T0327518

Design of Integrated Circuits for Optical Communications

Design of Integrated Circuits for Optical Communications

Second Edition

Behzad Razavi

A JOHN WILEY & SONS, INC., PUBLICATION

Published by John Wiley & Sons, Inc., Hoboken, New Jersey
Published simultaneously in Canada

For general information on our other products and services or for technical support, please contact our Customer Care Department within the United States at (800) 762-2974, outside the United States at (317) 572-3993 or fax (317) 572-4002.

Wiley also publishes its books in a variety of electronic formats. Some content that appears in print may not be available in electronic formats. For more information about Wiley products, visit our web site at www.wiley.com.

Library of Congress Cataloging-in-Publication Data:

Razavi, Behzad.
Design of integrated circuits for optical communications / Behzad Razavi. — Second edition.
pages cm
Includes index.
ISBN 978-1-118-33694-6 (hardback)
1. Optoelectronic devices. 2. Optical communications—Equipment and supplies. 3. Integrated optics. 4. Integrated circuits—Design and construction. I. Title.
TK8320.R39 2012
621.382'7—dc23 2012013971

Printed in the United States of America.

10 9 8 7 6 5 4 3 2 1

To Angelina

Contents

Preface to First Edition

The increasing demand for high-speed transport of data has revitalized optical communications, leading to extensive work on high-speed device and circuit design. This book has been written to address the need for a tutorial text dealing with the analysis and design of integrated circuits (ICs) for optical communication systems and will prove useful to both graduate students and practicing engineers. The book assumes a solid understanding of analog design, e.g., at the level of *Design of Analog CMOS Integrated Circuits* by B. Razavi or *Analysis and Design of Analog Integrated Circuits* by P. Gray, P. Hurst, S. Lewis, and R. Meyer.

The book comprises ten chapters. Chapter 1 provides an introduction to optical communications, setting the stage for subsequent developments. Chapter 2 describes basic concepts, building the foundation for analysis and design of circuits. Chapter 3 deals with optical devices and systems, bridging the gap between optics and electronics.

Chapter 4 addresses the design of transimpedance amplifiers, focusing on low-noise broadband topologies and their trade-offs. Chapter 5 extends these concepts to limiting amplifiers and output buffers, introducing methods of achieving a high gain with a broad bandwidth.

Chapter 6 presents oscillator fundamentals, and Chapter 7 focuses on LC oscillators. Chapter 8 describes the design of phase-locked loops, and Chapter 9 applies the idea of phase locking to clock and data recovery circuits. Chapter 10 deals with high-speed transmitter circuits such as multiplexers and laser drivers.

The book can be adopted for a graduate course on high-speed IC design. In a quarter system, parts of Chapters 3, 4, and 10 may be skipped. In a semester system, all chapters can be covered.

A website for the book provides additional resources for the reader, including an image set and web links. Visit **www.mhhe.com/razavi** for more information.

I would like to express my gratitude to the reviewers who provided invaluable feedback on all aspects of the book. Specifically, I am thankful to Lawrence Der (Transpectrum), Larry DeVito (Analog Devices), Val Garuts (TDK Semiconductor), Michael Green (University of California, Irvine), Yuriy Greshishchev (Nortel Networks), Qiuting Huang (Swiss Federal Institute of Technology), Jaime Kardontchik (TDK Semiconductor), Tai-Cheng Lee (National Taiwan University), Howard Luong (Hong Kong University of Science and Technology), Bradley Minch (Cornell University), Hakki Ozuc (TDK Semiconductor), Ken Pedrotti (University of California, Santa Cruz), Gabor Temes (Oregon State University), and Barry Thompson (TDK Semoconductor). I also wish to thank Michelle Flomenthoft, Betsy Jones, and Gloria Schiesl of McGraw-Hill for their kind support.

My wife, Angelina, encouraged me to start writing this book soon after we were married. She typed the entire text and endured my late work hours—always with a smile. I am very grateful to her.

Behzad Razavi
July 2002

Preface

The field of optical communications has experienced some change since the first edition of this book was published. While the fundemantals remain the same, the field has tried to find a place in mass markets and, specifically, spawned "passive optical networks." In addition, many new circuit techniques have been introduced for broadband applications, including optical systems.

This second edition reflects the new developments in the field. Recently reported circuit techniques for transimpedance amplifiers, broadband amplifiers, laser drivers, and clock and data recovery circuits have been described. Moreover, a new chapter dedicated to "burst-mode" circuits, i.e., building blocks required in passive optical networks, has been added.

Behzad Razavi
April 2012

About the Author

Behzad Razavi is an award-winning teacher, researcher, and author. He holds a PhD from Stanford University and is Professor of Electrical Engineering at University of California, Los Angeles. His current research includes wireless transceivers, frequency synthesizers, phase-locking and clock recovery for high-speed data communications, and data converters.

Prof. Razavi served on the Technical Program Committees of the International Solid-State Circuits Conference (ISSCC) from 1993 to 2002 and VLSI Circuits Symposium from 1998 to 2002. He has also served as Guest Editor and Associate Editor of the IEEE Journal of Solid-State Circuits, IEEE Transactions on Circuits and Systems, and International Journal of High Speed Electronics.

Professor Razavi received the Beatrice Winner Award for Editorial Excellence at the 1994 ISSCC, the best paper award at the 1994 European Solid-State Circuits Conference, the best panel award at the 1995 and 1997 ISSCC, the TRW Innovative Teaching Award in 1997, and the best paper award at the IEEE Custom Integrated Circuits Conference in 1998. He was the co-recipient of both the Jack Kilby Outstanding Student Paper Award and the Beatrice Winner Award for Editorial Excellence at the 2001 ISSCC. He received the Lockheed Martin Excellence in Teaching Award in 2006, the UCLA Faculty Senate Teaching Award in 2007, and the CICC Best Invited Paper Award in 2009. He was also recognized as one of the top 10 authors in the 50-year history of ISSCC. For his pioneering contributions to high-speed communication circuits, Prof. Razavi received the IEEE Donald Pederson Award in Solid-State Circuits in 2012.

Professor Razavi has served as an IEEE Distinguished Lecturer and is a Fellow of IEEE. He is the author of *Principles of Data Conversion System Design, RF Microelectronics* (translated to Chinese, Japanese, and Korean), *Design of Analog CMOS Integrated Circuits* (translated to Chinese, Japanese, and Korean), *Design of Integrated Circuits for Optical Communications*, and *Fundamentals of Microelectronics* (translated to Korean and Portuguese), and the editor of *Monolithic Phase-Locked Loops and Clock Recovery Circuits* and *Phase-Locking in High-Performance Systems*.

Chapter 1

Introduction to Optical Communications

The rapidly-growing volumes of data in telecommunication networks have rekindled interest in high-speed optical and electronic devices and systems. With the proliferation of the Internet and the rise in the speed of microprocessors and memories, the transport of data continues to be the bottleneck, motivating work on faster communication channels.

The idea of using light as a carrier for signals has been around for more than a century, but it was not until the mid-1950s that researchers demonstrated the utility of the optical fiber as a medium for light propagation [1]. Even though early fibers suffered from a high loss, the prospect of guided transmission of light with a very wide modulation band ignited extensive research in the area of optical communications, leading to the practical realization of optical networks in the 1970s.

This chapter provides an overview of optical communications, helping the reader understand how the concepts introduced in subsequent chapters fit into the "big picture." We begin with a brief history and study a generic optical system, describing its principal functions. Next, we present the challenges in the design of modern optical transceivers. Finally, we review the state of the art and the trends in transceiver design.

1.1 Brief History

Attempts to "guide" light go back to the 1840s, when a French physicist named Jacque Babinet demonstrated that light could be "bent" along a jet of water. By the late 1800s, researchers had discovered that light could travel inside bent rods made of quartz. The "fiber" was thus born as a flexible, transparent rod of glass or plastic.

In 1954, Abraham van Heel of the Technical University of Delft (Holland) and Harold Hopkins and Narinder Kapany of the Imperial College (Britain) independently published the idea of using a bundle of fibers to transmit images. Around the same time, Brian O'Brien of the American Optical Company recognized that "bare" fibers lost energy to the surrounding air, motivating van Heel to enclose the fiber core in a coating and hence lower the loss. Fiber loss was still very high, about 1,000 dB/km, limiting the usage to endoscopy applications.

The introduction of the laser as an intense light source in the 1950s and 1960s played a crucial role in fiber optics. The broadband modulation capability of lasers offered great potential for carrying information, although no suitable propagation medium seemed available. In 1966, Charles Ko and Charles Hockem of the Standard Telecommunication Laboratory (Britain) proposed that the optical fiber could be utilized as a signal transmission medium if the loss was lowered to 20 dB/km. They also postulated that such a low loss would be obtained if the impurities in the fiber material were reduced substantially.

Four years later, Robert Mauer and two of his colleagues at Corning Glass Works demonstrated silica fibers having a loss of less than 20 dB/km. With advances in semiconductor industry, the art of reducing impurities and dislocations in fibers improved as well, leading to a loss of 4 dB/km in 1975 and 0.2 dB/km in 1979. The dream of carrying massive volumes of information over long distances was thus fulfilled: in 1977, AT&T and GTE deployed the first fiber optic telephone system.

The widespread usage of optical communication for the transport of high-speed data stems from (1) the large bandwidth of fibers (roughly 25 to 50 GHz) and (2) the low loss of fibers (0.15 to 0.2 dB/km). By comparison, the loss reaches 200 dB/km at 100 MHz for twisted-pair cables and 500 dB/km at 1 GHz for low-cost coaxial cables. Also, wireless propagation with carrier frequencies of several gigahertz incurs an attenuation of tens of decibels across a few meters while supporting data rates lower than 100 Mb/s.

The large (and free) bandwidth provided by fibers has led to another important development: the use of multiple wavelengths (frequencies) to carry several channels on a single fiber. For example, it has been demonstrated that 100 wavelengths, each carrying data at 10 Gb/s, allow communication at an overall rate of 1 Tb/s across 400 km.

1.2 Generic Optical System

The goal of an optical communication (OC) system is to carry large volumes of data across a long distance. For example, the telephone traffic in Europe is connected to that in the United States through a fiber system installed across the Atlantic Ocean.

Depicted in Fig. 1.1(a), a simple OC system consists of three components: (1) an electro-optical transducer (e.g., a laser diode), which converts the electrical data to optical form (i.e., it produces light for logical ONEs and remains off for logical ZEROs); (2) a fiber, which carries the light produced by the laser; and (3) a photodetector (e.g., a photodiode), which senses the light at the end of the fiber and converts it to an electrical signal. We call the transmit and receive sides the "near end" and the "far end," respectively. As explained in Chapter 3, lasers are driven by electrical currents, and photodiodes generate an output current.

With long or low-cost fibers, the light experiences considerable attenuation as it travels from the near end to the far end. Thus, (1) the laser must produce a high light intensity, e.g., tens of milliwatts; (2) the photodiode must exhibit a high sensitivity to light; and (3) the electrical signal generated by the photodiode must be amplified with low noise. These observations lead to the more complete system shown in Fig. 1.1(b), where a "laser driver" delivers large currents to the laser and a "transimpedance amplifier" (TIA) amplifies the photodiode output with low noise and sufficient bandwidth, converting it to a voltage. For

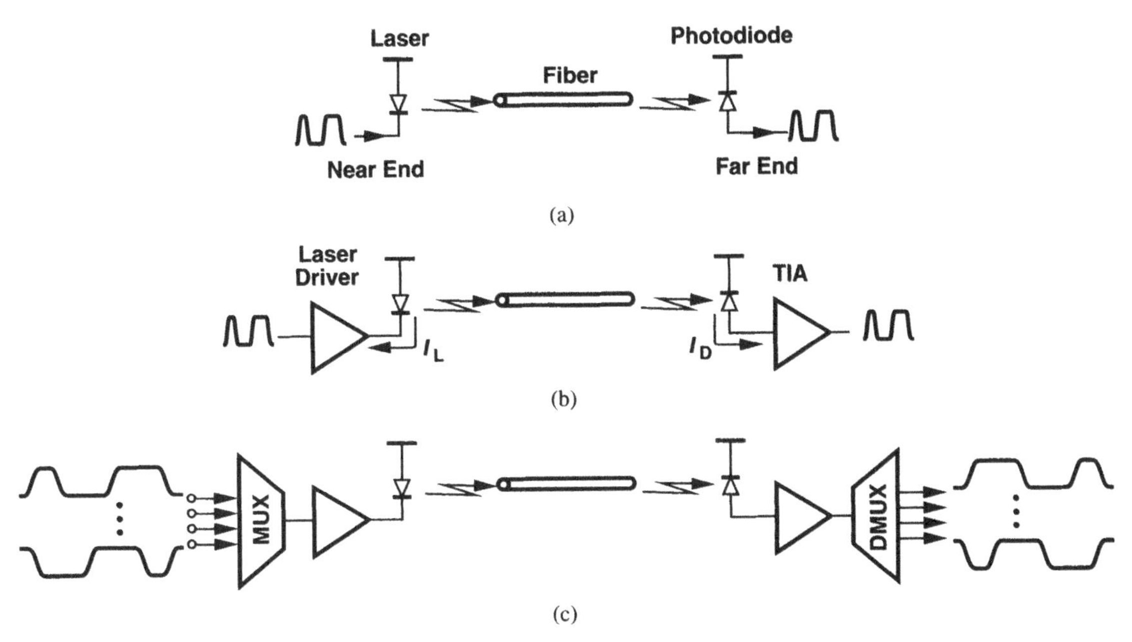

Figure 1.1 (a) Simple optical system, (b) addition of driver and amplifier, (c) addition of MUX and DMUX.

example, data at a rate of 10 Gb/s may be applied to the laser driver, modulate the laser light at a wavelength of 1.55 μm, and emerge at the output of the TIA with an amplitude of 10 mV.

The transmit and receive operations in Fig. 1.1(b) process high-speed "serial" data, e.g., a single stream of data at 10 Gb/s. However, the actual data provided to the transmitter (TX) is in the form of many low-speed channels ("parallel" data) because it is generated by multiple users. The task of parallel-to-serial conversion is performed by a "multiplexer" (MUX). Similarly, the receiver (RX) must incorporate a "demultiplexer" (DMUX) to reproduce the original parallel channels. The resulting system is shown in Fig. 1.1(c).

The topology of Fig. 1.1(c) is still incomplete. Let us first consider the transmit end. The multiplexer requires a number of clock frequencies with precise edge alignment. These clocks are generated by a phase-locked loop (PLL). Furthermore, in practice, the MUX output suffers from nonidealities such as "jitter" and "intersymbol interference" (ISI), mandating the use of a "clean-up" flipflop before the laser driver. These modifications lead to the transmitter illustrated in Fig. 1.2(a).

The receive end also requires additional functions. Since the TIA output swing may not be large enough to provide logical levels, a high-gain amplifier (called a "limiting amplifier") must follow the TIA. Moreover, since the received data may exhibit substantial noise, a clean-up flipflop (called a "decision circuit") is interposed between the limiting amplifier and the DMUX. The receiver thus appears as shown in Fig. 1.2(b).

The receiver of Fig. 1.2(b) lacks a means of generating the clock necessary for the de-

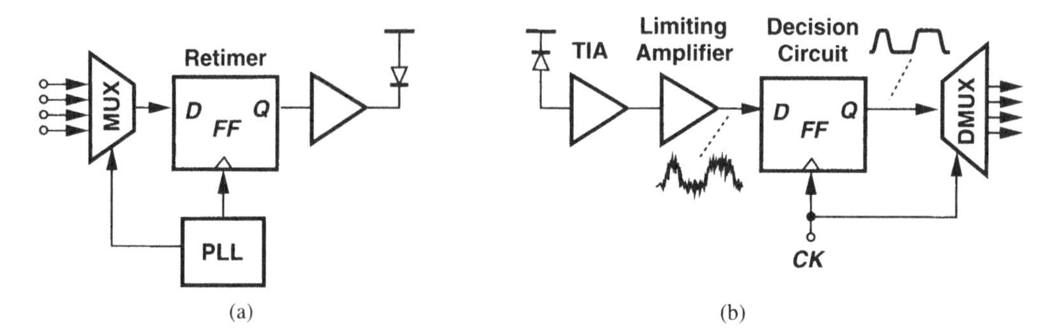

Figure 1.2 Modified (a) transmitter and (b) receiver.

cision circuit and the DMUX. This clock must bear a well-defined phase relationship with respect to the received data so that the flipflop samples the high and low levels "optimally," i.e., at the midpoint of each bit. The task of generating such a clock from the incoming data is called "clock recovery." The overall operation of clock recovery and data cleanup is called "clock and data recovery" (CDR). Figure 1.3 shows the complete system. Note that the laser driver incorporates power control (Chapter 10) and the TIA employs automatic gain control (AGC) (Chapter 4).

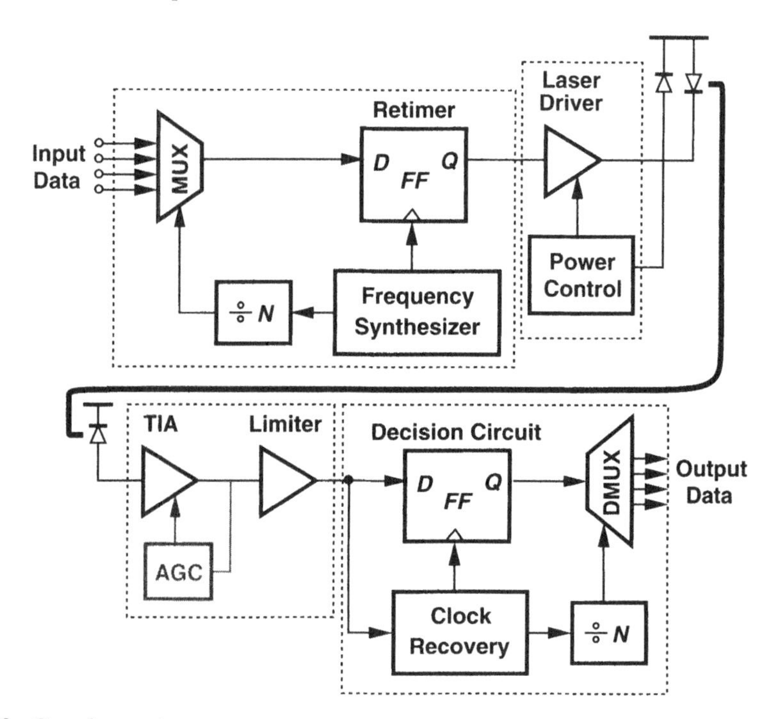

Figure 1.3 Complete system.

1.3 Design Challenges

While the system topology of Fig. 1.3 has not changed much over the past several decades, the design of its building blocks and the levels of integration have. Motivated by the evolution and affordability of IC technologies as well as the demand for higher performance, this change has created new challenges, necessitating new circuit and architecture techniques. We review some of the challenges here.

The transmitter of Fig. 1.3 entails several issues that manifest themselves at high speeds and/or in scaled IC technologies. Since the jitter of the transmitted data is determined primarily by that of the PLL, a robust, low-noise design with high supply and substrate rejection becomes essential. Furthermore, the design of skew-free multiplexers proves difficult at high data rates.

Another critical challenge arises from the laser driver, a circuit that must deliver tens of milliamperes of current with very short rise and fall times. Since laser diodes may experience large voltage swings between on and off states, the driver design becomes more difficult as scaled technologies impose lower supply voltages. The package parasitics also severely limit the speed with which such high currents can be switched to the laser [2].

The optical components in Fig. 1.3, namely, the laser diode, the fiber, and the photodiode, introduce their own nonidealities, requiring close interaction between electronic and optical design. Effects such as chirp, dispersion, attenuation, and efficiency play a major role in the overall link budget.

The receiver of Fig. 1.3 also presents many problems. The noise, gain, and bandwidth of the TIA and the limiter directly impact both the sensitivity and the speed of the overall system, raising additional issues as the supply voltage scales down. Moreover, the clock and data recovery functions must provide a high speed, tolerate long runs (sequences of identical bits), and satisfy stringent jitter and bandwidth requirements.

Full integration of the transceiver shown in Fig. 1.3 on a single chip raises a number of concerns. The high-speed digital signals in the MUX and DMUX may corrupt the receiver input or the oscillators used in the PLL and the CDR circuit. The high slew rates produced by the laser driver may lead to similar corruptions and also desensitize the TIA. Finally, since the oscillators in the transmit PLL and the receive CDR circuit operate at slightly different frequencies (with the difference given by the mismatch between the crystal frequencies in two communicating transceivers), they may "pull" each other, generating substantial jitter.

The above issues have resulted in multichip solutions that integrate the noisy and sensitive functions on different substrates. The dashed boxes in Fig. 1.3 indicate a typical partitioning, suggesting the following single-chip blocks: the PLL/MUX circuit (also called the "serializer"), the laser driver along with its power control circuitry, the TIA/limiter combination, and the CDR/MUX circuit (also called the "deserializer"). Recent work has integrated the serializer and deserializer (producing a "SERDES") but the TX and RX amplifiers may remain in isolation.

1.4 State of the Art

The new optical revolution is reminiscent of the monumental change that radio-frequency (RF) design began to experience in the early 1990s. This resurgence entails three important trends: (1) Modular, general-purpose building blocks are gradually replaced by end-to-end solutions that benefit from device/circuit/architecture codesign. (2) Greater levels of integration on a single chip provide higher performance and lower cost. (3) Mainstream VLSI technologies such as CMOS and BiCMOS continue to take over the territories thus far claimed by GaAs and InP devices. Modern OC transceiver applications continue to challenge designers along many dimensions.

Realization in CMOS Technology The cost and integration advantages of CMOS technology have motivated extensive work on high-speed CMOS design. Issues such as noise, speed, voltage headroom, and substrate coupling pose many difficulties in the design of CMOS transceivers. Research on 10-Gb/s CMOS CDR circuits yielded results in 2000 [3], and CMOS serializers and deserializers operating at this rate were reported in 2002 [4, 5].

Speed With the increasing volume of data transported in the backplane of the Internet, optical communication at rates of 40 Gb/s has become attractive. Such high speeds emerge as a new territory for IC design because prior work at these frequencies ("millimeter-wave frequencies") has been limited to narrowband, low-complexity circuits for wireless applications. Pushing bipolar and, preferably, CMOS technologies to such speeds, designers must cope with broadband characterization of active and passive devices, transmission-line behavior of on-chip interconnects, and high-speed packaging issues. A 40-Gb/s SiGe CDR circuit has been reported in [6].

Level of Integration Integrating a complete SERDES on a single CMOS chip serves as the first step toward much greater sophistication in OC transceiver design. Two important trends particularly suited to CMOS technology are: (1) integration of the SERDES along with the large digital processor that interfaces with the network (the "framer"); such integration eliminates a large number of high-speed printed-circuit board (PCB) lines between the two, simplifying the package design and saving substantial power. (2) integration of multiple SERDES on one chip; since the total data rate can be increased through the use of multiple light *wavelengths* on a single fiber, an important thrust is to integrate several SERDES on the same substrate, thereby increasing the "port density."

Power Dissipation At high speeds and/or high port densities, the power dissipation of optical transceivers becomes critical as it determines the type and size of the package in which the entire module is housed. Today's 10-Gb/s SERDES consume about 1 W of power, leading to serious packaging issues if four must be integrated on one chip. Interestingly, the low supply voltage required for deep-submicron CMOS technologies does reduce the overall power dissipation (e.g., in the output buffers) while making circuit design more difficult.

References

1. D. G. Goff, *Fiber Optic Reference Guide,* Boston: Focal Press, 1999.
2. H.-M. Rein and M. Moller, "Design Considerations for Very High Speed Si Bipolar ICs Operating up to 50 Gb/s," *IEEE Journal of Solid-State Circuits,* vol. 31, pp. 1076–1090, August 1996.
3. J. Savoj and B. Razavi, "A 10-Gb/s CMOS Clock and Data Recovery Circuit," *Symp. on VLSI Circuits Dig. of Tech. Papers*, pp. 136–139, June 2000.
4. M. M. Green et al., "OC-192 Transmitter in Standard 0.18-μm CMOS," *ISSCC Dig. of Tech. Papers,* pp. 186–187, Feb. 2002.
5. J. Cao et al., "OC-192 Receiver in Standard 0.18-μm CMOS," *ISSCC Dig. of Tech. Papers,* pp. 187–188, Feb. 2002.
6. M. Reinhold et al., "A Fully Integrated 40-Gb/s Clock and Data Recovery IC with 1:4 DMUX in SiGe Technology," *IEEE Journal of Solid-State Circuits,* vol. 36, pp. 1937–1945, Dec. 2001.

CHAPTER 2

Basic Concepts

This chapter forms the background necessary for the analysis and design of optical communication circuits and systems. We first review the properties of random binary data and consider methods of generating pseudo-random sequences. Next, we study the effect of bandwidth limitation and noise on random data. Finally, we introduce the concepts of phase noise and jitter and review transmission lines.

2.1 Properties of Random Binary Data

Most optical communication systems employ simple binary amplitude modulation of the lightwave for ease of detection. The random binary sequence (RBS) experiences various imperfections in the optical and electrical domains, raising important design issues. In this section, we study properties of random data to the extent necessary for circuit and system analysis.

A random binary sequence consists of logical ONEs and ZEROs that carry the information and usually occur with equal probabilities [Fig. 2.1(a)]. If each bit period is T_b seconds, we say the bit rate, R_b, is equal to $1/T_b$ bits per second.[1] The sequence depicted

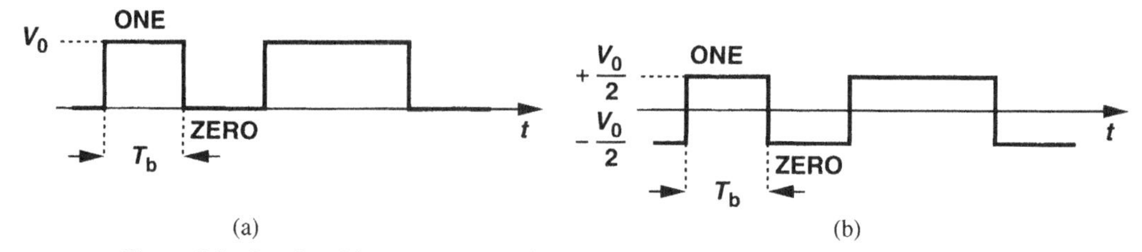

Figure 2.1 Random binary sequence with (a) finite and (b) zero dc content.

in Fig. 2.1(a) contains a nonzero average value because the logical ZEROs are represented

[1] We use b/s for random data and Hz for periodic waveforms, e.g., clocks.

by a zero voltage (or current). In some cases, it is simpler to view the waveform as shown in Fig. 2.1(b), where the ONEs and ZEROs assume equal and opposite values, thereby yielding a zero average.

The random nature of data implies that a binary sequence may contain arbitrarily long strings of consecutive ONEs or ZEROs (also called "runs") (Fig. 2.2). We say the data

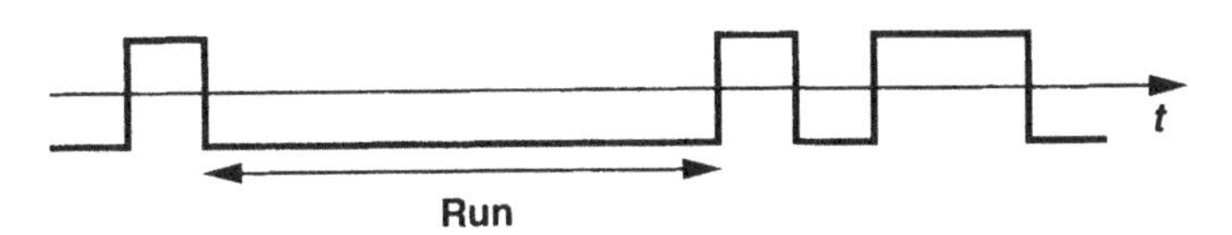

Figure 2.2 Random binary sequence with long run.

exhibits a low "transition density." Such strings create difficulties in the design of many optical transceiver circuits. In particular, operations such as ac coupling, offset cancellation, and clock recovery are sensitive to low transition densities, failing completely if a run becomes arbitrarily long. For this reason, optical communication standards typically specify the maximum "run length,"[2] i.e., the maximum number of consecutive ONEs or ZEROs allowed in the data. A typical run may be as long as 72 bits. To avoid exceeding such a run length, the data is encoded properly in the transmitter.

It is also instructive to examine random binary data in the frequency domain. How is the spectrum of a random sequence obtained?[3] Let us represent the random binary sequence by

$$x(t) = \sum_k b_k p(t - kT_b), \tag{2.1}$$

where $b_k = \pm 1$ and $p(t)$ denotes the pulse shape. That is, the RBS is viewed as positive and negative replicas of a basic pulse that are repeated every T_b seconds (Fig. 2.3). While we can assume that $p(t)$ is simply a rectangular pulse of width T_b, it is still useful to obtain

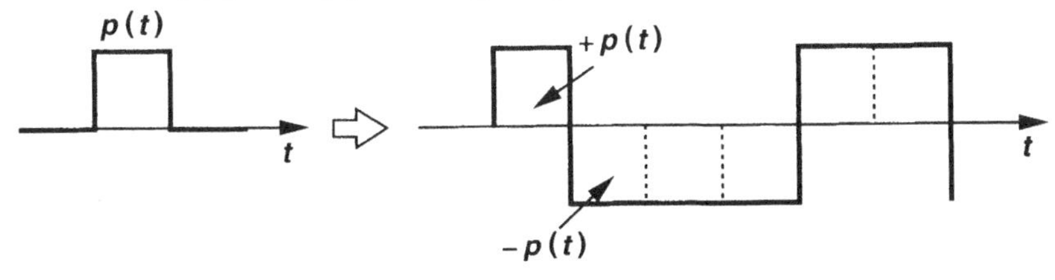

Figure 2.3 Random sequence viewed as random repetition of a pulse.

the spectrum of $x(t)$ for an arbitrary $p(t)$.

---

[2] Also called "consecutive identical digits" (CIDs).

[3] The spectrum of a waveform indicates how much power the signal carries in a 1-Hz bandwidth at each frequency.

It can be proved that, if the positive and negative pulses in (2.1) occur with equal probabilities, then the power spectral density of $x(t)$ is given by

$$S_x(f) = \frac{1}{T_b}|P(f)|^2, \tag{2.2}$$

where $P(f)$ represents the Fourier transform of $p(t)$ [1]. Equation (2.2) reveals many interesting properties of various data formats used in communications. As we will see throughout this book, many of these properties directly impact the design of transceivers.

Let us now compute $S_x(f)$ if $p(t)$ is a rectangular pulse T_b seconds wide and repeated every T_b seconds. Since the Fourier transform of such a pulse is equal to

$$P(f) = T_b\frac{\sin(\pi f T_b)}{\pi f T_b}, \tag{2.3}$$

the spectrum of the random sequence is expressed as

$$S_x(f) = T_b\left[\frac{\sin(\pi f T_b)}{\pi f T_b}\right]^2. \tag{2.4}$$

Noting that $\sin(\pi f T_b)$ vanishes at $f = n/T_b$ for integer values of n, we construct the spectrum as shown in Fig. 2.4(a). To show a wider magnitude range of $S_x(f)$, it is common

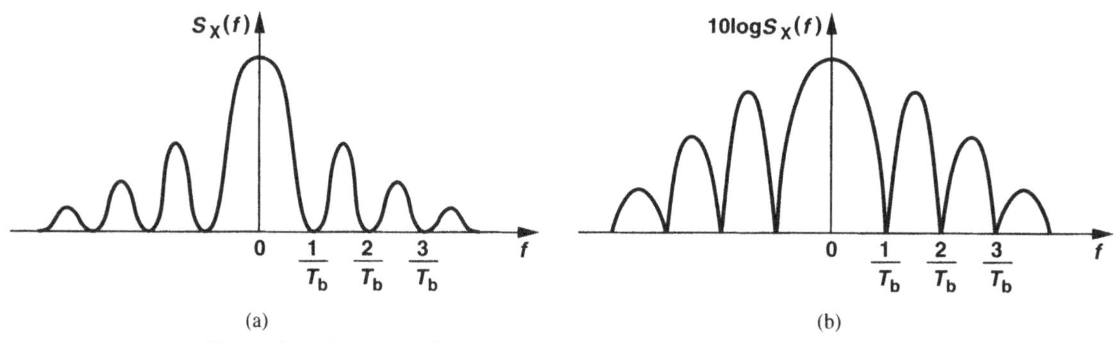

Figure 2.4 Spectrum of random binary data with (a) linear and (b) logarithmic vertical scale.

to use a logarithmic scale for the vertical axis [Fig. 2.4(b)].

The above analysis yields an important attribute of random binary sequences. For a bit rate of $1/T_b$, the spectrum exhibits no power at frequencies equal to $1/T_b$, $2/T_b$, etc. In other words, if the waveform is applied to a 1-Hz bandpass filter centered at $f = 1/T_b, 2/T_b, \cdots$, very little energy is observed. For example, a 10-Gb/s sequence does not contain a 10-GHz component [Fig. 2.5(a)]. This somewhat surprising result is better understood if we note that the fastest waveform at 10 Gb/s consists of a 1010 sequence with each bit 100 ps wide [Fig. 2.5(b)]. Such a signal is a 5-GHz square wave, containing only odd harmonics at 5 GHz, 15 GHz, etc. Another method of proving the existence of the nulls is described in Section 2.2.

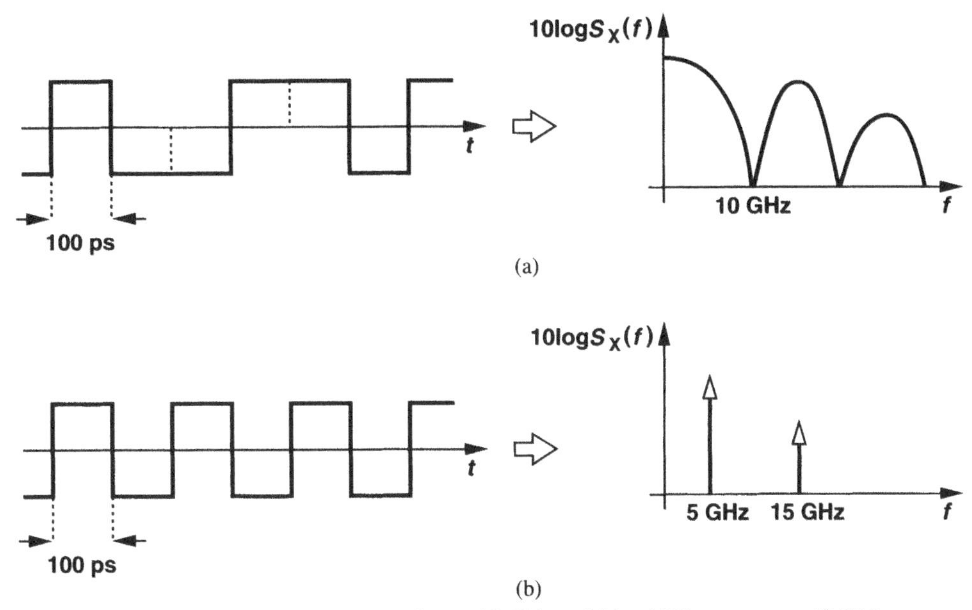

Figure 2.5 Spectra of (a) random binary data at 10 Gb/s and (b) a 1010 sequence at 10 Gb/s.

The foregoing observation proves critical in the task of clock recovery. As explained in Chapter 9, for clock recovery by phase locking, the waveform must contain a periodic component at the bit rate, i.e., the spectrum must display an impulse (sometimes called a "spectral line") at $f = 1/T_b$. To understand intuitively why random binary data does not provide such an impulse, we multiply the waveform by a sinusoid of frequency $1/T_b$ and compute the average of the result.[4] As illustrated in Fig. 2.6, the average is zero if the

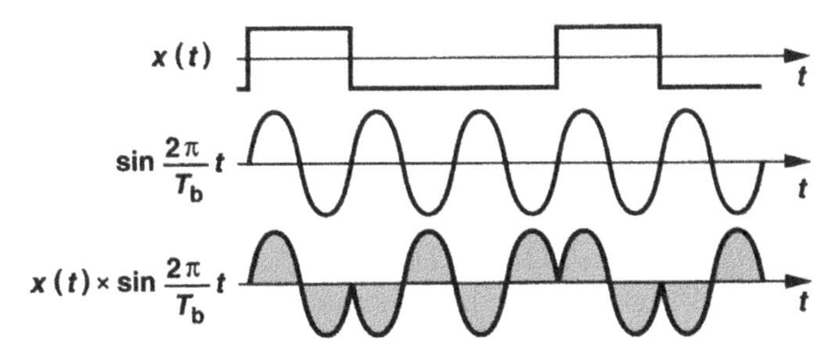

Figure 2.6 Correlation of random binary data with a sinusoid.

ONEs and ZEROs occur with equal probabilities.

Even though we have assumed that the ONEs and ZEROs are represented by equal and opposite pulses, the spectrum of Fig. 2.4 exhibits a finite power in the vicinity of zero frequency. Is this a contradiction to the assumption of zero dc content? No, if a waveform

[4]This is equivalent to calculating the coefficient in the Fourier series of a periodic waveform.

has a nonzero average, then its spectrum contains an *impulse* at zero. The finite power in Fig. 2.4 simply means that the signal is likely to have *arbitrarily* low frequencies if it is observed for a very long time. This property is akin to the long runs that appear in random data.

2.2 Generation of Random Data

In simulation and characterization, it is difficult to generate completely random binary waveforms because for randomness to manifest itself, the sequence must be very long. For this reason, it is common to employ standard "pseudo-random" binary sequences (PRBSs). Each PRBS is in fact a repetition of a pattern that itself consists of a random sequence of a number of bits (Fig. 2.7). As an example of PRBS generation, consider the circuit shown in

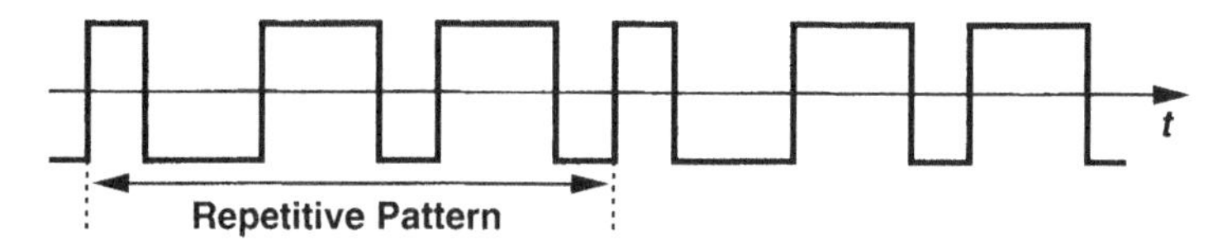

Figure 2.7 Pseudorandom binary sequence.

Fig. 2.8(a), where three master-slave flipflops form a shift register and an XOR gate senses

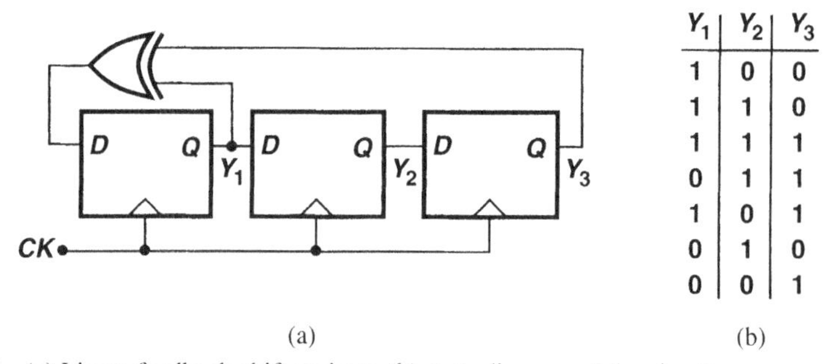

Y_1	Y_2	Y_3
1	0	0
1	1	0
1	1	1
0	1	1
1	0	1
0	1	0
0	0	1

Figure 2.8 (a) Linear feedback shift register, (b) state diagram of the circuit.

Y_1 and Y_3, returning the result to the input of the first flipflop.

Suppose the register begins with the initial state $Y_1Y_2Y_3 = 100$. The circuit then goes through the states shown in Fig. 2.8(b), producing $Y_3 = 0011101$. Also appearing in Y_1 and Y_2, this pattern repeats every $2^3 - 1 = 7$ clock cycles. We also note that if the initial condition is 000, the register remains in a degenerate state. Thus, some means of initialization is necessary. In signal processing literature, the circuit of Fig. 2.8(a) is called a "linear feedback" shift register [2, 3], and the operation of the circuit is expressed by the following polynomial:

$$p(y) = y^3 + y^1 + 1. \tag{2.5}$$

Here, the power in each term indicates the delay (e.g., Y_3 is a replica of the XOR gate output but delayed by three clock cycles) and the last term signifies the reset necessary to avoid the degenerate state. The polynomial may also be written as $y^3 \oplus y^2 \oplus 1$ to show the XOR operation explicitly.

The waveform produced by the above circuit is an example of a relatively random sequence. One attribute of randomness is "dc balance," i.e., the total number of ONEs in each period differs from that of ZEROs by only one. Note also that the maximum run length is equal to 3. Other properties of PRBSs are described in [3].

The linear feedback shift register of Fig. 2.8 creates a pattern 7 bits long. This technique can be extended to an m-bit system so as to produce a sequence of length $2^m - 1$. For example, many optical communication circuits are tested with a PRBS of length $2^7 - 1, 2^{15} - 1, 2^{23} - 1$, or $2^{31} - 1$, i.e., maximum run lengths of 7, 15, 23, or 31, respectively. The polynomials used for generating $2^{15} - 1$ and $2^{23} - 1$ are expressed as $y^{15} \oplus y^{14} \oplus 1$ and $y^{23} \oplus y^{18} \oplus 1$, respectively.

It is important to note that the spectrum of pseudo-random data sequences is quite different from that given by Eq. (2.4). Since the random pattern is repeated periodically, we expect the spectrum to contain only impulses. As an example, consider the PRBS illustrated in Fig. 2.9(a), where a 7-bit pattern is repeated. (For simplicity, the ZEROs are represented

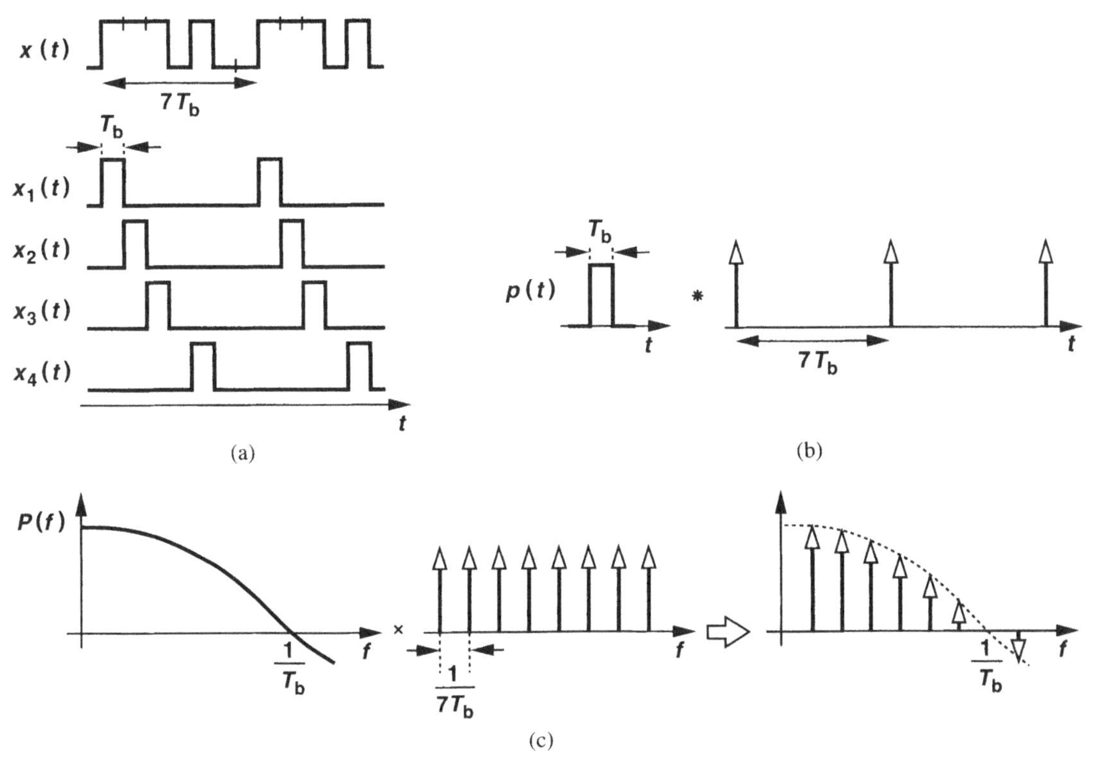

Figure 2.9 (a) PRBS data viewed as superposition of periodic pulses, (b) decomposition of periodic pulses as a basic pulse and train of impulses, (c) resulting spectra.

by a zero level rather than a negative voltage.) The signal can be decomposed into a number of simple, identical periodic waveforms, $x_1(t), \cdots, x_4(t)$, that differ by only a phase shift. Each waveform consists of a pulse of width T_b that is repeated every $7T_b$ seconds. Alternatively, as shown in Fig. 2.9(b), each waveform can be viewed as the convolution of a single pulse of width T_b with a train of impulses having a period $7T_b$. The Fourier transform of one such waveform is therefore given by the product of those of the pulses and the train [Fig. 2.9(c)]. Since $x_1(t), \cdots, x_4(t)$ have the same magnitude but different phase shifts, the Fourier transform of $x(t)$ still resembles that shown in Fig. 2.9(c), that is, it consists of only spectral lines. Of course, for long random patterns, the impulses are very closely spaced, creating an almost continuous spectrum. For example, with a PRBS of length $(2^{23} - 1)T_b$ at 1 Gb/s, the impulses appear at integer multiples of $[(2^{23} - 1)T_b]^{-1} \approx 119$ Hz. Note, however, that the nulls at n/T_b persist regardless of the length of the sequence. This is another approach to proving that random binary data lacks a spectral line at the bit rate.

2.3 Data Formats

2.3.1 NRZ and RZ Data

The binary sequences studied thus far are called "non-return to zero" (NRZ) data, a format employed in most high-speed applications. This type of waveform is called NRZ to distinguish it from "return to zero" (RZ) data, in which each bit consists of *two* sections: the first section assumes a value that represents the bit value, and the second section is always equal to a logical zero [Fig. 2.10(a)]. In other words, every two symbols carrying information are separated by a redundant zero symbol.

In contrast to NRZ data, RZ waveforms exhibit a spectral line at a frequency equal to the bit rate, thereby simplifying the task of clock recovery (Chapter 9). As shown in Fig. 2.10(b), this property is derived by viewing the RZ waveform as the product of an NRZ sequence, $x_1(t)$, and a periodic square wave, $x_2(t)$. Note that $x_1(t)$ and $x_2(t)$ toggle between 0 and 1, exhibiting impulses at dc in their spectrum. Since the spectrum of the square wave contains impulses at $1/T_b, 3/T_b$, etc., the spectrum of the RZ signal, $S_{x1}(f) * S_{x2}(f)$, appears as shown in Fig. 2.10(d). It can be shown that the overall spectrum still has a sinc squared shape [1].

Comparison of the spectra in Figs. 2.4(b) and 2.10(c) reveals the drawback of RZ data: it occupies about twice as much bandwidth as NRZ data does, intensifying the trade-offs in circuit design. Other data formats such as Manchester and Miller codes also exist [1] but, due to speed or modulation and detection difficulties, they are rarely used in high-speed optical systems.

2.3.2 8B/10B Coding

The long runs present in random data streams create difficulties in clock and data recovery as well as in ac coupling between stages. In CDR design, the lack of data transitions during a long run allows the oscillator to drift and hence generate jitter. In ac coupling, a long run exhibits a nonzero "running average" (i.e., a nonzero "local" dc content), which is blocked by high-pass filtering.

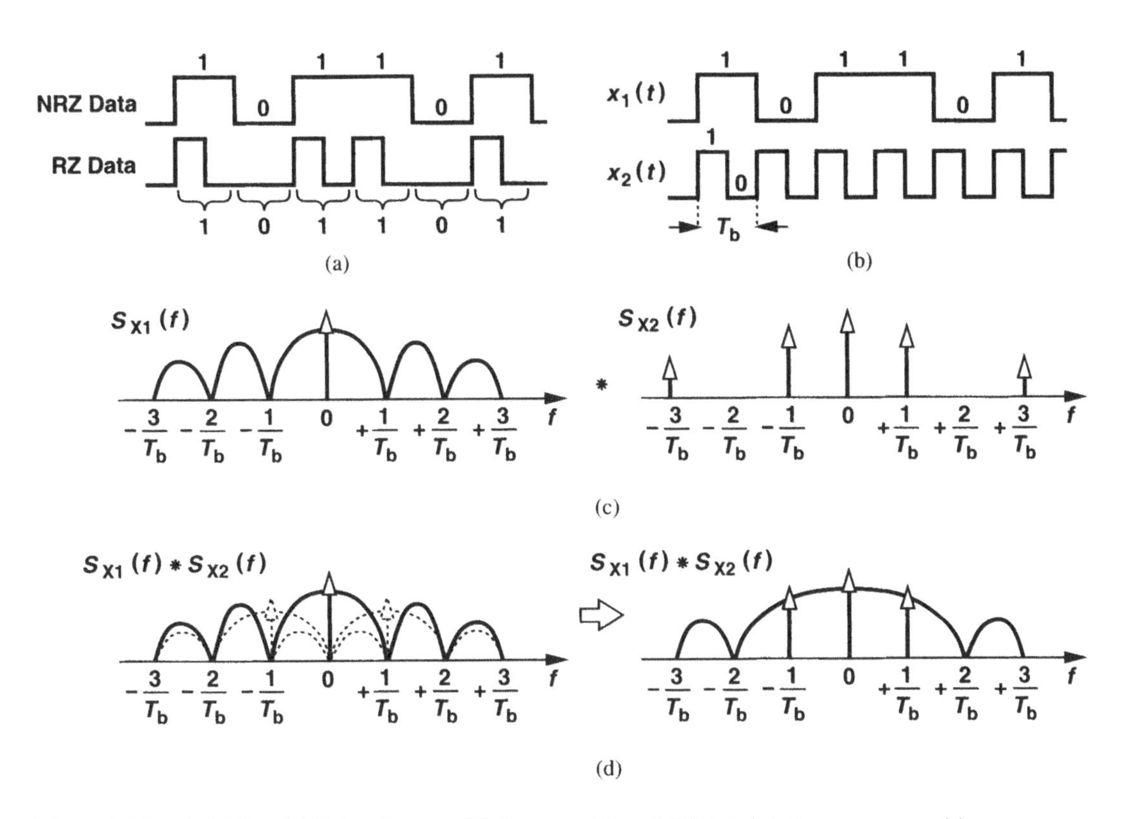

Figure 2.10 (a) NRZ and RZ data formats, (b) decomposition of RZ data into two sequences, (c) spectra of x_1 and x_2, (d) spectrum of RZ data.

In order to alleviate the above issues, random data may be encoded so as to limit the maximum run length. The result is loosely called a "dc-balanced" code. For example, 8-bit/10-bit (8B/10B) coding [4] converts a sequence of 8 bits to a 10-bit word while guaranteeing a maximum run length of 5 bits. Thus, the data rate rises by 25% (called the "overhead") but many aspects of the design are relaxed.

Described in [4] in detail, 8B/10B coding decomposes each byte into two blocks of 5 bits and 3 bits, converting them to 6-bit and 4-bit equivalents, respectively. Two important attributes of this coding scheme are (1) the difference between the number of ONEs and the number of ZEROs in each output block is equal to 0 or 2; (2) if one block contains two excess ONEs (ZEROs), then it is followed by another that includes two excess ZEROs (ONEs). As a result, each two consecutive blocks exhibit a total average of zero, hence the term "dc-balanced."

8B/10B coding is employed in a number of standards, including Fibre Channel and 10-Gigabit Ethernet. A more efficient version, called 64-bit/66-bit coding has also been employed to lower the overhead [5].

2.4 Effect of Bandwidth Limitation on Random Data

2.4.1 Effect of Low-Pass Filtering

In the design of high-speed circuits, the bandwidth trades with gain and power dissipation. Furthermore, in low-noise applications, the bandwidth must be chosen carefully so that the signal is processed with high fidelity while the total integrated noise is minimized.

In order to determine the bandwidth required of a circuit, we must examine the signal quality as the bandwidth is reduced. For a periodic square wave, a low-pass filter attenuates the high-frequency components, yielding finite rise and fall times [Fig. 2.11(a)]. But how about random binary data? As illustrated in Fig. 2.11(b), for a single ONE followed by

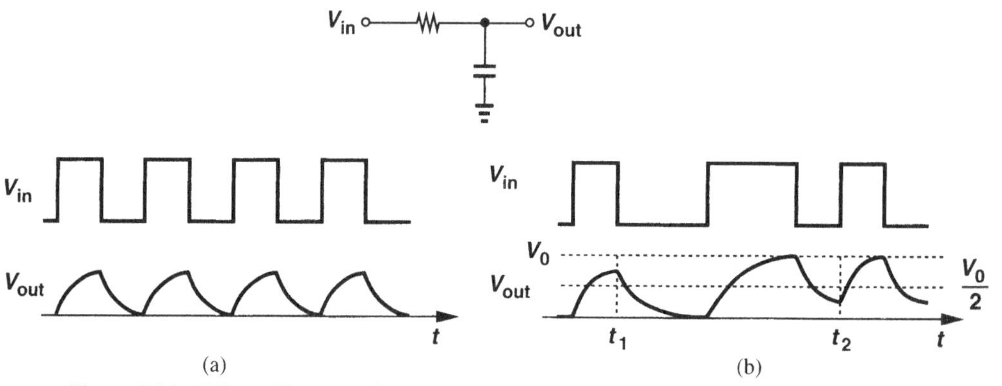

Figure 2.11 Effect of low-pass filtering on (a) periodic data and (b) random data.

a ZERO, the output does not come close to V_0, but for two consecutive ONEs, it does. A similar effect occurs for ZEROs as well. This phenomenon is undesirable because the output voltage levels corresponding to ONEs and ZEROs vary with time, making it difficult to define a decision threshold. For example, if the threshold is set at $V_0/2$, then the levels at $t = t_1$, and $t = t_2$ are very susceptible to noise and can be misinterpreted by the detector. Also, the time instants at which the waveform crosses $V_0/2$ experience random variations.

The above phenomenon is called "intersymbol interference" (ISI) because it can also be viewed as follows. Suppose, as shown in Fig. 2.12(a), the input data sequence is represented by a superposition of positive and negative steps. Each step gives rise to an exponential response at the output [Fig. 2.12(b)], thereby corrupting the levels produced for subsequent bits; the narrower the bandwidth, the longer the exponential tails and the greater the ISI.

2.4.2 Eye Diagrams

Since ISI manifests itself differently for different bit patterns, long sequences of random waveforms must be examined for ISI effects, a tedious task. A common tool for visualizing the nonidealities in random data is the "eye diagram." Such a diagram "folds" all of the bits into a short interval, e.g., two bits wide, thereby displaying an accumulation of distorted edges and levels. As an example, consider the relatively "clean" waveform depicted in Fig. 2.13(a) and suppose an oscilloscope displays the data across a horizontal span of 2 ns. If triggered every 2 ns, the oscilloscope displays the waveform first from t_1 to t_2, next from

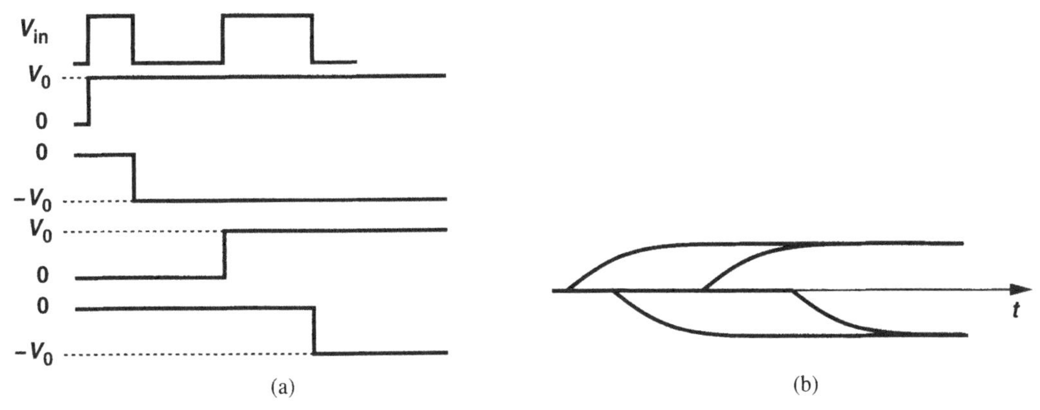

Figure 2.12 Use of superposition to view ISI.

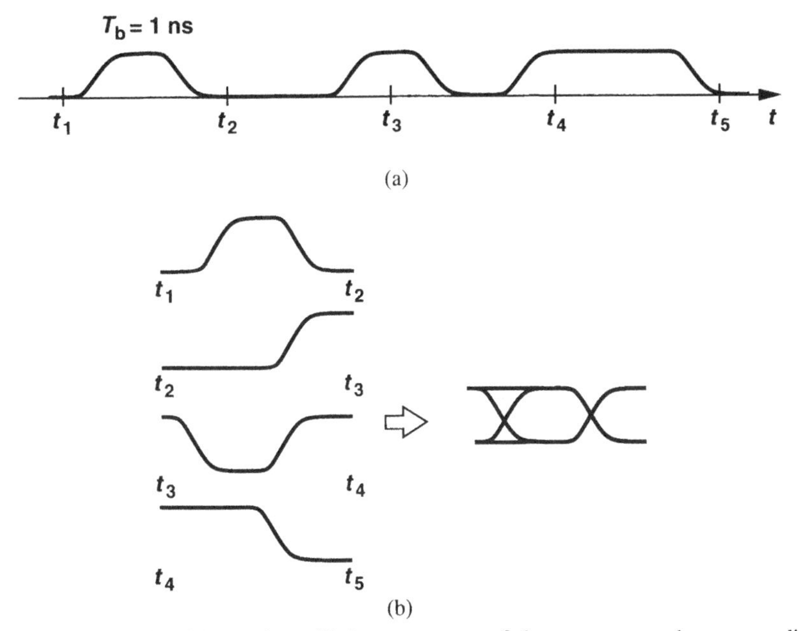

Figure 2.13 (a) Random binary data, (b) 2-ns segments of the sequence and corresponding eye diagram.

t_2 to t_3, then from t_3 to t_4, etc., each time starting on the left side of the screen and ending on the right. Thus, the segments are overlapped on the display, yielding the "eye" shown in Fig. 2.13(b).

To examine the effect of limited bandwidth on the eye diagram, let us consider a 1-Gb/s random binary sequence applied to a first-order low-pass filter having a -3-dB bandwidth of 300 MHz. Since the filter introduces a time constant $\tau \approx 530$ ps, we note that the step response, $V_0[1 - \exp(-t/\tau)]$, reaches within 15% of V_0 in one bit period and within 2.3% of

V_0 in two bit periods [Fig. 2.14(a)]. Thus, the worst-case degradation in high or low levels is approximately equal to 15%. To construct the eye diagram, we first plot the response to a

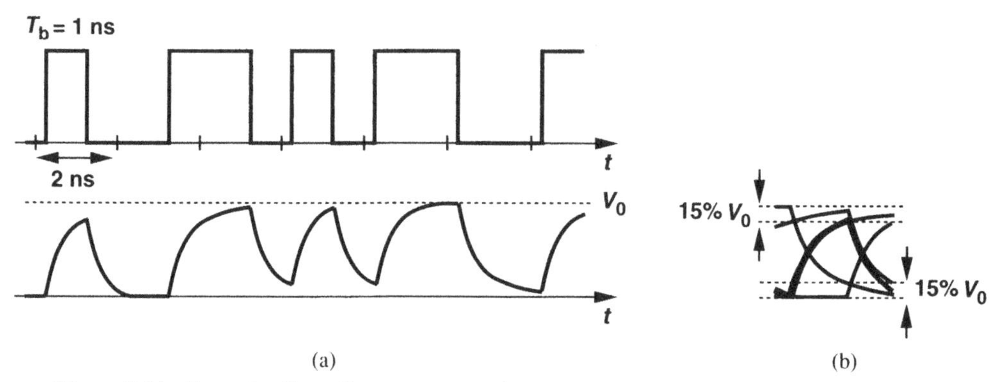

Figure 2.14 Example of eye diagram construction.

relatively random sequence and subsequently superimpose all 2-ns intervals [Fig. 2.14(b)]. We say the "eye closure" is about 30%. We also observe that the limited bandwidth degrades the *time resolution* as well because the edges experience random aberrations with respect to their ideal position. This effect is an example of "jitter," to be studied in Section 2.6.

In addition to limited circuit bandwidth, any imperfection that affects the magnitude or phase response of a system may result in ISI. Figure 2.15 illustrates two examples in high-speed applications. In Fig. 2.15(a), transistor M_1 is driven by random data and delivers a large current to R_L. Due to the finite inductance of the bond wire (and the package pin), each time M_1 turns on, the output voltage experiences ringing, corrupting subsequent bits. Figure 2.15(b) shows an eye diagram obtained from simulation of this circuit.

Another example of ISI generation is depicted in Fig. 2.15(c), where the output of a circuit travels through a trace on PCB, driving a termination resistor, R_T, and another circuit. Due to various nonidealities such as bond wire inductances, parasitic capacitances, and mismatches between the trace impedance and R_T, the signal received at the far end may suffer from significant ISI [6]. Figure 2.15(d) shows a simulated eye diagram for this case.

The foregoing examples indicate that eye diagrams provide an efficient means of studying the effect of circuit and system nonidealities. In order to construct an eye diagram in time-domain simulations, a periodic ramp waveform is generated to emulate the behavior of an oscilloscope. Consider the sequence $x(t)$ depicted in Fig. 2.16(a) and suppose $r(t)$, with a period equal to $2T_b$, is generated as well. Now, let us plot each $2T_b$ interval of $x(t)$ as a function of $r(t)$, obtaining the segments shown in Fig. 2.16(b). Superimposing the segments, we arrive at the eye diagram of Fig. 2.16(c).

2.4.3 Effect of High-Pass Filtering

We have thus far studied the effect of *low*-pass filtering on random data. It is also important to examine the effect of *high*-pass filtering. For example, to remove dc offsets in ampli-

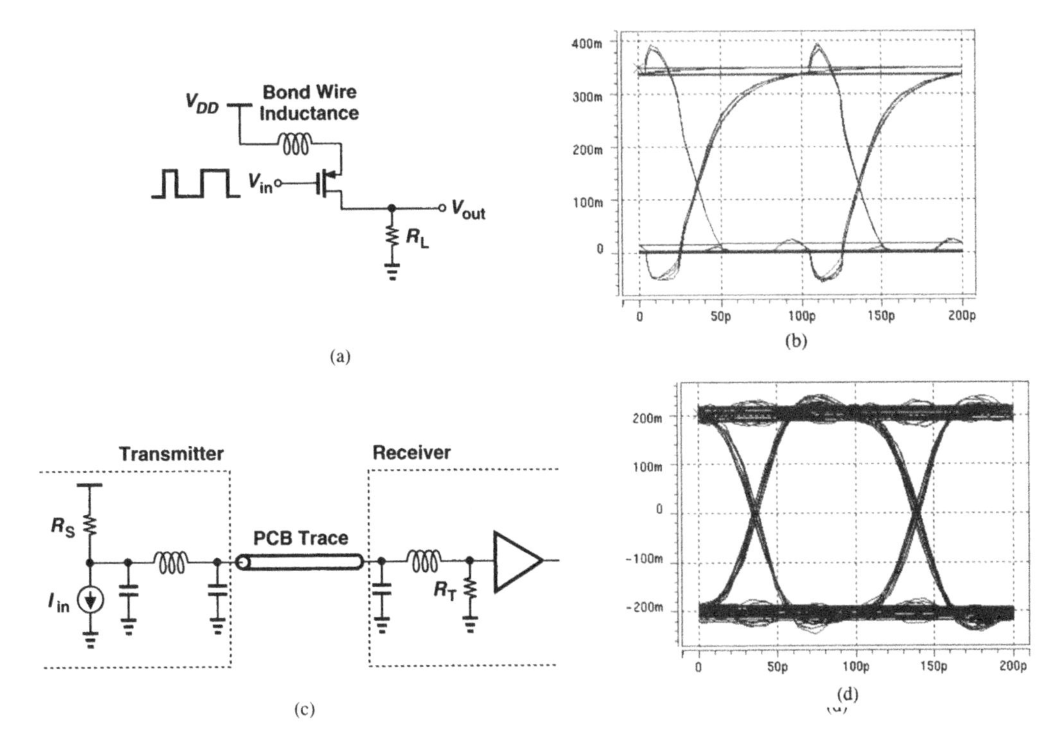

Figure 2.15 (a) Typical output stage, (b) ISI resulting from package parasitics, (c) typical point-to-point connection, (d) ISI resulting from transmitter and receiver nonidealities.

fiers, we may employ capacitive coupling between stages. Moreover, cascaded stages may require different dc voltages at their interfaces, making capacitive coupling an attractive choice, especially at low supply voltages.

To understand the effect of high-pass filtering, suppose a random binary sequence is applied to a first-order RC section (Fig. 2.17). Each transition at the input immediately appears at the output, but the output *levels* may "droop" significantly for long runs. As a result, the bits after each long run suffer from a large (temporary) dc shift, making it difficult to set a decision threshold. This phenomenon can also be viewed as ISI because each bit level depends on the preceding pattern.

The above effect is called "dc wander" (or "baseline wander") because the "instantaneous" dc value of the output waveform continues to change randomly. To minimize dc wander, $\tau_1 = R_1C_1$ must be sufficiently greater than the longest permissible run to ensure negligible droop. Some standards simply specify the lower -3-dB end of the bandwidth to avoid dc wander. For example, in multigigahertz systems, the lower end may be as small as a few tens of kilohertz.

As an example, suppose a 10-Gb/s optical standard requires a high-pass corner frequency of less than 250 kHz. What is the longest run that degrades the subsequent bit level by 0.2 dB if a first-order high-pass filter is assumed? As illustrated in Fig. 2.18, the output level must degrade by less than 0.2 dB after mT_b, where m is the maximum run length.

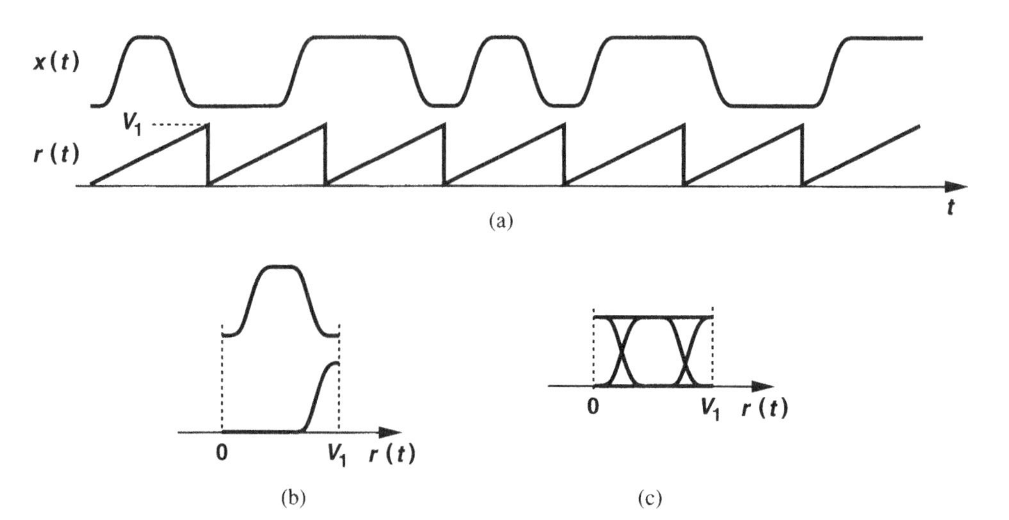

Figure 2.16 (a) Random sequence and a synchronous ramp, (b) two-bit segments of random sequence plotted versus the ramp, (c) resulting eye diagram.

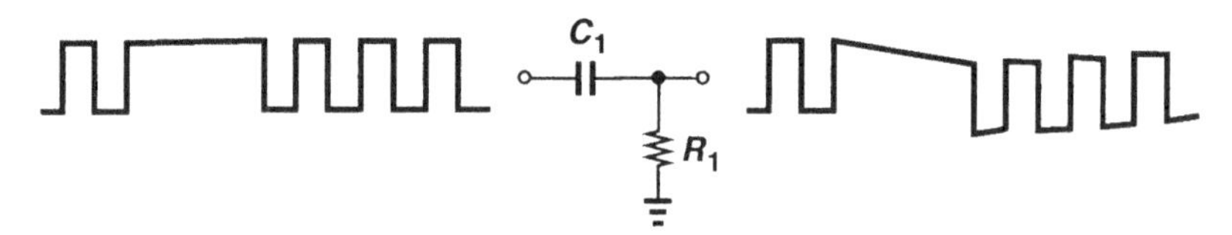

Figure 2.17 Effect of high-pass filtering on random data.

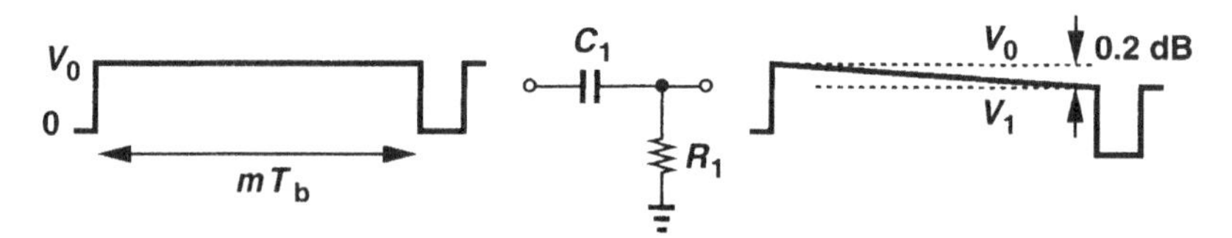

Figure 2.18 Example of droop in random data due to high-pass filtering.

The output voltage is given by

$$V_1 = V_0 \exp \frac{-mT_b}{\tau_1}. \tag{2.6}$$

Since a 0.2-dB droop translates to $20 \log V_1 = 20 \log V_0 - 0.2$ dB, we have $V_1/V_0 \approx 0.9772$ and hence,

$$0.9772 = \exp \frac{-mT_b}{\tau_1}. \tag{2.7}$$

Thus, $mT_b \approx 0.023\tau_1$ and, with $\tau_1 = (2\pi \times 250\text{ kHz})^{-1} \approx 636.6$ ns and $T_b = 100$ ps, we obtain $m \approx 147$.

2.5 Effect of Noise on Random Data

Random data propagating through an optical fiber may experience considerable attenuation. For example, low-cost plastic fibers exhibit losses as high as 3 dB/m (Chapter 3). Thus, the noise of the receiver can significantly impact the detection of the data. Since noise directly trades with gain, bandwidth, and power dissipation, it is important to determine how much noise can be tolerated for a given bit error rate (BER).

Recall from Chapter 1 that, as shown in Fig. 2.19(a), the data bits must ideally be sampled by the clock at their midpoints so as to provide maximum distance from the decision

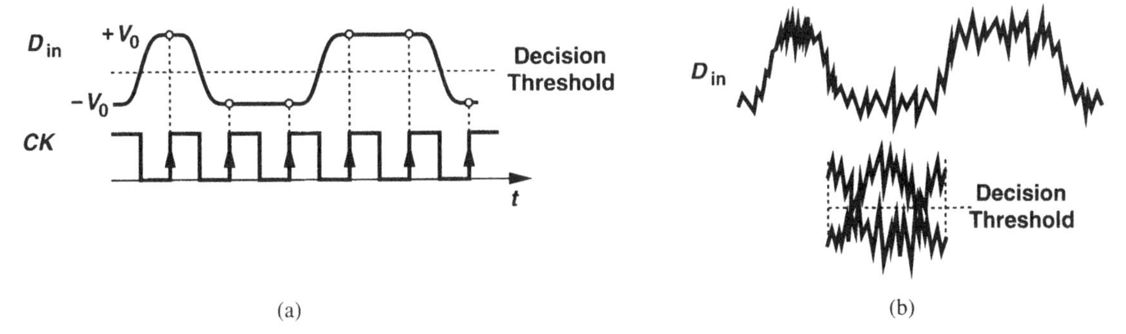

Figure 2.19 (a) Noiseless data sampled by clock, (b) noisy data and corresponding eye diagram.

threshold. The noise, $n(t)$, added to the signal degrades both the amplitude and the time resolution [Fig. 2.19(b)], closing the eye and increasing the BER. More specifically, with a greater noise amplitude, the logical levels shown in Fig. 2.19(b) cross the threshold erroneously more often.

To derive the error rate in terms of the additive noise amplitude, we follow three steps: (1) assume a noise model, e.g., an amplitude distribution ["probability density function" (PDF)] for $n(t)$; (2) obtain the amplitude distribution of signal + noise, i.e., if the binary sequence toggles between $-V_0$ and $+V_0$, calculate the PDFs of $-V_0+n(t)$ and $+V_0+n(t)$; (3) determine the probability of erroneous decision, i.e., $-V_0+n(t) > 0$ or $+V_0+n(t) < 0$.

Assuming that the noise amplitude exhibits a Gaussian distribution with zero mean, we write the PDF of $n(t)$ as:

$$P_n = \frac{1}{\sigma_n\sqrt{2\pi}} \exp\frac{-n^2}{2\sigma_n^2}, \tag{2.8}$$

where σ_n denotes the root mean square (rms) value of the noise. This equation simply means that if the noise amplitude is sampled many many times, the resulting values follow a Gaussian histogram (Fig. 2.20), with 68% of the samples falling in the range $[-\sigma_n\ \ +\sigma_n]$, 95% of the samples falling in the range $[-2\sigma_n\ \ +2\sigma_n]$, etc. As a rule of thumb, we assume the amplitude of random Gaussian noise rarely exceeds $\pm 4\sigma_n$.

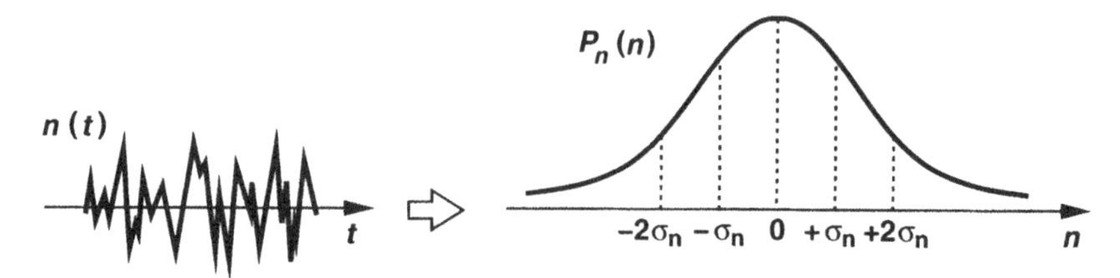

Figure 2.20 Gaussian noise distribution.

The next step is to obtain the PDF of signal + noise. What is the PDF of the binary sequence itself? If ONEs and ZEROs occur with equal probabilities, then the PDF consists of two impulses at $x = -V_0$ and $x = +V_0$, each having a weight of $1/2$ [Fig. 2.21(a)]. This is because $x(t)$ assumes no other value with a nonzero probability.[5] How is the amplitude

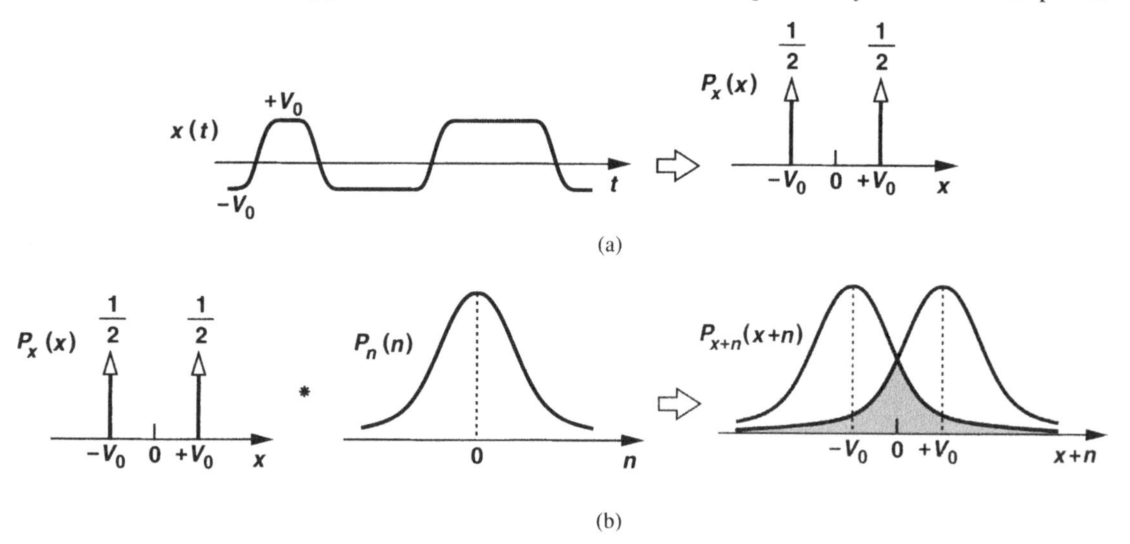

Figure 2.21 (a) PDF of noiseless random data, (b) effect of additive noise on PDF.

distribution of $x(t) + n(t)$ determined? We recall from the theory of probability that if two independent random variables are added, their PDFs are convolved. Thus, as illustrated in Fig. 2.21(b), $x(t) + n(t)$ displays a PDF consisting of two Gaussian distributions centered around $-V_0$ and $+V_0$, each having an rms value of σ_n. The shaded tails represent samples of $-V_0 + n(t)$ that are positive and samples of $+V_0 + n(t)$ that are negative, i.e., erroneous bits.

The last step is to calculate the probability of error. The probability that the actual bit is a logical ZERO but the received level, $-V_0 + n(t)$, is positive is given by the shaded area

[5]We assume the detector samples the midpoint of each bit, ignoring rising and falling edges of the data.

in Fig. 2.21(b) from 0 to $+\infty$:

$$P_{0\to 1} = \frac{1}{2}\int_0^{+\infty} \frac{1}{\sigma_n\sqrt{2\pi}} \exp\frac{-(u+V_0)^2}{2\sigma_n^2} du, \tag{2.9}$$

where the exponential term represents a Gaussian function shifted to the left by V_0. Similarly,

$$P_{1\to 0} = \frac{1}{2}\int_{-\infty}^{0} \frac{1}{\sigma_n\sqrt{2\pi}} \exp\frac{-(u-V_0)^2}{2\sigma_n^2} du. \tag{2.10}$$

If ONEs and ZEROs arrive with equal probabilities and the noise corrupts high and low levels equally, then $P_{0\to 1} = P_{1\to 0}$ and we need calculate only one. A change of variable, $z = (u+V_0)/\sigma_n$, simplifies Eq. (2.9) to

$$P_{0\to 1} = \frac{1}{2}\int_{V_0/(\sigma_n)}^{\infty} \frac{1}{\sqrt{2\pi}} \exp\frac{-z^2}{2} dz \tag{2.11}$$

$$= \frac{1}{2} Q\left(\frac{V_0}{\sigma_n}\right), \tag{2.12}$$

where $Q(\cdot)$ is called the "Q function" and defined as

$$Q(x) = \int_x^{\infty} \frac{1}{\sqrt{2\pi}} \exp\frac{-u^2}{2} du. \tag{2.13}$$

The total probability of error is therefore equal to

$$P_{tot} = Q\left(\frac{V_0}{\sigma_n}\right). \tag{2.14}$$

Note that V_0 is *half* of the peak-to-peak signal swing and σ_n is the rms value of noise. For simplicity, we may also write:

$$P_{tot} = Q\left(\frac{V_{pp}}{2\sigma_n}\right), \tag{2.15}$$

where $V_{pp} = 2V_0$. The dependence of P_{tot} upon V_{pp}/σ_n is expected because the latter is a measure of the signal-to-noise ratio (SNR) in the system. As explained in Chapter 4, σ_n is obtained by integrating the noise across the entire bandwidth, BW, e.g., $\sigma_n \propto \sqrt{\text{BW}}$.

The Q function is not available in closed form, but for $x > 3$, it can be approximated with high accuracy by

$$Q(x) \approx \frac{1}{x\sqrt{2\pi}} \exp\frac{-x^2}{2}. \tag{2.16}$$

As an example, suppose a transimpedance amplifier produces an output noise voltage of $\sigma_n = 1\ \text{mV}_{rms}$. If the peak-to-peak signal swing at the output is equal to 10 mV, what is the error rate? Since $V_0 = 10\ \text{mV}/2 = 5\ \text{mV}$ and $\sigma_n = 1\ \text{mV}_{rms}$, we can substitute $x = V_0/\sigma_n = 5$ in Eq. (2.16), obtaining $Q(5) = 2.97 \times 10^{-7}$. This result implies that nearly 3 bits out of every 10 million bits are likely to be excessively corrupted.

In practice, we often know the required bit error rate and must determine the corresponding V_{pp}/σ_n ratio. This is possible by iteration in Eq. (2.16). Figure 2.22 plots $Q(x)$ from both Eqs. (2.13) and (2.16). Note the strong dependence of the error rate upon x. As

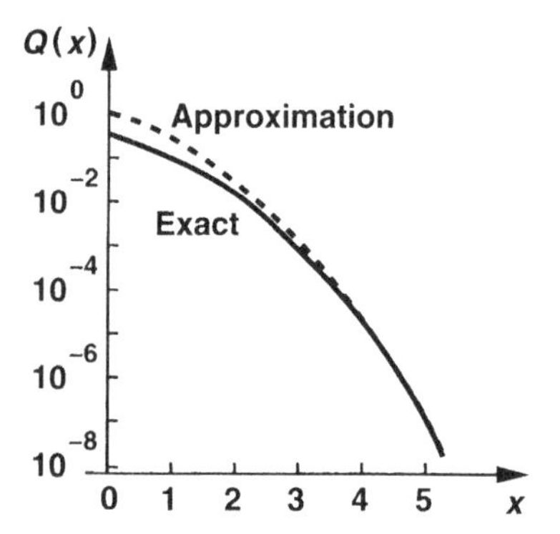

Figure 2.22 Q function.

a rule of thumb, BER drops by two orders of magnitude for each decibel increase in SNR. In hand calculations, it is helpful to remember that $Q(6) \approx 10^{-9}$, $Q(7) \approx 10^{-12}$, and $Q(8) \approx 10^{-15}$.

Our studies of ISI and noise point to an important trade-off: if the bandwidth of an amplifier is excessively narrow, random data experiences substantial ISI. On the other hand, if the bandwidth is excessively wide, the signal suffers from large noise. As we will see in Chapter 4, a reasonable compromise results if the bandwidth is approximately equal to 0.7 times the data rate.

2.6 Phase Noise and Jitter

The design of communication systems and circuits must often deal with phase noise and jitter. These two effects are closely related and must be studied in the context of oscillators and phase-locked loops.

2.6.1 Phase Noise

Consider a hypothetical oscillator producing an ideal sinusoid at a frequency $\omega_0 = 2\pi f_0 = 2\pi/T_0$. The output waveform can be expressed as $V_{out} = V_0 \cos \omega_0 t$ and the spectrum (Fourier transform) consists of two impulses at $\omega = \pm\omega_0$ [Fig. 2.23(a)]. Note that the period is constant and the zero-crossing points occur at exactly integer multiples of T_0. Also, the spectrum indicates that the signal carries no energy at any frequency other than ω_0. For example, if the waveform is applied to a very narrowband filter centered around $\omega_0 + \Delta\omega$, then the output is zero for $\Delta\omega \neq 0$.

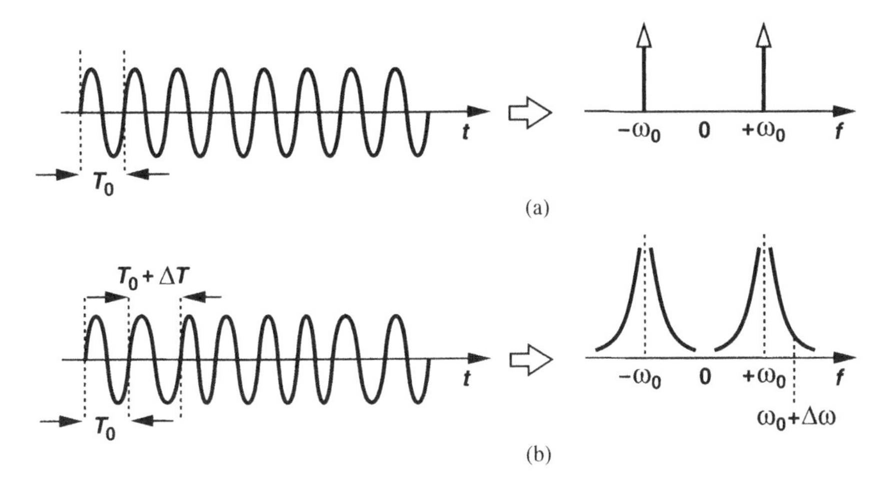

Figure 2.23 (a) Spectrum of a noiseless sinusoid, (b) effect of jitter.

In actual oscillators, circuit and system noise varies the period of oscillation randomly—as if the oscillator occasionally operates at frequencies other than ω_0 [Fig. 2.23(b)]. Here, the zero crossings may not occur at integer multiples of T_0 and the output spectrum exhibits "skirts" around the original impulses, revealing that the signal carries a finite energy at $\omega_0 + \Delta\omega$. For now, we assume that the amplitude of the output waveform is constant and unaffected by noise.

How is phase noise expressed mathematically? Since the instantaneous frequency varies randomly, we may be tempted to write $V_{out} = V_0 \cos\{[\omega_0 + \omega_n(t)]t\}$, where $\omega_n(t)$ is a small random component that represents the frequency variations. However, as explained in Chapter 6, this expression is incorrect and must be rewritten as

$$V_{out} = V_0 \cos[\omega_0 t + \phi_n(t)], \tag{2.17}$$

where $\phi_n(t)$ is a small random phase component with zero average. Intuitively, we note that the zero crossings are "modulated" by $\phi_n(t)$ because they appear at $\omega_0 t + \phi_n(t) = k\pi/2$ for odd k, that is, $t = k(\pi/2)/\omega_0 - \phi_n(t)/\omega_0$. Equivalently, the period of oscillation varies from one cycle to the next.

The quantity $\phi_n(t)$ in Eq. (2.17) (more specifically, its spectrum, $S_{\phi n}$) is called phase noise. Since ϕ_n is typically very small, we can assume $\phi_n(t) \ll 1$ rad, i.e., $\cos\phi_n \approx 1$ and $\sin\phi_n \approx \phi_n$, thus simplifying Eq. (2.17) to

$$V_{out}(t) = V_0 \cos\omega_0 t \cos[\phi_n(t)] - V_0 \sin\omega_0 t \sin[\phi_n(t)] \tag{2.18}$$

$$\approx V_0 \cos\omega_0 t - V_0 \phi_n(t) \sin\omega_0 t. \tag{2.19}$$

It follows that the spectrum of V_{out} consists of (a) impulses at $\pm\omega_0$ and (b) the spectrum of $\phi_n(t)$ translated to $\pm\omega_0$ due to multiplication by $\sin\omega_0 t$, i.e., $S_{\phi n}(\omega \pm \omega_0)$. This result agrees with the illustration in Fig. 2.23(b). The exact shape of $S_{\phi n}$ is derived in [7], but it can often be approximated by $1/(\Delta\omega)^2$.

To quantify $S_{\phi n}$, we measure the average power carried in 1 Hz in the phase noise skirts of Fig. 2.23(b). Since the spectrum of ϕ_n is not flat, the power must be measured at a specified frequency "offset," $\Delta\omega$, from ω_0 (Fig. 2.24). Also, to allow a fair comparison of different oscillators, the measured power in 1 Hz at $\Delta\omega$ must be normalized to the "carrier"

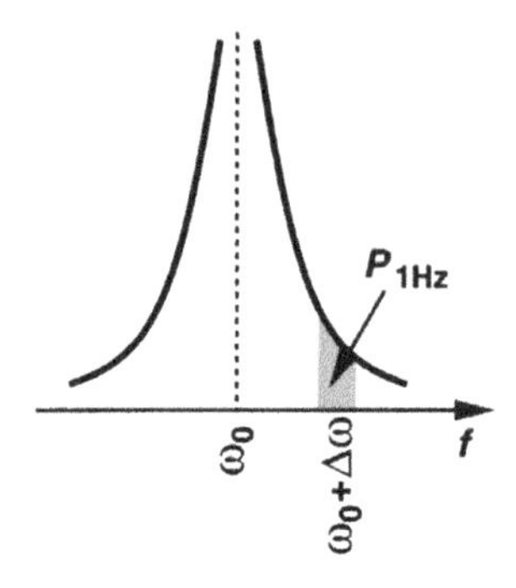

Figure 2.24 Definition of phase noise.

power, P_c, i.e., the power carried by the impulse at ω_0. In Eq. (2.19), the carrier power is approximately equal to $V_{rms}^2 = V_0^2/2$. The result is expressed as

$$\text{Relative Phase Noise}|_{\Delta\omega} = 10\log\frac{P_{1\text{Hz}}|_{\Delta\omega}}{P_c}\ \text{dBc/Hz}, \tag{2.20}$$

where the unit dBc/Hz denotes decibels with respect to the carrier, emphasizing the normalization.

As an example, suppose the phase noise spectrum of an oscillator is given by

$$P_{1\text{Hz}}|_{\Delta\omega} = \frac{(50\ \text{mV}_{\text{rms}})^2}{\Delta\omega^2}. \tag{2.21}$$

If the oscillation amplitude is equal to 0.5 V_{rms}, what is the relative phase noise at 100-kHz offset? At $\omega = 2\pi \times 100$ kHz, we have

$$S_{\phi n}(2\pi \times 100\ \text{kHz}) = 6.333 \times 10^{-15}\ \text{V}^2/\text{Hz}. \tag{2.22}$$

Normalizing this value to the carrier power, $(0.5\text{V}_{\text{rms}})^2$, we obtain $S_{\phi n}(2\pi \times 100\ \text{kHz})/(0.5\ \text{V}_{\text{rms}})^2 = 2.533 \times 10^{-14}$. Thus,

$$\text{Relative Phase Noise} = 10\log(2.533 \times 10^{-14}) \tag{2.23}$$

$$\approx -136\ \text{dBc/Hz}. \tag{2.24}$$

It is important to note that phase noise must be specified at a frequency offset. Typical numbers used in practice are $\Delta\omega = 2\pi \times (100\ \text{kHz})$ or $\Delta\omega = 2\pi \times (1\ \text{MHz})$.

2.6.2 Jitter

In optical communications, the time-domain behavior of phase noise is of great interest because it represents the extent to which the zero crossings of a waveform are corrupted. The deviation of the zero crossings from their ideal position in time is called jitter. Alternatively, the deviation of each period from the ideal value can be called jitter. These views bear subtle differences, leading to several definitions of jitter [7].

Suppose a noisy oscillator operates at a nominal frequency $\omega_0 = 2\pi f_0 = 2\pi/T_0$ and its output is plotted against an ideal square wave with period T_0 [Fig. 2.25(a)].[6] To quantify jitter, we can measure the deviation of each positive (or negative) transition point of $x_2(t)$

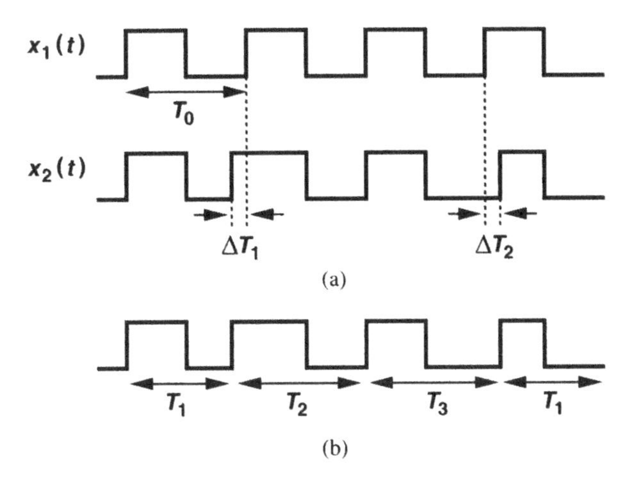

Figure 2.25 (a) Cycle-to-cycle jitter, (b) variable cycles.

from its corresponding point in $x_1(t)$, i.e., ΔT_1, ΔT_2, etc. This type of jitter is called "absolute jitter" because it results from comparison with an ideal reference, i.e., absolute points in time. Since the deviations are random, we measure a very large number of ΔT's and determine the root mean square value of absolute jitter as:

$$\Delta T_{abs,rms} = \lim_{N\to\infty} \frac{1}{N}\sqrt{\Delta T_1^2 + \Delta T_2^2 + \cdots + \Delta T_N^2}. \tag{2.25}$$

Another type of jitter is "cycle-to-cycle jitter," obtained by measuring the difference between every two consecutive cycles of the waveform and taking the rms value [Fig. 2.25(b)]:

$$\Delta T_{cc,rms} \approx \lim_{N\to\infty} \frac{1}{N}\sqrt{(T_2 - T_1)^2 + (T_3 - T_2)^2 + \cdots + (T_N - T_{N-1})^2}. \tag{2.26}$$

Note that cycle-to-cycle jitter does not require a reference signal.

---

[6]To visualize jitter more easily, we consider square waveforms rather than sinusoids.

Absolute and cycle-to-cycle jitter are commonly used to characterize the quality of signals in the time domain. A third type of jitter that may also prove useful is "period jitter" [7], defined as the deviation of each cycle from the *average* period of the waveform, $\overline{T}$:

$$T_{p,rms} = \lim_{N\to\infty} \frac{1}{N}\sqrt{(\overline{T}-T_1)^2 + (\overline{T}-T_2)^2 + \cdots + (\overline{T}-T_N)^2}. \tag{2.27}$$

For a phase-locked oscillator, absolute jitter and period jitter are equal.

2.6.3 Relationship Between Phase Noise and Jitter

The phase noise of oscillators can be measured in simulations or in the laboratory much more readily than the jitter can. It is therefore desirable to establish a relationship between the two quantities. Here, we develop a simple expression that provides insight and practical utility. The reader is referred to [7, 8, 9] for more details.

For absolute jitter, we compare the actual signal with an ideal reference, noting that the deviation of each zero crossing is $\Delta T_j = (2\pi/T_0)\phi_{n,j}$, where $\phi_{n,j}$ denotes the value of ϕ_n (in radians) in the vicinity of zero crossing number j. Thus,

$$\Delta T^2_{abs,rms} = \lim_{N\to\infty}\frac{1}{N}\sum_{j=1}^{N}\Delta T_j^2 \tag{2.28}$$

$$= \left(\frac{2\pi}{T_0}\right)^2 \lim_{N\to\infty}\frac{1}{N}\sum_{j=1}^{N}\phi^2_{n,j}. \tag{2.29}$$

The summation can be approximated by an integral:

$$\Delta T^2_{abs,rms} = \left(\frac{2\pi}{T_0}\right)^2 \lim_{T\to\infty}\frac{1}{T}\int_{-T/2}^{T/2}\phi_n^2(t)dt. \tag{2.30}$$

The limit represents the average power of ϕ_n and, from Parceval's theorem [1], is equivalent to the area under the spectrum of ϕ_n:

$$\Delta T^2_{abs,rms} = \left(\frac{2\pi}{T_0}\right)^2 \int_{-\infty}^{+\infty} S_{\phi n}(f)df. \tag{2.31}$$

The cycle-to-cycle jitter and the phase noise can also be related by an expression [7]. This is described in Chapter 9.

2.6.4 Jitter Due to Additive Noise

We observed in Section 2.4 that the bandwidth of an amplifier must be chosen properly so as to achieve a reasonable balance between noise and ISI. We also quantified the effect of noise on the *amplitude* of data bits and hence the bit error rate. In this section, we examine the effect of additive noise on the *transitions* of data, arriving at the resulting jitter.

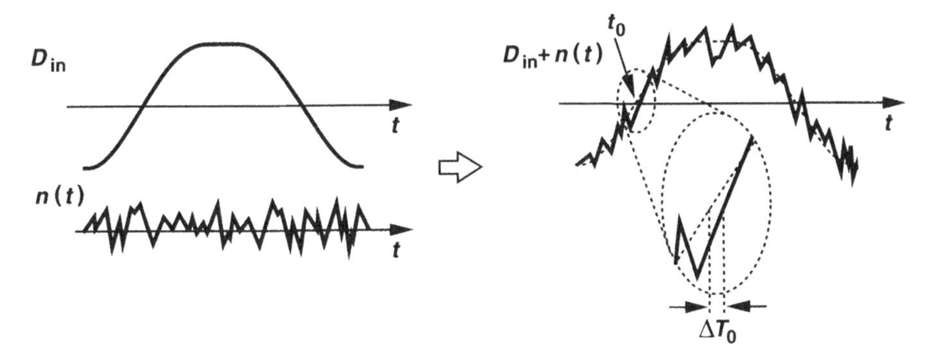

Figure 2.26 Effect of additive noise on jitter.

Suppose, as shown in Fig. 2.26, a random binary sequence D_{in} is corrupted by additive noise $n(t)$. Focusing on the zero crossings of $D_{in}+n(t)$, we recognize that if, in the vicinity of $t = t_0$, the noise assumes a value of $n(t_0)$, then the zero crossing of $D_{in}+n(t_0)$ deviates from the ideal value by

$$\Delta T_0 = \frac{n(t_0)}{S}, \tag{2.32}$$

where S denotes the slope (slew rate) of the transition. Thus, the sharper the edge, the less the effect of additive noise.

To gain more insight, let us assume an ideal random binary sequence with period T_b is applied to an amplifier having a first-order low-pass transfer function with $f_{-3\text{dB}} = 0.7/T_b$ (Fig. 2.27). We wish to compute the jitter resulting from the amplifier noise. Since each

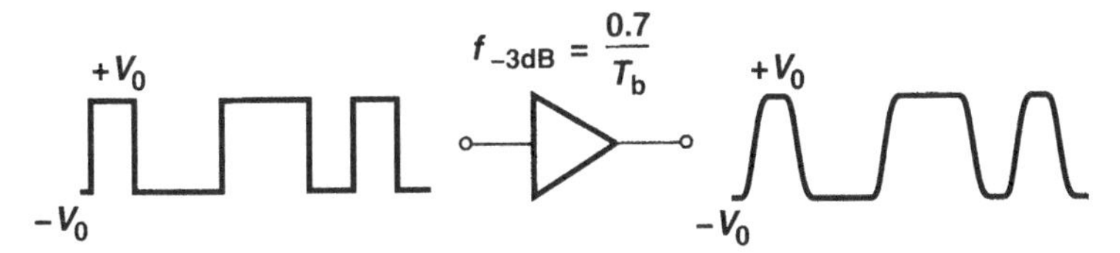

Figure 2.27 Effect of low-pass filtering on jitter.

edge of D_{out} experiences an exponential change of the form $-V_0 + 2V_0[1 - \exp(-t/\tau)]$, where $\tau = T_b/(2\pi \times 0.7)$, we can write the slope as:

$$|S| = \frac{2V_0}{\tau} \exp \frac{-t}{\tau}. \tag{2.33}$$

Each edge crosses zero when $\exp(-t_0/\tau) = 0.5$. The slope at the zero crossing is therefore equal to:

$$|S_{t_0}| = \frac{V_0}{\tau} \tag{2.34}$$

$$= \frac{1.4\pi}{T_b} V_0. \tag{2.35}$$

Substituting for S in Eq. (2.32), we now have

$$\Delta T_0 = \frac{n(t_0)}{1.4\pi V_0} T_b. \tag{2.36}$$

As an example, recall from Section 2.5 that for BER $= 10^{-12}$, we must have $V_0/n_{rms}(t_0) = 7$. Thus,

$$\Delta T_{0,rms} = \frac{T_b}{1.4\pi \times 7} \tag{2.37}$$

yielding a normalized jitter of

$$\frac{\Delta T_{0,rms}}{T_b} = \frac{1}{9.8\pi} \tag{2.38}$$

$$\approx 3.25\%. \tag{2.39}$$

This amount of jitter may not be acceptable in some applications.

2.7 Transmission Lines

A good understanding of transmission lines (T lines) is essential to the design of high-speed circuits. We briefly review the basic properties of T lines here and return to more advanced topics in Chapter 5.

2.7.1 Ideal Transmission Lines

A transmission line (T line) is an interconnect whose length is a significant fraction of the wavelength of interest or, equivalently, whose end-to-end delay is not negligible with respect to other time scales in the environment. For example, consider the simple structure shown in Fig. 2.28(a), where an ideal interconnect (with zero resistance) and a ground plane form a T line. A step applied to one end experiences delay as it propagates through the line. As depicted in Fig. 2.28(b), the line can be modeled by a distributed LC ladder, where L_0 and C_0 denote the inductance and capacitance of the line per unit length, respectively. The delay is calculated from the velocity of the wave, given by

$$v = \frac{1}{\sqrt{L_0 C_0}}. \tag{2.40}$$

Note that the propagation behavior and the delay in this case are quite different from those of long on-chip interconnects exhibiting substantial series *resistance*.

The propagation delay in T lines yields an interesting property: if a step is applied to one end, the instantaneous behavior at that end does not depend on the length of the line or

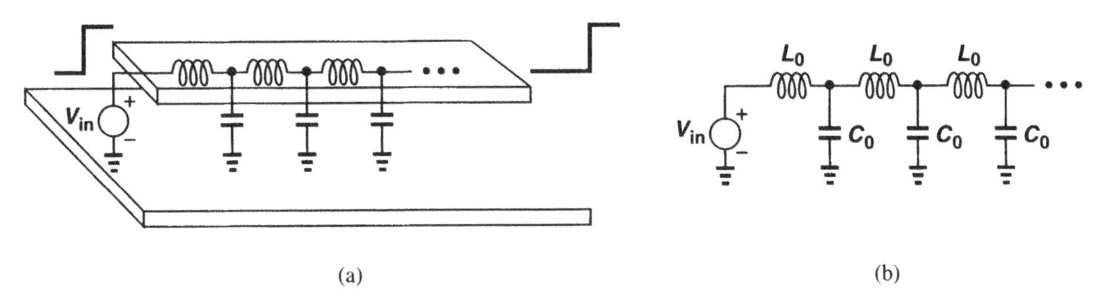

Figure 2.28 Distributed LC model of a T line.

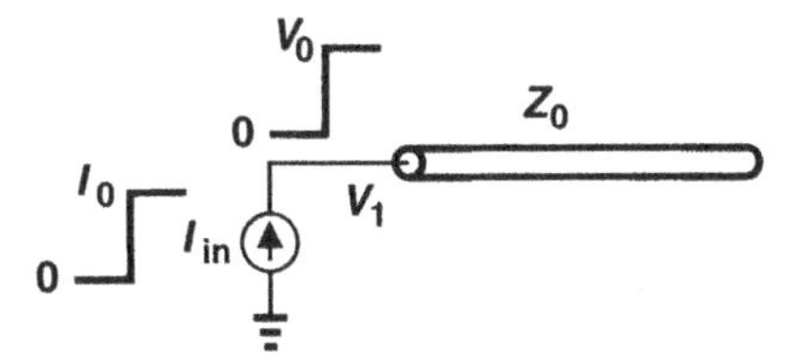

Figure 2.29 Current step at the input of an infinitely long T line.

the load seen at the other end. For example, an input current step gives rise to a voltage step whose height is a *fundamental* property of the line (Fig. 2.29). In fact, we have $V_0/I_0 = Z_0$, where Z_0 represents the "characteristic impedance" of the line and is equal to

$$Z_0 = \sqrt{\frac{L_0}{C_0}}. \tag{2.41}$$

Note that Z_0 in this case is a real number, containing no reactance. This example suggests that the characteristic impedance can be viewed as the "instantaneous" input impedance of a T line.

The voltage step generated in Fig. 2.29 propagates on the line, establishing a current of I_0 at each point as it travels. If the far end is terminated by a resistance equal to Z_0 (Fig. 2.30), then the relationship $V_0/I_0 = Z_0$ remains valid as the wave reaches the load,

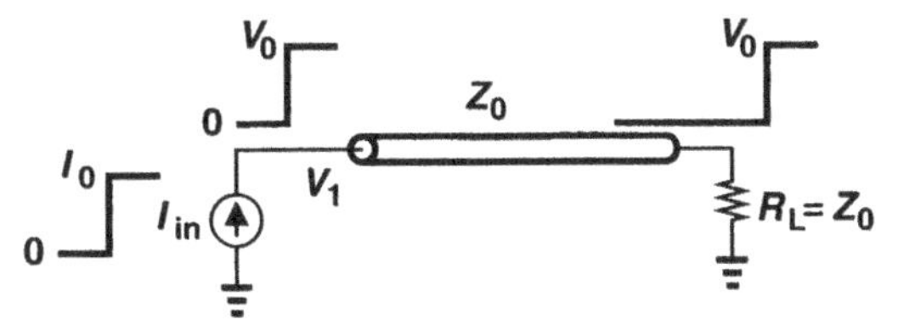

Figure 2.30 T line with matched load.

and the transient ceases thereafter. On the other hand, if $R_L \neq Z_0$, the voltage and current waveforms approaching the load violate Ohm's law, requiring that a *reflection* be generated.

For example, if $R_L = \infty$ (Fig. 2.31), then when the I_0 wave reaches the end, a current wave equal to I_0 must start from the load and propagate to the left so that the total current through $R_L = \infty$ is still zero. The reflected current wave produces a voltage step of height

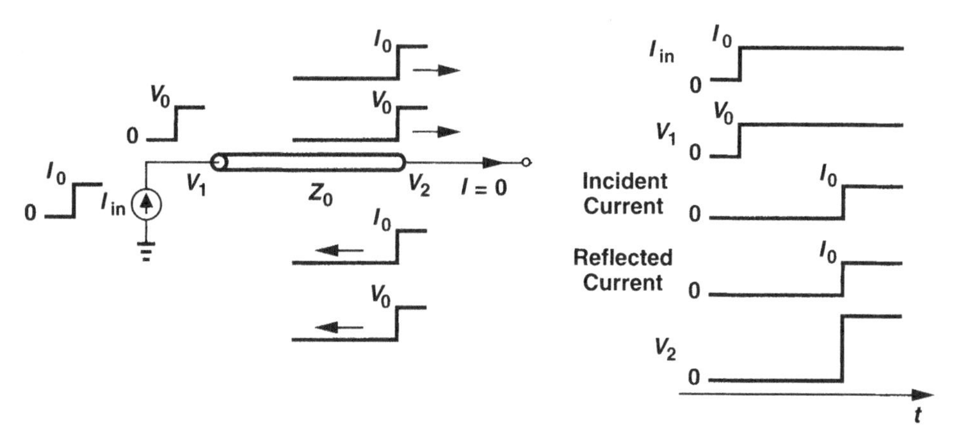

Figure 2.31 T line with open load.

V_0, raising the local potential on the line to $2V_0$ as it travels toward the source.

It is also instructive to study the case $R_L = 0$. As illustrated in Fig. 2.32, the V_0 wave reaching the end demands a reflection of $-V_0$ so as to maintain zero potential across the

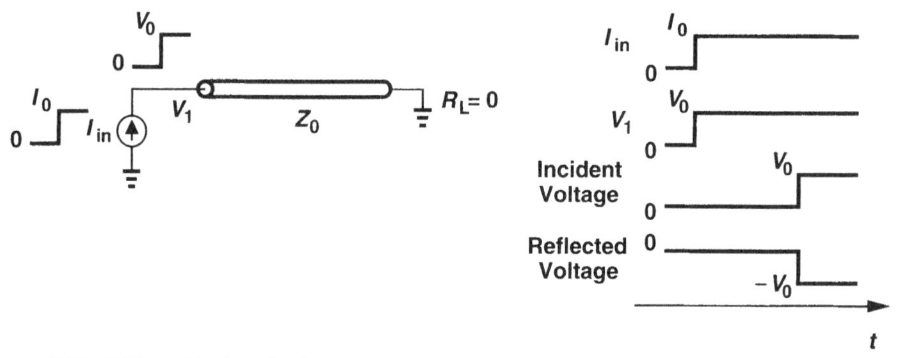

Figure 2.32 T line with short load.

load. The $-V_0$ component then propagates to the left, forcing the local voltage to zero as it travels on the line. The reader is encouraged to follow the transients in Figs. 2.31 and 2.32 as they go back and forth.

The above analysis indicates that if the termination at the far end deviates from Z_0, then reflection is inevitable. The reflection can nevertheless be absorbed if the *source* impedance is equal to Z_0. For example, consider the arrangement shown in Fig. 2.33, where a voltage source having an output resistance of $R_S = Z_0$ provides a step of height V_0. The wave amplitude is immediately halved at the near end because of the voltage division between R_S and the instantaneous input impedance of the T line. The resulting step travels to the far end, sees $R_L = \infty$ and generates a reflection of amplitude $V_0/2$, thereby raising the po-

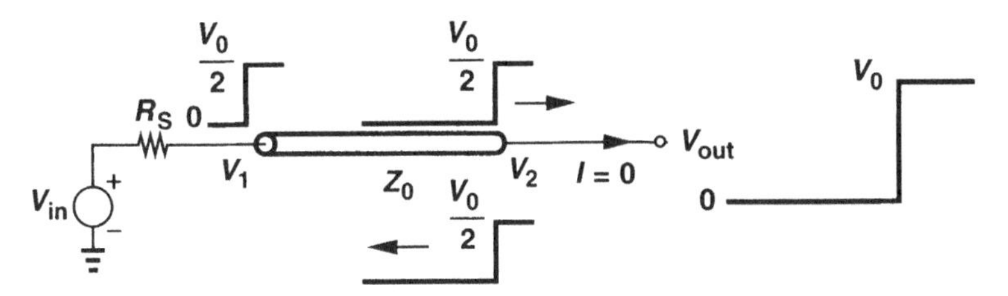

Figure 2.33 Absorption of secondary reflection by source impedance.

tential at the load to V_0. The reflection propagates to the left, reaches the source, and, with $R_S = Z_0$, produces no "secondary" reflections. This important observation is exploited in the design of high-speed output buffers.

For sinusoidal excitation of a matched T line, we express the wave propagation as:

$$V(t, x) = V_0 \cos(\frac{2\pi t}{T} t - \frac{2\pi}{\lambda} x), \tag{2.42}$$

where T denotes the temporal period and λ the spatial period (wavelength). Note that $\lambda = v \cdot T$. It is common to write $2\pi/\lambda = \beta = 2\pi/(v \cdot T) = \omega\sqrt{L_0 C_0}$. As depicted in Fig. 2.34, the waveform repeats along the x axis every λ meters. Also, a snapshot of the

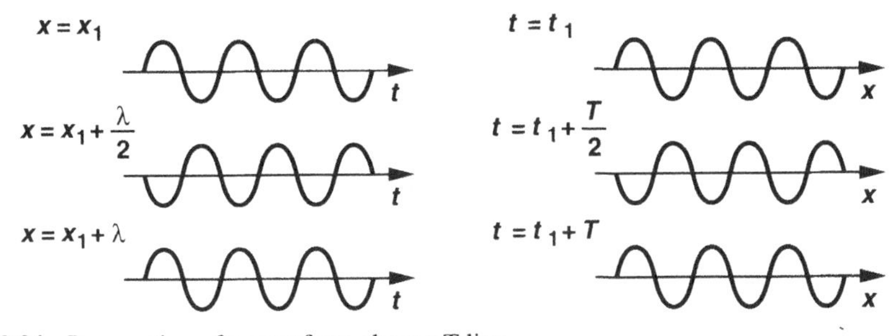

Figure 2.34 Propagation of a waveform along a T line.

waveform taken at $t = t_1$ is identical to that at $t = t_1 + T, t = t_1 + 2T$, etc.

Our study of T lines thus far has neglected the resistance of the conductors. If a lossless uniform line is matched at the far end, then a step propagates and reaches the load with no distortion or attenuation while experiencing a frequency-independent delay. We therefore say lossless T lines exhibit an infinite bandwidth and a linear phase shift, i.e., a phase shift linearly proportional to the frequency. If the T line suffers from substantial resistance (or, equivalently, is very long), then the resulting loss must be taken into account.

2.7.2 Lossy Transmission Lines

T lines suffer from losses in the conductors and in the dielectric material. The distributed model of Fig. 2.28(b) is therefore modified as shown in Fig. 2.35 to represent the losses. Here, the resistance R_S embodies both low-frequency and high-frequency conductor losses

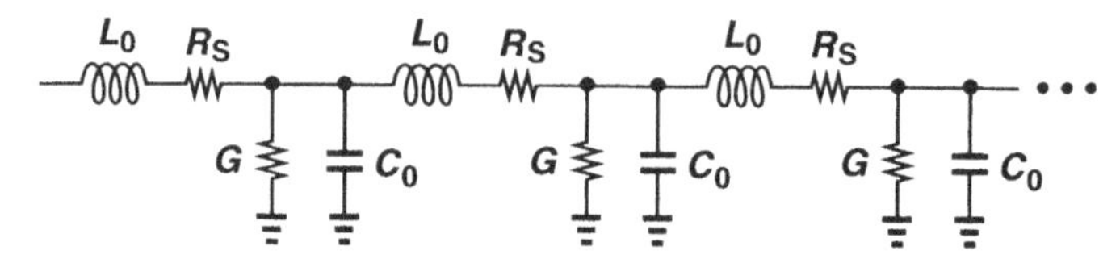

Figure 2.35 Model of lossy T line.

while conductance G models the dielectric loss. The characteristic impedance of such a line is equal to [10]:

$$Z_0 = \sqrt{\frac{j\omega L_0 + R_S}{j\omega C_0 + G}}, \tag{2.43}$$

a complex quantity.[7] While it may appear that at high frequencies, Eq. (2.43) reduces to $Z_0 = \sqrt{L_0/C_0}$, we must note that line losses also become significant as ω rises. Thus, R_S and G must still be taken into account. We limit our study to lines having a small loss here.

It is instructive to consider the propagation behavior expressed by Eq. (2.42) for lossy lines. We have [10]:

$$V(t, x) = V_0 \exp(-\alpha x) \cos(\frac{2\pi t}{T} t - \beta x), \tag{2.44}$$

where α denotes the attenuation factor,

$$\alpha \approx \frac{1}{2}(R_S\sqrt{\frac{C_0}{L_0}} + G\sqrt{\frac{L_0}{C_0}}), \tag{2.45}$$

and $\beta \approx \omega\sqrt{L_0 C_0}$. Equation (2.44) is commonly abbreviated as

$$V(x) = V_0 \exp(-\alpha x) \exp(-j\beta x), \tag{2.46}$$

where it is understood that the time dependence is not shown and only the real part of $V(x)$ is of interest.

References

1. L. W. Couch, *Digital and Analog Communication Systems,* Fourth Ed., New York: Macmillan Co., 1993.
2. J. G. Proakis, *Digital Communications,* Fourth Ed., Boston: McGraw-Hill, 2001.
3. B. Sklar, *Digital Communications*, New Jersey: Prentice-Hall, 1988.
4. A. X. Widmer and P. A. Franaszek, "A DC-Balanced, Partitioned-Block, 8B/10B Transmission Code," *IBM J. Res. and Develop.,* vol. 27, pp. 440–451, Sept. 1983.

[7] Unlike Eq. (2.41), this expression is derived for a steady-state sinusoidal excitation rather than a step input.

5. R. C. Walker and R. Dugan, "Low Overhead Coding Proposal for 10Gb/s Serial Links," *IEEE 802.3 High-Speed Study Group*, Nov. 1999, http://grouper.ieee.org/groups /802/3/10G_study/public/nov99/walker_1_1199.pdf.
6. J. Hauenschild and H.-M. Rein, "Influence of transmission-line interconnections between Gbit/s ICs on time jitter and instabilities," *IEEE Journal of Solid-State Circuits,* vol. 25, pp. 763–766, June 1990.
7. F. Herzel and B. Razavi, "A Study of Oscillator Jitter Due to Supply and Substrate Noise," *IEEE Trans. Circuits and Systems, Part II,* vol. 46, pp. 56–62, Jan. 1999.
8. W. P. Robins, *Phase Noise in Signal Sources*, London: Peregrinus, Ltd., 1982.
9. A. Hajimiri, S. Limotyrakis, and T. H. Lee, "Jitter and Phase Noise in Ring Oscillators," *IEEE Journal of Solid-State Circuits,* vol. 34, pp. 790–804, June 1999.
10. S. Ramo, J. R. Whinnery, and T. van Duzer, *Fields and Waves in Communication Electronics,* Second Ed., New York: Wiley, 1984.

CHAPTER 3

Optical Devices

The design of integrated circuits for optical communications is heavily influenced by the limitations of optical components. For example, to determine the required input-referred noise current of a transimpedance amplifier, the designer must know the transmitter output power, the fiber loss, and the photodiode response. Also, it is desirable to electrically compensate for optical device nonidealities such as fiber dispersion to save cost and space.

This chapter deals with three main components of optical systems: laser diodes, optical fibers, and photodiodes. We first study the principles of laser action and present Fabry-Perot (FP) and distributed-feedback (DFB) lasers. Next, we review loss and dispersion characteristics of optical fibers and describe photodiodes such as PIN and avalanche structures. Finally, we consider the effect of nonidealities in a typical OC system.

For a more detailed study, the reader is referred to the numerous books written on the subject, e.g., [1, 2, 3, 4].

3.1 Laser Diodes

A laser diode is a semiconductor device that produces light (not necessarily visible) in response to a current. Unlike light-emitting diodes (LEDs), lasers generate an extremely sharp spectral line, i.e., the output light energy is heavily concentrated around one frequency (or wavelength). In this context, we say lasers are "monochromatic."[1] Furthermore, lasers produce extremely focussed beams that can travel a long distance with little spatial divergence. (Laser pointers are a low-cost example.)

Figure 3.1(a) shows the input-output characteristic of a typical laser diode. If the current is less than a threshold value, I_{TH}, the optical power is small and the device does not operate as a laser. As the current exceeds I_{TH}, P_{out} rises relatively linearly until it approaches a saturated level. In optical communications, the current may be switched between two levels [Fig. 3.1(b)], thereby modulating the output power.

[1] The term monochromatic means "having a single color."

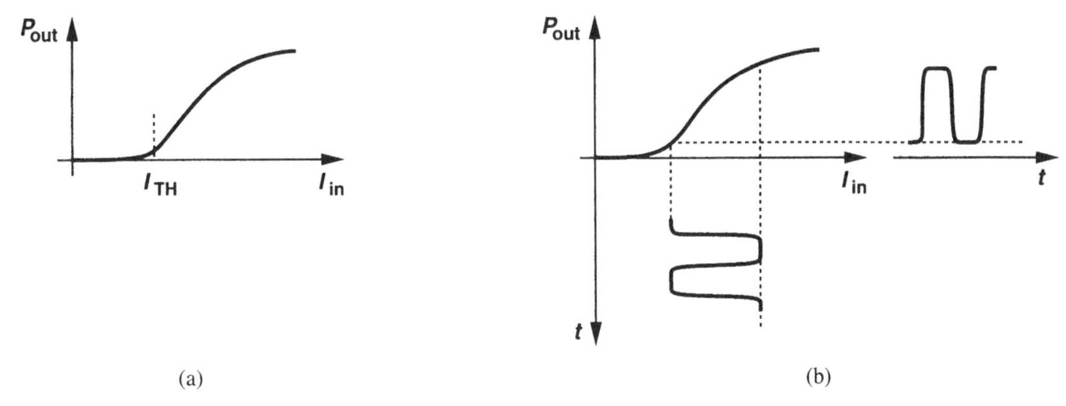

Figure 3.1 (a) Typical laser diode characteristic, (b) input and output waveforms.

What aspects of lasers are important in optical communications? We study a number of parameters that must be considered in transceiver design.

Efficiency Efficiency represents the optical power produced by the laser for a given input current. A more precise measure is the "quantum efficiency," defined as the number of output photons divided by the number of input electrons, both measured in a given time interval. Efficiency determines the current that the laser driver circuit must provide so as to generate a given optical power.

Spectral Purity A laser driven by a constant current ideally "lases" at a single wavelength, i.e., the output light is perfectly monochromatic. In reality, however, the light spectrum is smeared [Fig. 3.2(a)] or contains "side modes" [Fig. 3.2(b)].[2] As explained in Section 3.2.2, such nonidealities lead to other issues in propagation through fibers.

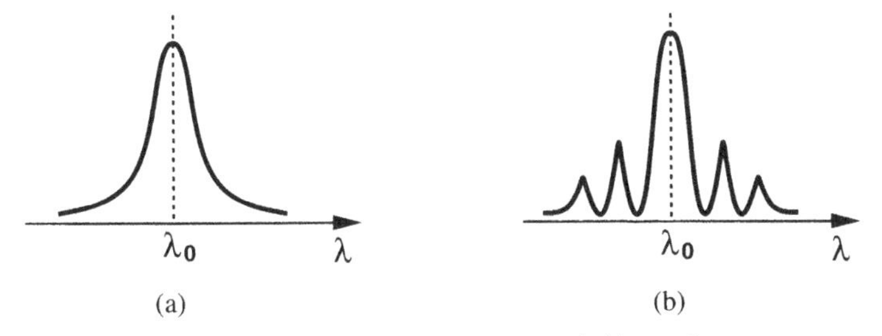

Figure 3.2 (a) Output spectrum of a laser, (b) appearance of side modes.

Threshold Current From the circuit design standpoint, it is desirable that I_{TH} in Fig. 3.1(a) be small so that the laser driver does not consume substantial idle power. More importantly, a low I_{TH} allows a high ratio between the logical ONEs and ZEROs, easing the task of detection.

[2]In the optical domain, the spectra are usually plotted as a function of the light wavelength, λ, rather than its frequency.

Voltage Drop The voltage across a laser diode includes two components: one due to I_{TH} and another resulting from current swings. Typical lasers exhibit voltage swings in the range of 1.5 to 3 V, requiring high supplies and large transistor breakdown voltages.

Switching Speed With optical communication rates approaching 40 to 100 Gb/s, lasers that can turn on and off rapidly become essential. As explained in Section 3.1.3, several phenomena limit the switching speed of lasers.

Output Power The laser output power determines both the maximum tolerable attenuation in the fiber and the sensitivity required of the receiver. High-power lasers must use large semiconductor devices and hence suffer from a high input capacitance, making the design of high-speed laser drivers difficult.

Lifetime Lasers must typically operate for about 10^5 hours (approximately 12 years). Due to aging, the laser characteristics change considerably during this time. In particular, the threshold current varies substantially with aging (and with temperature), requiring a monitor device and feedback circuitry to ensure a constant output power (Section 3.1.3).

Other parameters of lasers such as linearity and tunability may also become important in some applications.

3.1.1 Operation of Lasers

A laser is an optical oscillator. As explained in Chapter 6, a system oscillates if it provides both amplification and positive feedback at the frequency of interest. Thus, a laser can be viewed as an optical amplifier embedded in an optical feedback loop.

How can light be amplified as it travels through a medium? We must first understand the concepts of "spontaneous emission" and "stimulated emission." Suppose all of the atoms in a material exhibit two energy levels [Fig. 3.3(a)]. This means the outer electrons of the atoms can assume an energy level of E_1 or E_2. For example, in semiconductors, E_1 and E_2 correspond to the valence band and the conduction band, respectively. Recall from physics that an electron in the lower state may be excited to the upper state if it absorbs a photon having proper energy [Fig. 3.3(b)]. Similarly, an electron in the upper state may spontaneously fall to E_1 and emit a photon [Fig. 3.3(c)]. Called spontaneous emission, this effect is exploited in many familiar light sources, including fluorescent lamps and LEDs. More specifically, the operation of LEDs is based on the transition of electrons from the conduction band to the valence band (and recombination with holes) and emission of photons resulting from this transition.

Spontaneous emission is characterized by two properties. First, it entails a certain lifetime, i.e., the electrons falling to E_2 remain at this level for some time before returning to E_1. Second, the photons produced by spontaneous transitions assume arbitrary phases (and polarization), creating "incoherent" light.

The transition of electrons from a higher state need not be spontaneous. If a photon entering the material carries proper energy, it may *stimulate* an electron in E_2, forcing it to jump to the lower level and emit another photon [Fig. 3.3(d)]. The new photon is now *in phase* with the original one, and the two continue to travel through the material and stimulate more electrons, creating new photons. Called "stimulated emission," this effect

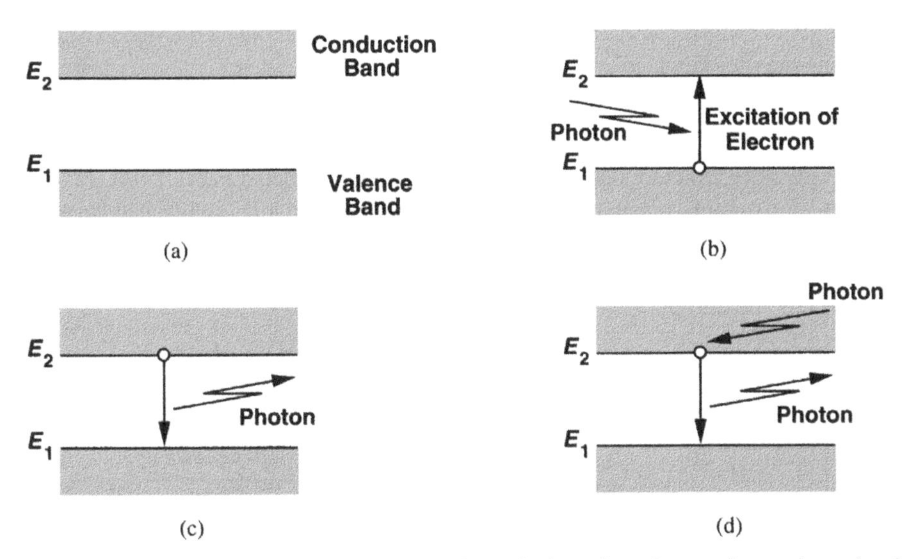

Figure 3.3 (a) Energy bands in a semiconductor, (b) excitation of an electron from valence band by a photon, (c) transition of an electron from conduction band, (d) excited transition of an electron from conduction band.

may continue indefinitely if electrons in the upper state are supplied steadily.[3] Of course, some of the photons are absorbed by the electrons that jump from E_1 to E_2.

If the density of electrons in the upper state in Fig. 3.3(d) somehow exceeds that in the lower state, the light passing through the material creates more stimulated emission than photon absorption, i.e., it is amplified. Called "population inversion," this condition occurs if the electrical current "pumped" through the laser is large enough to supply the high-energy electrons.

With light amplification available, we now seek methods of creating an optical oscillator. A simple approach is to confine the medium between two mirrors having a proper spacing such that the light reflects from each and experiences a total phase shift of zero in one round trip (Fig. 3.4). In practice, one of the mirrors is semitransparent, allowing most

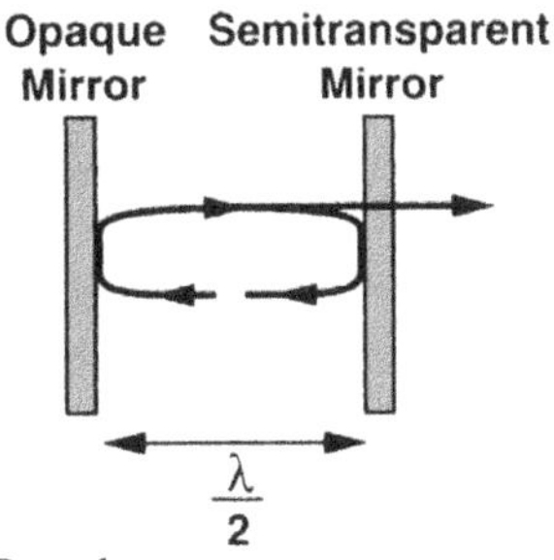

Figure 3.4 Simple view of a Fabry-Perot laser.

[3]The term "laser" is the acronym for "light amplification by stimulated emission of radiation."

of the light to exit the cavity (serving as the laser output) and only a fraction to be reflected to sustain the oscillation.

3.1.2 Types of Lasers

The simple laser structure described in Section 3.1.1 is called a Fabry-Perot (FP) laser. It consists of a suitable medium (e.g., a piece of semiconductor) sandwiched between two mirrors. Figure 3.5 depicts an example, where the electrical current flows from top to bottom and the light exits from the right facet. Each dimension is on the order of a few hundred microns. As the current forced through the medium increases, the structure first exhibits

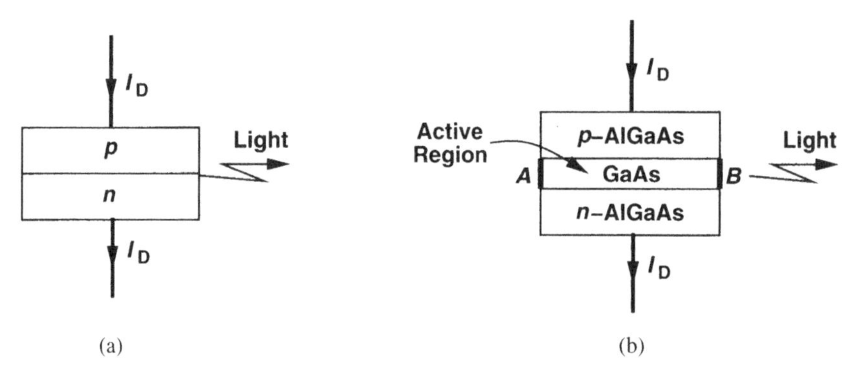

Figure 3.5 (a) Simple laser structure, (b) double heterostructure.

spontaneous emission, producing incoherent light with a relatively broad spectrum. If the current exceeds the threshold, the optical gain is sufficient to sustain oscillation, initiating laser action.

Lasers typically employ "direct-bandgap" semiconductors such as AlGaAs and InGaAsP. The bandgap energy of the material must be of such a value that the stimulated emission exhibits the required laser wavelength, e.g., 1.55 μm.

The geometry in Fig. 3.5(a) provides little confinement of carriers or light, displaying a very high threshold current (in the ampere range). In practice, the "double heterostructure" of Fig. 3.5(b) is used, where a thin layer of GaAs is sandwiched between two p and n regions of AlGaAs. With its higher bandgap energy, AlGaAs forms a barrier for the electrons and holes generated in the active region. Furthermore, owing to its lower refractive index, AlGaAs allows the active region to behave as an optical waveguide (i.e., a fiber), with the light traveling back and forth between the cleaved facets A and B. The laser thus achieves both carrier and optical confinement.

In reality, FP lasers may not produce a single light frequency. As the pumping current exceeds the threshold, high-order modes in the cavity between the mirrors are stimulated, generating other frequencies as well (Fig. 3.6). As explained in Section 3.2.2, these "sidemodes" produce undesirable effects while propagating through fibers. FP lasers are used for data rates as high as roughly 5 Gb/s.

To suppress the sidemodes observed in FP lasers, a *resonant* structure can be added to the simple cavity, thereby prohibiting unwanted components from growth in the feedback loop. To convert a uniform piece of material to a frequency-selective medium, we introduce

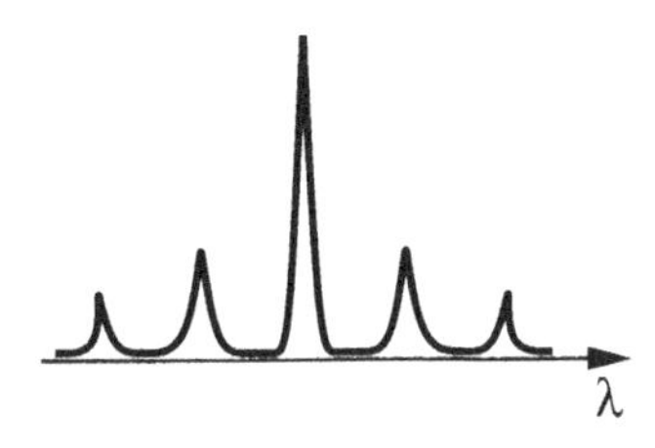

Figure 3.6 Effect of high-order modes in a Fabry-Perot laser.

a periodic perturbation in thickness or refractive index in the direction of propagation. Illustrated in Fig. 3.7(a), the idea is to allow one frequency to propagate through the medium with minimal loss whereas other frequencies are reflected from each interface. In optics, the above structure is called a "grating" and realized as in Fig. 3.7(b), where each "groove" modifies the effective value of n. If the light wavelength equals $2d$, then it resonates with

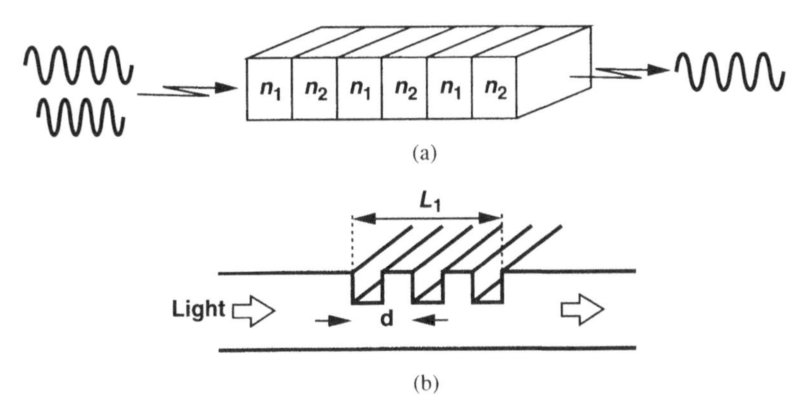

Figure 3.7 (a) Resonant structure using periodic variation of refractive index, (b) realization using a grating.

the grating and passes through with little attenuation; other wavelengths experience less gain. This phenomenon is analogous to resonance in LC circuits.

Figure 3.8 depicts a laser diode incorporating a resonant structure. The light traveling

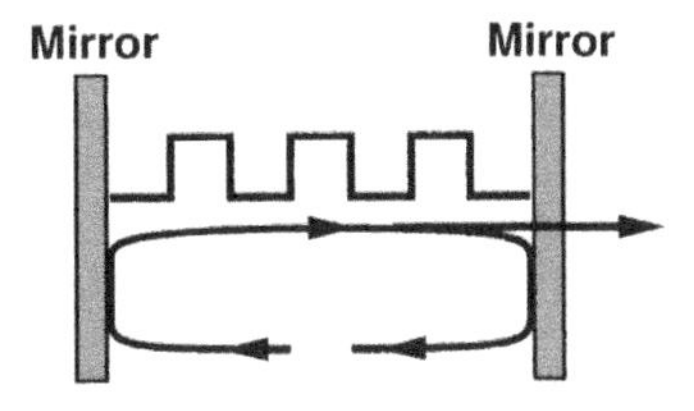

Figure 3.8 Distributed feedback laser.

around the loop between the two mirrors experiences the filtering selectivity provided by the grating, yielding an oscillation concentrated around the natural frequency of the resonator. Called a "distributed feedback" (DFB) laser, this device exhibits sidemodes about 30 dB below the fundamental and operates at rates as high as 20 Gb/s.

It is interesting to note that FP lasers can be viewed as "optical ring oscillators" (the light simply incurs a certain delay as it goes around the loop) whereas DFB lasers are similar to LC oscillators (the light experiences the resonant selectivity of the grating).

The laser geometries studied thus far are sometimes called "edge-emitting" structures to emphasize that, as illustrated in Fig. 3.5(b), the light exits from the edge of the device. Such structures suffer from a number of drawbacks: (a) the emitted light beam exhibits a noncircular cross section and is therefore difficult to couple to the fiber; (b) on-wafer testing of the laser diodes is not possible, requiring that they be diced first; (c) it is not possible to construct a two-dimensional array of edge-emitting devices on a wafer.

"Vertical cavity surface emitting lasers" (VCSELs[4]) are heterostructures designed to guide light in a direction *perpendicular* to the substrate. Approaching a true Fabry-Perot geometry, VCSELs employ mirrors *above* and *below* the active region so that the light travels up and down during amplification. Figure 3.9(a) depicts an example, where each mirror consists of a large number of thin layers with alternating refractive indices n_1 and

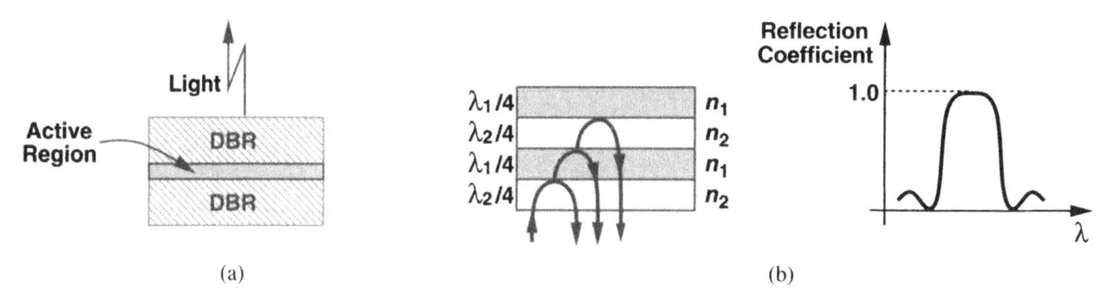

Figure 3.9 (a) VCSEL structure, (b) distributed Bragg reflector.

n_2. Called a "distributed Bragg reflector" (DBR), the mirror structure is shown in more detail in Fig. 3.9(b), revealing that proper choice of n_1, n_2, and the layer thicknesses ensures that most of the light at the wavelength of interest is reflected. Each layer may be only a few hundred angstroms thick.

Most early VCSELs produced light at a wavelength of 0.83 μm and also suffered from a high series resistance. With improvements in the technology, VCSELs operating at 1.3 μm and 1.5 μm have been demonstrated and their series resistance has been reduced. In contrast to edge-emitting devices, VCSELs typically require a lower threshold current with smaller voltage swings, but their output power is also smaller.

3.1.3 Properties of Lasers

With our basic understanding of lasers, we now examine some of their properties that impact the design of optical communication circuits and systems.

[4]Pronounced "vexels."

Turn-On Delay When a laser diode is turned on, the photon generation begins as spontaneous emission until the carrier density exceeds a threshold level. Thus, stimulated emission, i.e., true laser action, occurs after some delay.

If the laser current begins from below the threshold, then the turn-on delay experiences substantial *random* variations due to the random nature of spontaneous emission. Consequently, as shown in Fig. 3.10, the optical data suffers from jitter.

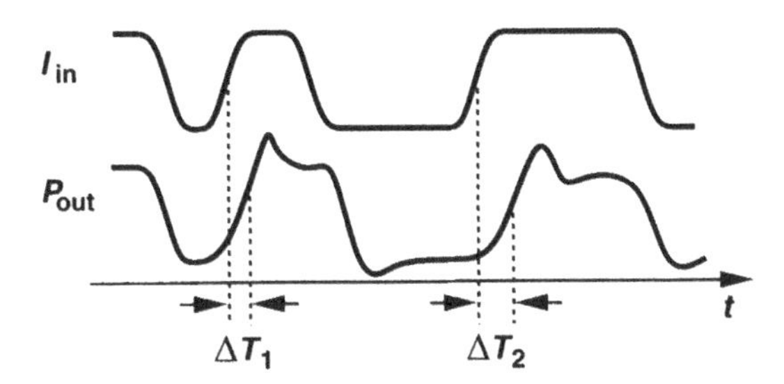

Figure 3.10 Effect of variable delay in lasers.

Frequency Chirping The oscillation frequency of lasers is a function of the medium's refractive index. This index in turn exhibits dependence on the carrier concentration in the medium. As a result, when the laser current is modulated by ONEs and ZEROs, the light *frequency* also experiences modulation. Illustrated in Fig. 3.11, this effect is called "frequency chirping." The principal consequence of chirping is the broadening of the light

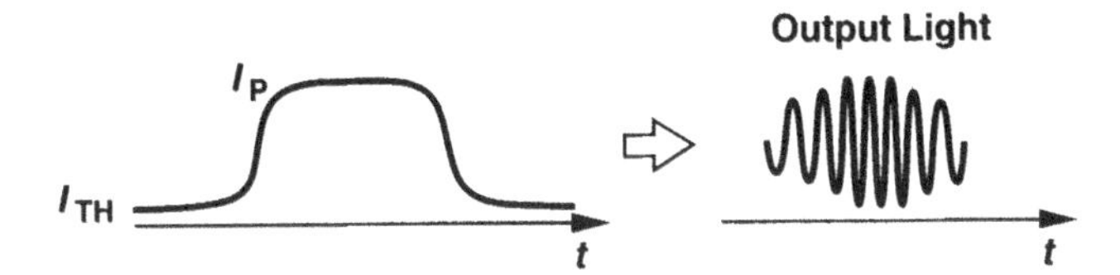

Figure 3.11 Chirping in lasers.

spectrum (by even more than one order of magnitude), leading to substantial dispersion in optical fibers carrying such signals (Section 3.2.2).

Relaxation Oscillation When the current flowing through a laser is stepped from a small value (below the threshold) to a large value, the laser output power does not vary in the form of a simple step. Depicted in Fig. 3.12(a), the output time-domain waveform suffers from "ringing," requiring some time to settle. This effect is called "relaxation oscillation." Equivalently, if the laser is driven by a sinusoidal current and P_{out} is plotted as a function of the input frequency, the behavior shown in Fig. 3.12(b) is observed, revealing peaking in the vicinity of f_1.

Relaxation oscillation results from the exchange of energy between photons and electrons. As the laser turns on, if stimulated emission raises the photon density above a certain level, then the electron density in the upper energy level begins to *fall*. This in turn *lowers* the photon density. Electrons then accumulate again to initiate strong light emission. The

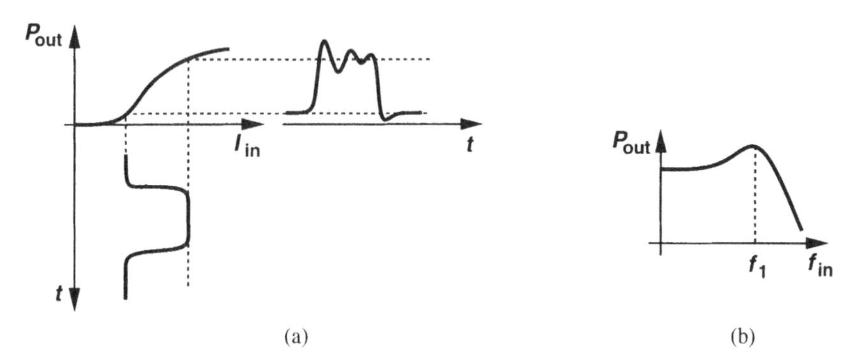

Figure 3.12 Relaxation oscillation in lasers.

rise and fall in the electron density and the output power continues (with decaying amplitude) until the system reaches equilibrium. The frequency of ringing typically ranges from 1 to 4 GHz.

Relaxation oscillation limits the speed of lasers. With random data, the ringing produced at each transition may affect the next, thus creating intersymbol interference and jitter. The system data rate is therefore typically limited to about 5 Gb/s for a fiber length of 5 km. Note that the carrier density fluctuation also results in frequency chirping, further widening the spectrum.

In order to alleviate the problems of random delay and relaxation oscillation, lasers are usually turned off only partially during logical ZEROs. Illustrated in Fig. 3.13, this method minimizes spontaneous emission at the cost of degrading the difference between

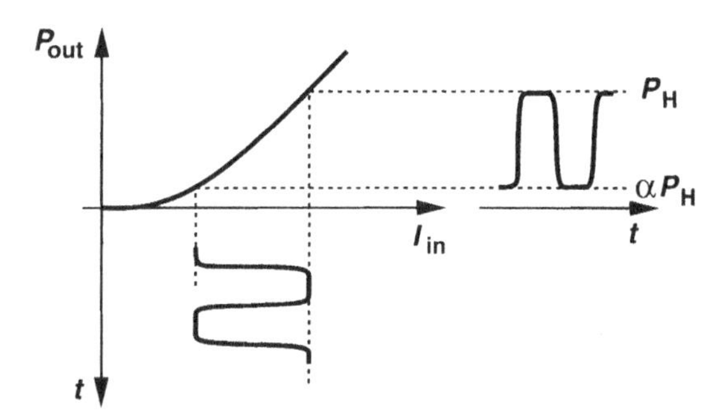

Figure 3.13 Choice of extinction ratio.

the logical ONEs and ZEROs (P_H and αP_H, respectively). Called the "extinction ratio" (ER), the quantity $P_H/(\alpha P_H) = 1/\alpha$ creates a trade-off between speed (i.e., random delay and relaxation oscillation) and detectability (the difference between highs and lows). In fact, the signal-to-noise ratio degrades by approximately $(1 - \alpha)$. As a compromise, the ER is typically set in the range of 10 to 15 dB.

Drift with Aging and Temperature The threshold current of laser diodes varies substantially with aging and temperature. For example, I_{TH} rises by 1 to 2% per degree centigrade. Thus, to maintain a constant extinction ratio, a servo loop is usually placed around the laser driver and the diode, controlling the low and high ends of the pump current. The servo loop requires monitoring the optical output, converting the result to an electrical component, comparing this value with a reference, and using the error to adjust the output current of the laser driver. Illustrated conceptually in Fig. 3.14, the loop employs a photodiode (Section 3.3) to measure the light generated by the laser, adjusting I_1 according to

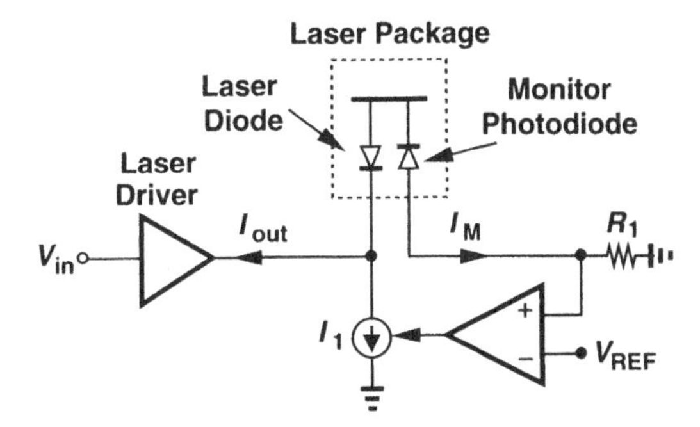

Figure 3.14 Servo loop to stabilize laser output power.

the difference between $I_M R_1$ and V_{REF}. The photodiode is typically included in the laser package. In some cases, the servo loop may also control the amplitude of the modulation current, I_{out}. We return to this topology in Chapter 10.

3.1.4 External Modulation

Turn-on delay, relaxation oscillation, and frequency chirping limit the switching rate of laser diodes and/or the length of the fiber carrying the information. For high-speed applications, e.g., at 10 Gb/s, it is common to separate the generation and modulation of the light into two independent functions. As shown in Fig. 3.15, a "continuous-wave" (CW) laser produces light with constant intensity, and an *external* modulator impresses the data on the light.

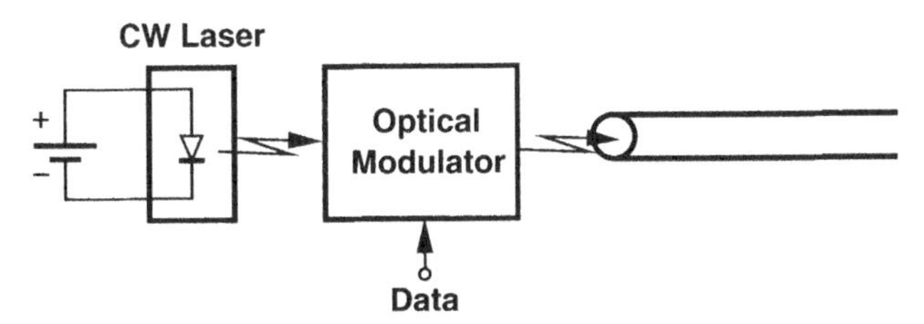

Figure 3.15 External modulation of lasers.

A common type of modulator used in high-speed applications is the "Mach-Zehnder" device. The Mach-Zehnder modulator employs a Lithium Niobate ($LiNbO_3$) crystal, in which the propagation velocity is a function of the electric field applied across it. This property allows the phase shift experienced by light to be modulated by an electrical signal. However, most systems require binary modulation of the light *amplitude* rather than the phase. The Mach-Zehnder modulator therefore controls the output amplitude through phase modulation and addition of *two* signals. Shown in Fig. 3.16, the structure consists of two signal paths, one of which experiences a variable phase shift controlled by V_{in}. If

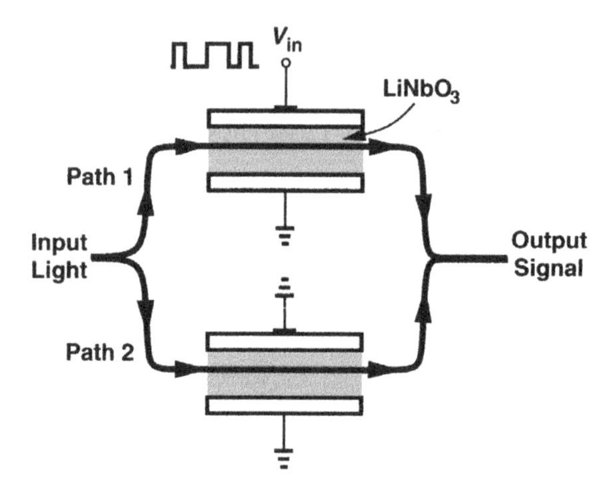

Figure 3.16 Mach-Zehnder modulator.

path 1 delays the light by π radians, then the output rays add destructively, creating no light. Thus, the peak value of V_{in} and the path length must be chosen so as to produce the required phase shift.

With the aid of Mach-Zehnder modulators, communication at rates of 10 Gb/s along 100 km of fiber has become possible. These devices, however, introduce difficulties in circuit design. First, they require the peak value of V_{in} in Fig. 3.16 to be as high as 4 to 6 V, creating breakdown issues in high-speed transistors operating in the driver circuitry. Second, since the modulator power/voltage characteristic drifts with aging and temperature, a control loop similar to that in Fig. 3.14 is necessary (Chapter 10).

3.2 Optical Fibers

The light produced and modulated by laser diodes and external modulators travels through an optical fiber. As mentioned in Chapter 1, fibers provide much greater bandwidths and lower losses than coaxial cables do, playing a critical role in high-speed long-distance communications.

Figure 3.17 shows the structure of a fiber. The "core" guides the light while the "cladding" confines the propagation and the "jacket" protects the fiber. This type is called a "single-mode fiber" (SMF). The core in low-loss fibers is made of "silica," a material derived from silicon dioxide (SiO_2). The small diameter of the core makes mechanical handling and alignment very difficult.

Two aspects of fibers play critical roles in the performance of OC systems: loss and dispersion.

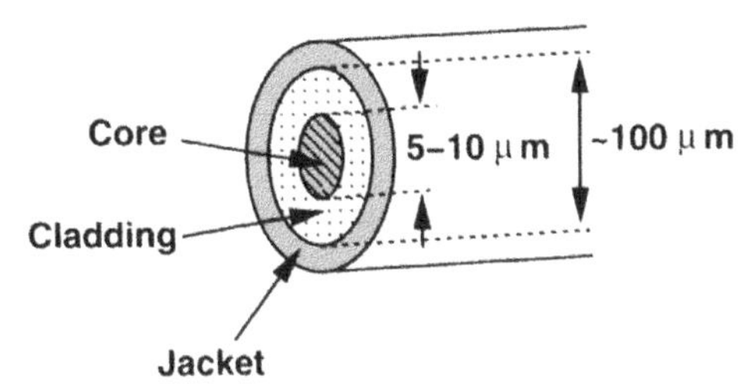

Figure 3.17 Structure of single-mode fiber.

3.2.1 Fiber Loss

The loss of fibers directly impacts the maximum allowable distance between the transmitter and the receiver. The light traveling through a fiber incurs two types of loss: absorptive and radiative.

Absorptive losses appear when the light wavelength stimulates atomic resonances in the fiber material itself ("intrinsic loss") or in the impurities in the fiber. Radiative losses arise when the light experiences scattering in the fiber, thereby exiting the core and radiating through the cladding. The most significant scattering mechanism is due to small dislocations (inhomogeneities) that occur in the fiber during fabrication and create fluctuations in composition and density. Called "Rayleigh scattering," this effect yields a loss that can be expressed as

$$\alpha_R = \frac{C_R}{\lambda^4}, \tag{3.1}$$

where C_R is a coefficient determined by the fiber uniformity, dimensions, and doping. The value of C_R typically falls in the range of 0.8 to 1 dB.μm^4/km, translating to a loss of about 0.15 dB/km at $\lambda = 1.55$ μm.

Figure 3.18 plots the loss characteristics of a typical silica SMF as a function of the light

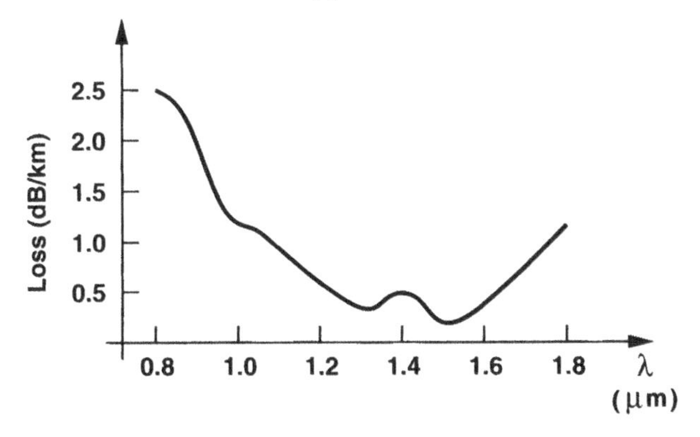

Figure 3.18 Loss profile of a typical SMF.

wavelength. The minimum attenuation occurs at $\lambda \approx 1.55$ μm, hence the widespread use of this wavelength in optical communications. The loss is relatively constant for about 200 nm around this wavelength, providing a flat bandwidth of roughly 25,000 GHz. In reality, however, fiber dispersion reduces the useful bandwidth by several orders of magnitude below this value.

The high manufacturing cost and difficult mechanical handling of silica fibers have encouraged the use of other materials. A common alternative are "plastic" fibers, found in three types: polymethylmethacrylate (PMMA), polystyrene (PS), and polycarbonate (PC). Plastic fibers have both a lower cost and a much greater core diameter ($\sim 1000\ \mu$m), obviating the need for precision connectors or cutters. Furthermore, they are soft and nonbrittle.

Plastic fibers, however, suffer from a number of drawbacks. First, they exhibit substantial loss, e.g., 55 dB/km for PMMA (at $\lambda \approx 0.76\ \mu$m). Second, their operation temperature cannot exceed 80°C. Nevertheless, plastic fibers are extensively used over short distances.

3.2.2 Fiber Dispersion

Dispersion in fibers has become a serious issue, motivating extensive work in both optical and electrical domains to reduce its effect. In this section, we study two important types of dispersion and describe how they degrade the performance in modern networks.

Dispersion is a familiar phenomenon in electronics. If different frequencies comprising a pulse propagate at different velocities in a medium, then the pulse shape changes. Equivalently, if the phase of the medium's transfer function does not vary linearly with frequency, then dispersion arises.

As a simple example, suppose the two sinusoids shown in Fig. 3.19 are summed and

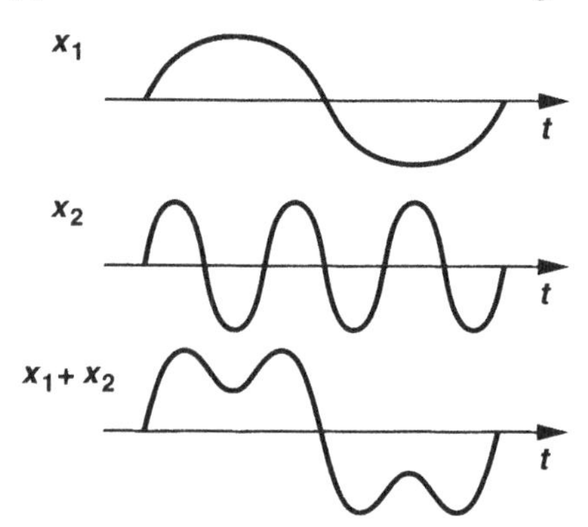

Figure 3.19 Summation of two sinusoids.

applied to a linear time-invariant system:

$$x(t) = x_1(t) + x_2(t) \tag{3.2}$$

$$= A_1 \cos \omega_1 t + A_2 \cos(3\omega_1 t). \tag{3.3}$$

We note that for $x(t)$ to retain its shape, the two components must incur equal delays (in seconds) as they propagate through the system[5]:

$$y_1(t) = A_1 \cos[\omega_1(t - \Delta t)] \tag{3.4}$$

$$y_2(t) = A_2 \cos[3\omega_1(t - \Delta t)]. \tag{3.5}$$

How do equal delays translate to phase shift? Expanding the argument in each output waveform, we write the phase shifts as:

$$\phi_1 = -\omega_1 \cdot \Delta t \tag{3.6}$$

$$\phi_2 = -3\omega_1 \cdot \Delta t. \tag{3.7}$$

Thus, in a nondispersive system, the phase shift (in radians) is a linear function of frequency.

As an example of a dispersive system, consider the first-order RC network shown in Fig. 3.20(a). As depicted in Fig. 3.20(b), the phase of the transfer function, $\angle H(s)$, is an

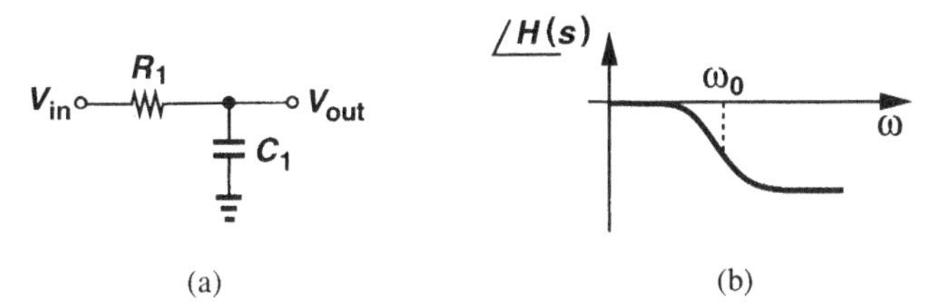

Figure 3.20 (a) First-order low-pass filter, (b) phase characteristic.

approximately linear function of frequency only for $0.1\omega_0 < \omega < 10\omega_0$.

As another example, suppose an all-pass system introduces a constant phase at all frequencies [Fig. 3.21(a)]. Here, higher frequencies incur less *delay*, in a sense "falling behind" lower frequencies. In particular, if the waveform $x(t)$ shown in Fig. 3.19 is applied to such a system, we have:

$$y_1(t) = A_1 \cos(\omega_1 t + \phi_1) \tag{3.8}$$

$$y_2(t) = A_2 \cos(3\omega_1 t + \phi_1), \tag{3.9}$$

and hence,

$$y_1(t) = A_1 \cos[\omega_1(t + \frac{\phi_1}{\omega_1})] \tag{3.10}$$

$$y_2(t) = A_2 \cos[3\omega_1(t + \frac{\phi_1}{3\omega_1})]. \tag{3.11}$$

[5]We assume an "all-pass" system, i.e., the magnitude of the transfer function remains equal to unity at all frequencies.

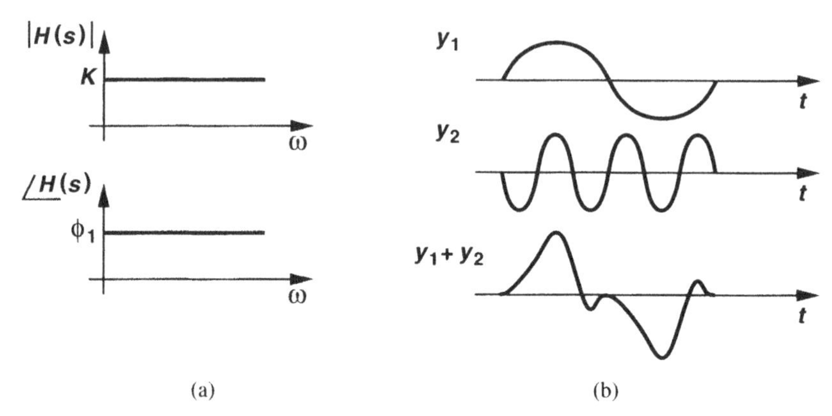

Figure 3.21 (a) All-pass system with flat phase response, (b) input and output signals.

Shown in Fig. 3.21(b), $y_1(t) + y_2(t)$ displays a different shape from $x_1(t) + x_2(t)$.

In digital data communication, dispersion becomes an issue because it typically *broadens* the pulses, thereby creating ISI. We now study two types of dispersion. For more details the reader is referred to [2, 6].

Chromatic Dispersion The propagation delay of light in a medium with a refractive index n is given by $v = c/n$, where $c \approx 3 \times 10^8$ m/s. In practice, the value of n varies with the light wavelength, leading to different velocities if the light contains different frequencies. For example, the refractive index of pure silica is given by [1]:

$$n^2(\lambda) = 1 + \frac{b_1\lambda^2}{\lambda^2 - a_1} + \frac{b_2\lambda^2}{\lambda^2 - a_2} + \frac{b_3\lambda^2}{\lambda^2 - a_3}, \tag{3.12}$$

where a_j and b_j are empirical constants. Since, as shown in Fig. 3.22, light modulated by random binary data exhibits a squared sinc spectrum, components at f_1 and f_2 reach the

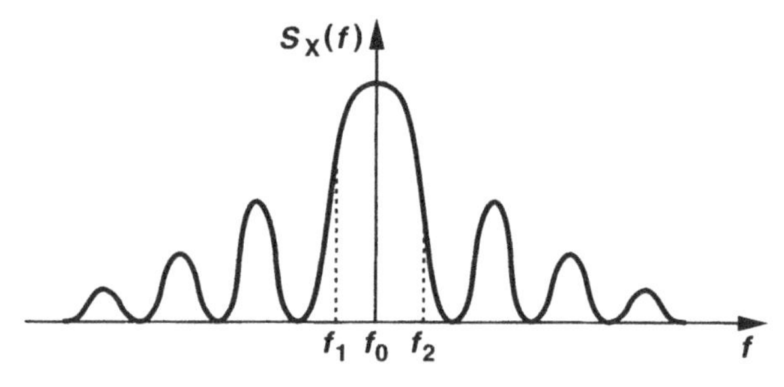

Figure 3.22 Spectrum of light modulated by random binary data.

far end at different times, thus creating chromatic dispersion.[6] Similarly, laser sidemodes (Fig. 3.6) experience dispersion, corrupting the data.

[6]"Chromatic" means related to color, referring to dispersion of different light colors (frequencies).

In optical communications, it is common to study the effect of dispersion by applying a Gaussian pulse to the system. Depicted in Fig. 3.23, such a pulse displays two useful

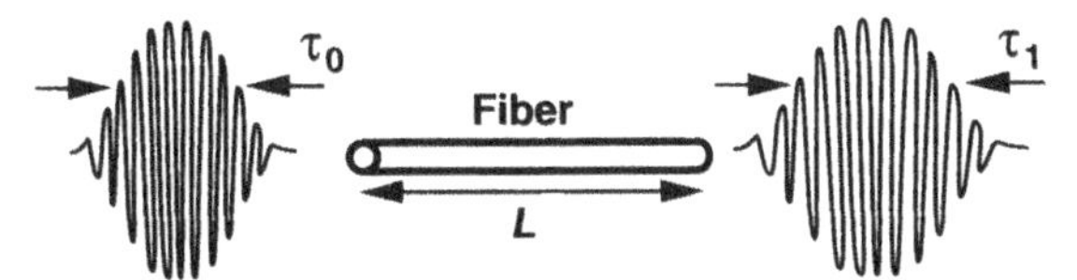

Figure 3.23 Broadening of Gaussian pulse due to dispersion.

properties: (a) due to its smooth edges, it occupies a small bandwidth; and (b) its Fourier transform is also a Gaussian and can be constructed easily. It can be shown that a Gaussian pulse $p(t) = A\exp(-t^2/\tau_0^2)$ modulating a light ray of wavelength λ_0 and traveling through a fiber of length L remains Gaussian and exhibits a width of [1]

$$\tau_1 = \tau_0\sqrt{1 + \frac{4L^2}{\tau_0^4}\left(\frac{\lambda_0^3}{2\pi c^2}\right)^2\left(\frac{d^2n}{d\lambda^2}\right)^2}. \tag{3.13}$$

For pulse broadening to be small, the second term under the square root must remain much less than unity and hence

$$\tau_1 \approx \tau_0 + \frac{2L^2}{\tau_0^4}\left(\frac{\lambda_0^3}{2\pi c^2}\right)^2\left(\frac{d^2n}{d\lambda^2}\right)^2. \tag{3.14}$$

Thus, the pulse is broadened by

$$\Delta\tau = \tau_1 - \tau_0 \tag{3.15}$$

$$\approx \frac{2L^2}{\tau_0^4}\left(\frac{\lambda_0^3}{2\pi c^2}\right)^2\left(\frac{d^2n}{d\lambda^2}\right)^2. \tag{3.16}$$

Note that, for small dispersion, $\Delta\tau \propto L^2$. In the optical fiber literature, on the other hand, it is common to define the broadening as [1]

$$\Delta\tau = \sqrt{\tau_1^2 - \tau_0^2}, \tag{3.17}$$

which, from (3.13), is equal to

$$\Delta\tau = \frac{2L}{\tau_0}\frac{\lambda_0^3}{2\pi c^2}\left|\frac{d^2n}{d\lambda^2}\right|. \tag{3.18}$$

With this definition, $\Delta\tau \propto L$ (!). This result is used to estimate the broadening for other pulses as well. Since a Gaussian pulse of width τ_0 occupies a bandwidth of roughly $\Delta\omega \approx 2/\tau_0$, we write Eq. (3.18) as

$$\Delta\tau = \Delta\omega L \cdot \frac{\lambda_0^3}{2\pi c^2}\left|\frac{d^2n}{d\lambda^2}\right|. \tag{3.19}$$

Thus, a well-behaved pulse having a bandwidth $\Delta\omega$ experiences a broadening given by Eq. (3.19). As a rule of thumb, a broadening of greater than one-fourth of the bit period is considered unacceptable [7].

While mathematically convenient, Gaussian pulses prove inadequate in representing random binary data. As illustrated in Fig. 3.24, two consecutive ONEs appear as one wide

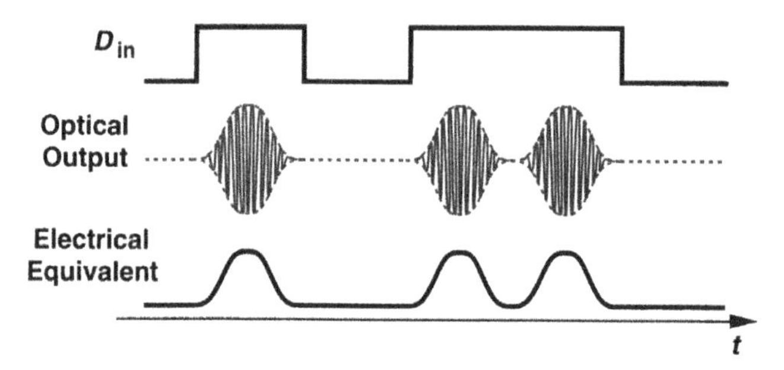

Figure 3.24 Use of Gaussian pulses to represent random data.

pulse in reality but as two narrow pulses in Gaussian modeling. Broadening therefore leads to unrealistic ISI between the ONEs.

Chromatic dispersion is usually quantified by an index, δ, that indicates the change in delay incurred by light per kilometer of fiber length for 1 nm of change in the wavelength. Figure 3.25 plots δ as a function of λ, suggesting that δ exceeds 10 ps/nm/km at $\lambda = 1.55$

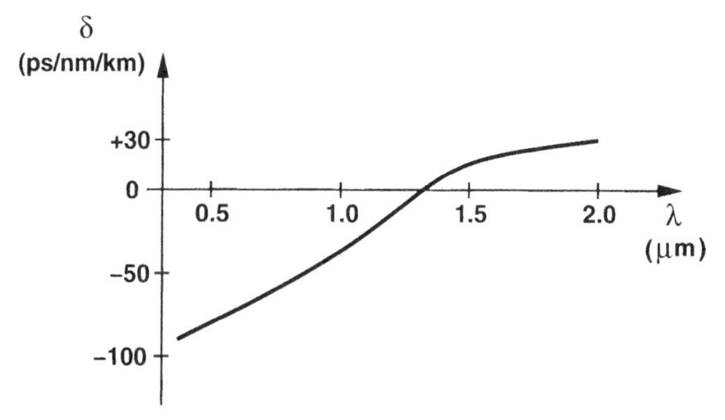

Figure 3.25 Chromatic dispersion profile of a typical SMF.

μm. This means wavelengths that are 1 nm (about 133 GHz) apart experience a delay difference of 10 ps across 1 km. While 133 GHz of bandwidth appears much greater than today's demand, we must note that (1) modulation of light amplitude at 40 Gb/s does produce significant energy at ± 100 GHz around the center frequency; (2) dispersion manifests itself even at 10 Gb/s if the fiber is hundreds of kilometers long.

Figure 3.25 also reveals that δ drops to zero in the vicinity of $\lambda = 1.3$ μm. Thus, the wavelengths of minimum loss and minimum dispersion do not coincide, requiring a compromise.

Some GaAs lasers produce a light wavelength of about 0.84 μm, at which the dispersion index reaches several tens of ps/nm/km. The removal of the fiber-induced dispersion in the electrical domain is under active research. Recent work has also produced "dispersion-shifted fibers" (DSFs), in which the wavelength of minimum dispersion is shifted to 1.55 μm. A third type, the "dispersion-flattened fiber" (DFF), is designed to have minimum dispersion in the range of 1.2 μm to 1.6 μm.

Polarization-Mode Dispersion During manufacturing, the cross section of optical fibers inevitably suffers from slight asymmetries, appearing as an ellipse rather than a circle [Fig. 3.26(a)]. This effect is exacerbated if the fiber senses asymmetric stresses after

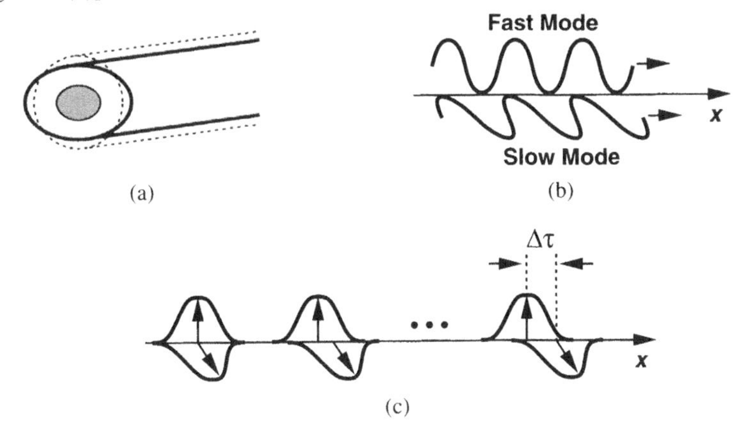

Figure 3.26 (a) Ellipticity in a fiber, (b) PMD, (c) pulse bifurcation.

installation. As depicted in Fig. 3.26(b), the resulting "ellipticity" yields different propagation velocities for different polarization modes, thereby leading to polarization-mode dispersion (PMD) [6].

The difference between propagation velocities is modeled by different refractive constants, n_s and n_f, for the slow and fast modes, respectively. Called the "birefringence," $\Delta n = n_s - n_f$ typically falls in the range of 10^{-7} to 10^{-5}.

Figure 3.26(c) illustrates the effect of PMD on a narrow pulse. As the waveform propagates in the fiber, one polarization mode falls behind, resulting in "pulse bifurcation." It can be shown that, for a fiber of length L, the bifurcation time ("differential delay") is given by [6]

$$\Delta\tau = L\left[\frac{\Delta n}{c} - \frac{\omega}{c}\cdot\frac{\partial(\Delta n)}{\partial\omega}\right], \tag{3.20}$$

where ω denotes the light frequency. The normalized quantity $\Delta\tau/L$ often serves as a measure of PMD in fibers, assuming a typical value of 1.5 ps/km.

Interestingly, the dependence of PMD upon the length of the fiber changes as the latter increases. In long fibers, the direction of ellipticity does not remain constant from one end to the other, leading to a slower accumulation of $\Delta\tau$ with length (Fig. 3.27). If the ellipticity appears randomly, then it can be assumed that $\Delta\tau \propto \sqrt{L}$ for $L > L_C$. Unfortunately, the

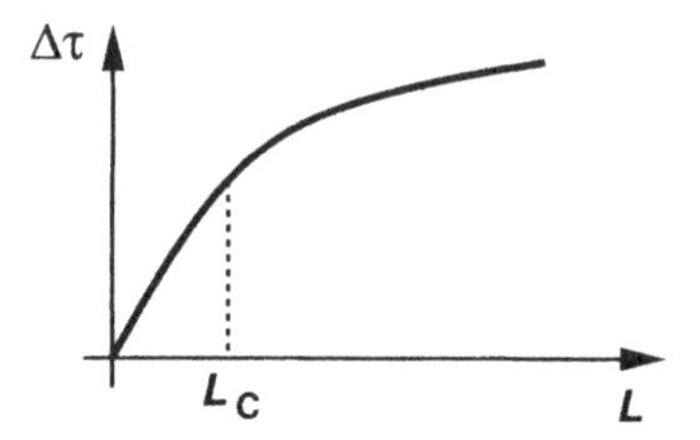

Figure 3.27 Dependence of PMD on fiber length.

value of L_C varies from a few meters to 1 km under different conditions, making it difficult to clearly distinguish the "short-fiber regime" from the "long-fiber regime." Nevertheless, since PMD is significant only for tens or hundreds of kilometers of length, the square-root dependence proves more relevant to practical applications.

An important issue related to PMD is its drift with temperature, necessitating *adaptive* cancellation at the far end. For long fibers, the drift is random and $\Delta\tau$ exhibits a Maxwellian distribution (Fig. 3.28) [6]:

$$P_X(x) = \sqrt{\frac{2}{\pi}} \frac{x^2}{\alpha^3} \exp \frac{-x^2}{2\alpha^3}. \tag{3.21}$$

It can be shown that for a 1-dB power penalty with a probability of 1/18,000 (30 minutes

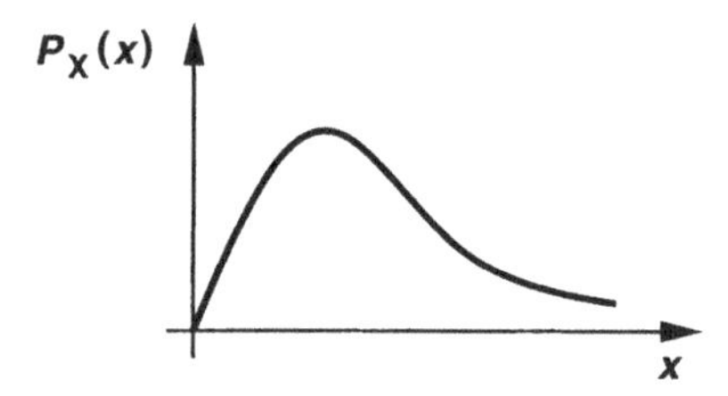

Figure 3.28 PMD distribution.

per year), the mean differential delay, $< \Delta\tau >$, must satisfy:

$$< \Delta\tau > \leq 0.14T_b, \tag{3.22}$$

where T_b denotes the bit rate. Also, for a given bit rate, R_b, the maximum tolerable PMD is given by

$$(PMD_{max})^2 = \frac{0.02}{R_b^2 \cdot L}. \tag{3.23}$$

For example, if $R_b = 10$ Gb/s and $L = 100$ km, then $PMD_{max} = 14$ ps, i.e., 1.4 ps/$\sqrt{\text{km}}$.

3.3 Photodiodes

The light carried by a fiber is converted to an electrical signal at the receive end by means of a photodiode. Various properties of photodiodes affect the sensitivity and speed of the receiver front end.

Photodiodes produce current in response to light, in a sense the reverse action of light-emitting diodes. If a pn junction is illuminated with light, the electrons in the valence band may be stimulated and raised to the conduction band. As a result, a photon is absorbed and an electron-hole pair capable of conducting current is generated (Fig. 3.29). This phenomenon occurs if the photon energy E_p exceeds the bandgap energy of the material, E_g.

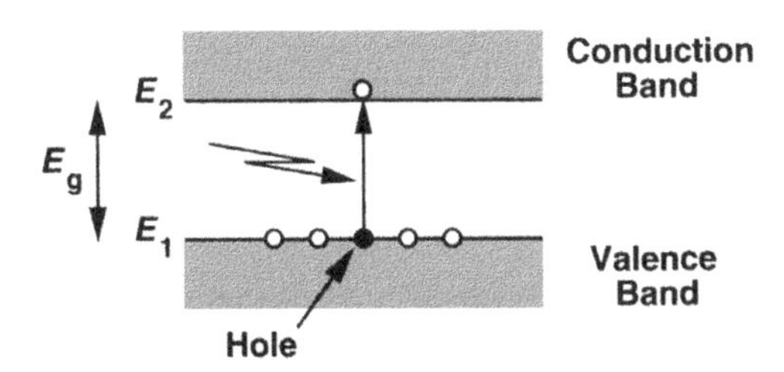

Figure 3.29 Generation of an electron-hole pair by means of a photon.

The photon energy is given by Plank's equation:

$$\lambda_p = \frac{hc}{\lambda}, \tag{3.24}$$

where $h = 6.63 \times 10^{-34}$ J.s denotes the Plank constant. Thus, the light wavelength must fall below a threshold, $\lambda_{TH} \approx 1.24/E_g$ μm·eV to stimulate carrier generation.

The generation of electron-hole pairs in a pn junction shifts the reverse "breakdown" characteristic toward the origin [Fig. 3.30(a)]. Indeed, if a reverse bias voltage is applied across the diode, the electrons and holes created by light are absorbed to opposite polarities, producing a continuous current [Fig. 3.30(b)].

It is important to note that photodiode action occurs mostly in the depletion region; the electric field in other regions of the diode is small, allowing the electrons and holes to recombine before they are swept away. For this reason, efficient photodiode operation demands both a reverse bias and a large depletion region.

3.3.1 Responsivity and Efficiency

The current generated by a photodiode, I_p, is linearly proportional to the optical power, P_{op}:

$$I_p = R_{ph} P_{op}, \tag{3.25}$$

where R_{ph} is called the "responsivity." For example, some photodiodes exhibit a responsivity of 1 A/W, i.e., they produce a current of 1 mA when illuminated by 1 mW of light

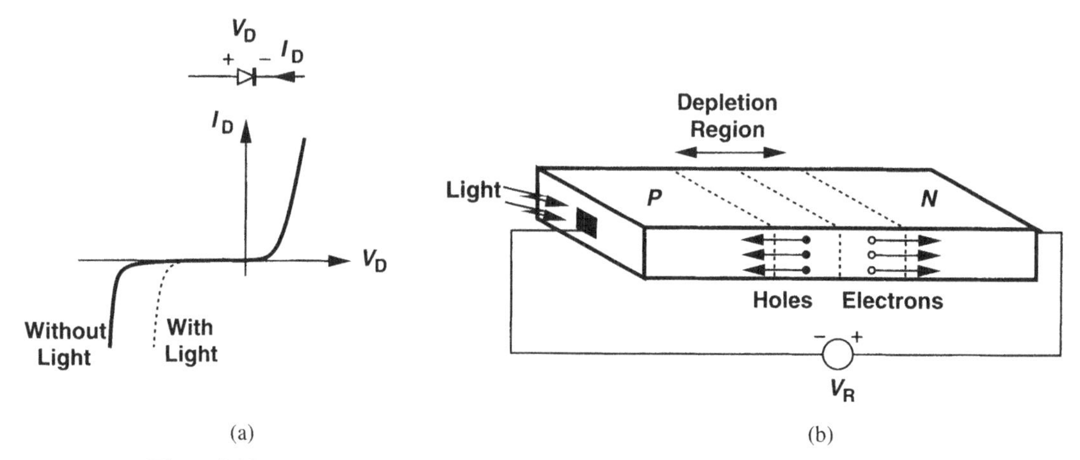

Figure 3.30 (a) Photodiode characteristic, (b) current conduction by electrons and holes.

with proper wavelength. The responsivity can be increased by widening the diode surface area that receives the light, but at the cost of greater junction capacitance.

In an ideal photodiode, every photon entering the device generates an electron-hole pair. In reality, however, some photons are reflected from the surface or absorbed by the material to produce heat. We therefore define the "quantum efficiency" of photodiodes as the number of electrons generated divided by the number of photons applied (in a given time interval):

$$\eta = \frac{I_p/q}{P_{op}/(hc/\lambda)}, \tag{3.26}$$

where q denotes the electron charge and the denominator represents the number of photons. From (3.25), we have

$$\eta = R_{ph}\frac{hc}{q\lambda} \tag{3.27}$$

$$= \frac{1.24R_{ph}}{\lambda}. \tag{3.28}$$

Typical values of η fall in the range of 0.8 to 0.9.

In this chapter, we study two types of photodiodes commonly used at the front end of optical receivers. We should mention, however, that many of today's high-performance systems interpose an *optical* amplifier between the fiber and the photodiode to boost the light level, thereby lowering the effect of the receiver noise.

3.3.2 PIN Diodes

Recall that photodiode action occurs primarily in the depletion region. To create a wide depletion region, an intrinsic (undoped) piece of semiconductor can be interposed between the p and n sections (Fig. 3.31), leading to a "PIN diode." With a reverse-bias voltage, the

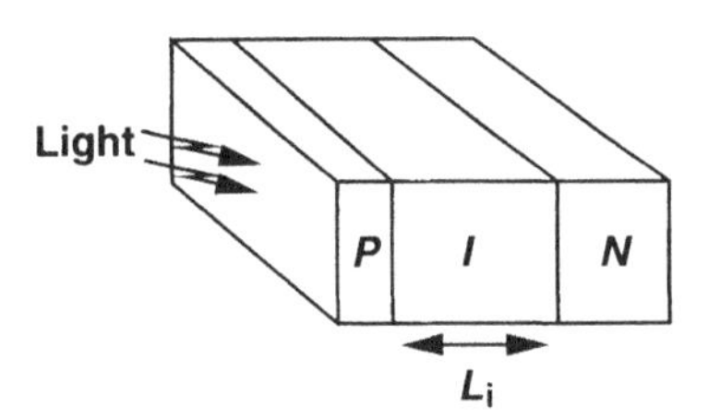

Figure 3.31 PIN structure.

intrinsic region is almost completely depleted, providing a depletion region as wide as L_i, typically about 20 μm. PIN diodes exhibit a quantum efficiency of approximately 0.8.

The switching speed of PIN diodes is limited by two effects. First, the finite transit time of electrons and holes through the intrinsic region limits the rate at which the current can vary in response to light. For a transit time of T_{tr}, the -3-dB bandwidth of the light-current transfer function is roughly equal to $0.44/T_{tr}$ [2]. For example, if carriers reach a saturated velocity of 8×10^6 cm/s, then $T_{tr} \approx 0.2$ ns for $L_i = 20$ μm, yielding a -3-dB bandwidth of 2.2 GHz.

The second factor limiting the speed arises from the junction capacitance, C_j, of the diode. As explained in Chapter 4, C_j greatly impacts the design of transimpedance amplifiers. Interestingly, if L_i in Fig. 3.31 is reduced to decrease the transit time, then the junction capacitance rises (and the quantum efficiency falls). Also, if the cross section of the diode is enlarged to increase the output current, then C_j increases as well. For photodiodes operating in the gigahertz range, C_j falls in the range of 0.1 to 0.5 pF.

PIN diodes display a low noise, making the subsequent amplifier the major noise contributor. Such diodes operate with wavelengths of 0.5 to 1 μm in Si and 1 to 1.6 μm in AlGaAs.

3.3.3 Avalanche Photodiodes

The operation of PIN diodes is based on the generation of *one* electron-hole pair for each photon entering the lattice. In avalanche photodiodes (APDs), on the other hand, the generated electrons and holes carry so much energy that they themselves can stimulate other electrons and holes, creating an avalanche effect.

To understand the principle of operation, we first review avalanche breakdown in pn junctions. At high electric fields, an electron may travel at such high velocities (even occasionally greater than the saturated velocity) that it can stimulate more electrons from the valence band into the conduction band. Called "impact ionization," this phenomenon can lead to an indefinite growth (multiplication) of carriers if each new conduction-band electron acquires enough energy to free more electrons.

APDs operate with controlled avalanche, that is, with a multiplication factor, M, of several hundred rather than infinity. Thus, each photon entering the device may create hundreds of electron-hole pairs, providing a large output current. Shown in Fig. 3.32 is the structure of a typical APD, consisting of a sandwich of n^+, p, and I layers atop a p^+ substrate. As in PIN diodes, the intrinsic region enables generation of electron-hole pairs. Grown as a very uniform and thin layer, the p region supports a high electric field to create

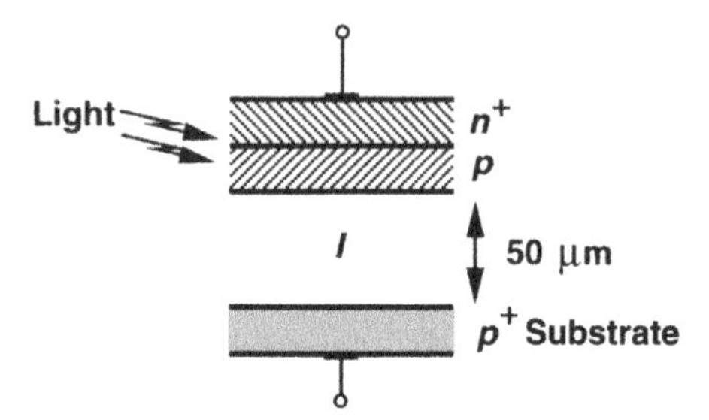

Figure 3.32 APD structure.

avalanche. For a light wavelength of 0.8 μm and $\eta \approx 0.9$, multiplication factors of several hundred and responsivities of about 10 A/W are achieved.

Interestingly, APDs suffer from a "gain-bandwidth" trade-off. The higher the multiplication factor is, the longer the avalanche persists, limiting the switching rate of the current. For this reason, APDs rarely achieve a speed higher than a few gigahertz.

The use of APDs in optical communications entails several other issues. First, the avalanche process is quite noisy, making the noise of the photodiode significant. Second, this avalanche noise appears only when light is applied, i.e., only for one logical level. We say the noise is greater on ONEs than on ZEROs. Third, the reverse-bias voltage applied to APDs must be controlled precisely so as to avoid full avalanche breakdown. Fourth, the high responsivity of APDs is obtained only at high reverse voltages, typically in the vicinity of 10 V, requiring additional power supplies in the system if the circuits themselves must operate with lower voltages. The last two properties are in contrast to those of PIN diodes.

3.4 Optical Systems

Following our study of optical devices, we now briefly describe general aspects of optical systems, concentrating on issues resulting from device and circuit limitations.

The light produced by an optical transmitter is modulated by the data. While a variety of schemes have been attempted, the most common method is direct modulation of the light intensity by binary data, known as "amplitude shift keying" (ASK) or "on-off keying" (OOK). To perform ASK, the laser diode or the modulator is directly driven by the data, producing substantial light in response to one logical level (Fig. 3.33).

The light traveling through the fiber experiences loss, requiring a highly-sensitive receiver to detect the signal with acceptable bit error rate. The sensitivity of the receiver is determined by (a) the responsivity of the photodiode, (b) the overall input-referred noise of the receiver, and (c) the detection method employed in the receiver. For a transmitted power of P_T, a fiber loss of L_F, and a photodiode responsivity of R_{PD}, the input current applied to the receiver is

$$I_{in} = P_T \cdot L_F \cdot R_{PD}. \tag{3.29}$$

For example, suppose an optical transmitter injects a "high" output level of 10 mW into

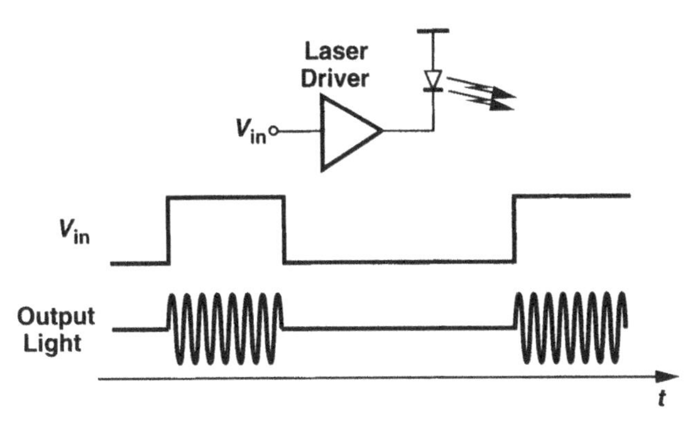

Figure 3.33 ASK modulation of light.

a plastic fiber that is 100 m long. Let us calculate the maximum tolerable receiver noise for BER $= 10^{-12}$ if the fiber exhibits a loss of 300 dB/km and the receiver photodiode has a responsivity of 1 A/W. The fiber attenuates the light power by 30 dB (a factor of 1000), yielding a peak-to-peak current $I_{in} = 10\ \mu$A. For BER $= 10^{-12}$, Eqs. (2.15) and (2.16) require that the total input-referred rms noise current of the receiver be $\sqrt{\overline{I_n^2}} = 10\ \mu A/14 \approx 714$ nA.

This example ignores two important nonidealities. First, the finite extinction ratio of the lasers brings the high and low levels closer to each other, degrading the detection. Second, for avalanche photodiodes or PIN diodes followed by optical amplifiers, the noise on logical ONEs may be greater than that on ZEROs.

The foregoing study provides the foundation for "link budget" calculations. Based on the transmitted power, the length (and hence loss) of the fiber, and the properties of the photodiode, the link budget determines the maximum tolerable noise of the receiver and/or the detection method for a given bit error rate. In this context, it is often said, "every dB counts." For example, if the fiber suffers from a loss of 0.2 dB/km, then a 1-dB improvement in the link budget allows a 5-km increase in the length of the fiber.

Most optical receivers employ "direct detection," i.e., they sample the signal level every bit period, compare the result to a threshold, and decide whether the bit is a ONE or a ZERO (Fig. 3.34). The error rate associated with direct detection is expressed by Eqs. (2.15) and

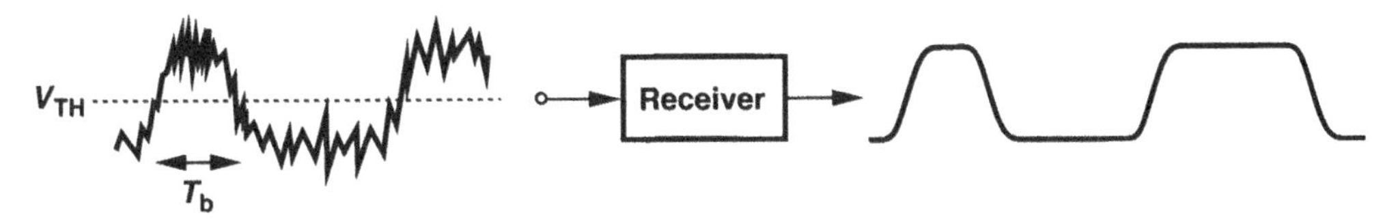

Figure 3.34 Direct detection of ASK-modulated light.

(2.16) for additive Gaussian noise.

The reader may wonder if direct detection is the optimum method of determining the logical levels in the presence of noise. After all, if each bit lasts for T_b seconds, sampling only one point in this interval seems to ignore a great deal of information. In fact, we intuitively expect that if single-point sampling is preceded by *integration*, the effect of noise can be reduced. Illustrated in Fig. 3.35, the idea is to integrate the waveform for each

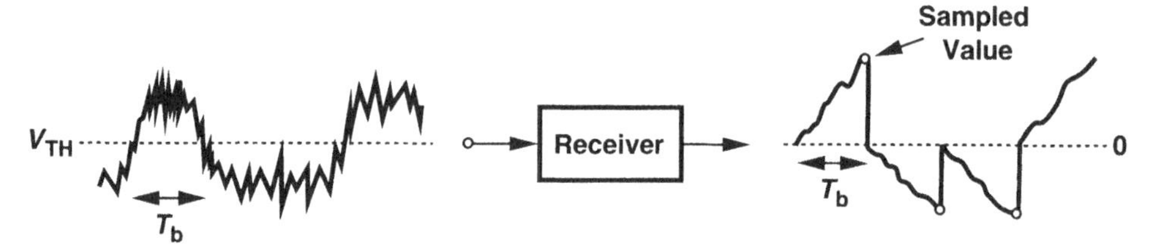

Figure 3.35 Use of integration before detection.

bit period, sample the final value, and reset the integrator before the next bit. The detection now "averages" the noise, producing a more reliable representation of the bit and lowering the bit error rate.

The integrate-and-dump technique described above is an example of "matched filtering," a common method of optimum detection in digital communications [5]. While improving the performance, the integrate-and-dump scheme nevertheless entails difficult issues in high-speed systems. First, the integration must exactly coincide with bit period, requiring phase alignment between data and clock while accounting for skews introduced by the integrator itself. Second, the reset operation must be extremely fast so as to prepare the integrator for the next bit. Interleaved matched filters may resolve the latter issue [8], but the alignment of data and clock remains a challenging task if the data carries significant noise.

A common measure of performance in optical systems is the number of photons required per bit. For example, 1 mW of optical power corresponds to 7.5×10^{15} photons per second. (With a responsivity of 1 A/W, a photodiode produces 1 mA of current from such a signal.) We may then ask: what is the *minimum* required number of photons per bit for a given error rate? More specifically, does the error rate go to zero if the transmission and reception introduce no noise? Recall from Section 3.3 that Plank's equation gives the *average* number of photons per unit time as $P_{op}/(hc/\lambda)$. However, the exact arrival time of each photon is random.[7] Now suppose a noiseless receiver can count each single photon, translating the arrival of the photon in the time interval T_b to a logical ONE. If during a T_b period, no photon arrives, then either a logical ZERO has been transmitted, or the photon corresponding to a logical ONE has been lost. From these observations, it can be proved that a noiseless, "photon-counting" receiver exhibits the following error rate [2]:

$$\text{BER} = \frac{1}{2}\exp^{\gamma T}, \tag{3.30}$$

where $\gamma = P_{op}/(hc/\lambda)$ denotes the average number of photons received per unit time. For example, 20 photons per bit yields an error rate of 10^{-9}. This is called the "quantum

[7]The arrival time of photons has a Poisson distribution [2], similar to that of trains.

limit" of the receivers. Actual receiver implementations display sensitivities about one to two orders of magnitude worse than the quantum limit.

References

1. A. K. Ghatak and K. Thyagarajan, *Introduction to Fiber Optics,* Cambridge: Cambridge University Press, 1998.
2. M. M.-K. Liu, *Principles and Applications of Optical Communication,* Chicago: McGraw-Hill, 1996.
3. G. P. Agrawal, *Fiber Optic Communication Systems*, Third Ed., New York: Wiley, 2002.
4. B. E. A. Saleh and M. C. Teich, *Fundamentals of Photonics*, New York: Wiley, 1991.
5. L. W. Couch, *Digital and Analog Communication Systems,* Fourth Ed., New York: Macmillan Co., 1993.
6. I. P. Kaminow and T. L. Koch, Editors, *Optical Fiber Telecommunications IIIA*, San Diego: Academic Press, 1997.
7. P. S. Henry, "Lightwave Primer," *IEEE J. of Quantum Electronics,* vol. 21, pp. 1862-1879, Dec. 1985.
8. J. Savoj and B. Razavi, "A CMOS Interface Circuit for Detection of 1.2-Gb/s RZ Data," *ISSCC Dig. of Tech. Papers*, pp. 278–279, Feb. 1999.

CHAPTER 4

Transimpedance Amplifiers

The light traveling through a fiber experiences loss before reaching a photodiode at the far end. The photodiode then transforms the light intensity to a proportional current, which is subsequently amplified and converted to voltage by a transimpedance amplifier. The design of TIAs entails many trade-offs between noise, bandwidth, gain, supply voltage, and power dissipation, presenting difficult challenges in both CMOS and bipolar technologies.

This chapter deals with the design of low-noise TIAs. Following a summary of performance requirements, we describe basic TIA topologies and their noise, gain, and speed trade-offs. Next, we study supply rejection and differential TIAs and review high-performance techniques such as gain boosting, feedback, and inductive peaking. Finally, we deal with automatic gain control (AGC) and present some case studies.

4.1 General Considerations

A transimpedance amplifier converts an input current, I_{in}, to an output voltage, V_{out}. The circuit is characterized by a "transimpedance gain," defined as $R_T = \delta V_{out}/\delta I_{in}$. For example, a gain of 1 kΩ means the TIA produces a 1-mV change in the output in response to a 1-μA change in the input. TIAs must satisfy stringent requirements imposed by link budget and speed considerations. Before describing TIA performance parameters, we examine a simple topology to arrive at some fundamental limitations.

Since photodiodes generate a small current and since most of the subsequent processing occurs in the voltage domain, the current must be converted to voltage. As depicted in Fig. 4.1(a), a single resistor can perform this function, providing a transimpedance gain equal to R_L. However, the time constant $R_L C_D$ leads to a severe trade-off between gain, noise, and bandwidth. Modeling the thermal noise of R_L as shown in Fig. 4.1(b), where $\overline{I_n^2} = 4kT/R_L$ (per unit bandwidth), we have

$$\overline{V_{n,out}^2} = \int_0^\infty \frac{4kT}{R_L} \left| R_L || \frac{1}{C_D j 2\pi f} \right|^2 df \tag{4.1}$$

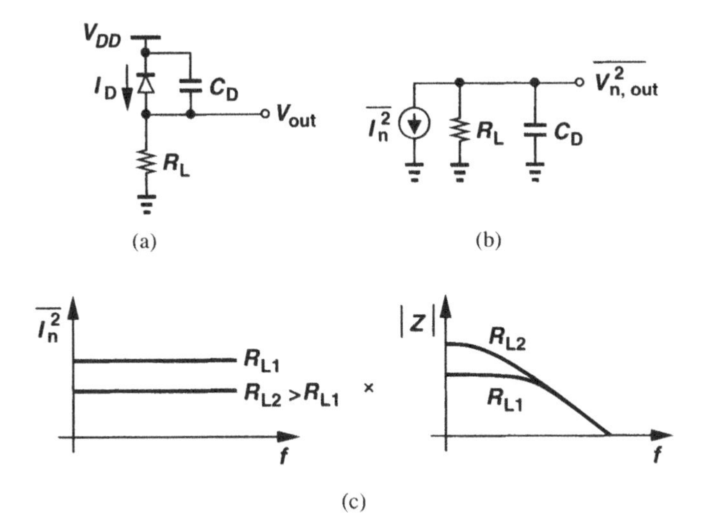

Figure 4.1 (a) Conversion of photodiode current to voltage by a resistor, (b) equivalent circuit for noise calculation, (c) effect of resistor value.

$$= \int_0^\infty \frac{4kT}{R_L} \frac{R_L^2 df}{R_L^2 C_D^2 4\pi^2 f^2 + 1}. \tag{4.2}$$

Noting that

$$\int \frac{du}{1+u^2} = \tan^{-1} u, \tag{4.3}$$

we can write

$$\overline{V_{n,out}^2} = \frac{kT}{C_D}. \tag{4.4}$$

This equation reveals that the total integrated noise is independent of R_L. We observe from Fig. 4.1(c) that, as R_L increases, $\overline{I_n^2}$ decreases but the area under $|Z|^2 = R_L^2/(R_L^2 C_D^2 4\pi^2 f^2 + 1)$ increases such that $\overline{V_{n,out}^2}$ remains constant. Of course, the circuits following the diode/resistor combination limit the bandwidth to less than infinity, reducing the noise contributed by R_L. Nonetheless, kT/C_D provides a rough estimate in initial calculations.

For fair comparison of different designs, the noise must be referred to the input so that it does not depend on the gain.[1] Since the circuit of Fig. 4.1(a) has a transimpedance gain

[1] Input-referred noise is the value that, if applied to the input of the equivalent noiseless circuit, produces an output noise equal to that of the original, noisy circuit. Note that input-referred noise is a fictitious quantity in that it cannot be observed in a circuit.

of R_L, its total input-referred noise current is equal to

$$\overline{I_{n,in}^2} = \frac{\overline{V_{n,out}^2}}{R_L^2} \quad (4.5)$$

$$= \frac{kT}{R_L^2 C_D}, \quad (4.6)$$

indicating that R_L must be maximized.

However, we must also apply a minimum bandwidth constraint proportional to the bit rate. Assuming for now that the −3-dB bandwidth, $(2\pi R_L C_D)^{-1}$, must be equal to the bit rate, R_b, we summarize the circuit's properties as

$$\text{Transimpedance Gain} = R_L \quad (4.7)$$

$$\frac{1}{2\pi R_L C_D} = R_b \quad (4.8)$$

$$\overline{I_{n,in}^2} = \frac{kT}{R_L^2 C_D}. \quad (4.9)$$

Equations (4.8) and (4.9) indicate a direct trade-off between speed and noise. More specifically, eliminating C_D or R_L from these equations yields the following results:

$$\overline{I_{n,in}^2} = 2\pi kT \frac{R_b}{R_L} \quad (4.10)$$

$$\overline{I_{n,in}^2} = 4\pi^2 kT C_D R_b^2. \quad (4.11)$$

For example, if $C_D = 0.2$ pF and $R_b = 10$ Gb/s, then $\sqrt{\overline{I_{n,in}^2}} = 1.81\ \mu A_{rms}$. In practice, the circuit following the diode and resistor may contribute significant noise as well.

The fundamental trade-offs expressed by Eqs. (4.10) and (4.11) suggest that the diode/resistor network of Fig. 4.1(a) is ill-suited to high-performance applications, and other circuit topologies must be sought that relax these constraints.

4.1.1 TIA Performance Parameters

Noise The input-referred noise current, $I_{n,in}$, of a TIA determines the minimum input current that yields a given bit error rate, directly impacting the link budget. (Recall from Section 2.5 that an error rate of, say, 10^{-12} requires that $I_{in,pp}/I_{n,in,rms} \approx 14$.) Thus, as mentioned in Chapter 3, a 1-dB decrease in $I_{n,in}$ allows a 5-km increase in the length of the fiber if the latter has a loss of 0.2 dB/km. The total noise current of high-performance TIAs typically falls in the range of 0.5 to 2 μA_{rms}.

The stringent TIA noise requirements limit the choice of circuit topologies and in particular the number of devices in the signal path. As will be seen throughout this chapter, the problem of noise becomes more severe as the supply voltage scales down.

Bandwidth Our study of noise in Section 2.5 suggests that the bandwidth of TIAs must be *minimized* so as to reduce the total integrated noise. On the other hand, the observations in Section 2.4 reveal that limited bandwidth introduces intersymbol interference in random data, closing the eye both vertically and horizontally. To this end, we must quantify the relationship between bandwidth and ISI.

TIA circuits typically contain multiple poles (and zeros), making it difficult to calculate the ISI. Here, we consider a single-pole low-pass filter to gain an intuitive understanding of the phenomenon. As shown in Fig. 4.2(a), for moderate ISI, the effect of only the previous bit need be taken into account, and the worst case occurs when two or more consecutive

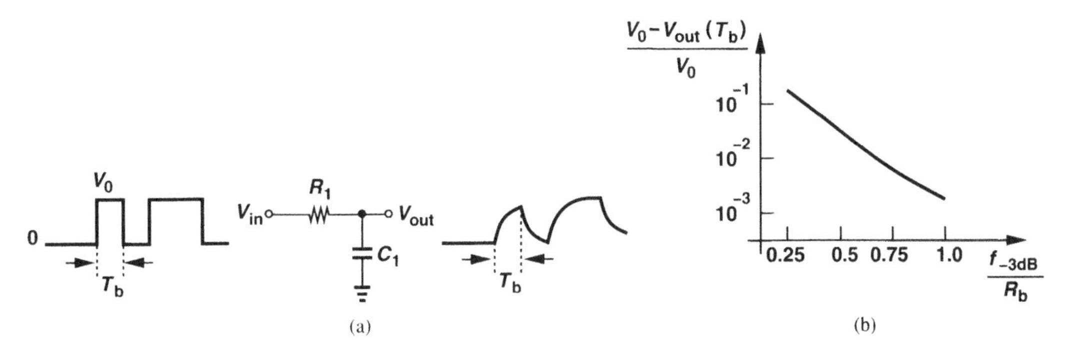

Figure 4.2 Response of an RC network to random data, (b) eye closure as a function of fractional bandwidth.

ZEROs (ONEs) are followed by a single ONE (ZERO). The settling for each bit is then expressed as

$$V_{out}(t) = V_0 \left(1 - \exp\frac{-t}{\tau}\right), \tag{4.12}$$

where $\tau = R_1C_1$. The error between V_{out} at $t = T_b$ and the final value is equal to

$$V_0 - V_{out}(T_b) = V_0 \exp\frac{-T_b}{\tau} \tag{4.13}$$

$$= V_0 \exp\frac{-2\pi f_{-3dB}}{R_b}, \tag{4.14}$$

where $f_{-3dB} = (2\pi R_1C_1)^{-1}$ and $R_b = 1/T_b$. The error grows exponentially as the bit rate increases. Figure 4.2(b) plots the normalized error as a function of f_{-3dB}/R_b. It is important to note that the eye closure is *twice* the value given by (4.14) and plotted in Fig. 4.2(b) because both the high and low levels experience ISI. The typical bandwidth chosen for TIAs is in the vicinity of $0.7R_b$, yielding a normalized error of 1.23% and hence a total eye closure of 2.46%, i.e., $|20\log(1 - 2.46\%)| = 0.216$ dB.

It is instructive to study the trade-off between noise and ISI as the bandwidth is reduced from R_b to $0.7R_b$ to $0.5R_b$. For $f_{-3dB} = R_b$, Eq. (4.14) yields a total eye closure of only 0.033 dB, but the noise is integrated across more than R_b hertz. If $f_{-3dB} = 0.7R_b$, then the total integrated noise current (or voltage) drops by a factor of approximately $\sqrt{0.7} \approx 0.84$, i.e., 1.55 dB. Thus, the choice $f_{-3dB} = 0.7R_b$ improves the performance by about 1.55 dB − 0.033 dB = 1.517 dB. If $f_{-3dB} = 0.5R_b$, then the eye closure due to ISI rises to 0.79 dB whereas the integrated noise current drops by about 3 dB.

Let us now examine the jitter resulting from limited bandwidth. Assuming that the decision threshold in Fig. 4.2(a) is equal to $V_0/2$, we wish to compute the time instants at which the output waveform crosses the threshold. First, consider the case shown in Fig. 4.3(a), where a logical ONE is preceded by several ZEROs. The time for V_{out} to reach

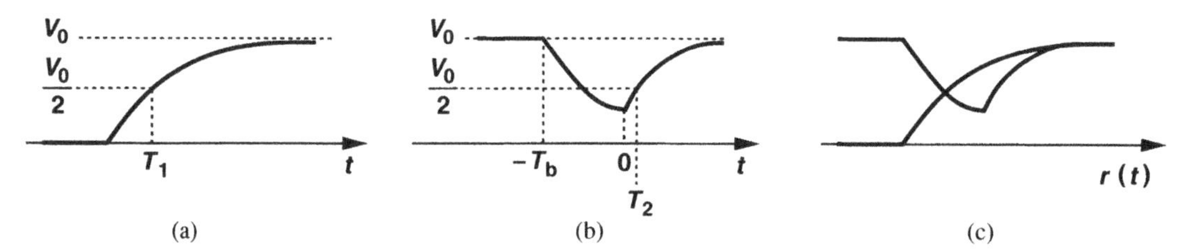

Figure 4.3 Response of an RC network to a long run followed by (a) a rising edge, (b) a falling edge, and (c) the resulting eye diagram. (As explained in Chapter 2, $r(t)$ is a periodic ramp.)

$V_0/2$ is therefore determined from

$$\frac{V_0}{2} = V_0(1 - \exp\frac{-T_1}{\tau}), \tag{4.15}$$

and hence

$$T_1 = \tau \ln 2 \tag{4.16}$$

$$\approx 0.693\tau. \tag{4.17}$$

Next, consider the sequence depicted in Fig. 4.3(b), where a single logical ZERO appears between two ONEs, forcing V_{out} to begin at a higher value at $t = 0$. Here, V_{out} reaches $V_0/2$ earlier than 0.693τ. We write the initial value of V_{out} at $t = 0$ as $V_{out}(0) = V_0 \exp(-T_b/\tau)$, arriving at

$$V_{out}(t) = V_0 - (V_0 - V_0 \exp\frac{-T_b}{\tau}) \exp\frac{-t}{\tau}. \tag{4.18}$$

It follows that V_{out} reaches $V_0/2$ at

$$T_2 = \tau \ln[2(1 - \exp\frac{-T_b}{\tau})]. \tag{4.19}$$

As expected, if $T_b \gg \tau$, then $T_2 \to T_1$. The difference between T_1 and T_2 can be viewed as the jitter in the zero crossings [Fig. 4.3(c)]. Normalizing the jitter to the bit period, we have

$$\frac{T_1 - T_2}{T_b} = \frac{-\tau}{T_b} \ln(1 - \exp\frac{-T_b}{\tau}), \tag{4.20}$$

or, expressing in terms of f_{-3dB}/R_b,

$$\frac{T_1 - T_2}{T_b} = \frac{-R_b}{2\pi f_{-3dB}} \ln(1 - \exp\frac{-2\pi f_{-3dB}}{R_b}). \tag{4.21}$$

For $f_{-3dB} > 0.3R_b$, the exponential term is relatively small, allowing the approximation $\ln(1 - \epsilon) \approx \epsilon$:

$$\frac{T_1 - T_2}{T_b} = \frac{R_b}{2\pi f_{-3dB}} \exp\frac{-2\pi f_{-3dB}}{R_b}. \tag{4.22}$$

For example, the jitter equals 1.38% for $f_{-3dB} = 0.5R_b$ and 0.28% for $f_{-3dB} = 0.7R_b$. Thus, the choice $f_{-3dB} = 0.7R_b$ introduces negligible jitter in a first-order system.

As mentioned earlier, practical TIAs contain multiple poles (and zeros), requiring careful simulations to determine the eye closure and the jitter resulting from the limited bandwidth. Figure 4.4 depicts the output of a second-order system as the bandwidth goes from R_b to $0.7R_b$ to $0.5R_b$. We note that the vertical closure of the eye is equal to 0.2 dB and 1 dB for the last two cases, respectively. While the noise-ISI trade-off appears to improve as the bandwidth goes from $0.7R_b$ to $0.5R_b$, the former choice is more common because actual circuits may contain more poles and process and temperature variations mandate additional margin.

Gain The transimpedance gain of TIAs must be large enough to overcome the noise of the subsequent stage, typically a 50-Ω driver or a limiting amplifier. Since the gain trades with bandwidth and voltage headroom, the two stages are designed together so as to optimize the overall performance. At high speeds (e.g., 10 Gb/s) and low supply voltages (e.g., 1.8 V), the gain of the TIA may be limited to a few hundred ohms, making the design of the following stage quite difficult.

As an example, suppose a limiting amplifier exhibits an input-referred noise voltage of 5 nV/$\sqrt{\text{Hz}}$. What transimpedance gain must precede the circuit to reduce its noise contribution to 1 pA/$\sqrt{\text{Hz}}$? We have $R_T = (5\ \text{nV}/\sqrt{\text{Hz}})/(1\ \text{pA}/\sqrt{\text{Hz}}) = 5\ \text{k}\Omega$.

Overload Response TIAs may receive large input currents, e.g., if they appear at the end of a short fiber. For example, with a 1-mW received power and a responsivity of 1 A/W, the input current reaches 1 mA. As with other analog circuits, TIAs introduce nonlinearity in the signal as the input level increases. While the binary nature of data may imply that a high nonlinearity can be tolerated, other issues must be taken into account. Some TIA topologies or some types of transistors may substantially distort the data waveform if the input current is large. For example, bipolar devices in the signal path may enter saturation for one bit, failing to respond properly to the next. Also, feedback circuits behave poorly if the signal drives the stages into saturation. As a result, both the high and low levels and the zero crossings of the waveforms may be corrupted.

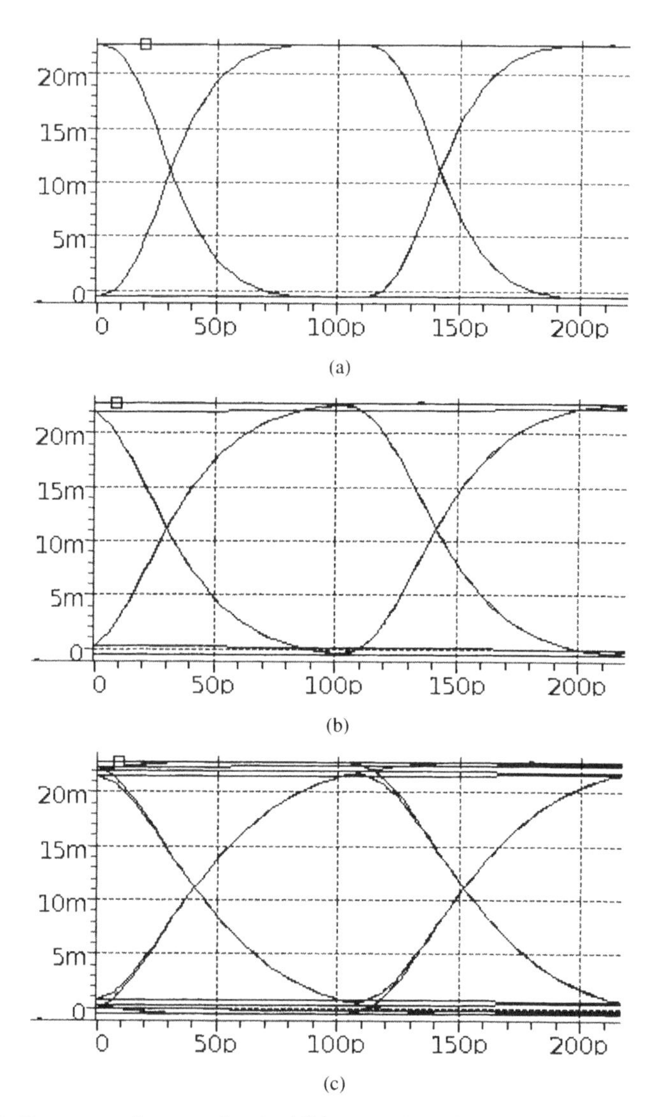

Figure 4.4 Response of a second-order TIA to random data for bandwidth equal to (a) R_b, (b) $0.7R_b$, and (c) $0.5R_b$.

The problem of overload often requires that the gain of TIA be adjusted according to the input level. Called "automatic gain control," the idea is to monitor the signal amplitude, compare it with a reference, and continuously adjust the transimpedance such that the output level remains relatively constant. AGC circuits are studied in Section 4.7.

In some applications, the TIA *linearity* may also mandate the use of AGC. For example, if the fiber introduces substantial dispersion (Chapter 3), then the TIA must amplify the signal with enough linearity to allow the subsequent equalizer to remove the ISI.

Output Impedance Stand-alone TIAs must drive a 50-Ω transmission line on a printed-circuit board to communicate with the subsequent stage. As explained in Chapter 5, 50-Ω drivers impose severe trade-offs between gain, power dissipation, and bandwidth. The trend, however, is to integrate the TIA and the limiting amplifier on one chip, thereby avoiding the interstage driver.

4.1.2 SNR Calculations

Our knowledge of analog design suggests that the noise of a circuit must be referred to the input so as to allow a fair comparison of different topologies. For example, in the common-source stage of Fig. 4.5(a), we represent the thermal noise of M_1 and R_D in unit bandwidth

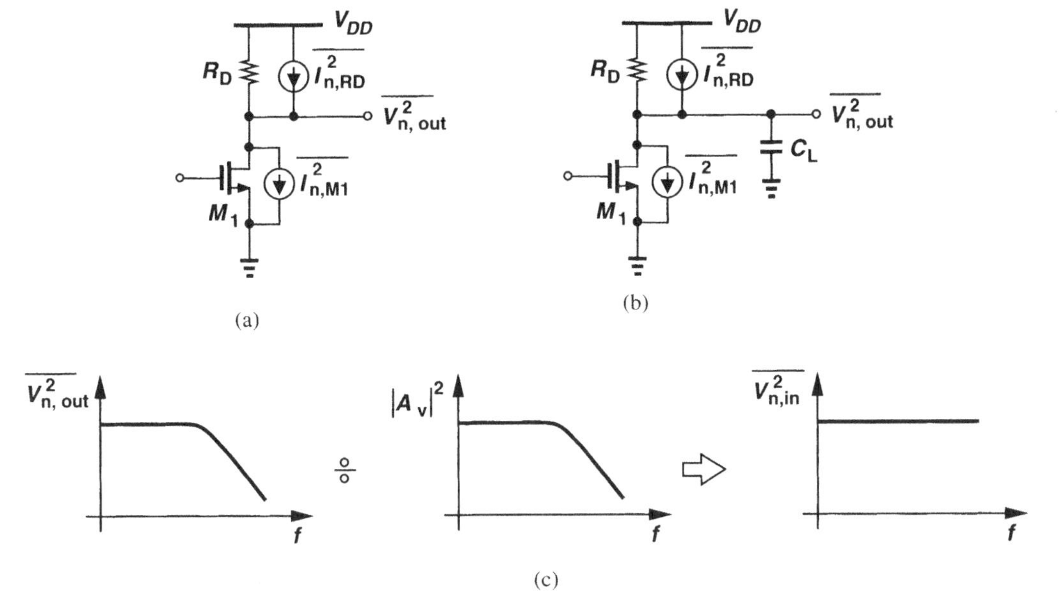

Figure 4.5 (a) Noise sources in a common-source stage, (b) CS stage driving a capacitive load, (c) input-referred noise of the stage.

by two current sources:

$$\overline{I_{n,M1}^2} = 4kT\gamma g_m \ \mathrm{A^2/Hz} \tag{4.23}$$

$$\overline{I_{n,RD}^2} = \frac{4kT}{R_D} \ \mathrm{A^2/Hz}, \tag{4.24}$$

where γ denotes the excess noise coefficient, equal to $2/3$ for long-channel devices and reaching as high as 2.5 in deep submicron technologies. Neglecting channel-length modulation in M_1, we express the output noise voltage as:

$$\overline{V_{n,out}^2} = (\overline{I_{n,M1}^2} + \overline{I_{n,RD}^2})R_D^2 \tag{4.25}$$

$$= 4kT(\gamma g_m + \frac{1}{R_D})R_D^2. \tag{4.26}$$

Thus, the input-referred noise voltage per unit bandwidth equals

$$\overline{V_{n,in}^2} = 4kT(\gamma g_m + \frac{1}{R_D})R_D^2 \div (g_m R_D)^2 \tag{4.27}$$

$$= 4kT(\frac{\gamma}{g_m} + \frac{1}{g_m^2 R_D}). \tag{4.28}$$

As a more complete example, let us repeat the above calculations for the circuit of Fig. 4.5(b), where a load capacitor C_L is added to the output node. We rewrite Eq. (4.26) as

$$\overline{V_{n,out}^2} = 4kT(\gamma g_m + \frac{1}{R_D})\left|R_D || \frac{1}{|C_L s|}\right|^2. \tag{4.29}$$

Since the voltage gain is equal to $g_m[R_D||(C_L s)^{-1}]$, we have

$$\overline{V_{n,in}^2} = \frac{\overline{V_{n,out}^2}}{g_m^2\left|R_D || \frac{1}{C_L s}\right|^2} \tag{4.30}$$

$$= 4kT(\frac{\gamma}{g_m} + \frac{1}{g_m^2 R_D}). \tag{4.31}$$

Interestingly, the input-referred noise voltage is independent of the load capacitance because, as depicted in Fig. 4.5(c), the gain and the output noise voltage fall at the same rate with frequency. Of course, the roll-off of the voltage gain at high frequencies makes the noise of the *subsequent* stage more significant.

The foregoing calculations become more complex and less intuitive if (1) the circuit contains additional poles and (2) the *total* integrated noise is of interest. (TIAs satisfy both conditions.) To appreciate the difficulty, let us return to the above example, where $\overline{V_{n,in}^2}$ is independent of frequency. How is the total noise voltage calculated? Across what bandwidth must $\overline{V_{n,in}^2}$ be integrated? To resolve this quandary, we recall that $\overline{V_{n,in}^2}$ is merely a fictitious measure and consider the actual operation of the circuit. As illustrated in Fig. 4.6, the common-source stage amplifies the signal by a factor of $g_m R_D$ while adding thermal noise. The total noise at the *output* is readily expressed as

$$\overline{V_{n,out,tot}^2} = \int_0^\infty \overline{V_{n,out}^2} df \tag{4.32}$$

$$= \int_0^\infty 4kT(\gamma g_m + \frac{1}{R_D})\frac{R_D^2 df}{R_D^2 C_L^2 (2\pi f)^2 + 1} \tag{4.33}$$

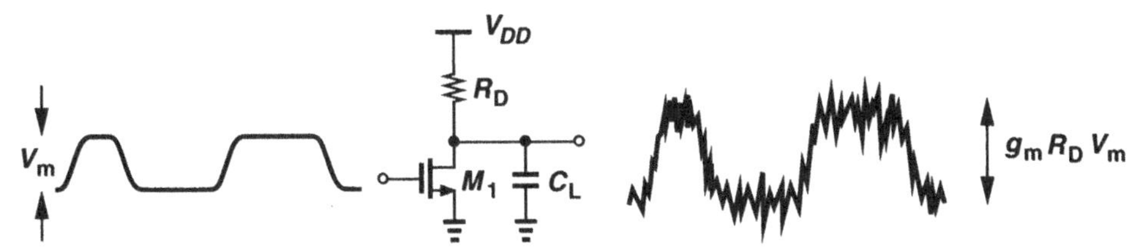

Figure 4.6 Amplification of random data in the presence of noise.

$$= (\gamma g_m + \frac{1}{R_D})R_D \int_0^\infty \frac{4kT}{R_D} \frac{R_D^2 df}{R_D^2 C_L^2 (2\pi f)^2 + 1} \tag{4.34}$$

which, from Eqs. (4.2) and (4.4), reduces to:

$$\overline{V_{n,out,tot}^2} = \frac{kT}{C_L}(\gamma g_m R_D + 1). \tag{4.35}$$

The above result suggests that the signal-to-noise ratio is more easily calculated at the *output* port:

$$SNR = \frac{(g_m R_D V_m)^2}{\overline{V_{n,out,tot}^2}} \tag{4.36}$$

$$= \frac{g_m^2 R_D^2 V_m^2}{\frac{kT}{C_L}(\gamma g_m R_D + 1)} \tag{4.37}$$

$$\approx \frac{C_L}{\gamma kT} g_m R_D V_m^2, \tag{4.38}$$

where it is assumed that $\gamma g_m R_D \gg 1$. For BER calculations, the square root of this result serves as the argument of the Q function (Chapter 2).

In summary, for SNR calculations, it is the *output* noise that must be integrated across $0 < f < \infty$ to represent the total noise corrupting the signal. Thus, the SNR is more easily obtained at the output port—just as carried out in simulations or measurements. We can of course rewrite Eq. (4.37) as

$$SNR = \frac{V_m^2}{\frac{kT}{C_L}\frac{(\gamma g_m R_D + 1)}{g_m^2 R_D^2}}, \tag{4.39}$$

and, since the numerator denotes the input signal, view the denominator as the total input-referred noise, $\overline{V_{n,in,tot}^2}$. In other words,

$$\overline{V_{n,in,tot}^2} = \frac{1}{g_m^2 R_D^2}\overline{V_{n,out,tot}^2}. \tag{4.40}$$

Note that the output noise is divided by the *midband* gain (rather than by the frequency-dependent gain) because the output signal amplitude is approximately equal to $g_m R_D V_m$.

It is sometimes necessary to compare the noise performance of circuits having different bandwidths and different gains. In such cases, the total noise does not provide a fair measure as it is not integrated across the same bandwidth. We therefore define the "average input noise" as:

$$\overline{V^2_{n,in,avg}} = \frac{\overline{V^2_{n,in,tot}}}{f_{-3dB}}, \tag{4.41}$$

where f_{-3dB} is the -3-dB bandwidth of the overall circuit. Expressed per unit bandwidth, $\overline{V^2_{n,in,avg}}$ allows direct comparison of different designs.

As an example, let us compute the average input noise current of the circuit shown in Fig. 4.7. Recall from Eq. (4.6) that the total input-referred noise current of this topology is

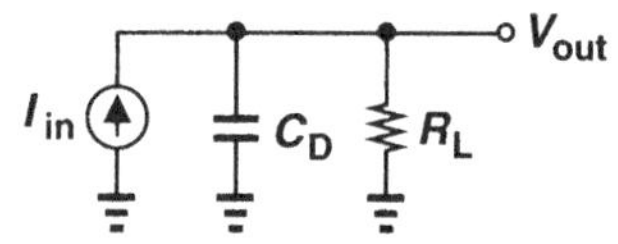

Figure 4.7 Current source driving an RC load.

equal to $kT/(R_L^2 C_D)$. Thus,

$$\overline{I^2_{n,in,avg}} = \frac{kT}{R_L^2 C_D} \div \frac{1}{2\pi R_L C_D} \tag{4.42}$$

$$= \frac{2\pi kT}{R_L} \ \mathrm{A}^2/\mathrm{Hz}. \tag{4.43}$$

4.1.3 Noise Bandwidth

The total noise corrupting a signal in a circuit results from all of the frequency components that fall in the bandwidth of the circuit. Consider a multipole circuit having the output noise spectrum shown in Fig. 4.8(a). Since the noise components above ω_{p1} are not negligible,

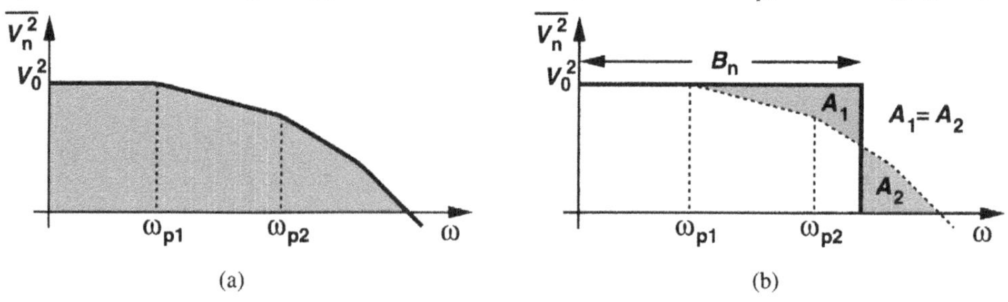

Figure 4.8 (a) Output noise spectrum of a circuit, (b) concept of noise bandwidth.

the total output noise must be evaluated by calculating the total area under the spectral density:

$$\overline{V_{n,out,tot}^2} = \int_0^\infty \overline{V_{n,out}^2} df. \tag{4.44}$$

However, as depicted in Fig. 4.8(b), it is sometimes helpful to represent the total noise simply as $V_0^2 \cdot B_n$, where the bandwidth B_n is chosen such that

$$V_0^2 \cdot B_n = \int_0^\infty \overline{V_{n,out}^2} df. \tag{4.45}$$

Called the "noise bandwidth," B_n allows a fair comparison of circuits that exhibit the same low-frequency noise, V_0^2, but different high-frequency transfer functions. As an exercise, the reader can prove that the noise bandwidth of a one-pole system is equal to $\pi/2$ times the pole frequency.

4.2 Open-Loop TIAs

The example of Fig. 4.1(a) and its stringent noise-bandwidth trade-off point to the need for TIAs that provide higher performance. The problem of noise prohibits complex configurations that introduce many devices in the signal path, constraining TIA circuits to primarily two topologies: open-loop (common-gate/base) stages and current-feedback amplifiers.

Before studying TIAs, we make an observation. Recall from the example of Fig. 4.1 that R_L directly couples the noise and bandwidth equations. Thus, for a given photodiode capacitance, the (input) bandwidth of a TIA is maximized by minimizing the input resistance of the circuit. We must therefore seek circuits that provide both a low input resistance and a high gain.

4.2.1 Low-Frequency Behavior

An amplifier stage that exhibits a low input impedance is the common-gate (CG) (for field-effect devices) or common-base (CB) (for bipolar devices) topology (Fig. 4.9). Neglecting second-order effects in the transistors for now, we note that the input resistance of each stage is approximately equal to $1/g_m$, where g_m denotes the transconductance of the input transistor. Proper choice of the bias current (and, for M_1, device dimensions) yields a relatively low input resistance, maximizing the input bandwidth. Let us first carefully examine the low-frequency behavior of the two circuits shown in Fig. 4.9.

CG Low-Frequency Behavior Including both body effect and channel-length modulation and assuming I_B is ideal, we construct the small-signal equivalent circuit of the CG stage (Fig. 4.10). Since all of I_{in} flows through R_D, the transimpedance gain is

$$R_T = R_D. \tag{4.46}$$

To obtain the input resistance, which is equal to $-V_1/I_{in}$, we recognize that the current through r_O is equal to $I_{in} + g_m V_1 + g_{mb} V_1$. Adding the voltage drops across r_O and R_D

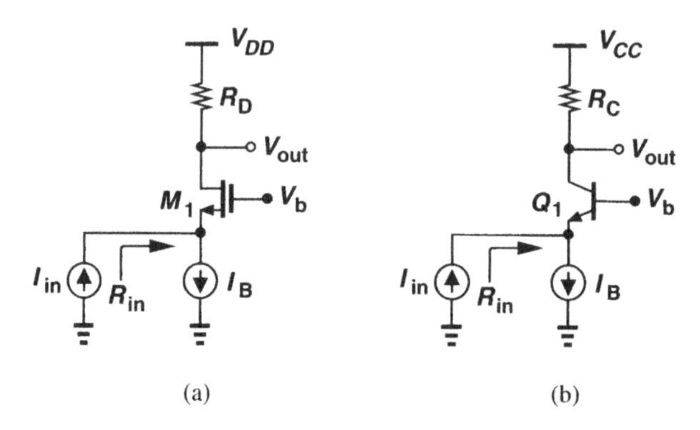

Figure 4.9 (a) Common-gate and (b) common-base topologies.

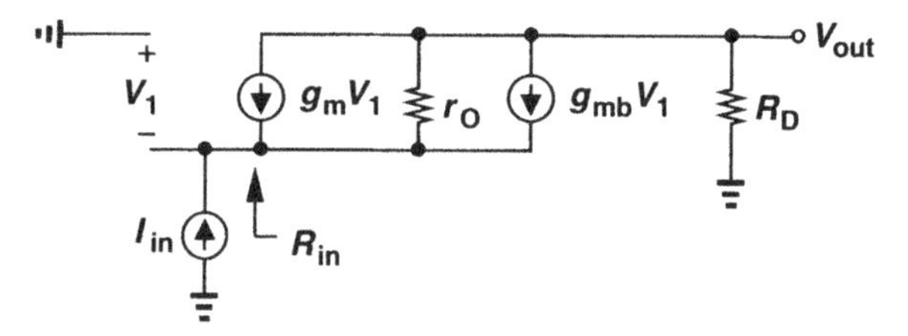

Figure 4.10 Small-signal model of common-gate stage.

and equating the result to $-V_1$, we have

$$(I_{in} + g_m V_1 + g_{mb} V_1) r_O + I_{in} R_D = -V_1. \tag{4.47}$$

Thus,

$$R_{in} = \frac{-V_1}{I_{in}} \tag{4.48}$$

$$= \frac{r_O + R_D}{1 + (g_m + g_{mb}) r_O}. \tag{4.49}$$

Since typically, $(g_m + g_{mb}) r_O \gg 1$,

$$R_{in} \approx \frac{1}{g_m + g_{mb}} + \frac{R_D}{(g_m + g_{mb}) r_O}. \tag{4.50}$$

Note that the positive feedback through r_O *increases* the input resistance. If $r_O = \infty$, R_{in} is simply equal to $1/(g_m + g_{mb})$, reflecting the impact of body effect. With short-channel devices, on the other hand, the second term in Eq. (4.50) may not be negligible.

It is instructive to analyze the low-frequency noise performance of the CG stage.[2] Figure 4.11(a) depicts a more realistic implementation, with M_2 operating as the bias current

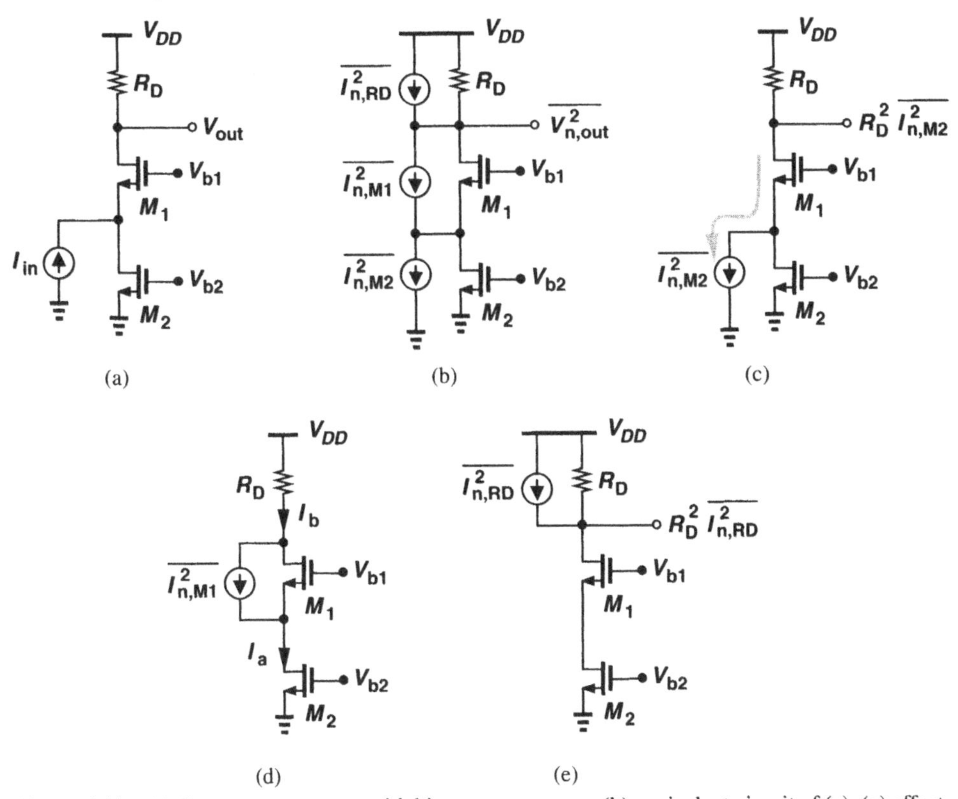

Figure 4.11 (a) Common-gate stage with bias current source, (b) equivalent circuit of (a), (c) effect of noise current of M_2, (d) effect of noise current of M_1, (e) noise contribution of R_D.

source. Adding the noise sources as in Fig. 4.11(b) and neglecting channel-length modulation and body effect for simplicity, we compute the contribution of each source by superposition. (1) All of $\overline{I_{n,M2}^2}$ flows through R_D, generating an output noise of $R_D^2\overline{I_{n,M2}^2}$ [Fig. 4.11(c)]; (2) No part of $\overline{I_{n,M1}^2}$ flows through R_D [1] because in Fig. 4.11(d), I_a is zero ($r_{O2} = \infty$) and I_b must be equal to I_a; (3) All of $\overline{I_{n,RD}^2}$ flows through R_D because the impedance seen looking into the the drain of M_2 is infinite [Fig. 4.11(e)]. It follows that the output noise per unit bandwidth is

$$\overline{V_{n,out}^2} = (\overline{I_{n,M2}^2} + \overline{I_{n,RD}^2})R_D^2 \tag{4.51}$$

$$= 4kT\left(\gamma g_{m2} + \frac{1}{R_D}\right)R_D^2. \tag{4.52}$$

[2] We neglect flicker noise in broadband amplifiers because the bandwidths of interest in today's technology are much greater than typical flicker noise corner frequencies.

Dividing this quantity by the transimpedance gain yields the input-referred noise current:

$$\overline{I_{n,in}^2} = 4kT(\gamma g_{m2} + \frac{1}{R_D}) \tag{4.53}$$

$$= \overline{I_{n,M2}^2} + \overline{I_{n,RD}^2}. \tag{4.54}$$

Equation (4.54) is a remarkable result, suggesting that the noise currents of M_2 and R_D are referred to the input with a unity factor. This is the principal drawback of common-gate TIAs.

To extend our understanding, let us determine the contribution of $\overline{I_{n,M1}^2}$ in Fig. 4.11(a) if channel-length modulation is not neglected. Using the equivalent circuit shown in Fig. 4.12 and noting that the currents through R_D and r_{O2} are equal, we write $V_1 =$

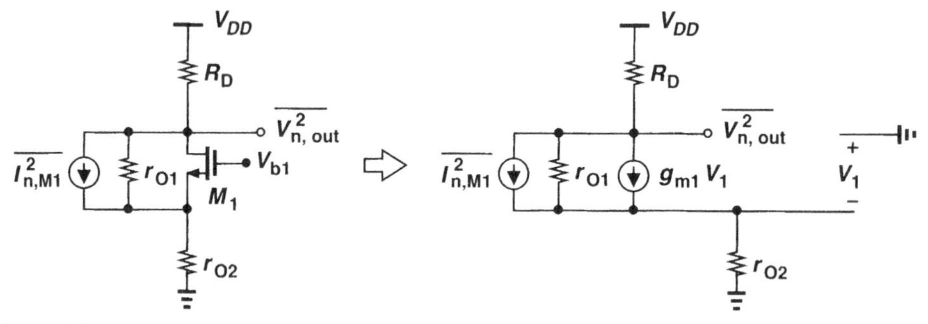

Figure 4.12 Equivalent circuit of common-gate stage.

$(V_{n,out}/R_D)r_{O2}$. Since the current through r_{O1} is given by $-V_{n,out}/R_D - g_{m1}V_1 - I_{n,M1}$, the voltages across r_{O1}, R_D, and r_{O2} can be summed as

$$\left(\frac{-V_{n,out}}{R_D} - g_{m1}V_1 - I_{n,M1}\right) r_{O1} - \frac{V_{n,out}}{R_D} r_{O2} - V_{n,out} = 0. \tag{4.55}$$

Thus,

$$V_{n,out} = \frac{-r_{O1}R_D}{r_{O2}(1 + g_{m1}r_{O1}) + r_{O1} + R_D} I_{n,M1}. \tag{4.56}$$

Typically, $g_{m1}r_{O1} \gg 1$ and $g_{m1}r_{O1}r_{O2} \gg r_{O1} + R_D$, yielding

$$V_{n,out} \approx \frac{-R_D}{g_{m1}r_{O2}} I_{n,M1}. \tag{4.57}$$

That is,

$$\overline{V_{n,out}^2} = 4kTR_D \frac{\gamma R_D}{g_{m1}r_{O2}^2}. \tag{4.58}$$

In other words, if $g_{m1}r_{O2}^2$ is comparable with γR_D, then the noise contributed by M_1 becomes comparable with that of R_D.

Let us now return to Eq. (4.54) and attempt to minimize the noise currents of M_2 and R_D. Interestingly, $\overline{I_{n,M2}^2}$ and $\overline{I_{n,RD}^2}$ trade *with each other*. This is quantified as follows. Since $g_{m2} = 2I_{D2}/(V_{GS2} - V_{TH2})$, where I_{D2} and $V_{GS2} - V_{TH2}$ are the drain current and gate-source overdrive voltage of M_2, respectively, and since $\overline{I_{n,M2}^2} = 4kT\gamma g_{m2}$, we have

$$\overline{I_{n,M2}^2} = 4kT\gamma \frac{2I_{D2}}{V_{GS2} - V_{TH2}}. \tag{4.59}$$

That is, for a given bias current, the overdrive voltage of M_2 and hence the minimum tolerable drain-source voltage of this transistor must be maximized. On the other hand, the noise current of R_D is minimized by maximizing R_D. Both of these trends require greater voltage headroom. In fact, since $R_D I_{D2} + V_{DS2}$ must remain less than V_{DD} and V_{DS2} must exceed $V_{GS2} - V_{TH2}$, we have

$$R_D I_{D2} + \frac{2I_{D2}}{\dfrac{\overline{I_{n,M2}^2}}{4kT\gamma}} < V_{DD} \tag{4.60}$$

and hence

$$\frac{4kT}{\overline{I_{n,RD}^2}} + \frac{8kT\gamma}{\overline{I_{n,M2}^2}} < \frac{V_{DD}}{I_{D2}}. \tag{4.61}$$

Thus, $\overline{I_{n,M2}^2} + \overline{I_{n,RD}^2}$ must be minimized subject to the above constraint. This limitation makes low-voltage design more difficult.

CB Low-Frequency Behavior We now consider the common-base stage of Fig. 4.13(a), seeking the differences between this circuit and the CG topology. With the aid of the equivalent circuit shown in Fig. 4.13(b), where r_b represents the transistor base re-

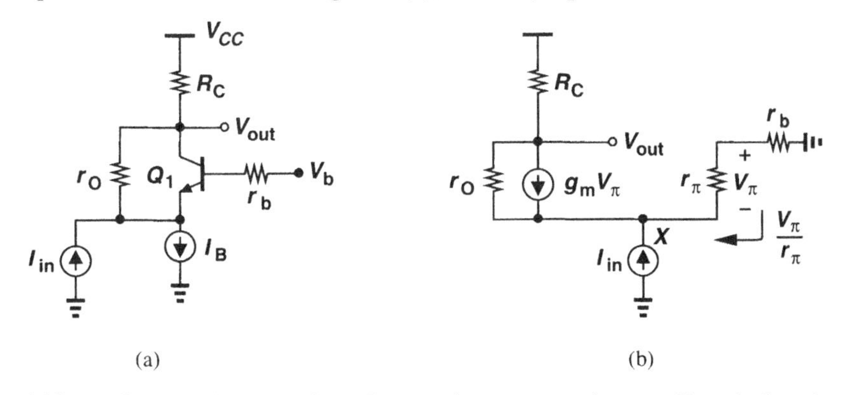

Figure 4.13 (a) Common-base stage including transistor output resistance, (b) equivalent circuit.

sistance, we note that $V_X = -V_\pi(1 + r_b/r_\pi)$. Also, $(I_{in} + V_\pi/r_\pi)R_C = V_{out}$. Adding

V_X and the voltage drop across r_O and equating the result to V_{out}, we obtain the transimpedance gain:

$$\frac{V_{out}}{I_{in}} = \frac{(1+g_m r_O)r_\pi + r_b}{(1+g_m r_O)r_\pi + r_b + R_C + r_O} R_C. \tag{4.62}$$

It is reasonable to assume that $(1+g_m r_O)r_\pi + r_b \gg R_C + r_O$ because $(1+g_m r_O)r_\pi \approx \beta r_O$. That is,

$$R_T \approx R_C. \tag{4.63}$$

Using the above expressions for V_X and V_π, we can also derive the input impedance:

$$R_{in} = \frac{V_X}{I_{in}} \tag{4.64}$$

$$\approx \frac{(\frac{1}{g_m} + \frac{r_b}{\beta})(r_O + R_C)}{r_O + \frac{r_b + R_C}{\beta}}. \tag{4.65}$$

For large β, (4.65) reduces to

$$R_{in} \approx \frac{1}{g_m} + \frac{R_C}{g_m r_O}, \tag{4.66}$$

similar to the case of the CG stage.

Let us analyze the noise performance of the CB stage, including the contribution of the bias current source [Fig. 4.14(a)] as well. Shown in Fig. 4.14(b), the equivalent circuit

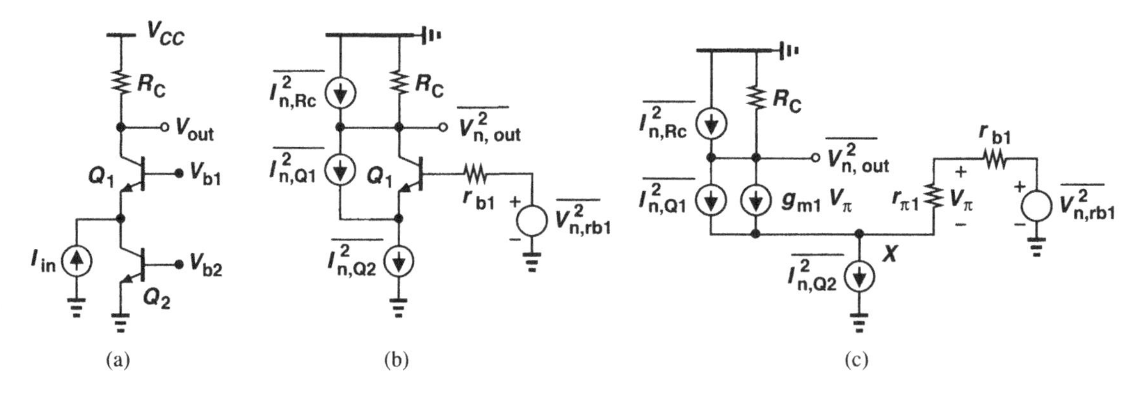

Figure 4.14 (a) CB stage with bias current source, (b) equivalent circuit including noise sources, (c) small-signal equivalent.

contains (1) the collector current shot noise of Q_1, $\overline{I^2_{n,Q1}} = 4kT(g_{m1}/2)$; (2) the base resistance thermal noise of Q_1, $\overline{V^2_{n,rb1}} = 4kTr_{b1}$; (3) the same components for Q_2 and represented by $\overline{I_{n,Q2}^2} = 4kT(g_{m2}/2) + (4kTr_{b2})g_{m2}^2$; (4) the thermal noise of R_C, $\overline{I_{n,Rc}^2} = 4kT/R_C$.[3] Neglecting the Early effect and assuming a large β, we use the equivalent circuit of Fig. 4.14(c) and write

$$-\frac{V_{n,out}}{R_C} + I_{n,Rc} = I_{n,Q1} + g_{m1}V_\pi \tag{4.67}$$

and hence

$$V_\pi = \frac{1}{g_{m1}}\left(-\frac{V_{n,out}}{R_C} + I_{n,Rc} - I_{n,Q1}\right). \tag{4.68}$$

Also, summing the currents at node X gives

$$\frac{V_\pi}{r_{\pi 1}} + g_{m1}V_\pi + I_{n,Q1} = I_{n,Q2}. \tag{4.69}$$

Substituting for V_π from (4.68), we have

$$-\left(1+\frac{1}{\beta}\right)\frac{V_{n,out}}{R_C} + \left(1+\frac{1}{\beta}\right)I_{n,RC} - \frac{I_{n,Q1}}{\beta} = I_{n,Q2}. \tag{4.70}$$

If $\beta \gg 1$, then

$$V_{n,out} \approx R_C I_{n,Rc} - \frac{R_C I_{n,Q1}}{\beta} - R_C I_{n,Q2} \tag{4.71}$$

or

$$\overline{V_{n,out}^2} \approx 4kTR_C + \frac{R_C^2}{\beta^2}4kT\frac{g_{m1}}{2} + 4kTR_C^2\left(\frac{1}{2g_{m2}} + r_{b2}\right). \tag{4.72}$$

In a typical design, the second term in (4.72) is negligible and

$$\overline{V_{n,out}^2} \approx 4kTR_C + 4kTR_C^2\left(\frac{1}{2g_{m2}} + r_{b2}\right). \tag{4.73}$$

Thus, the output noise results from primarily Q_2 and R_C, similar to the case of the CG stage. Also, the noise voltage $V_{n,rb1}$ does not contribute to $V_{n,out}$ because, with no Early effect, the voltage gain from the base of Q_1 to the output is zero.

Dividing both sides of (4.73) by $R_T^2 = R_C^2$ yields the input-referred noise current per unit bandwidth as

$$\overline{I_{n,in}^2} = \frac{4kT}{R_C} + 4kT\left(\frac{1}{2g_{m2}} + r_{b2}\right). \tag{4.74}$$

[3] The effect of the base shot noise current is neglected.

Thus, the noise currents of R_C and Q_2 are referred to the input with unity gain.

It is possible to reduce the noise contribution of Q_2 by emitter degeneration, but at the cost of voltage headroom. The idea can be studied with the aid of the circuit shown in Fig. 4.15(a). To determine the output noise current, we use the equivalent circuit of Fig. 4.15(b)

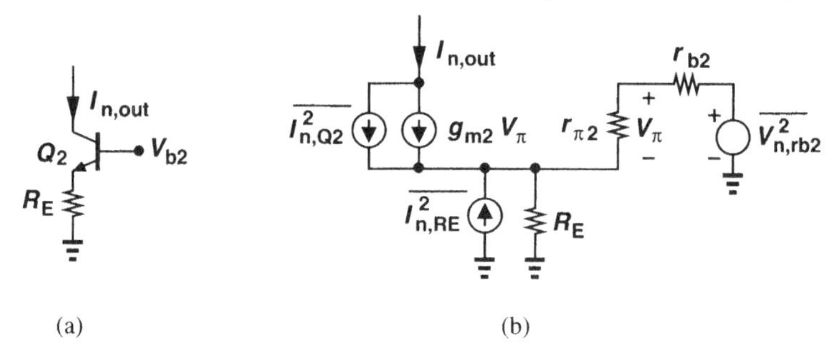

Figure 4.15 (a) Current source with resistive degeneration, (b) equivalent circuit for noise calculation.

and write $g_{m2}V_\pi + I_{n,Q2} = I_{n,out}$ and hence $V_\pi = (I_{n,out} - I_{n,Q2})/g_{m2}$. Summing the voltage drops across $R_E, r_{\pi 2}$, and r_{b2} yields

$$(I_{n,RE} + I_{n,out} + \frac{I_{n,out} - I_{n,Q2}}{g_{m2}r_{\pi 2}})R_E + \frac{I_{n,out} - I_{n,Q2}}{g_{m2}}(1 + \frac{r_{b2}}{r_{\pi 2}}) = V_{n,rb2}. \tag{4.75}$$

It follows that

$$I_{n,out} \approx \frac{-I_{n,RE}R_E + I_{n,Q2}(\frac{R_E}{\beta_2} + \frac{1}{g_{m2}} + \frac{r_{b2}}{\beta_2}) + V_{n,rb2}}{R_E + \frac{1}{g_{m2}} + \frac{r_{b2}}{\beta_2}}. \tag{4.76}$$

Voltage headroom issues typically limit the drop across R_E to a few hundred millivolts, resulting in $R_E/\beta_2 \ll 1/g_{m2}$ because $R_E I_E/\beta_2 \ll V_T$. Thus,

$$\overline{I^2_{n,out}} \approx \frac{4kTR_E + 4kT\frac{g_{m2}}{2}(\frac{1}{g_{m2}} + \frac{r_{b2}}{\beta_2})^2 + 4kTr_{b2}}{(R_E + \frac{1}{g_{m2}} + \frac{r_{b2}}{\beta_2})^2}. \tag{4.77}$$

To arrive at a simple expression, let us assume a practical case, $r_{b2} \ll R_E, \beta_2/g_{m2}$ (to minimize the noise of r_{b2}) and :

$$\overline{I^2_{n,out}} \approx 4kT\frac{R_E + \frac{1}{2g_{m2}}}{(R_E + \frac{1}{g_{m2}})^2} \tag{4.78}$$

$$\approx 4kT\frac{1 + 2g_{m2}R_E}{(1 + g_{m2}R_E)^2}\frac{g_{m2}}{2}. \tag{4.79}$$

Equation (4.79) suggests that resistive degeneration lowers the output noise current of a bipolar current source by a factor of $(1 + 2g_{m2}R_E)/(1 + g_{m2}R_E)^2$, a monotonically decreasing function of R_E. For example, if a drop of $10V_T$ is allocated to R_E, then $I_{E2}R_E \approx 10V_T$ and $g_{m2}R_E \approx 10$, reducing $\overline{I^2_{n,out}}$ by approximately a factor of 6.

4.2.2 High-Frequency Behavior

CG High-Frequency Behavior Let us first consider the common-gate stage at high frequencies. In the circuit shown in Fig. 4.16, $C_{in} = C_D + C_{GS1} + C_{SB1} + C_{GD2} + C_{DB2}$, where C_D denotes the photodiode capacitance. Also, C_{out} includes C_{GD1}, C_{DB1}, and

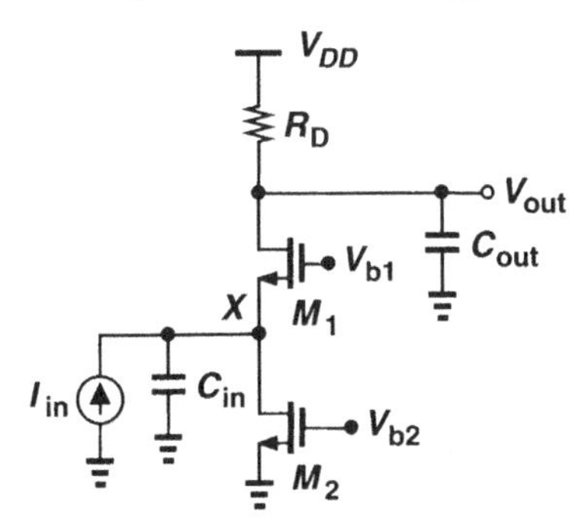

Figure 4.16 CG stage at high frequencies.

the input capacitance of the following stage. Neglecting channel-length modulation for simplicity and noting that the resistance seen at the source of M_1 to ground is equal to $(g_{m1} + g_{mb1})^{-1}$, we have

$$I_{in}\left(\frac{1}{C_{in}s}\Big|\Big|\frac{1}{g_{m1} + g_{mb1}}\right) = V_X. \tag{4.80}$$

Since the drain current of M_1 is equal to $(g_{m1} + g_{mb1})V_X$, the output voltage is given by

$$V_{out} = (g_{m1} + g_{mb1})V_X\left(R_D\Big|\Big|\frac{1}{C_{out}s}\right). \tag{4.81}$$

It follows from (4.80) and (4.81) that

$$\frac{V_{out}}{I_{in}} = (g_{m1} + g_{mb1})\left(\frac{1}{C_{in}s}\Big|\Big|\frac{1}{g_{m1} + g_{mb1}}\right)\left(R_D\Big|\Big|\frac{1}{C_{out}s}\right) \tag{4.82}$$

$$= \frac{(g_{m1} + g_{mb1})R_D}{(g_{m1} + g_{mb1} + C_{in}s)(R_D C_{out}s + 1)}. \tag{4.83}$$

As expected, $V_{out}/I_{in} = R_D$ for $s = 0$. In a typical design, the input pole, $(g_{m1} + g_{mb1})/C_{in}$, may be closer to the origin than is the output pole, $(R_D C_{out})^{-1}$, because the photodiode capacitance is quite large (in the range of 100 to 500 fF).

It is instructive to examine the CG stage for trade-offs between noise, bandwidth, and supply voltage. To maximize the magnitude of the input pole, the quantity $g_{m1} + g_{mb1}$

must be maximized—by increasing the width or the bias current of M_1. However, as the width increases, C_{GS1} rises more rapidly than $g_{m1} + g_{mb1}$ does, eventually limiting the bandwidth. If the bias current is increased, then V_{GS1}, the voltage drop across R_D, and the minimum allowable V_{DS2} must increase, requiring a greater supply voltage. If R_D is decreased to accommodate a greater bias current, then its noise current increases and the transimpedance gain falls. If M_2 is made wider to allow a smaller V_{DS2}, then both its noise current and drain capacitance increase.

The foregoing observations reveal that it becomes increasingly more difficult to achieve a broad band and a reasonable transimpedance gain at low supply voltages. In fact, owing to the low available gain in the first stage, the noise contributed by the subsequent stage is typically quite significant as well.

We now analyze the noise behavior of the CG stage at high frequencies (Fig. 4.17). Since the noise current of M_2, $I_{n,M2}$, is directly added to the input signal, we ignore

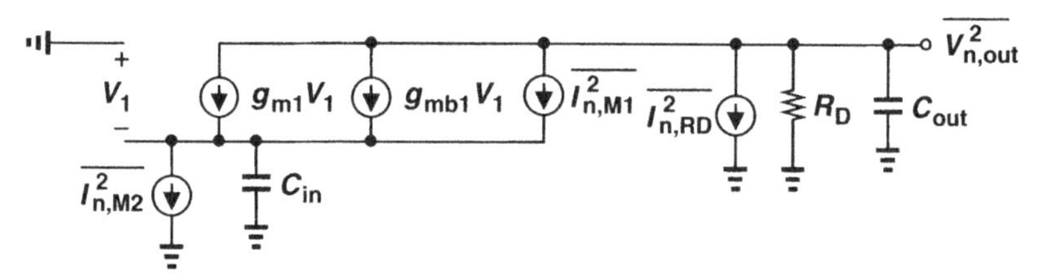

Figure 4.17 High-frequency noise model of CG stage.

it for the moment. Neglecting channel-length modulation, we have $(I_{n,M1} + g_{mb1}V_1 + g_{m1}V_1)/(C_{in}s) = -V_1$, i.e., $V_1 = -I_{n,M1}/(C_{in}s + g_{mb1} + g_{m1})$. Also, summing $I_{n,RD}$ and the currents through C_{out} and R_D yields

$$[-V_{n,out}(C_{out}s + \frac{1}{R_D}) + I_{n,RD}]\frac{1}{C_{in}s} = -V_1 \tag{4.84}$$

$$= \frac{I_{n,M1}}{C_{in}s + g_{mb1} + g_{m1}}. \tag{4.85}$$

Thus,

$$V_{n,out} = \frac{-R_D C_{in} s I_{n,M1}}{(C_{in}s + g_{m1} + g_{mb1})(R_D C_{out}s + 1)} + \frac{I_{n,RD}R_D}{R_D C_{out}s + 1}. \tag{4.86}$$

Dividing (4.86) by the transimpedance gain given by (4.83), we obtain

$$I_{in,in} = \frac{-C_{in}sI_{n,M1}}{g_{m1} + g_{mb1}} + \frac{I_{n,RD}(C_{in}s + g_{m1} + g_{mb1})}{g_{m1} + g_{mb1}} \tag{4.87}$$

$$= -\frac{C_{in}s}{g_{m1} + g_{mb1}}I_{n,M1} + (\frac{C_{in}s}{g_{m1} + g_{mb1}} + 1)I_{n,RD}. \tag{4.88}$$

Equation (4.88) reveals two important properties of the CG stage.[4] First, the noise contributed by M_1 directly scales with C_{in} and the frequency; as $|C_{in}s|$ increases, a greater fraction of $I_{n,M1}$ flows from the output node (rather than circulate inside M_1 [1]). Second, the noise contributed by R_D to the input also rises as $|C_{in}s|$ becomes comparable with $g_{m1} + g_{mb1}$. This is because the transimpedance gain falls as the frequency approaches the input pole.

While providing intuitive results, the above analysis does not readily yield the total *integrated* noise current. As explained in Section 4.1.2, the total noise must be computed at the output and subsequently divided by the midband gain. We must therefore integrate the following function from $f = 0$ to $f = \infty$:

$$\overline{V_{n,out}^2} = \frac{(2\pi R_D C_{in} f)^2 \overline{I_{n,M1}^2}}{[C_{in}^2(2\pi f)^2 + (g_{m1} + g_{mb1})^2][R_D^2 C_{out}^2(2\pi f)^2 + 1]} + \frac{R_D^2 \overline{I_{n,RD}^2}}{R_D^2 C_{out}^2(2\pi f)^2 + 1} + \frac{(g_{m1} + g_{mb1})^2 R_D^2 \overline{I_{n,M2}^2}}{[C_{in}^2(2\pi f)^2 + (g_{m1} + g_{mb1})^2][R_D^2 C_{out}^2(2\pi f)^2 + 1]}, \tag{4.89}$$

where the last term represents the product of the noise current of M_2 and the gain. The algebra may appear formidable, but the final result will be simple and intuitive.

Before integrating (4.89), it is instructive to examine each term qualitatively. We assume a typical case, where the input pole is much closer to the origin than the output pole is. As plotted in Fig. 4.18, the first term rises from zero, reaches a value of $R_D^2 \overline{I_{n,M1}^2}$ in midband,[5] and begins to fall in the vicinity of the output pole, $\omega_{p,out} = (R_D C_{out})^{-1}$. The second term corresponds to the noise of a first-order RC section similar to that studied in Section 4.1. The third term begins as the noise in a first-order circuit but experiences sharper fall above the output pole.[6]

In order to integrate (4.89), we note that

$$\frac{x^2}{(a^2x^2 + b^2)(c^2x^2 + d^2)} = \frac{1}{a^2d^2 - b^2c^2}\left(\frac{-b^2}{a^2x^2 + b^2} + \frac{d^2}{c^2x^2 + d^2}\right) \tag{4.90}$$

and

$$\frac{1}{(a^2x^2 + b^2)(c^2x^2 + d^2)} = \frac{1}{a^2d^2 - b^2c^2}\left(\frac{a^2}{a^2x^2 + b^2} + \frac{-c^2}{c^2x^2 + d^2}\right). \tag{4.91}$$

[4] Recall that the total noise current must also include $I_{n,M2}$.

[5] In midband, C_{in} exhibits a low impedance to ground, allowing most of $I_{n,M1}$ to flow from R_C.

[6] In practice, we may choose the output pole to be only a few times the input pole so as to minimize the area under the three plots in Fig. 4.18.

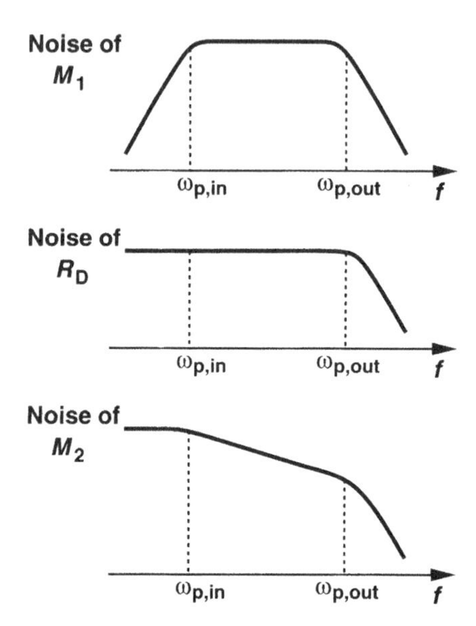

Figure 4.18 Noise contributions in a CG stage as a function of frequency.

To deal with the second-order fractions, we recall from Section 4.1 that

$$\int_0^{\infty} \frac{4kTRdf}{(2\pi RCf)^2 + 1} = \frac{kT}{C}. \tag{4.92}$$

Decomposing the first and third terms of (4.89) into partial fractions and carrying out the integration with the aid of (4.92), we obtain

$$\overline{V_{n,out,tot}^2} = \frac{kTg_{m1}R_DC_{in}\gamma}{C_{out}[C_{in} + (g_{m1} + g_{mb1})R_DC_{out}]} + \frac{kT}{C_{out}} + \frac{2kT(g_{m1} + g_{mb1})g_{m2}R_D^2\gamma}{C_{in} + (g_{m1} + g_{mb1})R_DC_{out}}. \tag{4.93}$$

If the input pole is assumed to be dominant, i.e., $C_{in}/(g_{m1} + g_{mb1}) \gg R_DC_{out}$, then

$$\overline{V_{n,out,tot}^2} \approx \frac{kT(g_{m1}R_D\gamma + 1)}{C_{out}} + \frac{2kT(g_{m1} + g_{mb1})g_{m2}R_D^2\gamma}{C_{in}}. \tag{4.94}$$

Since for short-channel MOSFETs, $\gamma > 1$, it is reasonable to assume $g_{m1}R_D\gamma \gg 1$. Dividing both sides by R_D^2 yields the total input-referred noise current:

$$\overline{I_{n,in,tot}^2} \approx \frac{kTg_{m1}\gamma}{R_DC_{out}} + \frac{2kT(g_{m1} + g_{mb1})g_{m2}\gamma}{C_{in}} \tag{4.95}$$

$$\approx 4kT\gamma(\frac{1}{4}g_{m1}\omega_{p,out} + \frac{1}{2}g_{m2}\omega_{p,in}). \tag{4.96}$$

This result can be interpreted as the noise of M_1 integrated across a bandwidth of $\omega_{p,out}/4$ plus the noise of M_2 integrated across a bandwidth of $\omega_{p,in}/2$. Since $\omega_{p,out} = (R_D C_{out})^{-1}$ and $\omega_{p,in} = (g_{m1} + g_{mb1})/C_{in}$ determine the bandwidth, they are defined by the optical standard and cannot be changed significantly. Thus, the only parameter under the designer's control is g_{m2}. However, for a given bias current, g_{m2} can be lowered only at the cost of voltage headroom.

Equation (4.96) predicts little flexibility in the design of common-gate TIAs, implying that this topology is ill-suited to low-noise applications. However, since CG stages exhibit a broad band with high stability, they are utilized in short-haul applications.

CB High-Frequency Behavior Let us now consider the common-base topology at high frequencies. As shown in Fig. 4.19(a), the circuit contains the following capacitances: (1) C_{in}, which includes the photodiode capacitance, the collector-base and collector-substrate capacitances of Q_2, and $C_{\pi 1}$; (2) C_{out}, which represents the collector-base and collector-substrate capacitances of Q_1 and the input capacitance of the subsequent stage. For gain calculations, we neglect the Early effect and the base resistance, arriving at the equivalent circuit shown in Fig. 4.19(b). The topology is now similar to a common-gate stage, exhibiting the following transfer function (if $\beta \gg 1$):

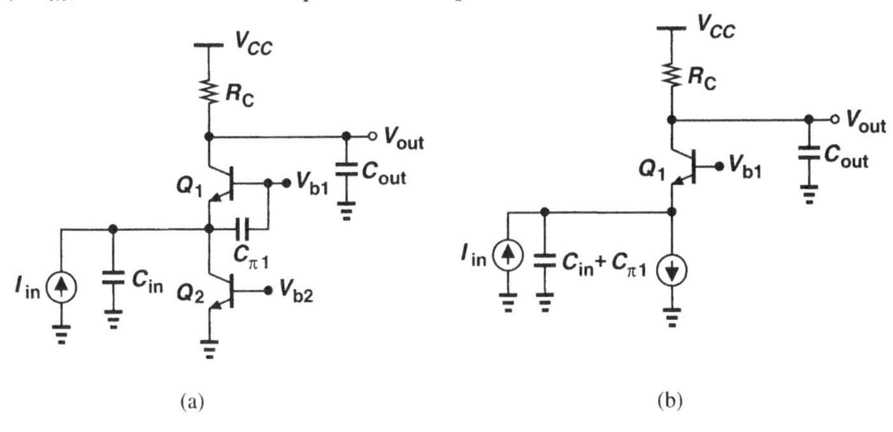

Figure 4.19 (a) CB stage at high frequencies, (b) simplified model of (a).

$$\frac{V_{out}}{I_{in}} = \frac{g_{m1} R_C}{[g_{m1} + (C_{in} + C_{\pi 1})s](R_C C_{out} s + 1)}. \tag{4.97}$$

The CB and CG stages suffer from somewhat similar trade-offs. To achieve a broad band, $g_{m1}/(C_{in} + C_{\pi 1})$ and $(R_C C_{out})^{-1}$ must be maximized. For a given C_{in}, a higher collector bias current gives a higher input pole (even though $C_{\pi 1}$ increases as well), but it requires a greater voltage drop across R_C. If R_C is decreased, the gain drops and the noise current contributed by R_C rises.

For noise calculations, we neglect $C_{\pi 1}$. This may not lead to accurate results, necessitating careful simulations, but it provides an intuitive understanding of the trade-offs in CB stages. Shown in Fig. 4.20(a), the noise equivalent resembles that in Fig. 4.17, but with

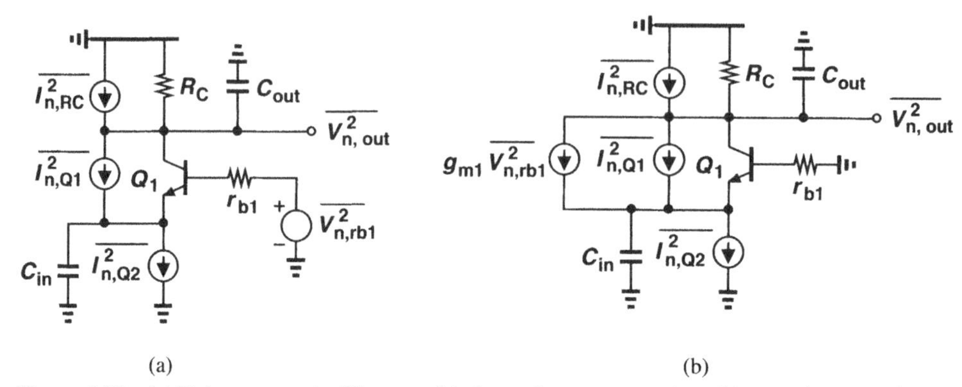

Figure 4.20 (a) Noise sources in CB stage, (b) alternative representation of base resistance noise.

r_{b1} and its noise voltage added to the circuit. Multiplying the noise voltage $\overline{V^2_{n,rb1}}$ by g^2_{m1}, we convert it to a current source connected between the collector and emitter of Q_1 [Fig. 4.20(b)] [1]. If $\beta \gg 1$, so that the effect of r_{b1} is negligible, then the results derived for the common-gate stage can be directly applied to this circuit as well. More specifically, we note that $\overline{I^2_{n,Q2}}$ and $\overline{I^2_{n,Q1}} + g^2_{m1}\overline{V^2_{n,rb1}}$ in Fig. 4.20(b) play the same roles as $\overline{I^2_{n,M2}}$ and $\overline{I^2_{n,M1}}$ in Fig. 4.17, respectively.

We now utilize Eq. (4.93) to estimate the output noise of the CB stage. Since $\overline{I^2_{n,Q2}} = 2kTg_{m2} + 4kT/r_{b2}$, we must replace $g_{m2}\gamma$ in (4.93) with $g_{m2} + 2/r_{b2}$. Similarly, since $\overline{I^2_{n,Q1}} + g^2_{m1}\overline{V^2_{n,rb1}} = 2kTg_{m1} + 4kTg^2_{m1}r_{b1}$, we replace $g_{m1}\gamma$ in the numerator of the first term of (4.93) with $g_{m1} + g^2_{m1}r_{b1}/2$. Thus,

$$\overline{V^2_{n,out}} = \frac{kT(g_{m1} + 2g^2_{m1}r_{b1})R_C C_{in}}{C_{out}(C_{in} + g_{m1}R_C C_{out})} + \frac{kT}{C_{out}} + \frac{2kTg_{m1}(g_{m1} + 2/r_{b2})R^2_C}{C_{in} + g_{m1}R_C C_{out}}. \tag{4.98}$$

Dividing both sides by the square of the midband gain, R^2_C, yields the total input-referred noise current:

$$\overline{I^2_{in,tot}} = \frac{kT(g_{m1} + 2g^2_{m1}r_{b1})C_{in}}{C_{out}(C_{in} + g_{m1}R_C C_{out})R_C} + \frac{kT}{C_{out}}\frac{1}{R^2_C} + \frac{2kTg_{m1}(g_{m1} + 2/r_{b2})}{C_{in} + g_{m1}R_C C_{out}}. \tag{4.99}$$

Note that the base resistance thermal noise appears directly along with the collector current shot noise, raising the overall input-referred noise. Larger transistors exhibit a smaller base resistance but greater junction capacitances.

In the above derivations, $C_{\pi 1}$ was neglected. To improve the approximation, C_{in} can be replaced with $C_{in} + C_\pi$ (as if r_{b1} were zero) in Eq. (4.99).

4.3 Feedback TIAs

The tight trade-offs in common-gate and common-base circuits make it difficult to achieve low noise. As mentioned in Section 4.2.1, the noise current of the load resistor and the bias current source are directly referred to the input, leading to a high noise at low supply voltages. For this reason, we must seek other TIA topologies that ameliorate these trade-offs.

The most common TIA configuration is the "voltage-current" or "shunt-shunt" feedback topology, where a negative feedback network senses the voltage at the output and returns a proportional current to the input. This type of feedback is chosen because it lowers both the input resistance—thus increasing the input pole magnitude and allowing the amplifier to absorb the photodiode current—and the output resistance, thereby yielding better drive capability.

4.3.1 First-Order TIA

Let us begin with the simple topology shown in Fig. 4.21, where R_F provides feedback around an ideal voltage amplifier, A. Since $V_X = V_{out}/(-A)$, we have $(V_{out} +$

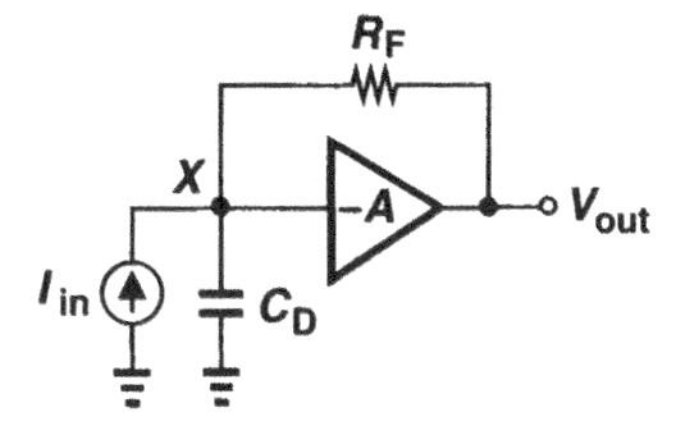

Figure 4.21 Feedback TIA.

$V_{out}/A)/R_F = -I_{in} - (V_{out}/A)C_D s$. That is,

$$\frac{V_{out}}{I_{in}} = -\frac{A}{A+1}\frac{R_F}{1+\frac{R_F C_D}{A+1}s}. \tag{4.100}$$

As expected, the feedback amplifier provides a midband transimpedance gain of approximately R_F but with a time constant of $R_F C_D/(A+1)$. The -3-dB bandwidth is thus equal to

$$f_{-3dB} \approx \frac{1}{2\pi}\frac{A}{R_F C_D}, \tag{4.101}$$

a factor of A greater than that of the simple resistive network of Fig. 4.1(a).

It is also instructive to examine the noise behavior of the circuit. Modeling the input-referred noise of the amplifier by a voltage source $\overline{V_{n,A}^2}$ and neglecting its input noise current, we arrive at the equivalent shown in Fig. 4.22. Since $V_X = V_{n,out}/(-A) + V_{n,A}$, we have

$$(V_{n,out} + \frac{V_{n,out}}{A} - V_{n,A} - V_{n,RF})\frac{1}{R_F} = (-\frac{V_{n,out}}{A} + V_{n,A})C_D s, \tag{4.102}$$

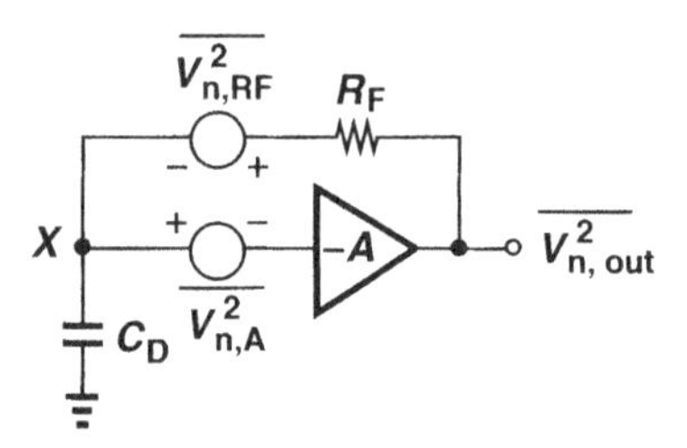

Figure 4.22 Noise sources in feedback TIA.

and hence

$$V_{n,out} = \frac{V_{n,RF} + (R_F C_D s + 1)V_{n,A}}{1 + R_F C_D s/A}, \tag{4.103}$$

where it is assumed $A \gg 1$. Before attempting to integrate the noise across the bandwidth, let us consider (4.103) in extreme cases. If $C_D = 0$, then $V_{n,out} = V_{n,RF} + V_{n,A}$, yielding an input noise current (per unit bandwidth) of

$$\overline{I^2_{n,in}} = \frac{\overline{V^2_{n,RF}} + \overline{V^2_{n,A}}}{R_F^2} \tag{4.104}$$

$$= \frac{4kT}{R_F} + \frac{\overline{V^2_{n,A}}}{R_F^2}. \tag{4.105}$$

Thus, the noise of R_F is directly referred to the input, and the noise voltage of A is divided by R_F. While the first term in Eq. (4.105) may appear to play the same role as the term $4kT/R_D$ in (4.53) for the common-gate/base structure, the critical difference is that in the topology of Fig. 4.21, R_F need not carry a bias current and its value does not limit the voltage headroom. Also, if R_F is large, the second term in (4.105) may be quite smaller than the contribution of the bias current source in CG and CB circuits.

In the other extreme case, we let C_D or s approach infinity. Consequently, (4.103) approaches $AV_{n,A}$. This is because C_D provides a low impedance from node X in Fig. 4.22 to ground, allowing $V_{n,A}$ to be amplified by A.

To compute the total output noise, we must integrate the squared magnitude of (4.103) from $f = 0$ to $f = \infty$. However, the observation that $V_{n,out} \to AV_{n,A}$ as $s \to \infty$ reveals that, if $AV_{n,A}$ does not fall with frequency, then its integral diverges. In other words, the assumption of infinite bandwidth for A yields an infinite output noise power! Noise calculations therefore mandate a more realistic model for the core amplifier, leading to second-order configurations.

4.3.2 Second-Order TIA

The core amplifier in Fig. 4.21 exhibits at least one pole, altering the response of the TIA in both the frequency and time domains. Assuming a transfer function of

$$A(s) = \frac{A_0}{1 + s/\omega_0} \tag{4.106}$$

for the voltage amplifier and $A_0 \gg 1$, we rewrite (4.100) as

$$\frac{V_{out}}{I_{in}} = -\frac{A(s)R_F}{A(s) + 1 + R_F C_D s} \tag{4.107}$$

$$= -\frac{A_0 R_F}{\frac{R_F C_D}{\omega_0}s^2 + (R_F C_D + \frac{1}{\omega_0})s + A_0 + 1}. \tag{4.108}$$

We examine this result for two cases. First, suppose the second pole of the closed-loop system is much higher in magnitude than the first pole. Noting that

$$(\frac{s}{\omega_{p1}} + 1)(\frac{s}{\omega_{p2}} + 1) = \frac{s^2}{\omega_{p1}\omega_{p2}} + (\frac{1}{\omega_{p1}} + \frac{1}{\omega_{p2}})s + 1, \tag{4.109}$$

and $1/\omega_{p1} + 1/\omega_{p2} \approx 1/\omega_{p1}$ if $\omega_{p2} \gg \omega_{p1}$, we have

$$\omega_{p1} \approx \frac{A_0 + 1}{R_F C_D + 1/\omega_0}. \tag{4.110}$$

Thus, the first pole is slightly lower than that of the first-order TIA topology. The second pole is obtained by recognizing that $(\omega_{p1}\omega_{p2})^{-1} = R_F C_D/[(A_0 + 1)\omega_0]$:

$$\omega_{p2} \approx \omega_0 + \frac{1}{R_F C_D}. \tag{4.111}$$

Thus, the second pole is equal to the sum of the open-loop poles of the circuit (so long as the assumption $\omega_{p2} \gg \omega_1$ remains valid).

Under what conditions can we assume $\omega_{p2} \gg \omega_{p1}$? We require that

$$\omega_0 + \frac{1}{R_F C_D} \gg \frac{A_0 + 1}{R_F C_D + \frac{1}{\omega_0}}, \tag{4.112}$$

and hence

$$(R_F C_D \omega_0 + 1)^2 \gg R_F C_D \omega_0 (A_0 + 1). \tag{4.113}$$

If $\omega_0 \gg (R_F C_D)^{-1}$, then $R_F C_D \omega_0 + 1 \approx R_F C_D \omega_0$ and

$$\omega_0 \gg \frac{A_0 + 1}{R_F C_D}. \tag{4.114}$$

In other words, the *open-loop* pole of the amplifier must be much higher than the *closed-loop* pole resulting from R_F and C_D.

In practice, the above condition may not hold. We then treat the amplifier as a typical second-order system, seeking means of ensuring a well-behaved time response, i.e., critically-damped behavior. Recall from basic control theory that if the denominator of a second-order transfer function is expressed as $s^2 + 2\zeta\omega_n s + \omega_n^2$, then ζ, the "damping factor," must be equal to $\sqrt{2}/2$ for critical damping. If $\zeta < \sqrt{2}/2$, then the step response exhibits ringing, both creating ISI and corrupting the high and low levels of the data.[7] Rewriting (4.108), we have

$$\frac{V_{out}}{I_{in}} = -\frac{\dfrac{A_0\omega_0}{C_D}}{s^2 + \dfrac{R_F C_D + 1/\omega_0}{R_F C_D/\omega_0}s + \dfrac{(A_0+1)\omega_0}{R_F C_D}}, \tag{4.115}$$

concluding that $\omega_n^2 = (A_0 + 1)\omega_0/(R_F C_D)$ and hence

$$\zeta = \frac{1}{2}\frac{R_F C_D \omega_0 + 1}{\sqrt{(A_0+1)\omega_0 R_F C_D}}. \tag{4.116}$$

If $\zeta = \sqrt{2}/2$, then

$$(R_F C_D \omega_0)^2 - 2A_0 R_F C_D \omega_0 + 1 = 0 \tag{4.117}$$

and

$$\omega_0 = \frac{A_0 \pm \sqrt{A_0^2 - 1}}{R_F C_D}. \tag{4.118}$$

Recognizing that only the sum of the two terms gives a valid solution, we obtain

$$\omega_0 \approx \frac{2A_0}{R_F C_D}. \tag{4.119}$$

Thus, the -3-dB bandwidth of the core amplifier must be chosen equal to *twice* the closed-loop bandwidth of the first-order TIA to ensure a critically-damped response. Intuitively, we expect that as ω_0 exceeds this value, the feedback becomes overdamped, a trend indeed observed in (4.116). Conversely, if $\omega_0 < 2A_0/(R_F C_D)$, then the step response suffers from ringing. For example, if $\omega_0 = A_0/(R_F C_D)$, then $\zeta \cong 0.5$.

[7] The choice $\zeta = \sqrt{2}/2$ is somewhat conservative for TIA design. Even with $\zeta \approx 0.5$, the circuit exhibits acceptable ringing. However, process and temperature variations often necessitate such conservatism.

With the above condition for critical damping, let us now determine other parameters of the circuit. In particular, we have

$$\omega_n = \sqrt{\frac{(A_0+1)\omega_0}{R_F C_D}} \tag{4.120}$$

$$\approx \frac{\sqrt{2}A_0}{R_F C_D}. \tag{4.121}$$

More importantly, the -3-dB bandwidth of the TIA is obtained by setting the magnitude of (4.115) to $\sqrt{2}/2$ times its low-frequency value. Writing (4.115) in a more general form,

$$\frac{V_{out}}{I_{in}} = -\frac{R_0\omega_n^2}{s^2 + 2\zeta\omega_n s + \omega_n^2}, \tag{4.122}$$

where $R_0 = [A_0/(A_0+1)]R_F$, we have

$$\left|\frac{R_0\omega_n^2}{-\omega_{-3\text{dB}}^2 + 2\zeta\omega_n\omega_{-3\text{dB}} + \omega_n^2}\right| = \frac{\sqrt{2}}{2}R_0. \tag{4.123}$$

That is,

$$(\omega_n^2 - \omega_{-3\text{dB}}^2)^2 + 4\zeta^2\omega_n^2\omega_{-3\text{db}}^2 = 2\omega_n^4, \tag{4.124}$$

which, along with $\zeta = \sqrt{2}/2$, yields

$$\omega_{-3\text{dB}} = \omega_n. \tag{4.125}$$

Thus, the -3-dB bandwidth of the second-order TIA is equal to

$$f_{-3dB} = \frac{1}{2\pi}\frac{\sqrt{2}A_0}{R_F C_D}. \tag{4.126}$$

Interestingly, the bandwidth is *greater* than that of the first-order topology by about 41%. This is because the pole introduced by the core amplifier creates an inductive behavior in the input impedance of the TIA, partially cancelling the roll-off due to the input capacitance.[8] Plotted in Fig. 4.23 is the -3-dB bandwidth as a function of ω_0, revealing that the bandwidth can be increased considerably.

The noise behavior of second-order TIA heavily depends on the actual implementation of the core amplifier. We analyze the noise performance of a typical implementation below.

Typical Implementation Let us now study a typical transistor implementation of the feedback TIA. Shown in Fig. 4.24 is a MOS example, with the core amplifier consisting

[8]Note that, despite the resonance between the capacitance and the effective inductance, proper choice of circuit parameters leads to two coincident *real* poles.

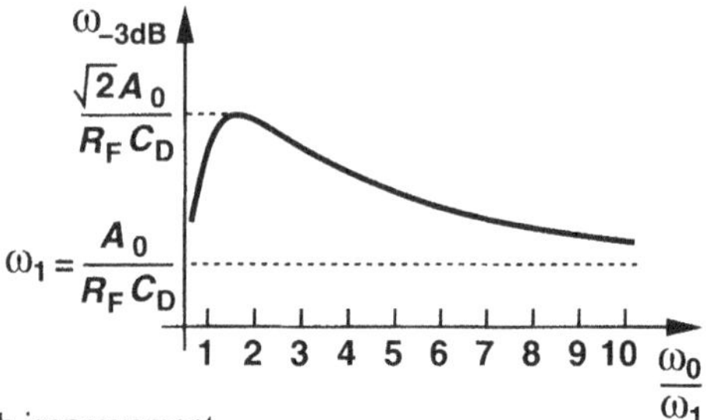

Figure 4.23 TIA bandwidth improvement.

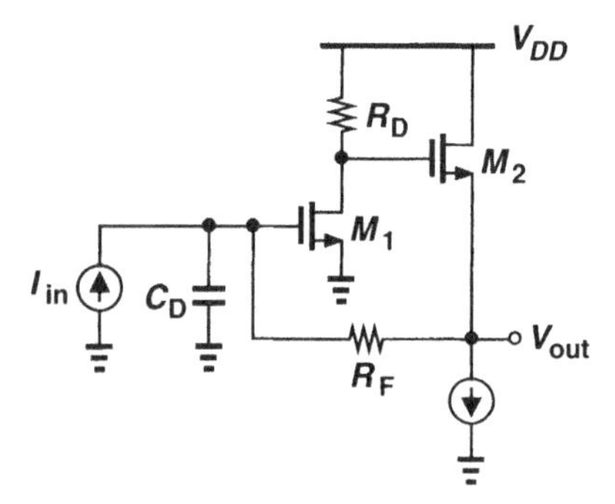

Figure 4.24 Implementation of feedback TIA.

of a common-source stage and a source follower. The source follower isolates R_D from the loading effect of both R_F and the input capacitance of the subsequent stage. In our analysis, we neglect channel-length modulation and body effect for simplicity.

If the output impedance of the source follower, g_{m2}^{-1}, is much less than R_F, the open-loop voltage gain of the core amplifier is approximately equal to $g_{m1}R_D$ and the closed-loop transimpedance gain is

$$R_T = \frac{g_{m1}R_D}{1 + g_{m1}R_D} R_F. \tag{4.127}$$

Since the loop gain is equal to $g_{m1}R_D$, the closed-loop input and output impedances at low frequencies are respectively given by

$$R_{in} \approx \frac{R_F}{1 + g_{m1}R_D} \tag{4.128}$$

and

$$R_{out} \approx \frac{1}{1 + g_{m1}R_D} \cdot \frac{1}{g_{m2}}. \tag{4.129}$$

It is important to note the voltage headroom issues in the circuit of Fig. 4.24. Since the voltage drop across R_D, V_{RD}, is limited to $V_{RD} = V_{DD} - V_{GS2} - V_{GS1}$, it is difficult to achieve a high open-loop gain by increasing R_D (for a given bias current). Thus, g_{m1} must be maximized by choosing a larger width for M_1, but at the cost of increasing the capacitances of M_1. At supply voltages below 2 V, this limitation may simply rule out the use of the source follower.

We now compute the input-referred noise current of the TIA using the method applied to the topology of Fig. 4.22. Viewing M_1, R_D, and M_2 as the core voltage amplifier and neglecting channel-length modulation and body effect, we express the output noise voltage as:

$$\overline{V^2_{n,out,core}} = 4kTg_{m1}\gamma R_D^2 + 4kTR_D + 4kT\frac{\gamma}{g_{m2}}, \tag{4.130}$$

where the last term represents the gate-referred noise voltage of M_2. Dividing (4.130) by the square of the voltage gain yields the input-referred noise of the core:

$$\overline{V^2_{n,in,core}} = 4kT\frac{\gamma}{g_{m1}} + \frac{4kT}{g_{m1}^2 R_D} + 4kT\frac{\gamma}{g_{m2}g_{m1}^2 R_D^2}. \tag{4.131}$$

From (4.105), we have

$$\overline{I^2_{n,in}} = \frac{4kT}{R_F} + \frac{\overline{V^2_{n,in,core}}}{R_F^2} \tag{4.132}$$

$$= \frac{4kT}{R_F} + \frac{4kT}{R_F^2}\left(\frac{\gamma}{g_{m1}} + \frac{1}{g_{m1}^2 R_D} + \frac{\gamma}{g_{m2}g_{m1}^2 R_D^2}\right). \tag{4.133}$$

For a given transimpedance gain, only the noise contributed by the core can be minimized. This requires maximizing R_D and the transconductances of M_1 and M_2. We will consider the frequency-dependent noise performance later.

Frequency Response The frequency response of the TIA implementation of Fig. 4.24 is also of interest. From the open-loop circuit of Fig. 4.25, we identify three poles:

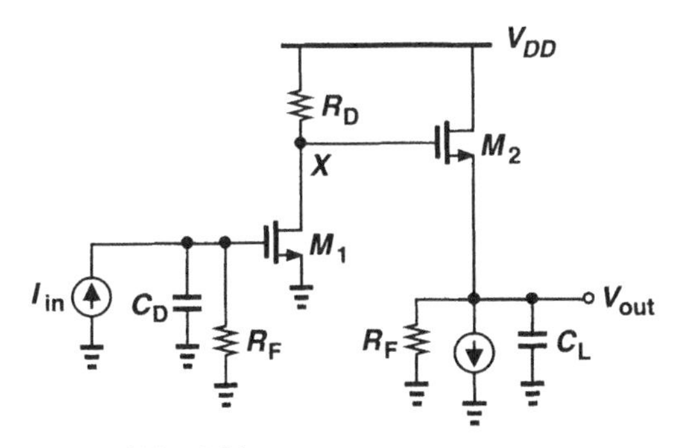

Figure 4.25 Open-loop equivalent of Fig. 4.24.

(1) R_F, C_D, C_{GS1}, and the Miller multiplication of C_{GD1} create a pole at the input node; (2) R_D, C_{DB1}, C_{GD2}, and a fraction of C_{GS2} yield a pole at node X; (3) C_{SB2} and the

input capacitance of the following stage, C_L, along with the output resistance of the source follower, $g_{m2}^{-1}||R_F \approx g_{m2}^{-1}$, form a pole at the output.

With three poles around the feedback loop, the TIA can even oscillate. Since the magnitudes of the poles may be comparable, it is difficult to ensure that the closed-loop step response exhibits negligible ringing. For example, for a large R_F and C_D (or a wide input transistor) the input pole may be critical. Similarly, since the stage following the TIA typically incorporates wide devices to achieve a reasonable noise and gain, the effect of the output pole may also be significant.[9]

It is also important to recognize that the TIA of Fig. 4.24 exhibits an *inductive* output impedance. This is because, as the frequency increases, the feedback becomes weaker, lowering the loop gain. For example, if only the input pole is considered, we can write:

$$Z_{out} \approx \frac{\frac{1}{g_{m2}}}{1 + g_{m1}R_D \frac{1}{R_F C_{in} s + 1}}, \tag{4.134}$$

$$\approx \frac{R_F C_{in} s + 1}{R_F C_{in} s + 1 + g_{m1} R_D} \cdot \frac{1}{g_{m2}}. \tag{4.135}$$

where C_{in} denotes the total capacitance from the input node to ground. Figure 4.26 plots the magnitude of Z_{out} as a function of frequency, revealing that the inductive behav-

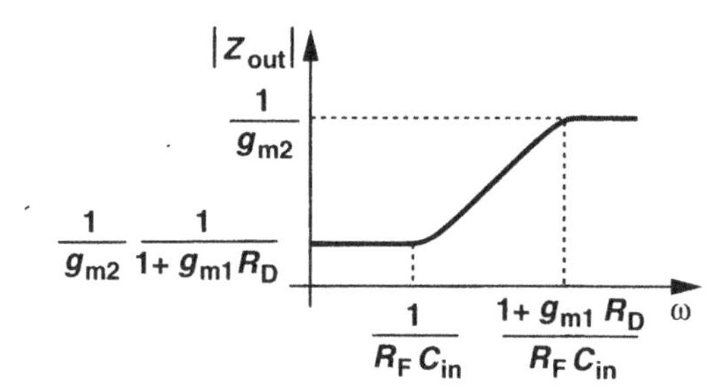

Figure 4.26 Output impedance of feedback TIA as a function of frequency.

ior begins in the vicinity of the dominant pole, $1/(R_F C_{in})$. Note that $Z_{out}(s = 0) = g_{m2}^{-1}/(1 + g_{m1}R_D)$ and $Z_{out}(s = \infty) = g_{m2}^{-1}$. As an exercise, the reader can determine the equivalent RL circuit that displays this behavior. The inductive output impedance may lead to ringing if the circuit drives substantial load capacitance.

To alleviate the stability issue, the circuit can be modified as shown in Fig. 4.27. Here, the output stage employs two source followers, one to drive the load capacitance and another to serve in the feedback loop. Now, the pole associated with the source of M_3 appears at very high frequencies, and C_L affects the phase margin negligibly. The circuit still contains two poles at the input and at the drain of M_1 but proper positioning of the poles

[9]Addition of a small capacitor in parallel with the feedback resistor may reduce ringing.

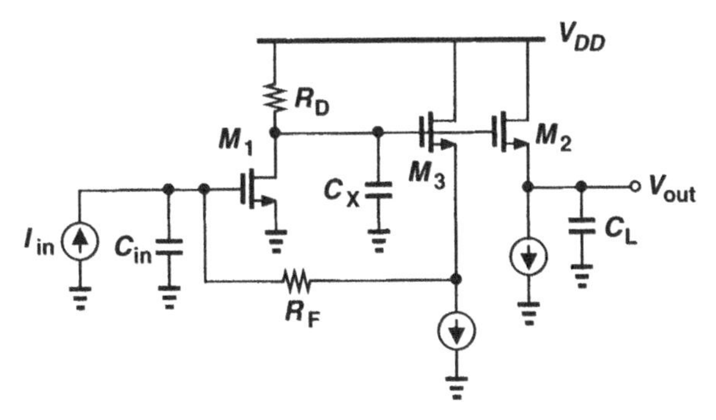

Figure 4.27 Modified feedback TIA to improve stability.

can yield an acceptable time response. In fact, we recall from (4.119) that for optimal response, the open-loop pole of the core amplifier, $\omega_0 = 1/(R_D C_X)$, must be equal to twice the closed-loop pole of the first-order circuit, $A_0/(R_F C_{in}) = g_{m1} R_D/(R_F C_{in})$. That is,

$$R_F C_{in} = 2 g_{m1} R_D \cdot (R_D C_X). \tag{4.136}$$

Thus, for given values of C_{in}, g_{m1}, R_D, and C_X, the feedback resistor can be chosen so as to provide a well-behaved step response.

Placing the source follower M_2 outside the feedback loop in Fig. 4.27 raises the output resistance, R_{out}, of the TIA, but it also reduces the inductive component. The output impedance of a source follower driven by a resistance R_S behaves as shown in Fig. 4.28 [1]. Since for M_2 in Fig. 4.27, R_S equals $R_D/(1 + g_{m1} R_D) \approx 1/g_{m1}$ at low frequencies

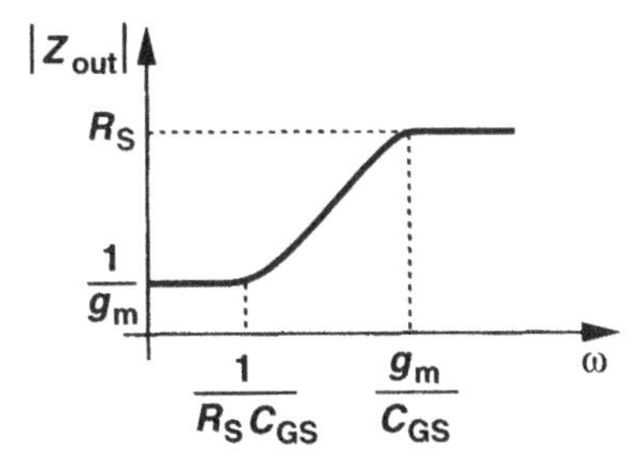

Figure 4.28 Output impedance of source follower.

and R_D at high frequencies (if R_D is less than R_F), and since C_{GS2} is typically smaller than C_{in}, the zero in Fig. 4.28 [$= 1/(R_S C_{GS})$] has a higher magnitude than that in Fig. 4.26 [$= 1/(R_F C_{in})$]. In other words, the inductive component manifests itself at higher frequencies.

High-Frequency Noise Performance The high-frequency noise performance of the TIAs in Figs. 4.24 and 4.27 must be analyzed by integrating the output noise voltage and dividing it by R_F. Let us model the circuit of Fig. 4.27 by the simplified equivalent shown in Fig. 4.29. Here, the feedback source follower is represented by a unity-gain voltage buffer and the noise of the output source follower is ignored. Summing the currents at

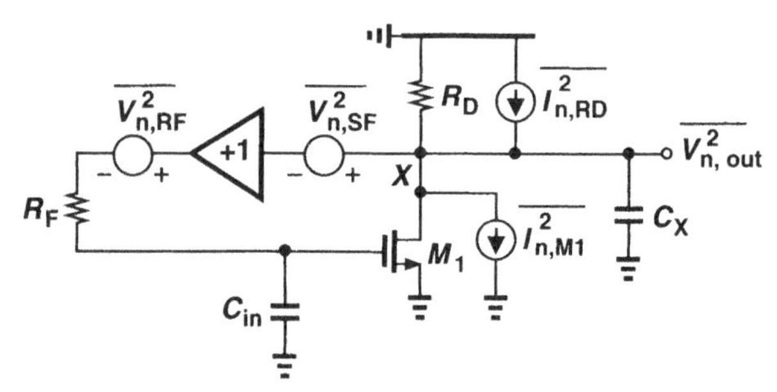

Figure 4.29 Noise sources in feedback TIA.

node X yields

$$-V_{n,out}(C_X s + \frac{1}{R_D}) + I_{n,RD} - I_{n,M1} = I_{D1}, \tag{4.137}$$

where I_{D1} denotes the small-signal drain current of M_1. Also, since the voltage across R_F is equal to $V_{GS1}C_{in}sR_F$, we have

$$V_{GS1} + V_{GS1}R_F C_{in} s + V_{n,RF} + V_{n,SF} = V_{n,out}. \tag{4.138}$$

Noting that $g_{m1}I_{D1} = V_{GS1}$, we obtain from (4.137) and (4.138)

$$\left[-V_{n,out}(C_X s + \frac{1}{R_D}) + I_{n,RD} - I_{n,M1}\right]\frac{1}{g_{m1}}(1+R_F C_{in} s)+V_{n,RF}+V_{n,SF} = V_{n,out}. \tag{4.139}$$

It follows that

$$V_{n,out} = \frac{(I_{n,RD} - I_{n,M1})\dfrac{1 + R_F C_{in} s}{g_{m1}} + V_{n,RF} + V_{n,SF}}{\dfrac{C_X}{g_{m1}} R_F C_{in} s^2 + (R_F C_{in} + \dfrac{C_X}{g_{m1}})s + 1}, \tag{4.140}$$

where it is assumed $g_{m1}R_D \gg 1$. We note that the noise currents of R_D and M_1 experience a first-order roll-off at high frequencies whereas the noise contributed by the feedback devices, R_F and the source follower, falls at a higher rate.

The second-order denominator of (4.140) makes the integration of noise by hand difficult, necessitating simulations. Nevertheless, applying the condition given by (4.136) simplifies the denominator to

$$D(s) = 2R_D^2 C_X^2 s^2 + (2g_{m1}R_D^2 C_X + \frac{C_X}{g_{m1}})s + 1. \tag{4.141}$$

If $g_{m1}^2 R_D^2 \gg 1$, then

$$D(s) = 2R_D^2 C_X^2 s^2 + 2g_{m1}R_D^2 C_X s + 1, \tag{4.142}$$

yielding the pole locations

$$s_{1,2} = -\frac{g_{m1}R_D^2C_X \pm \sqrt{g_{m1}^2R_D^4C_X^2 - 2R_D^2C_X^2}}{2R_D^2C_X^2} \tag{4.143}$$

$$= +\frac{g_{m1}R_D^2C_X}{2R_D^2C_X^2}\left[-1 \pm \sqrt{1 - \frac{2}{g_{m1}^2R_D^2}}\right]. \tag{4.144}$$

Assuming $g_{m1}^2R_D^2 \gg 2$ and $\sqrt{1-\epsilon} \approx 1 - \epsilon/2$ for small ϵ, we obtain

$$s_1 \approx \frac{-1}{(2g_{m1}R_D)(R_DC_X)} \tag{4.145}$$

$$s_2 \approx \frac{-g_{m1}}{C_X}. \tag{4.146}$$

Now, using integration techniques described in Section 4.2.1, the total noise can be calculated. The details are left as an exercise for the reader.

4.4 Supply Rejection

Most TIAs are designed as single-ended circuits because photodiodes provide a single-ended current. As a consequence, they suffer from poor supply rejection and even supply-dependent biasing. In this section, we study this issue.

In the common-gate and common-base topologies of Fig. 4.9, the bias current can be defined accurately by means of supply-independent bandgap techniques [1]. However, the output signal, V_{out}, is single-ended and prone to supply noise. In Fig. 4.9(a), for example, $V_{out} = V_{DD} - I_DR_D$. Since V_{out} is referenced to V_{DD} rather than ground, it is essential that the following stage sense its input with respect to V_{DD} as well.

To understand the above concept, let us consider the topologies shown in Fig. 4.30. The NMOS common-source stage in Fig. 4.30(a) produces an output proportional to $V_{GS1} = V_X$,[10] thus amplifying the noise on V_{DD} along with the signal. On the other hand, the PMOS common-source stage of Fig. 4.30(b) senses its input with respect to V_{DD}, thereby suppressing the effect of supply noise. That is, as V_{DD} rises and falls, the gate and source voltages of M_2 vary by the same amount, leading to a constant drain current. (In reality, channel-length modulation and device capacitances degrade the supply rejection.)

If the first stage of a TIA is followed by a differential pair [Fig. 4.31(a)], then the reference voltage V_{b2} must be generated such that it experiences the same amount of supply noise as does V_X. Thus, as shown in Fig. 4.31 (b), a replica of the common-gate stage generates V_{b2}. As explained in Section 4.5, the input-referred noise current of the overall circuit is $\sqrt{2}$ times that of the single-ended counterpart. The noise of the replica stage can be suppressed by adding C_b, but at the cost of a slight asymmetry with respect to the effect of supply noise at nodes X and Y.

---

[10]These are small-signal quantities.

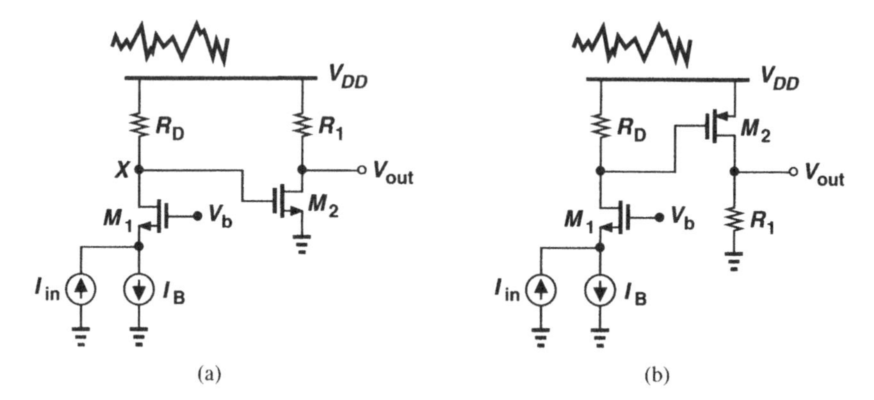

Figure 4.30 Effect of supply noise on CG stage with (a) NMOS and (b) PMOS common-source stages.

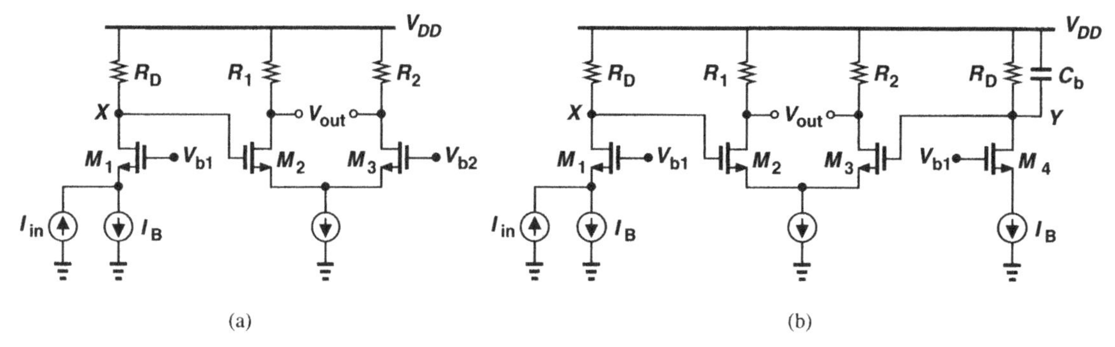

Figure 4.31 (a) Common-gate stage followed by differential pair, (b) addition of replica stage to improve symmetry.

The problem of supply dependence is more pronounced in single-ended feedback TIAs. Consider, for example, the simple topology of Fig. 4.24 again, depicted in Fig. 4.32(a) to analyze its supply rejection. The gate-source voltage of M_1 is given by

$$V_{DD} - R_D I_{D1} - V_{GS2} = V_{GS1}, \tag{4.147}$$

displaying direct dependence on V_{DD}. Neglecting channel-length modulation and body effect, we calculate the small-signal "gain" from V_{DD} to V_{out} by differentiating both sides of (4.147) with respect to V_{DD}:

$$1 - R_D \frac{\partial I_{D1}}{\partial V_{DD}} = \frac{\partial V_{GS1}}{\partial V_{DD}} \tag{4.148}$$

$$= \frac{\partial V_{GS1}}{\partial I_{D1}} \cdot \frac{\partial I_{D1}}{\partial V_{DD}} \tag{4.149}$$

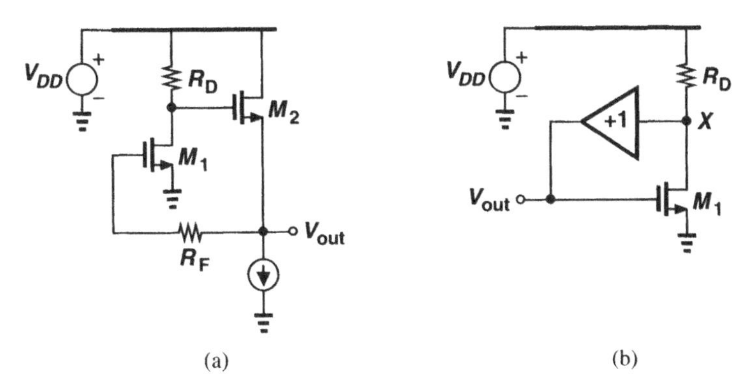

Figure 4.32 (a) Supply noise in feedback TIA, (b) equivalent circuit.

$$= \frac{1}{g_{m1}} \cdot \frac{\partial I_{D1}}{\partial V_{DD}}. \tag{4.150}$$

Thus,

$$\frac{\partial I_{D1}}{\partial V_{DD}} = \frac{1}{\frac{1}{g_{m1}} + R_D}. \tag{4.151}$$

We also note that $V_{DD} - R_D I_{D1} - V_{GS2} = V_{out}$ and hence $\partial V_{out}/\partial V_{DD} = 1 - R_D \partial I_{D1}/\partial V_{DD}$. It follows that:

$$\frac{\partial V_{out}}{\partial V_{DD}} = \frac{1}{1 + g_{m1} R_D}. \tag{4.152}$$

Equation (4.152) can also be obtained by viewing the circuit as shown in Fig. 4.32(b) and recognizing that M_1 operates as a diode-connected device in this test, i.e., it introduces a small-signal resistance equal to $1/g_{m1}$ from node X to ground. Thus, to the first order, supply variations are attenuated by only the resistive divider formed by R_D and $1/g_{m1}$.

The supply dependence of single-ended feedback TIAs poses a difficult challenge. To appreciate the issue, let us consider a remedy. Illustrated in Fig. 4.33, the idea is to bias the input stage by a constant current source and add a large capacitor C_B to provide a flat gain in the band of interest. But how large must C_B be? Recall from Chapter 2 that some optical standards specify a cut-off frequency of only a few tens of kilohertz for the lower end of the band. Thus, if, for example, $g_{m1} = (100\ \Omega)^{-1}$, then C_B must exceed 30 nF to yield a corner frequency of 50 kHz. Such a large capacitor must be placed off-chip, requiring a bond wire and a package pin that may experience noise coupling from other connections to the chip. Furthermore, if C_B acts as a short for such a broad bandwidth, then it fails to reject supply noise frequencies in the range of interest.

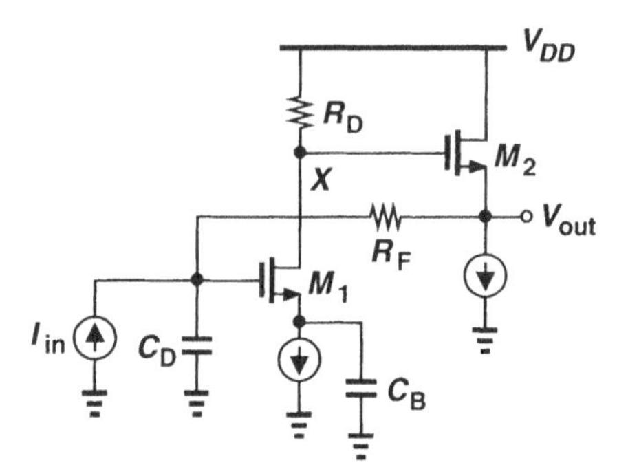

Figure 4.33 Feedback TIA with current-biased input stage.

4.5 Differential TIAs

It is desirable to employ differential topologies for TIAs, thus suppressing the effect of supply and substrate noise. However, since photodiodes produce a single-ended current and since their cathode is connected to a high voltage to allow a high quantum efficiency, the input signal is not differential, leading to several difficulties.

Let us begin with a common-gate TIA, convert it to differential form, and add a differential pair at the output [Fig. 4.34(a)]. We identify three issues in this circuit. First, the

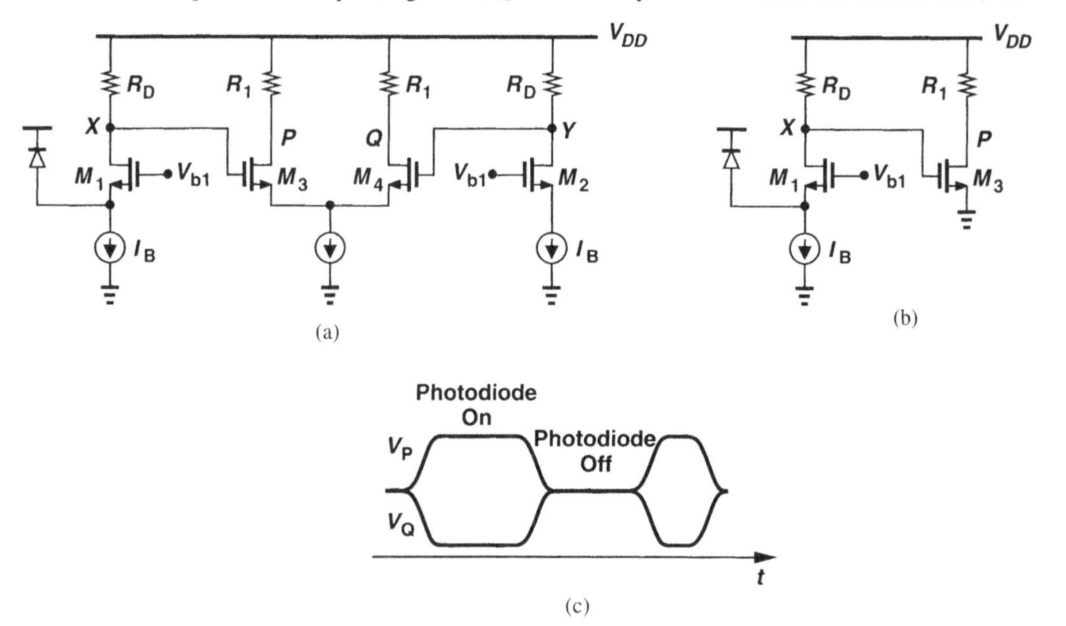

Figure 4.34 (a) Pseudo-differential CG stage, (b) half-circuit equivalent, (c) output waveforms.

signal generated at node X propagates to the output through two different paths, with M_3 operating as a common-source stage in the X-P path and M_3-M_4 as a cascade of a source follower and common-gate stage in the X-Q path. As a result, the high-frequency components of the signal experience unequal gains and phase shifts in the two paths, producing an asymmetric output waveform. Second, as the half circuit of Fig. 4.34(b) suggests, the input-referred noise current of the circuit is $\sqrt{2}$ times that of its single-ended counterpart, degrading the sensitivity by 3 dB. Third, if the circuit is perfectly symmetric, the output swings are *not* fully differential. As illustrated in Fig. 4.34(c), when the photodiode is off, V_{out1} and V_{out2} are *equal* and when the diode turns on, V_{out1} and V_{out2} change in opposite directions. This is a serious issue because it makes the choice of the decision threshold difficult.

The feedback TIA of Section 4.3 can also be converted to differential form. Depicted in Fig. 4.35, the circuit suffers from the same three issues as the above example. How-

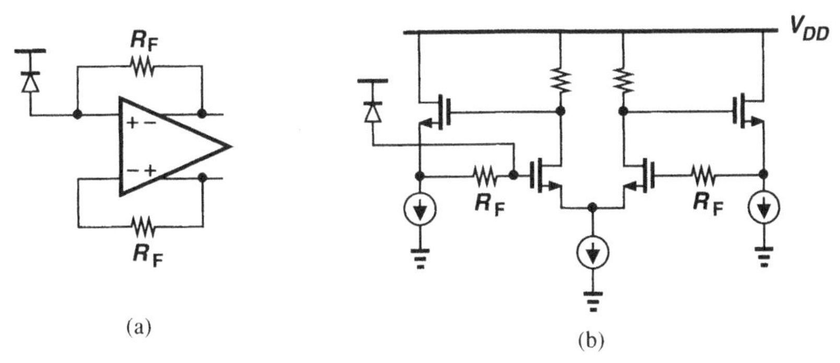

Figure 4.35 Differential feedback TIA.

ever, in contrast to its single-ended version, the circuit of Fig. 4.35 benefits from supply-independent biasing.

Figure 4.36 illustrates a common approach to converting the output of a single-ended

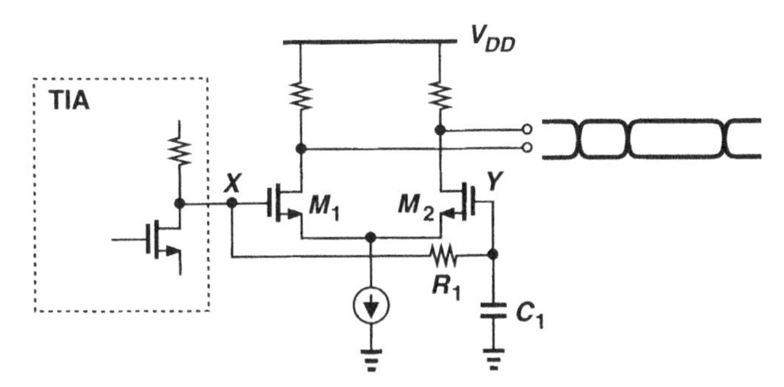

Figure 4.36 Single-ended to differential conversion.

TIA to a truly-differential signal. Here, the low-pass filter consisting of R_1 and C_1 extracts the dc level of the TIA output, applying the result to the gate of M_2. Since $V_X - V_Y$ displays a zero average, the output of the differential pair is also free from offset.

Note that the topology of Fig. 4.36 acts as a high-pass filter: at sufficiently low frequencies, the signals at nodes X and Y are identical, leading to a zero output. The time constant R_1C_1 must therefore reach tens of microseconds if the lower corner frequency of the high-pass transfer function is to fall below a few tens of kilohertz. For this reason, this technique often demands a large external capacitor.

The bond wire connecting the photodiode to the input of a TIA can experience substantial capacitive and inductive coupling from other bond wires. As a result, the signal may be heavily corrupted if the chip receives or delivers other large, high-speed waveforms. For example, as illustrated in Fig. 4.37, if the TIA is integrated along with a clock recovery

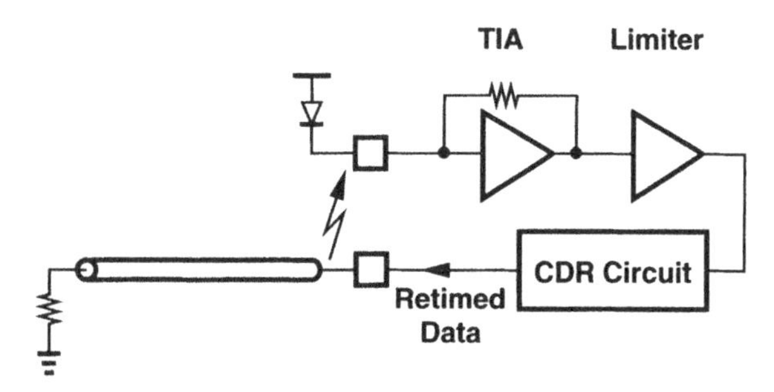

Figure 4.37 Noise coupling from CDR output to TIA input.

circuit, the retimed data, while exiting the chip, couples to the input current.

Another consequence of coupling to the input is instability. For a stand-alone TIA or a TIA/limiter cascade, the output signal delivered to an off-chip load couples to the TIA input through bondwires, the substrate, and the supply, thus creating a feedback loop (Fig. 4.38). The high loop gain and the existence of multiple poles and zeros in the loop may

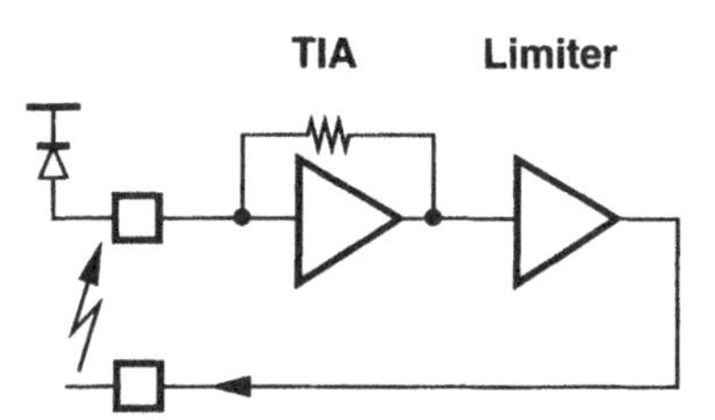

Figure 4.38 Coupling from limiter output to TIA input.

therefore give rise to oscillation or at least considerable ringing in the step response.

It is important to make two observations at this point. (1) If the high-speed outputs are differential, then the coupling to the input is determined by the *mismatches* between two nominally identical paths and is therefore greatly reduced. As shown in Fig. 4.39(a), the input signal now experiences the *sum* of the two outputs. Of course, in reality, the pads are arranged as shown in Fig. 4.39(b) to ensure minimal coupling. (2) For the differential TIAs of Fig. 4.35, the unused input must still be brought out through a bond wire so as to create partial symmetry at the input. Depicted in Fig. 4.40, the idea is that coupling from other wires disturbs the two inputs approximately equally, introducing mostly a common-mode change.

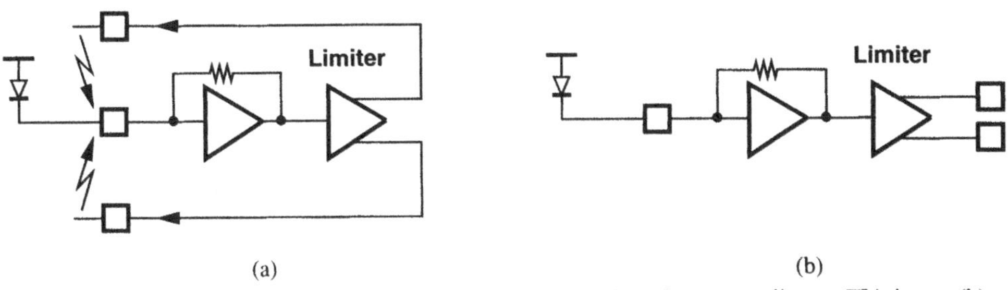

Figure 4.39 (a) Symmetric placement of limiter output pads to lower coupling to TIA input, (b) placement of limiter outputs far from TIA input.

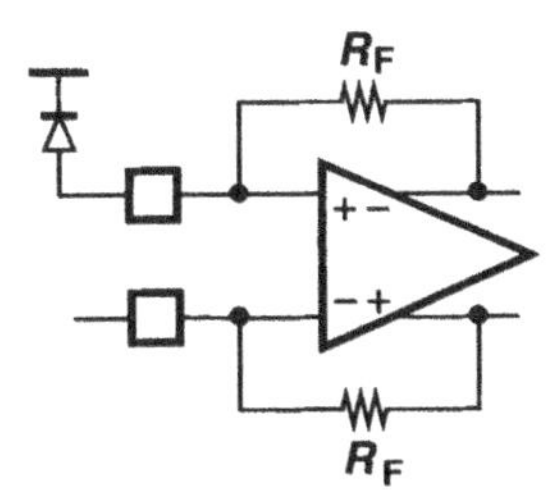

Figure 4.40 Connection of bond wire to unused TIA input to improve symmetry.

4.6 High-Performance Techniques

The topologies studied thus far in this chapter form the foundation for TIA design. In this section, we introduce a number of techniques that improve the performance along various dimensions: gain, noise, speed, and voltage headroom.

4.6.1 Gain Boosting

As mentioned in Sections 4.2.1 and 4.3, low supply voltages impose a small dc voltage drop across the load resistor in the input stage, thereby limiting the gain and yielding a high-input referred noise. For example, in the CG stage of Fig. 4.9(a) or the feedback TIA of Fig. 4.24, $I_{D1}R_D$ is constrained by the supply voltage. In CMOS technology, it is possible to add a PMOS current source in parallel with R_D, thus providing part of the bias current. Illustrated in Fig. 4.41 for both types of TIAs, this technique employs a current mirror consisting of M_3 and M_4 to deliver a large portion of the bias current drawn by the input transistor, M_1. For example, if M_4 carries $0.8I_{D1}$, then the value of R_D can be increased by a factor of five while maintaining the same voltage drop as the case without the PMOS current source. As a result, the gain of the input stage increases and the noise contributed by R_D decreases [as expressed by Eqs. (4.53) and (4.133)].[11]

[11]The finite value of R_D in Fig. 4.41(a) allows the definition of the output bias voltage. In Fig. 4.41(b), R_D may be infinite.

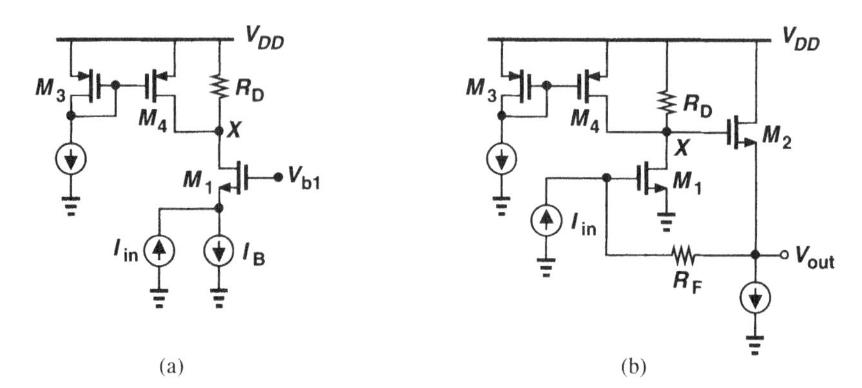

Figure 4.41 Addition of PMOS current source to (a) CG stage, (b) feedback TIA.

The addition of M_4 in the circuit of Fig. 4.41 nonetheless leads to other difficulties. First, the drain-bulk and drain-gate capacitances of M_4 may substantially increase the time constant at node X because the PMOS transistor suffers from low mobility and must therefore be quite wide to carry a large current with a reasonable drain-source voltage. This capacitance is especially problematic in the feedback TIA of Fig. 4.41(b) as it affects the phase margin and the step response. Second, the thermal noise current of M_4 directly adds to that of R_D, thus raising the input-referred noise. To quantify this effect, let us consider the two cases depicted in Fig. 4.42 and compare their input-referred noise currents. We

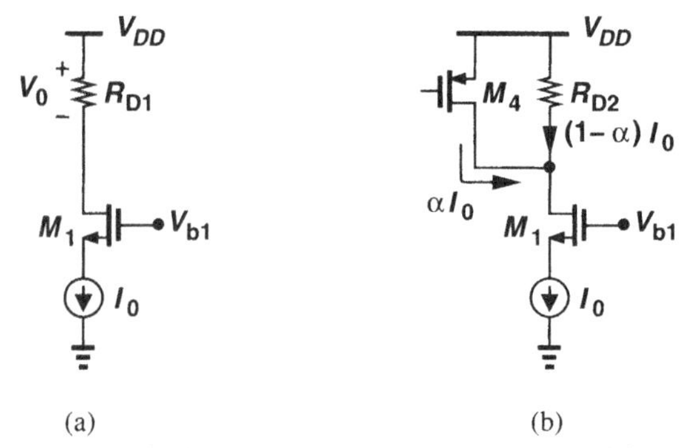

Figure 4.42 Noise performance of CG stage (a) without and (b) with PMOS current source.

assume the maximum allowable voltage across R_{D1} or R_{D2} is equal to V_0. In case (a), $R_{D1} = V_0/I_0$ contributes a noise current of $4kT/R_{D1}$. In case (b), suppose $I_{D4} = \alpha I_0$ and hence $I_{RD2} = (1-\alpha)I_0$. Consequently, R_{D2} can be as large as $V_0/[(1-\alpha)I_0]$. On the other hand, M_4 exhibits a transconductance of

$$g_{m4} = \frac{2I_{D4}}{V_{GS4} - V_{TH4}} \tag{4.153}$$

$$= \frac{2\alpha I_0}{V_0}, \tag{4.154}$$

where it is assumed the transistor is biased at the edge of saturation, i.e., $V_{GS4} - V_{TH4} = V_{DS4}$. Adding the thermal noise currents of R_{D2} and M_4, we have

$$\overline{I_{n,RD2}^2} + \overline{I_{n,M4}^2} = \frac{4kT}{R_{D2}} + 4kT\gamma g_{m4} \tag{4.155}$$

$$= 4kT\frac{(1-\alpha)I_0}{V_0} + 4kT\frac{2\gamma\alpha I_0}{V_0} \tag{4.156}$$

$$= 4kT\frac{I_0}{V_0}[1 + \alpha(2\gamma - 1)]. \tag{4.157}$$

How does this result compare with the thermal noise of R_{D1} in Fig. 4.42(a)? If $\gamma < 0.5$, then (4.157) yields a *lower* value for any $\alpha > 0$. If $\gamma > 0.5$, the noise of R_{D2} and M_4 is greater than that of R_{D1}. Since γ is equal to 2/3 for long-channel devices and greater than 2 for deep-submicron transistors, the addition of M_4 inevitably gives a somewhat higher noise.

The choice of α in Fig. 4.42(b) is determined by three factors: the desired increase in gain, the increase in noise [Eq. (4.157)], and the matching between I_0 and I_{D4}. Arising from mismatches in the current mirror and other sources of error, the last issue dictates an upper limit on α, e.g., $\alpha = 0.8$ or 0.9.

In pure bipolar technologies, addition of a *pnp* current source may not be feasible. Typical *pnp* transistors suffer from substantial collector-base and collector-substrate capacitance, degrading the speed.

4.6.2 Capacitive Coupling

The source followers in feedback TIAs constrain the voltage headroom considerably. With body effect and typical current levels and device dimensions, the source follower in Fig. 4.24 may require a gate-source voltage of 0.6 to 0.7 V in 0.18-μm technology. A possible remedy is to employ capacitive coupling, thereby isolating the dc levels. Illustrated in Fig. 4.43 are two embodiments of the idea. In Fig. 4.43(a), the input stage is coupled to the source follower through C_C and the gate bias voltage of M_2 is set at V_{DD} through R_B. Now R_D can sustain a greater voltage drop, thus providing a higher gain and a lower input-referred noise. However, the bottom-plate parasitic capacitance of C_C increases the time constant at node X.

In Fig. 4.43(b), the feedback incorporates capacitive coupling, I_2 and M_3 define the bias current of M_1, and R_B is large enough to isolate the signal path from the low impedance introduced by M_3. Capacitor C_B shunts the noise of I_2 and M_3 to ground. Note that the minimum supply voltage is now determined by $I_{D1}R_D + V_{GS2} + V_{I1}$, where V_{I1} denotes the minimum allowable voltage across I_1.

The principal issue in the ac-coupled circuits of Fig. 4.43 relates to the lower corner frequency of the passband. The closed-loop gain drops for $f < f_1 = (1 + g_{m1}R_D)/(2\pi R_B C_C)$ in the circuit of Fig. 4.43(a) and rises for $f < f_2 = 1/(2\pi R_F C_C)$ in the circuit of Fig. 4.43(b).[12] Since some standards require this frequency to be as low as

[12] We assume R_B can be sufficiently large.

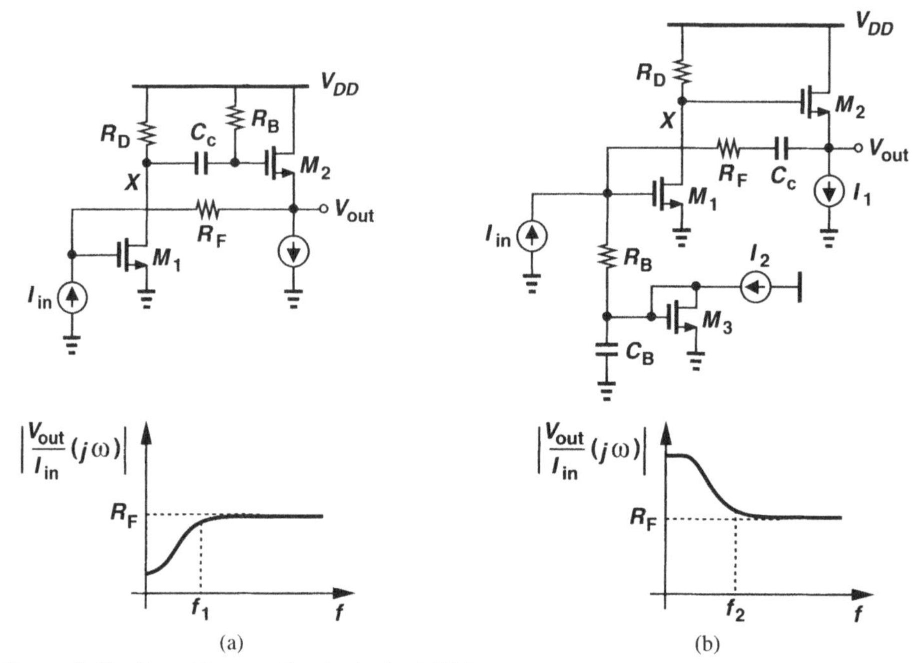

Figure 4.43 Capacitive coupling in feedback TIA.

a few tens of kilohertz, the time constants must reach tens of microseconds, necessitating very large resistor and capacitor values. For example, if $C_C = 1$ pF, the required value of R_F may exceed 10 MΩ. Such a large resistor exhibits substantial capacitance to the substrate, degrading the speed and/or the loop gain. Nonetheless, standards such as 10 Gigabit Ethernet and Fibre Channel, which employ 8B/10B encoding (Chapter 2), can benefit from these techniques.

4.6.3 Feedback TIAs

The feedback TIAs studied thus far incorporate a source/emitter follower as a buffer. In CMOS technology, however, source followers suffer from poor drive capability while consuming a large voltage headroom. We may then consider eliminating the source follower, arriving at the circuit of Fig. 4.44(a). In a typical design, the loading of R_F on R_D may not be negligible, and from the equivalent circuit of Fig. 4.44(b), we can write $V_1 = I_{in}R_F + V_{out}$ and $g_m(I_{in}R_F + V_{out}) + V_{out}/R_D = I_{in}$. Thus,

$$\frac{V_{out}}{I_{in}} = -\frac{g_m R_F - 1}{g_m R_D + 1} R_D. \tag{4.158}$$

(Note that the output resistance of M_1 can be included in parallel with R_D.) If $g_m R_F$, $g_m R_D \gg 1$, then

$$\frac{V_{out}}{I_{in}} \approx -R_F. \tag{4.159}$$

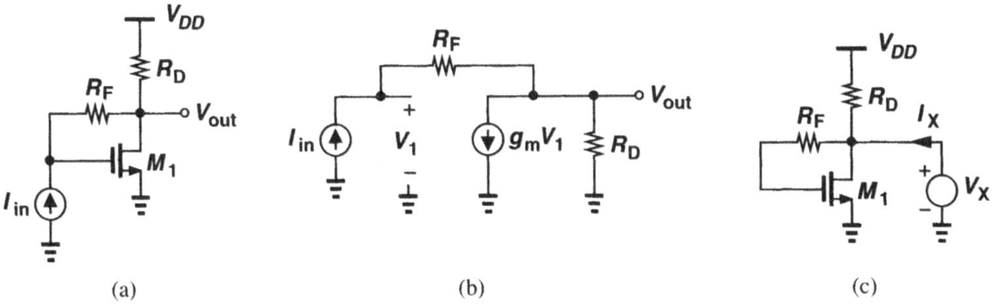

Figure 4.44 (a) TIA with no buffer in feedback path, (b) equivalent circuit of (a), (c) circuit to calculate the output impedance.

We also note that the input resistance of the circuit is given by

$$R_{in} = \frac{V_1}{I_{in}} \tag{4.160}$$

$$= R_F - \frac{g_m R_F - 1}{g_m R_D + 1} R_D \tag{4.161}$$

$$= \frac{R_F + R_D}{g_m R_D + 1}. \tag{4.162}$$

For the output resistance, the circuit of Fig. 4.44(c) gives

$$R_{out} = R_D || \frac{1}{g_m} \tag{4.163}$$

because R_F carries no current.

Comparing the above results with those expressed by (4.127), (4.128), and (4.129), we observe that the TIA of Fig. 4.44(a) provides approximately the same transimpedance gain but with higher input and output resistances. The advantage of this topology lies in the voltage headroom consumption, as is evident from $V_{DD} = R_D I_{D1} + V_{GS1}$. For a given current, increasing R_D leads to a lower input noise and increasing V_{GS1} yields a lower input capacitance.

Let us analyze the noise behavior of the circuit as well. From the equivalent circuit of Fig. 4.45, we note that the drain current of M_1 is equal to $g_m(V_{n,out} - V_{n,RF})$ and sum

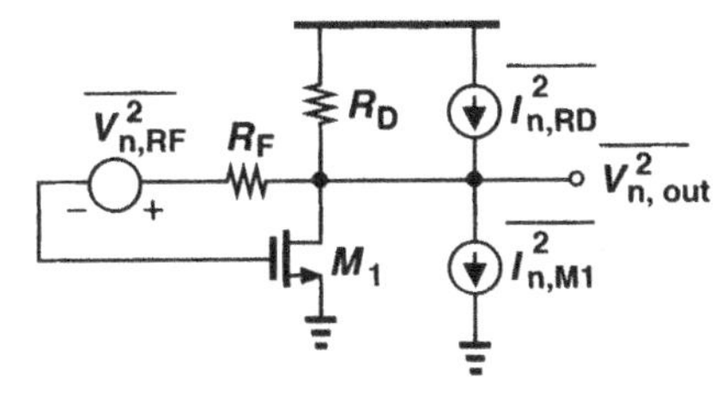

Figure 4.45 Noise equivalent of the TIA shown in Fig. 4.44(a).

the currents at the output node as

$$\frac{V_{n,out}}{R_D} + I_{n,M1} - I_{n,RD} + g_m(V_{n,out} - V_{n,RF}) = 0. \tag{4.164}$$

It follows that

$$V_{n,out} = \frac{-I_{n,M1} + I_{n,RD} + g_m V_{n,RF}}{g_m R_D + 1} R_D. \tag{4.165}$$

Dividing both sides by the gain, R_F, assuming $g_m R_D \gg 1$, and computing the mean-square values, we have

$$\overline{I^2_{n,in}} = \frac{\overline{I^2_{n,M1}} + \overline{I^2_{n,RD}} + g_m^2 \overline{V^2_{n,RF}}}{g_m^2 R_F^2} \tag{4.166}$$

$$= \frac{4kT\gamma}{g_m R_F^2} + \frac{4kT}{g_m^2 R_F^2 R_D} + \frac{4kT}{R_F}. \tag{4.167}$$

This quantity is similar to that in (4.133) but lacking the contribution of the source follower.

It is instructive to examine the frequency response of the TIA in Fig. 4.44(a). Consider the circuit shown in Fig. 4.46(a), where C_{in} is equal to the sum of the photodiode capacitance and C_{GS1}; $C_F = C_{GD1}$; and C_L is equal to the sum of C_{DB1} and the input

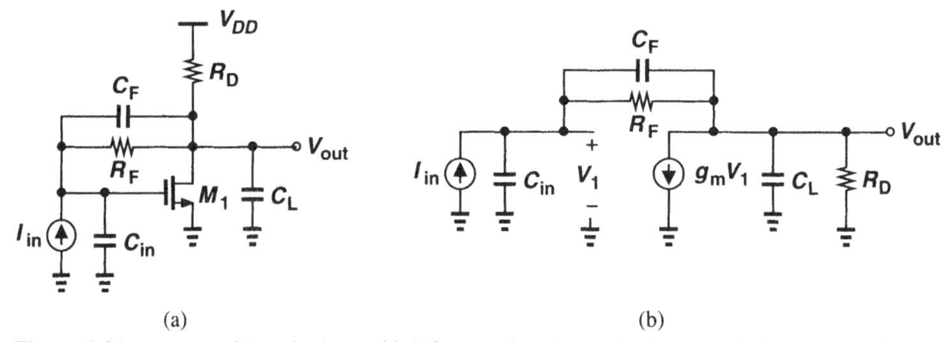

Figure 4.46 (a) TIA of Fig. 4.44(a) at high frequencies, (b) small-signal equivalent circuit of (a).

capacitance of the following stage. Using the equivalent circuit in Fig. 4.46(b), we have

$$I_{in} = V_1 C_{in} s + V_{out}(C_L s + \frac{1}{R_D}) + g_m V_1, \tag{4.168}$$

and hence

$$V_1 = \frac{I_{in} - V_{out}(C_L s + \frac{1}{R_D})}{g_m + C_{in} s}. \tag{4.169}$$

Also, V_1 and the voltage drop across $R_F||(C_F s)^{-1}$ must add up to V_{out}:

$$\frac{I_{in} - V_{out}(C_L s + \frac{1}{R_D})}{C_{in}s + g_m} + \frac{R_F}{R_F C_F s + 1}\left[-V_{out}(C_L s + \frac{1}{R_D}) - g_m \frac{I_{in} - V_{out}(C_L s + \frac{1}{R_D})}{C_{in}s + g_m}\right] = V_{out}. \tag{4.170}$$

It follows that

$$\frac{V_{out}}{I_{in}} = \frac{(R_F C_F s + 1 - g_m R_F)R_D}{R_F R_D \xi s^2 + [R_F(1 + g_m R_D)C_F + R_D C_L + (R_F + R_D)C_{in}]s + 1 + g_m R_D}, \tag{4.171}$$

where $\xi = C_F C_L + C_L C_{in} + C_{in} C_F$. As expected, (4.171) reduces to (4.158) for $s = 0$. The circuit contains one zero and two poles.

We must now determine the conditions for optimum step response. Following the analysis in Section 4.3 and neglecting the effect of the zero for now, we obtain the natural frequency and the damping factor as

$$\omega_n^2 = \frac{1 + g_m R_D}{R_F R_D (C_F C_L + C_L C_{in} + C_{in} C_F)} \tag{4.172}$$

$$\zeta = \frac{1}{2}\frac{R_F(1 + g_m R_D)C_F + R_D C_L + (R_F + R_D)C_{in}}{\sqrt{R_F R_D (1 + g_m R_D)(C_F C_L + C_L C_{in} + C_{in} C_F)}}. \tag{4.173}$$

The circuit parameters can be chosen to yield $\zeta = \sqrt{2}/2$ and, from (4.125), the −3-dB bandwidth is equal to ω_n. Note that if $g_m R_D \gg 1$, then

$$\omega_n \approx \sqrt{\frac{g_m}{R_F(C_F C_L + C_L C_{in} + C_{in} C_F)}}. \tag{4.174}$$

Interestingly, feedback yields a −3-dB bandwidth relatively independent of R_D. Note that the zero of the transfer function is given by $\omega_z = (1 - g_m R_F)/R_F C_F \approx -g_m/C_F$, which typically exhibits a much greater magnitude than ω_n. Thus, the effect of the zero is usually negligible.

The low-voltage topology of Fig. 4.44(a) can also be realized in bipolar technology. The analysis is similar to that described above for the MOS implementation.

The TIA of Fig. 4.44(a) still suffers from supply-dependent biasing. To alleviate this issue, the load resistor can be replaced with a current source whose value is defined by supply-independent biasing techniques [Fig. 4.47(a)] [1]. Due to the low output impedance of PMOS devices, transistor M_2 must be relatively long, and due to the limited voltage headroom, it must be wide. Thus, M_2 contributes substantial junction and overlap capacitance at the output node.

Let us compute the performance parameters of the above circuit. If R_D goes to infinity in Eqs. (4.158), (4.159) and (4.162), then $V_{out}/I_{in} = -R_F$, $R_{in} = 1/g_m$, and $R_{out} = 1/g_m$. To calculate the input-referred noise current, we write from Fig. 4.47(b):

$$g_{m1}(V_{n,out} - V_{n,RF}) + I_{n,M1} = I_{n,M2} \tag{4.175}$$

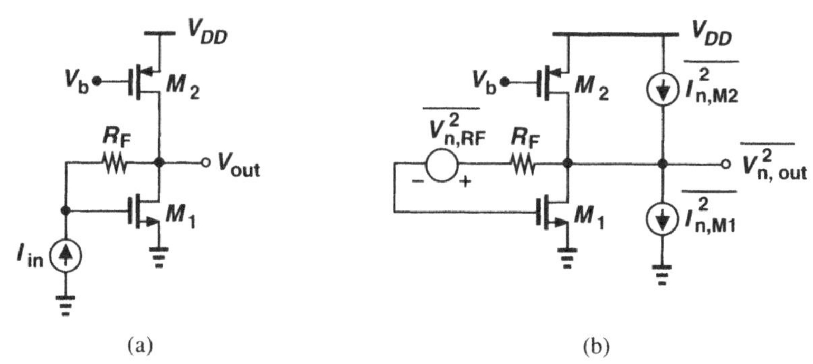

Figure 4.47 (a) Feedback TIA using a current-source load, (b) implementation of (a) along with noise sources.

and hence

$$V_{n,out} = \frac{-I_{n,M1} + I_{n,M2}}{g_{m1}} + V_{n,RF}. \tag{4.176}$$

Dividing by the transimpedance gain and calculating the mean-square values, we have

$$\overline{I^2_{n,in}} = \frac{4kT\gamma_n}{g_{m1}R_F^2} + \frac{4kT\gamma_p g_{m2}}{g_{m1}^2 R_F^2} + \frac{4kT}{R_F}, \tag{4.177}$$

where γ_n and γ_p denote the excess noise coefficients of NMOS and PMOS devices, respectively. As expected, the second term in (4.177) is greater than that in (4.167) because, for a given allowable dc voltage, $\gamma_p g_{m2} > 1/R_D$.

The frequency response of the circuit can be derived by allowing R_D to approach infinity in Eq. (4.171). The optimum -3-dB bandwidth is still given by (4.174), but ζ is simplified to

$$\zeta = \frac{1}{2}\frac{g_m R_F C_F + C_L + C_{in}}{\sqrt{g_m R_F (C_F C_L + C_L C_{in} + C_{in} C_F)}}. \tag{4.178}$$

4.6.4 Inductive Peaking

With the advent of monolithic inductors, inductive peaking techniques have become feasible in integrated circuits. The idea is to allow the capacitance that limits the bandwidth to resonate with an inductor, thereby improving the speed. The resonance must of course occur with minimal peaking and overshoot so as to provide a well-behaved response to random data.

Let us study inductive peaking with the aid of the example illustrated in Fig. 4.48. The common-source stage of Fig. 4.48(a) is simplified as shown in Fig. 4.48(b), revealing a -3-dB bandwidth equal to $(2\pi R_D C_L)^{-1}$. Now suppose an inductor is placed in series with the load resistor [Fig. 4.48(c)]. Using the equivalent circuit of Fig. 4.48(d), we seek conditions

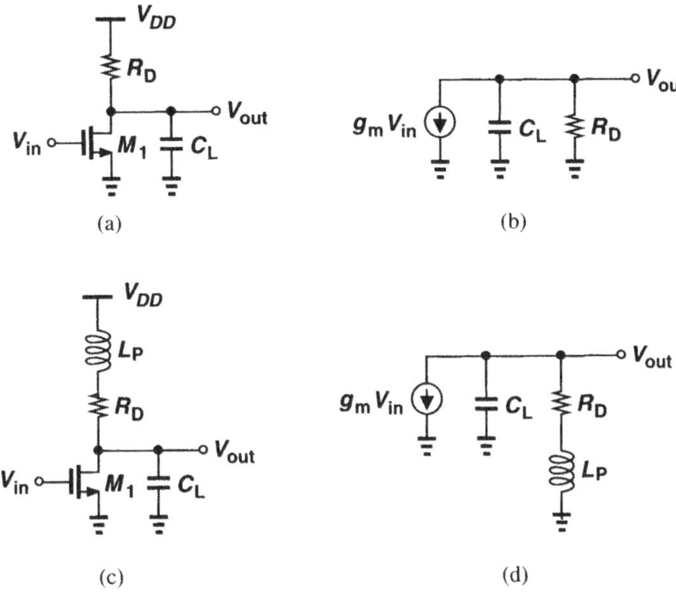

Figure 4.48 (a) CS stage with load capacitance, (b) small-signal equivalent of (a), (c) addition of inductor peaking, (d) small-signal equivalent of (c).

for a well-behaved response.[13] Comparing the circuits of Figs. 4.48(b) and 4.48(d), we note intuitively that, with an input step waveform, the inductor initially serves as an open circuit, allowing all of the current to flow through the capacitor rather than through the resistor. As a result, the output voltage changes faster in Fig. 4.48(d) than in Fig. 4.48(b). If the inductor value is excessively large, then V_{out} experiences overshoot before settling. We must therefore determine the amount of overshoot as a function of the circuit parameters.

Summing the currents flowing through the capacitor and the RL branch, we have

$$V_{out}C_L s + \frac{V_{out}}{L_P s + R_D} = -g_m V_{in}. \tag{4.179}$$

Thus,

$$\frac{V_{out}}{V_{in}} = -g_m \frac{L_P s + R_D}{L_P C_L s^2 + R_D C_L s + 1} \tag{4.180}$$

$$= -g_m R_D \cdot \frac{s + 2\zeta\omega_n}{s^2 + 2\zeta\omega_n s + \omega_n^2} \cdot \frac{\omega_n}{2\zeta}, \tag{4.181}$$

where $\zeta = (R_D/2)\sqrt{C_L/L_P}$ and $\omega_n^2 = (L_P C_L)^{-1}$. Neglecting the effect of the zero and assuming $\zeta = \sqrt{2}/2$, we can write from Eq. (4.125): $\omega_{-3-dB} = \omega_n = \sqrt{2}/(R_D C_L)$.

[13] It is interesting to note that, for broadband circuits, inductive "peaking" must exhibit minimal peaking in the frequency domain (although there may be some overshoot in the time domain).

However, with this choice, the zero frequency is equal to $R_D/L_P = 2/(R_D C_L)$, only slightly higher than the −3-dB bandwidth. For this reason, the zero must be included in the calculation of the bandwidth and time response.

To obtain the −3-dB bandwidth in the presence of the zero, we equate the squared magnitude of (4.181) to $[\omega_n R_D/(2\zeta)]^2/2$, arriving at

$$\omega^2_{-3-dB} = \left[\frac{1}{4\zeta^2} + 1 - 2\zeta^2 + \sqrt{(\frac{1}{4\zeta^2} + 1 - 2\zeta^2)^2 + 1}\right]\omega_n^2. \tag{4.182}$$

Since $\zeta = (R_D/2)\sqrt{C_L/L_P}$, we have $\omega_n^2 = (L_P C_L)^{-1} = 4\zeta^2/(R_D^2 C_L^2)$ and hence

$$\omega^2_{-3-dB} = \left[\frac{1}{4\zeta^2} + 1 - 2\zeta^2 + \sqrt{(\frac{1}{4\zeta^2} + 1 - 2\zeta^2) + 1}\right]\frac{4\zeta^2}{R_D^2 C_L^2}. \tag{4.183}$$

This equation expresses the bandwidth improvement beyond the value $1/(R_D C_L)$. But how should the value of ζ be chosen? Owing to the zero in the transfer function, the step response of the circuit exhibits more overshoot than the second-order systems of Section 4.3 do. Table 4.1 displays the characteristics for various overshoots with and without a zero in the transfer function.

Overshoot	5%	7.5%	10%
ζ (with zero)	0.73	0.69	0.65
ζ (without zero)	0.69	0.64	0.59
Bandwidth Improvement (with zero)	78%	82%	84%

Table 4.1 Characteristics of inductive peaking.

Let us interpret the above results carefully. In a typical design, R_D and C_L are determined by gain and voltage headroom requirements and the value of L_P is chosen to increase the speed. With the margin needed for process and temperature variations, a nominal overshoot of 7.5% provides a reasonable compromise between control over the step response and the improvement in bandwidth. Thus, in principle, inductive peaking can enhance the speed by 82%.[14] Note that the sharper (second-order) roll-off of the gain beyond the resonance frequency reduces the effect of the out-of-band noise as well.

In practice, since monolithic inductors suffer from parasitic capacitance and a low quality factor, Q, inductive peaking improves the speed to a lesser extent. Figure 4.49 shows the

[14] However, inductive peaking *within* a feedback loop may increase the ringing.

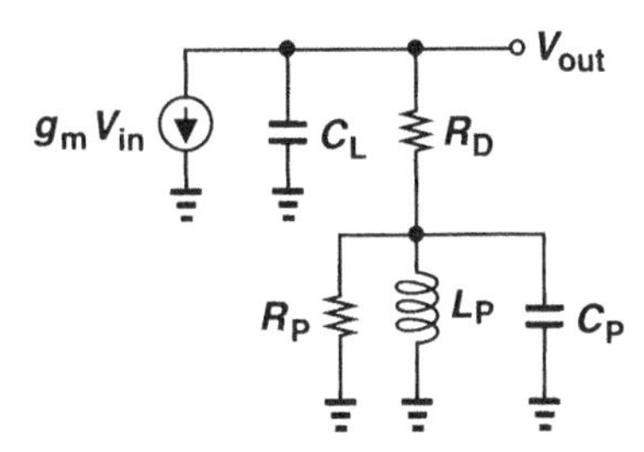

Figure 4.49 Inductive peaking including more realistic model of inductor.

circuit of Fig. 4.48(d) but with the inductor modeled by a simple parallel network, where R_P represents the effect of finite Q, i.e., $R_P = L_P\omega/Q$, and C_P denotes the parasitic capacitance of the inductor to ground. The circuit is now of third order and, due to the shunting effect of R_P, L_P has less impact on the output. Typical monolithic inductor characteristics limit the bandwidth improvement to roughly 50%. Accurate *broadband* inductor models are therefore necessary to predict the circuit's behavior (Chapter 7).

The inductive peaking concept studied thus far is called "shunt peaking" because the resistor/inductor combination appears in parallel with the output. This technique is usually applied to the *output* node(s) of amplifiers because most circuits of interest produce a *current* that subsequently flows through a load impedance. It is also possible to utilize inductive peaking at the input. For example, as shown in Fig. 4.50, an inductor can be interposed between a photodiode and the input of a transimpedance amplifier so as to

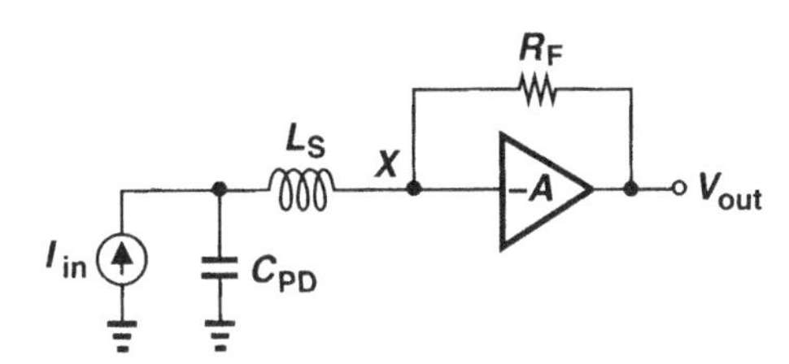

Figure 4.50 Series peaking at the input of a TIA.

increase the bandwidth. Called "series peaking" (because the inductor appears in series with the source), this method is similar to shunt peaking but it introduces no zero in the transfer function. Noting that the current through R_F is equal to $(V_{out} + V_{out}/A)/R_F$ and hence the voltage drop across L_S is $[(V_{out} + V_{out}/A)/R_F]L_S s$, we sum the currents at the input node:

$$\left[\frac{-V_{out}}{A} - (V_{out} + \frac{V_{out}}{A})\frac{L_S s}{R_F}\right] C_{PD} s = I_{in} + (V_{out} + \frac{V_{out}}{A})\frac{1}{R_F}. \tag{4.184}$$

It follows that

$$\frac{V_{out}}{I_{in}} = \frac{-1}{s^2 + \dfrac{R_F}{(A+1)L_S}s + \dfrac{1}{C_{PD}L_S}} \cdot \frac{AR_F}{(A+1)C_{PD}L_S}. \tag{4.185}$$

Following Eq. (4.125), we set $\zeta = \sqrt{2}/2$ and hence

$$\omega_{-3-dB} \approx \frac{\sqrt{2}A}{R_F C_{PD}}, \tag{4.186}$$

thereby increasing the bandwidth by approximately 41% with an overshoot of 4.3%.

We should point out two limitations of input series peaking. First, the loss and parasitic capacitance of monolithic inductors make them a poor candidate for L_S in Fig. 4.50. It is possible to employ a bond wire instead [2], but the length and shape of the wire and the capacitance of the photodiode must be controlled tightly. Second, the above analysis assumed infinite bandwidth for the core amplifier; even with one pole in A, the overall transfer function is of third order, possibly leading to ISI.

4.7 Automatic Gain Control

Optical receivers may experience vastly different input currents because the transmitted laser power, the length and hence loss of the fiber, and the efficiency of the photodiode may vary from one link to another. Thus, transimpedance amplifiers must accommodate a relatively wide dynamic range, typically from a few microamperes to a few milliamperes.

How does a TIA behave with a high input current? Consider the circuit shown in Fig. 4.51(a) and assume $R_F = 1$ kΩ and $V_{out} = 0.6$ V when $I_{in} = 0$. If a logical ONE is

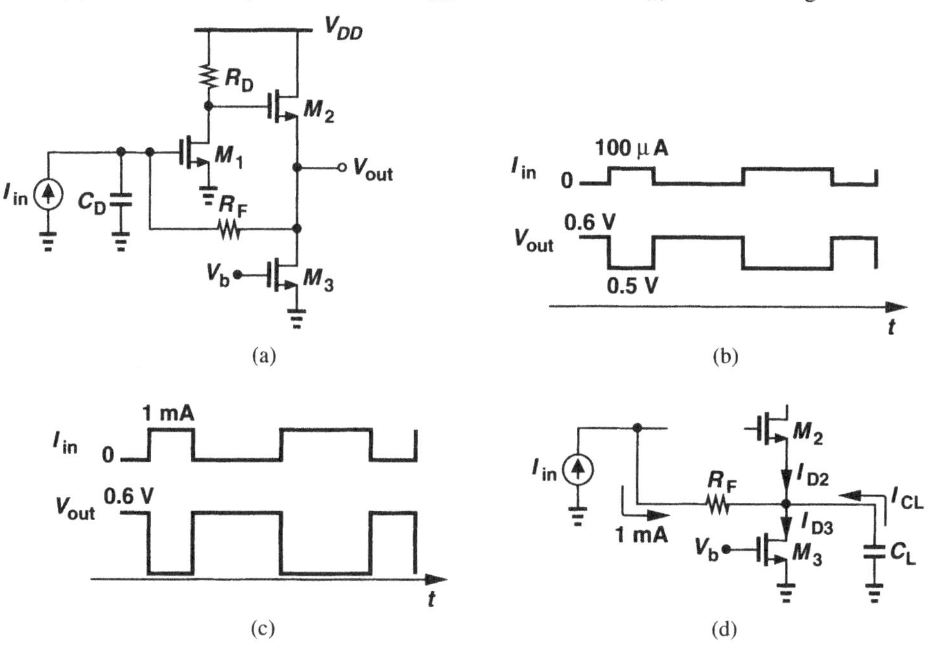

Figure 4.51 Effect of overload in a CMOS feedback TIA.

represented by $I_{in} = 100$ μA, then V_{out} goes *down* by 100 μA$\times$1 kΩ = 100 mV [Fig.

4.51(b)]. But what happens if I_{in} rises to 1 mA? Since a negative change of 1 mA×1 kΩ = 1 V cannot be accommodated in V_{out}, the circuit becomes heavily nonlinear. More specifically, as V_{out} falls, M_3 enters the triode region, reducing the gain of the source follower and hence the loop gain so that the transimpedance gain is much lower than R_F. Thus, V_{out} may seem to approach zero in this case [Fig. 4.51(c)]. However, as depicted in Fig. 4.51(d), the large input current flowing through R_F must also be absorbed by M_3, prohibiting V_{out} from reaching zero.

The above overload behavior appears benign: for large input currents, the TIA produces a reasonable voltage swing even though transistor M_3 enters the triode region. Unfortunately, at high speeds, other effects arise that make overload undesirable. First, with finite rise and fall times, clipping of the large signals may result in narrower widths for logical ZEROs or ONEs at the output, leading to pulsewidth distortion and hence degraded detection. Second, if the TIA drives a load capacitance, then, as illustrated in Fig. 4.51(d), the current available to discharge this capacitance, I_{CL}, is *reduced* at high input levels. Thus, the logical ZEROs at the output exhibit a long tail, introducing significant ISI.

The problem of overload is even more serious in bipolar TIAs as it may drive some of the transistors into saturation, degrading the speed considerably. For example, in the circuit of Fig. 4.52, a high input current forces both V_X and V_{out} to low values, saturating Q_1 and

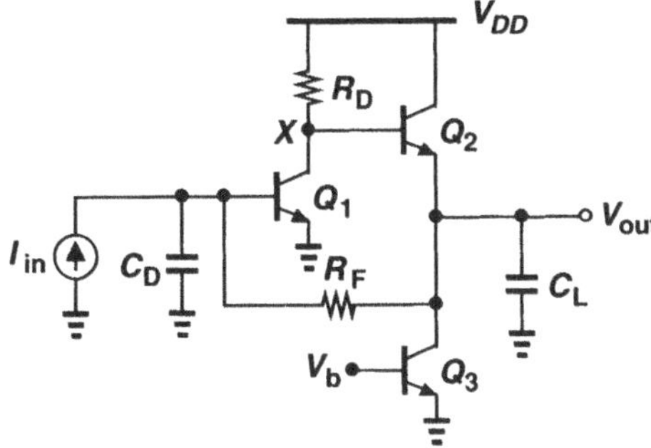

Figure 4.52 Effect of overload in a bipolar TIA.

Q_3.

The foregoing study suggests that the gain of TIAs must be automatically adjusted such that high input currents do not distort the output waveform. To this end, two functions must be added: a means of measuring the output swing and a method of reducing the gain. Let us examine these concepts in detail.

Interestingly, unlike typical voltage amplifiers, TIAs sense currents with only *one polarity* because the photodiode injects either electrons or no charge into the TIA. Illustrated in Fig. 4.53, this property implies that the *average* value of the output voltage varies in proportion to the *amplitude* of the input current. Thus, AGC can simply utilize the dc content of the output as a measure of the input swing. This is in contrast to AGC in amplifiers sensing both positive and negative inputs, where wave rectification or peak detection are required.

The above observation leads to the amplitude measurement circuit shown in Fig. 4.54. Here, a low-pass filter following the main TIA generates the output time average $V_{out,avg}$, and a replica of the TIA establishes a reference voltage corresponding to a zero input

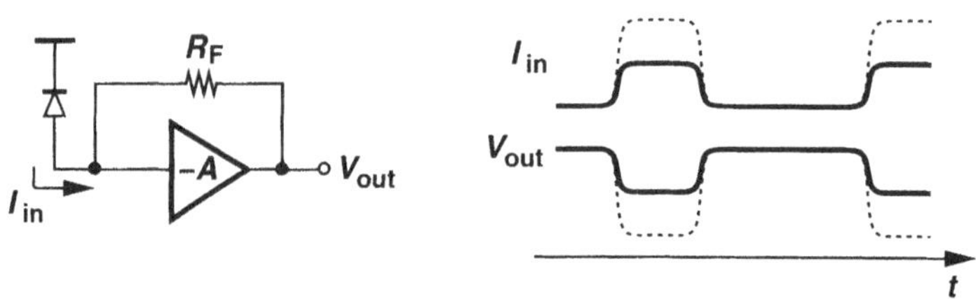

Figure 4.53 Conceptual illustration of AGC.

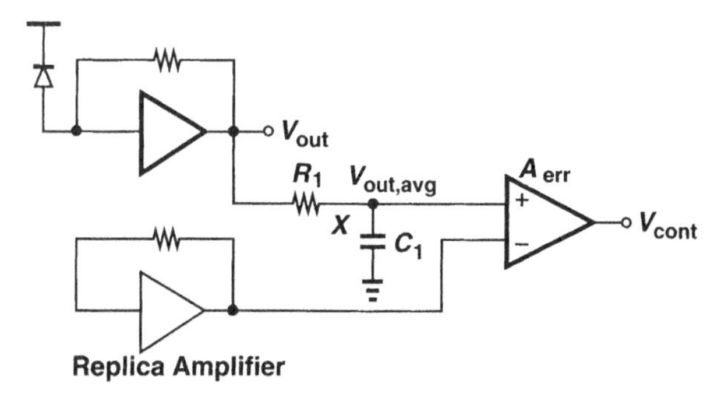

Figure 4.54 Generation of error voltage for AGC.

current. Amplifier A_{err} then produces an output, V_{cont}, proportional to the error and hence the input amplitude. Note that the replica amplifier can be scaled down with respect to the main TIA, saving power dissipation.

The principal issue in the topology of Fig. 4.54 stems from the required corner frequency of the low-pass filter, $\omega_{LPF} = 1/(R_1C_1)$. If ω_{LPF} is excessively large, then low-frequency components in the data waveform pass through the R_1-C_1 network unattenuated, forcing V_{out} to change with time and ultimately corrupt the data when the AGC loop is closed around the TIA. From another perspective, as depicted in Fig. 4.55, if a long run appears at the TIA output and the time constant R_1C_1 is not large enough, then

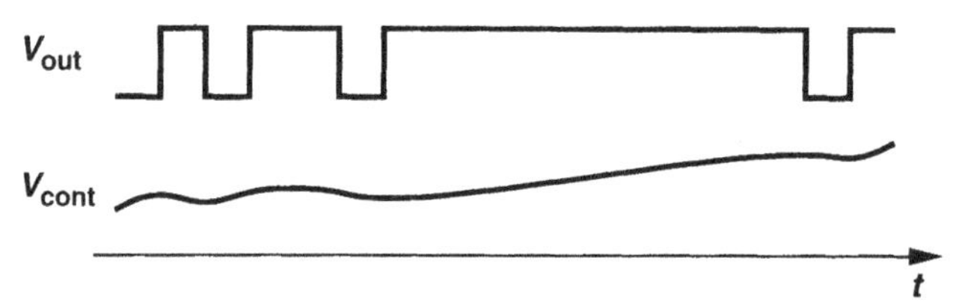

Figure 4.55 Effect of AGC for long runs.

$V_{out,avg}$ changes, exhibiting an incorrect average for *subsequent* bits.

Since some optical standards specify a low-frequency corner of only a few tens of kilohertz for the data, the time constant R_1C_1 in Fig. 4.54 must be on the order of a few hundred microseconds, requiring very large resistor and/or capacitor values. For this reason, capacitor C_1 is usually placed off-chip. However, the bond wire and package trace connecting node X to the external capacitor may experience substantial crosstalk from other high-speed signals. Thus, a reasonable fraction of C_1 is included on the chip to suppress the effect of coupled interference.

Having developed a method of measuring the output swing, we now address the second task: gain control. In feedback TIAs, the transimpedance gain is nearly equal to the feedback resistor, suggesting that varying the value of this resistor directly controls the gain. Shown in Fig. 4.56 is an example, where an NMOS device placed in parallel with R_F low-

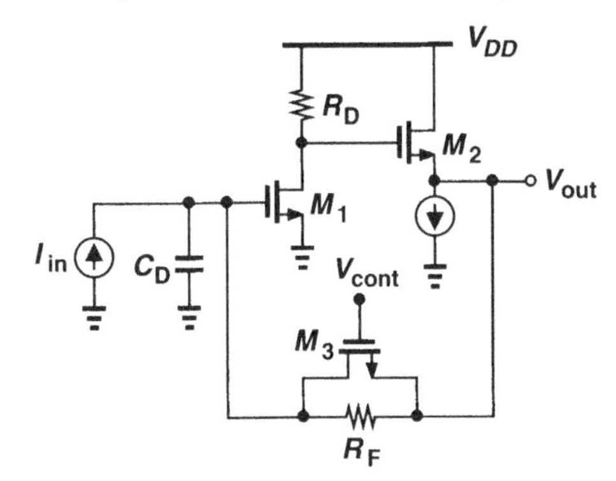

Figure 4.56 AGC in a feedback TIA.

ers the gain as V_{cont} becomes more positive. This is the most common approach to gain control in TIAs.

The above technique must nonetheless deal with several issues. First, recall from Section 4.3 that the feedback resistor affects the closed-loop poles and hence the step response. Thus, if R_F is varied over a wide range, the stability of the circuit suffers. In fact, considering Eq. (4.116) and noting that typically $R_F C_D \omega_0 \gg 1$ [for example, $R_F C_D \omega_0 = 2A_0$ from (4.119)], we have

$$\zeta \approx \frac{1}{2} \frac{R_F C_D \omega_0}{\sqrt{(A_0+1)R_F C_D \omega_0}} \tag{4.187}$$

$$\approx \frac{1}{2}\sqrt{\frac{R_F C_D \omega_0}{A_0+1}}. \tag{4.188}$$

That is, if R_F is lowered by, say, a factor of two, then ζ degrades by approximately 40%.

In order to maintain a relatively constant ζ, it is possible to lower A_0 in Eq. (4.188) along with R_F. For example, as shown in Fig. 4.57, a PMOS device acting as a variable resistor can be utilized to reduce the gain of the input common-source stage [3]. Here, V_{cont}^{+} and V_{cont}^{-} vary in opposite directions. A more complete version of this circuit is studied in Section 4.8.

The second issue related to scheme of Fig. 4.56 is the limited gain range at low supply voltages. Since M_3 turns on only for $V_{cont} > V_{GS1} + V_{TH3}$, its overdrive voltage cannot exceed $V_{DD} - (V_{GS1} + V_{TH3})$. In fact, the overdrive is even less than this value because amplifier A_{err} in Fig. 4.54 cannot produce a control voltage as high as V_{DD}.

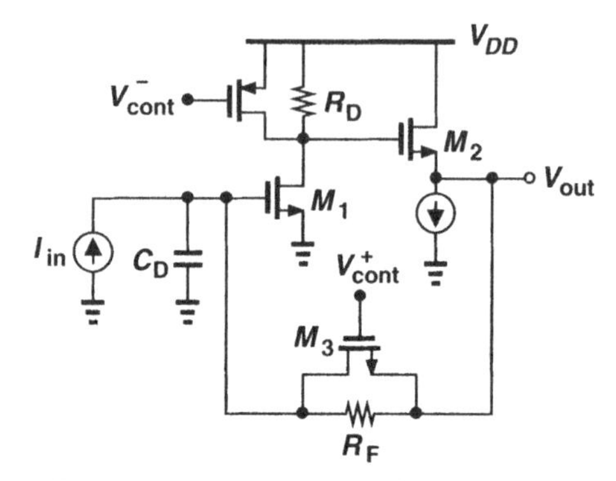

Figure 4.57 Open-loop and closed-loop gain control in a feedback TIA.

The limited overdrive of M_3 creates a trade-off between the gain dynamic range and the sensitivity of the gain control. To understand this effect, let us consider two extreme cases where M_3 is very wide or very narrow. As shown in Fig. 4.58, in each case, the

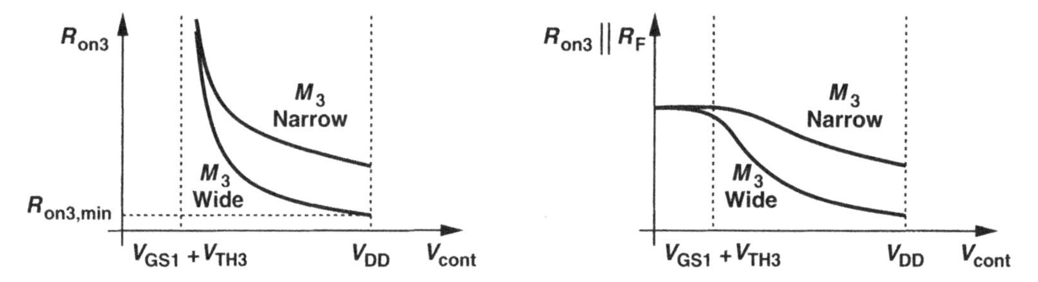

Figure 4.58 Variation of MOS device resistance and overall transimpedance.

on-resistance of the transistor, $R_{on3} = [\mu_n C_{ox}(W/L)_3(V_{GS3} - V_{TH3})]^{-1}$, varies from ∞ to a minimum value, $R_{on3,min}$, thereby changing the transimpedance gain from R_F to $R_F||R_{on3,min}$. If M_3 is very wide, $R_{on3,min}$ is small, yielding a wide gain range but at the cost of high sensitivity in R_{on}. Conversely, if M_3 is very narrow, the gain range is limited but the sensitivity of R_{on} is lower. The high gain sensitivity is undesirable because it may lead to instability in the AGC loop.

The third critical drawback of the technique shown in Fig. 4.56 relates to the capacitances introduced by M_3 itself. If $R_{on3,min}$ is to be low enough to provide a reasonable gain range, then M_3 must be a wide transistor because of its limited overdrive voltage. Thus, the channel and junction capacitances of M_3 may degrade the speed and raise the input-referred noise.

4.8 Case Studies

In order to solidify the concepts introduced in this chapter, we now study several examples of TIAs. While realized in bipolar and BiCMOS technologies, these circuit topologies are applicable to CMOS implementations as well.

Figure 4.59 shows the core circuit of a feedback TIA [4]. Operating from a 5-V supply,

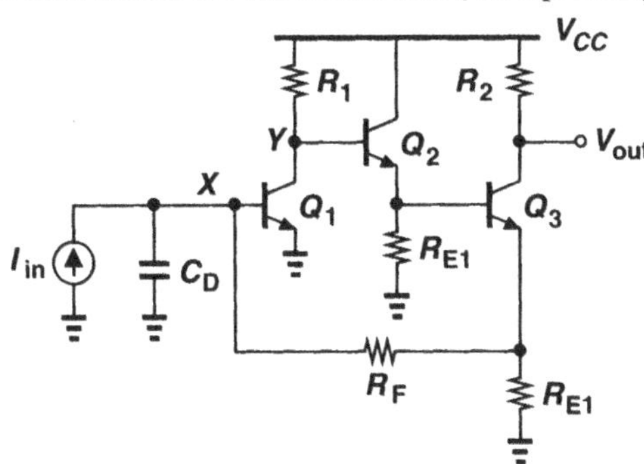

Figure 4.59 Bipolar feedback TIA.

the amplifier incorporates two cascaded emitter followers so as to allow a relatively high current in the output stage. Note that the output signal is sensed at the collector (rather than the emitter) of Q_3 for two reasons: (1) to isolate the feedback from the input capacitance of the subsequent stage; (2) to provide a dc level closer to V_{CC} and hence compatible with the subsequent stage. Note, however, that all three stages suffer from supply-dependent bias points.

In order to ensure a frequency response relatively independent of the photodiode capacitance, the above design employs a large device for Q_1 and a high voltage gain from X to Y, $A_{XY} \approx 100$, thereby making the Miller multiplication of the collector-base capacitance of Q_1, $A_{XY}C_{\mu 1}$, dominant. Since the voltage gain of the two emitter followers is near unity, the low-frequency loop gain is equal to A_{XY}, yielding a closed-loop −3-dB bandwidth of roughly $(2\pi R_F C_{\mu 1})^{-1}$. For the required bandwidth ($\sim$ 200 MHz), this relationship determines the value of R_F.

The bias current and collector load resistor of Q_1 are constrained by a trade-off between noise and stability. The large value of $R_F (\approx 7.2\ \text{k}\Omega)$ requires that the base shot noise current of Q_1 be minimized, necessitating a low bias current. On the hand, for a given voltage gain, $|A_{XY}| \approx g_{m1}R_1$, a lower bias current translates to a greater value for R_1, bringing the pole at node Y closer to the origin and degrading the phase margin. As a compromise, this design uses $I_{C1} = 0.5$ mA and $R_1 = 5\ \text{k}\Omega$.

Figure 4.60 depicts the core TIA along with the subsequent single-ended to differential converter. Resistors R_6 and R_8 provide level shift and the degenerated pair produces a differential output. As with the example of Fig. 4.34, the output voltages V_A and V_B are offset because the circuit generates *equal* voltages at A and B in the absence of light.

Another example of TIA design is shown in Fig. 4.61 in simplified form [5]. Implemented in GaAs MESFET technology, the circuit employs a cascode stage at the input of a feedback configuration. Resistor R_1 provides a path for the bias current of M_1, allowing a greater value for R_2 with a given supply voltage. The integrator consisting of R_3, C_2, and A_1 measures the average voltage at the *input* and adjusts the on-resistance of M_4, thereby performing AGC. Note that the time constant $R_G C_1$ must be very large to provide a low corner frequency for the passband.

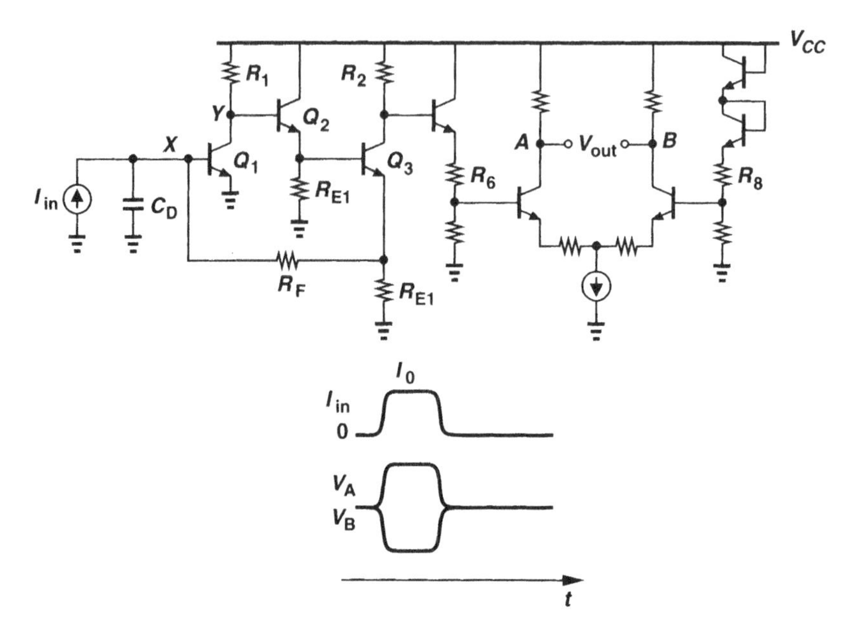

Figure 4.60 Circuit of Fig. 4.59 along with output stage.

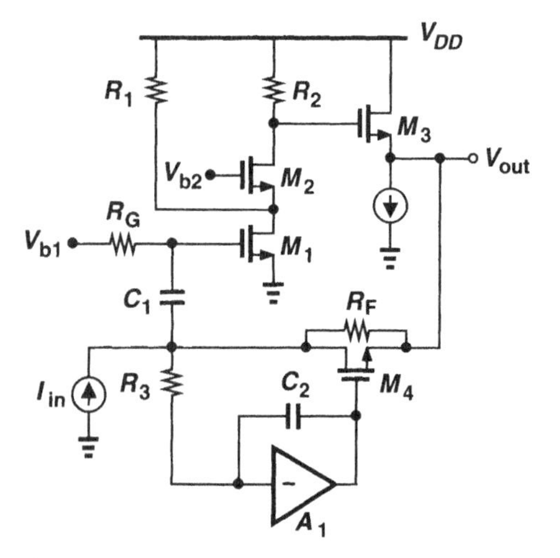

Figure 4.61 Feedback TIA using AGC.

Figure 4.62 shows a BiCMOS TIA with AGC [3]. Employing a feedback topology around a common-emitter core (Q_1, R_E, and R_C), the circuit senses the average output by means of amplifier A_1 and the low-pass network R_1 and C_1. The result controls the gate voltage of M_2, varying the transimpedance gain. Recall from Section 4.7 that if only the feedback resistance is varied, then the stability degrades. In this design, the open-loop

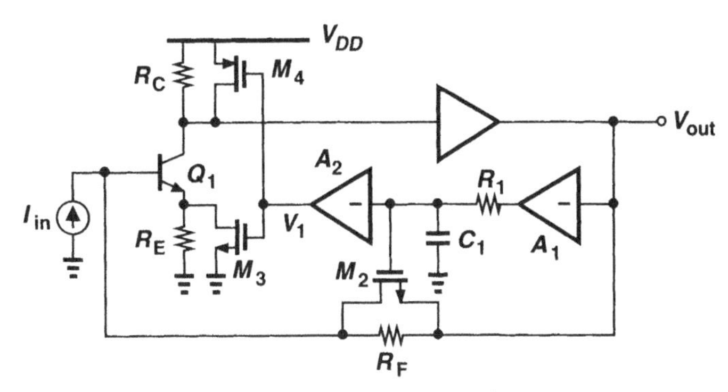

Figure 4.62 BiCMOS TIA applying gain control to emitter and collector networks.

gain of the amplifier is also controlled through A_2, M_3, and M_4, thereby maintaining $\zeta \approx \sqrt{R_F C_D \omega_0 / A_0}/2$ constant. The voltage gain of the common-emitter stage is adjusted by both M_3 and M_4. If V_1 is low, then R_E degenerates Q_1 while M_4 shunts R_C, yielding minimum gain. Conversely, if V_1 is high, M_3 provides a low resistance from the emitter of Q_1 to ground and M_4 is off, maximizing the gain.

Figure 4.63(a) illustrates an example of TIA design for 10-Gb/s operation [2]. The broad

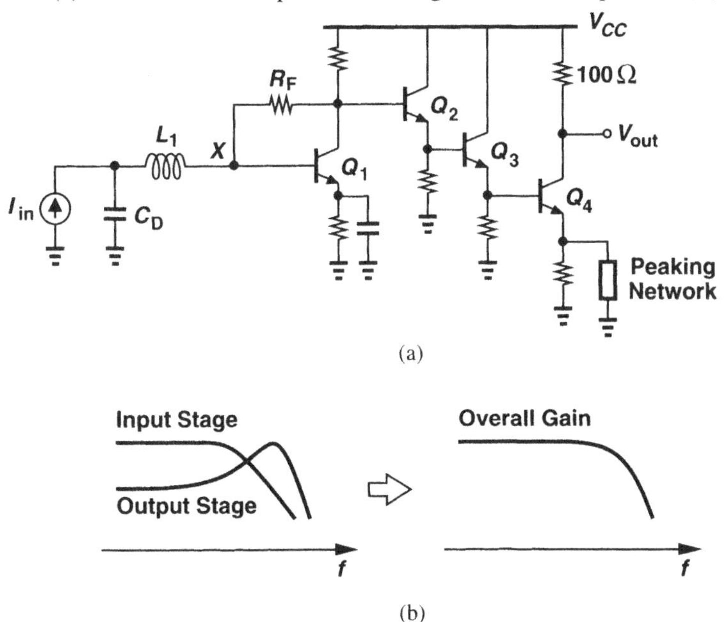

Figure 4.63 (a) Feedback TIA using series peaking at input, (b) effect of output peaking network.

bandwidth required here prohibits the use of feedback around more than one stage, reducing the TIA to a common-emitter stage with local feedback from the collector to the base. Two emitter followers are interposed between the input and output stages to avoid loading the former by the input capacitance of the latter.

The above design utilizes the bond wire inductance, L_1, to create inductive peaking at the input. Chosen to resonate with C_D near the -3-dB bandwidth of the amplifier, L_1 both reduces the input-referred noise current and increases the overall bandwidth. This design also employs a network in the emitter of Q_4 to raise the gain at high frequencies, compensating for the roll-off in the input stage and widening the band of the overall circuit [Fig. 4.63(b)] [2].

Another example of feedback TIA design is shown in Fig. 4.64 in simplified form [6].

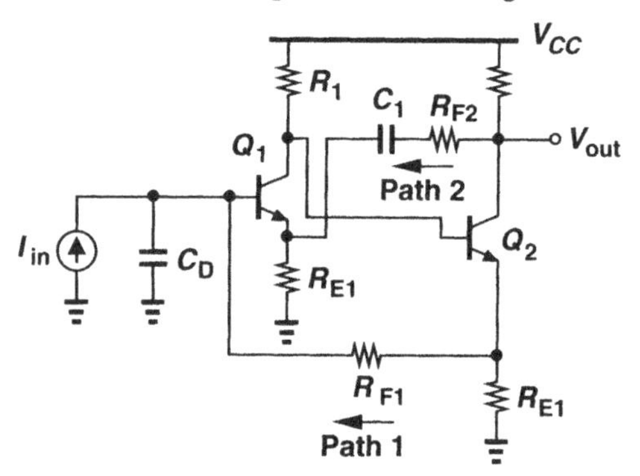

Figure 4.64 TIA using two feedback paths.

In addition to current feedback to the input, the circuit incorporates voltage feedback to the emitter of Q_1 through R_{F2} and C_1 [6]. Note that Q_2 serves as an emitter follower for R_{F1} and a common-emitter stage for R_{F2}. The additional feedback network is designed to compensate for the phase shift experienced by the signal through path 1, thereby reducing peaking in the frequency domain. The principal drawback of this approach is that feedback path 2 raises the input noise of the circuit considerably.

4.9 New Developments in TIA Design

In our study of TIAs in previous sections, we dealt with the common-gate/common-base stage as well as feedback topologies. In this section, we consider a number of new TIA circuits. Since the capacitance of photodiodes (around 100 to 200 fF) tends to limit the circuit's bandwidth at the input, a great deal of effort has been expended on reducing the input resistance of TIAs without raising their input noise.

Figure 4.65(a) depicts a CG stage driven by a photodiode. Recall from our previous analysis that the input-referred noise current of the circuit (at moderate frequencies) is given by

$$\overline{I^2_{n,in}} = \overline{I^2_{n,B}} + \frac{4kT}{R_D}, \tag{4.189}$$

where $\overline{I^2_{n,B}}$ denotes the noise current of the bias current source, I_B. The input bandwidth is equal to $g_m/(2\pi C_{PD})$ if channel-length modulation, body effect, and other capacitances are neglected. How can we increase this bandwidth? If the bias current is raised while

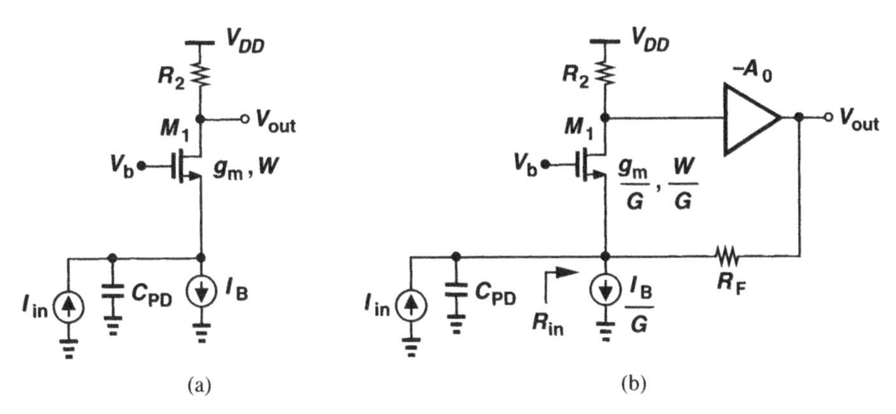

Figure 4.65 (a) Common-gate TIA, (b) feedback CG TIA.

the voltage headroom limitations remain constant, then (a) $\overline{I_{n,B}^2}$ rises because, e.g., the transconductance of the bias transistor providing I_B increases $[g_m = 2I_D/(V_{GS} - V_{TH})]$, and (b) the noise current contributed by R_D, $4kT/R_D$, also goes up because R_D must be reduced to accommodate the larger bias current.

Now consider the feedback CG circuit shown in Fig. 4.65(b) [7]. Here, another gain stage follows the CG amplifier, and resistor R_F senses the output voltage and returns a proportional current to the input. Also, the bias current, the load resistor, and the transistor width are scaled by a factor of $G > 1$ with respect to the CG circuit of Fig. 4.65(a).

Our observation is that the feedback around the circuit tends to lower the input resistance, R_{in}, broadening the input bandwidth, the principal benefit of this technique. It can be shown that [7]

$$R_{in} = \frac{1}{\dfrac{g_m}{G} + \dfrac{1 + g_m R_D A_0}{R_F}}. \tag{4.190}$$

But the reduction of R_{in} must be assessed with the input noise in mind. We have [7]

$$\overline{I_{n,in}^2} \approx \frac{4kT}{R_F} + \frac{\overline{I_{n,B}^2}}{G} + \frac{4kT}{GR_D} + \frac{\overline{V_{n,A0}^2}}{G^2 R_D^2}, \tag{4.191}$$

where $g_m R_F/G$ is assumed much greater than unity, and $\overline{V_{n,A0}^2}$ denotes the input-referred noise voltage of A_0. For a fair comparison of the circuits in Figs. 4.65(a) and (b), we follow the CG stage of Fig. 4.65(a) with an amplifier having an input-referred noise voltage of $\overline{V_{n,A0}^2}$. We now ask, if the two topologies display the same input noise, how much bandwidth improvement does the feedback circuit provide? Equating the input-noise currents yields

$$\overline{I_{n,B}^2} + \frac{4kT}{R_D} + \frac{\overline{V_{n,A0}^2}}{R_D^2} = \frac{\overline{I_{n,B}^2}}{G} + \frac{4kT}{GR_D} + \frac{\overline{V_{n,A0}^2}}{G^2 R_D^2} + \frac{4kT}{R_F}. \tag{4.192}$$

It is interesting that the feedback circuit exhibits lower noise contributions from I_B, R_D, and A_0 at the cost of an additional component from R_F. To arrive at a simple result, [7] pessimistically assumes that $4kT/(GR_D) + 4kT/R_F$ on the right-hand side must not exceed $4k/R_D$ on the left, obtaining

$$R_F \geq \frac{G}{G-1} R_D. \tag{4.193}$$

This is a sufficient condition for the feedback topology to be less noisy than the original CG stage. We can then ask, what minimum value of A_0 yields a lower input resistance than $1/g_m$? Using the above value of R_F, we have [7]

$$\frac{g_m}{G} + \frac{G-1}{GR_D}(1 + g_m R_D A_0) > g_m. \tag{4.194}$$

If $g_m R_D A_0 \gg 1$, then the condition reduces to

$$A_0 > 1. \tag{4.195}$$

In summary, the foregoing feedback technique broadens the input bandwidth without increasing the input-referred noise if $R_F > GR_D/(G-1)$ and $A_0 > 1$. For example, [7] chooses $G = 1.5$ and $A_0 \approx 2.2$, thus reducing the input resistance by 32%. Note that the power consumption of the CG stage itself also falls by a factor of 1.5.

A concern in the feedback TIA of Fig. 4.65(b) arises from the number of poles around the loop. With poles at the input, at the drain of M_1, and at the output of A_0 (especially the input capacitance of the next stage), the circuit may suffer from inadequate phase margin, exhibiting substantial ISI. This issue becomes more serious as A_0 increases (to extend the input bandwidth) but less serious as R_F increases (to lower its noise contribution).

We now turn our attention to another TIA topology [8]. Let us begin with the arrangement shown in Fig. 4.66(a), where M_2 serves as a source follower in the feedback loop

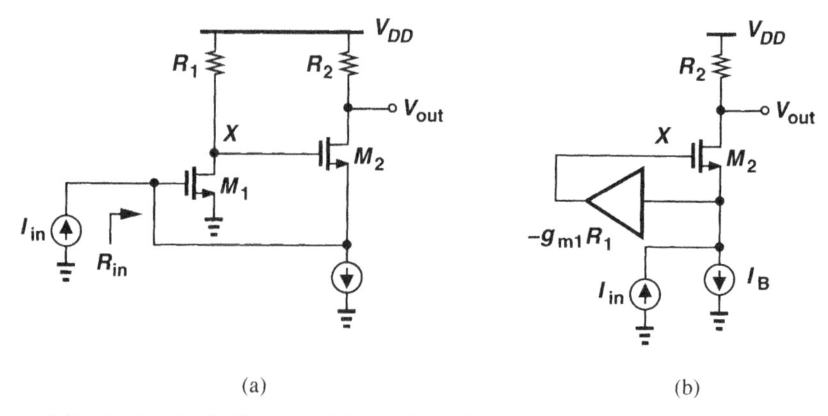

Figure 4.66 (a) Feedback TIA, (b) addition of amplifier in feedforward path.

and as a common-gate stage to generate the output voltage. At low frequencies, all of I_{in} flows through R_2 and the overall transimpedance gain is equal to $V_{out}/I_{in} = R_2$. The input impedance, R_{in}, is equal to the open-loop value ($\approx 1/g_{m2}$) divided by one plus the loop gain. If the loop is broken at the gate of M_2 and if channel-length modulation and body effect are neglected, the loop gain is simply equal to $g_{m1}R_1$. Thus,

$$R_{in} = \frac{1}{g_{m2}} \cdot \frac{1}{1 + g_{m1}R_1}. \tag{4.196}$$

In a manner similar to the topology of Fig. 4.65(b), the feedback loop in this circuit lowers the input resistance, thus extending the input bandwidth.

To determine the input-referred noise current, we redraw the TIA as shown in Fig. 4.66(b), where the common-source stage is represented as a *feedforward* amplifier driving the gate of M_2, i.e., as if M_2 were the main signal path. If I_B is ideal, then no noise current emerges from the drain of M_2 (why?), leaving R_2 as the only noise contributor to the output. This noise, $4kTR_2$, is divided by the square of the transimpedance gain, R_2, when referred to the input. In addition, the noise current of I_B, $\overline{I_{n,B}^2}$, directly corrupts the input. It follows that

$$\overline{I_{n,in}^2} = \frac{4kT}{R_2} + \overline{I_{n,B}^2}, \tag{4.197}$$

which is identical to that of a simple common-gate stage. In other words, the feedback (or feedforward, depending on how we view the circuit) lowers the input resistance without raising the input noise. Of course, the noise of a CG stage tends to be higher than that of a feedback CS stage. Nonetheless, scaling R_2 up and I_B down can reduce their noise contribution while feedforward can maintain a relatively low input resistance. For example, beginning with a CG stage, we scale R_2 and I_B (and W_2) by a factor of 4 and choose $1 + g_{m1}R_1 = 4$. As a result, $\overline{I_{n,in}^2}$, falls by the same factor while R_{in} is still equal to the initial value.

One drawback of the above TIA topology is that the bias voltage at node X in Fig. 4.66(a) is equal to $V_{GS1} + V_{GS2}$, limiting the drop across R_1 and hence the gain, $g_{m1}R_1$. For reasons mentioned previously, it is difficult to employ capacitive coupling in the loop. Alternatively, we can precede the feedforward amplifier in Fig. 4.66(b) with a level shift circuit. Illustrated in Fig. 4.67(a) [8], such a level-shift can be realized by a common-gate stage. Now, V_X can be low enough to accommodate only V_{GS2} and the headroom necessary for I_B.

Let us determine the input resistance, the transimpedance gain, and the input noise of this circuit while neglecting channel-length modulation and body effect. Transistor M_3 carries a small-signal current equal to $-g_{m3}V_{in}$, which upon flowing through R_3 and traveling through M_1 yields

$$V_X = g_{m3}V_{in}R_3(-g_{m1}R_1). \tag{4.198}$$

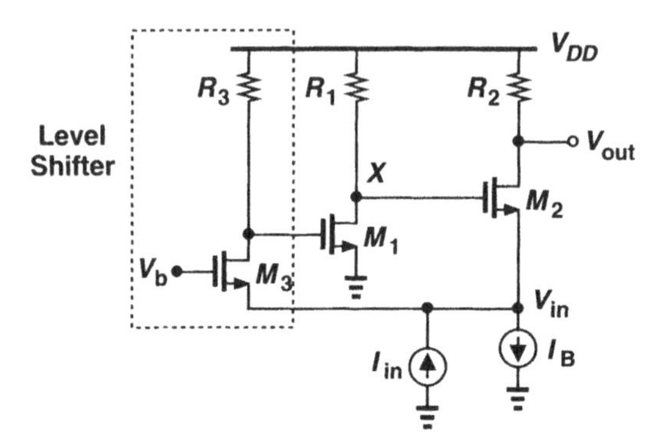

Figure 4.67 TIA using a two-stage feedforward path.

(For simplicity, we denote small-signal quantities by upper-case letters.) Since the drain current of M_2 is equal to $g_{m2}(V_X - V_{in})$, and since $I_{D3} + I_{D2} = -I_{in}$, we have

$$-g_{m3}V_{in} + g_{m2}(-g_{m3}V_{in}R_3g_{m1}R_1 - V_{in}) = I_{in}. \tag{4.199}$$

It follows that

$$\frac{I_{in}}{V_{in}} = g_{m3} + (1 + g_{m3}R_3g_{m1}R_1)g_{m2}. \tag{4.200}$$

Thus, the resistance presented by M_2 is reduced by a factor of $1 + g_{m3}R_3g_{m1}R_1$.

With V_{in} found in terms of I_{in}, we can now compute the drain current of M_2 and hence the transimpedance gain. We have

$$I_{D2} = g_{m2}(V_X - V_{in}) \tag{4.201}$$

$$= g_{m2}[g_{m3}V_{in}R_3(-g_{m1}R_1) - V_{in}] \tag{4.202}$$

$$= -g_{m2}(1 + g_{m3}R_3g_{m1}R_1) \times \frac{I_{in}}{g_{m3} + (1 + g_{m3}R_3g_{m1}R_1)g_{m2}}. \tag{4.203}$$

Since $V_{out} = -R_2I_{D2}$, the gain, R_0, is given by

$$R_0 = \frac{g_{m2}R_2(1 + g_{m3}R_3g_{m1}R_1)}{g_{m3} + (1 + g_{m3}R_3g_{m1}R_1)g_{m2}}. \tag{4.204}$$

It is reasonable to assume that the second term in the denominator is much greater than the first:

$$R_0 \approx R_2. \tag{4.205}$$

In other words, the circuit exhibits the transimpedance gain of a simple CG stage but with a much lower input resistance.

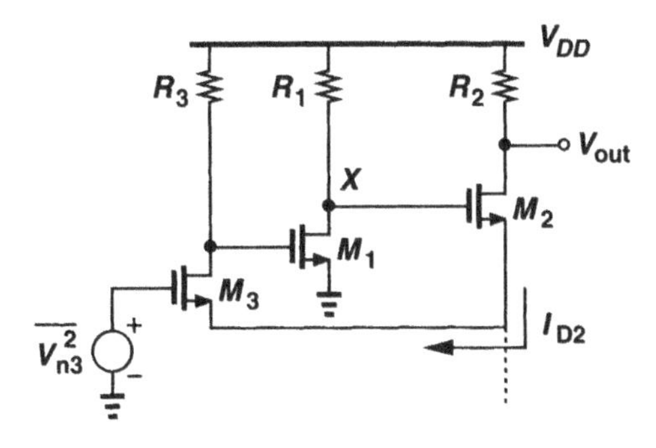

Figure 4.68 Circuit for analyzing noise in feedforward TIA.

To study the noise behavior of the TIA, we begin with the noise of M_3 and model it by a voltage source, $\overline{V_{n3}^2} = 4kT\gamma/g_{m3}$, in series with its gate (Fig. 4.68). To compute the noise current flowing through M_2 due to V_{n3}, we note that $I_{D3} = -I_{D2}$ and hence $V_{GS3} = -I_{D2}/g_{m3}$. Thus, the voltage at the source of M_2 is equal to $V_{n3} + I_{D2}/g_{m3}$. Also, the voltage at the drain of M_3 is given by $-I_{D3}R_3 = I_{D2}R_3/g_{m3}$. Multiplying this voltage by $-g_{m1}R_1$, subtracting from it the source voltage of M_2, and multiplying the result by g_{m2}, we obtain I_{D2}:

$$\left[+I_{D2}R_3(-g_{m1}R_1) - \left(V_{n3} + \frac{I_{D2}}{g_{m3}}\right)\right] g_{m2} = I_{D2}. \tag{4.206}$$

It follows that

$$I_{D2} = \frac{-V_{n3}}{\dfrac{1}{g_{m2}} + \dfrac{1}{g_{m3}} + g_{m1}R_1R_3} \tag{4.207}$$

$$= \frac{-g_{m2}g_{m3}V_{n3}}{g_{m3} + (1 + g_{m3}R_3g_{m1}R_1)g_{m2}}. \tag{4.208}$$

The output noise voltage is equal to $-I_{D2}R_2$.

Dividing the output noise voltage, $-I_{D2}R_2$, by the transimpedance gain given by (4.204), we obtain the input noise current due to M_3:

$$I_{n,in}|_{M3} = \frac{g_{m3}V_{n3}}{1 + g_{m3}R_3g_{m1}R_1} \tag{4.209}$$

$$\approx \frac{V_{n3}}{R_3g_{m1}R_1}. \tag{4.210}$$

The input contribution of M_3 is thus heavily attenuated.

Let us now consider the noise of R_3, M_1, R_1, and M_2. As illustrated in Fig. 4.69(a), all four noise sources can be referred to the gate of M_1 as

$$\overline{V_{n,eq}^2} = 4kTR_3 + \frac{4kT\gamma}{g_{m1}} + \frac{4kTR_1 + 4kT\gamma/g_{m2}}{g_{m1}^2R_1^2}. \tag{4.211}$$

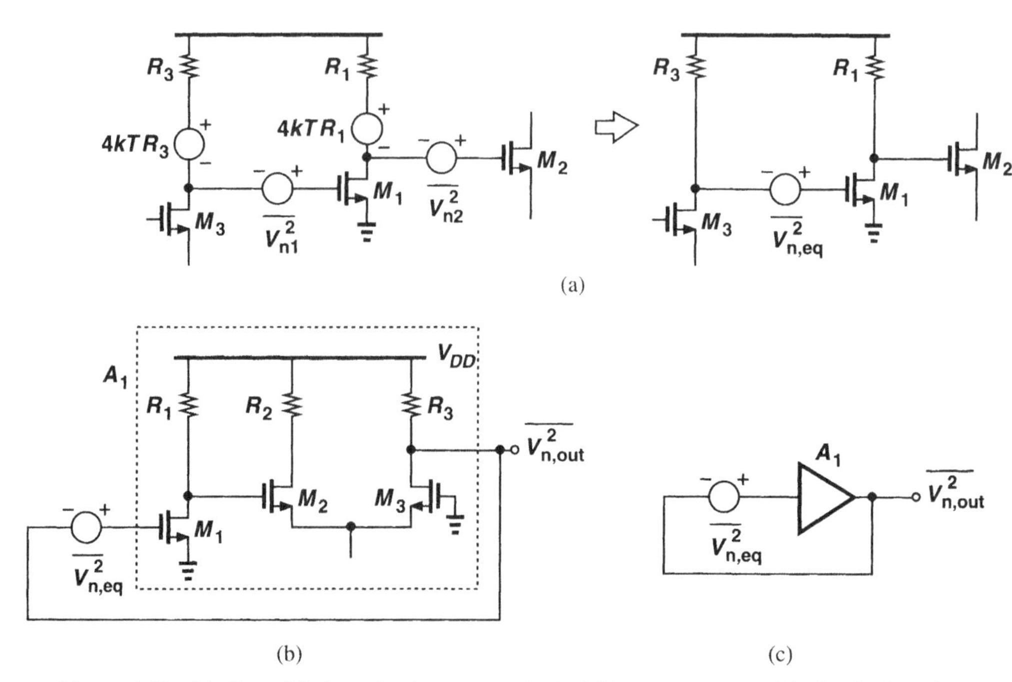

Figure 4.69 (a) Consolidation of noise sources in the TIA to one source, (b) circuit viewed as a feedback amplifier, (c) simplified model.

We must now see how this noise voltage contributes to the output. To this end, we redraw the TIA circuit as shown in Fig. 4.69(b), recognizing that the circuit in the dashed box is a voltage amplifier. Using the model in Fig. 4.69(c), we have $V_{n,out} \approx V_{n,eq}$ if $A_1 \gg 1$. Dividing this noise voltage by R_3 and noting that M_2 and M_3 carry equal and opposite small-signal currents, we obtain the noise voltage at the drain of M_2 as $(V_{n,eq}^2/R_3^2)R_2^2$. Dividing this noise by the transimpedance gain yields

$$\overline{I_{n,in}^2}|_{R3,M1,R1,M2} = \frac{4kTR_3}{R_2^2} + \frac{4kT\gamma}{g_{m1}R_2^2} + \frac{4kTR_1 + 4kT\gamma/g_{m2}}{g_{m1}^2 R_1^2 R_2^2} \times \frac{R_2^2}{R_3^2}. \tag{4.212}$$

We note from the first term that the noise of R_3 is directly referred to the input. Also, the second term may be significant if $g_{m1}R_3$ is not much greater than unity.

The TIA of Fig. 4.67 contains two other sources of noise. The noise current of R_2 is directly referred to the input, and the noise of the bias current source, I_B, is also added to the input. Since I_B must sink the currents of two common-gate stages, its noise current is greater than that of its counterpart in a simple CG stage.

References

1. B. Razavi, *Design of Analog CMOS Integrated Circuits,* Chicago: McGraw-Hill, 2001.

2. M. Neuhauser, H.-M. Rein, and H. Wernz, "Low-Noise High-Gain Si-Bipolar Preamplifiers for 10-Gb/s Optical Fiber Links – Design and Realization," *IEEE Journal of Solid-State Circuits,* vol. 31, pp. 24–29, Jan. 1996.
3. H. Korramabadi, L. D. Tzeng, and M. J. Tarsia, "A 1.05-Gb/s –31-dBm to 0-dBm BiCMOS Optical Preamplifier Featuring Adaptive Transimpedance," *ISSCC Dig. of Tech. Papers*, pp. 143–144, Feb. 1996.
4. R. G. Meyer and R. A. Blauschild, "A Wideband Low-Noise Monolithic Transimpedance Amplifier," *IEEE Journal of Solid-State Circuits,* vol. 21, pp. 530–533, Aug. 1986.
5. S. S. Taylor and T. P. Thomas, "A 2pA/$\sqrt{\text{Hz}}$ 622Mb/s GaAs MESFET Transimpedance Amplifier," *ISSCC Dig. of Tech. Papers*, pp. 254–255, Feb. 1994.
6. R. G. Meyer et al., "A 4-Terminal Wideband Monolithic Amplifier," *IEEE Journal of Solid-State Circuits,* vol. 16, pp. 634–639, Dec. 1981.
7. C.-F. Liao and S.-I. Liu, "40 Gb/s transimpedance-AGC amplifier and CDR circuit for b road-band data receivers in 90 nm CMOS," *IEEE Journal of Solid-State Circuits,* vol. 43, pp. 642-655, March 2008.
8. C. Kromer et al, "A low-power 20-GHz 52-dB. transimpedance amplifier in 80-nm CMOS," *IEEE Journal of Solid-State Circuits,* vol. 39, pp. 885-894, June 2004.

CHAPTER 5

Limiting Amplifiers and Output Buffers

The signal produced by the front-end transimpedance amplifier usually suffers from a small amplitude, on the order of a few tens of millivolts for the minimum input current level. The TIA must therefore be followed by additional amplifying stages that boost the signal swing to logical levels, e.g., on the order of 500 mV. Limiting amplifiers (also called "limiters")[1] provide both a high voltage gain and large output swings.

In this chapter, we study the design of low-noise, broadband limiters. Following an overview of limiter requirements, we examine basic cascaded amplifier stages and their shortcomings. Next, we study broadband techniques such as active-inductor peaking, capacitive degeneration, and f_T doublers. We then describe the design of output buffers and distributed amplifiers (DAs).

5.1 General Considerations

5.1.1 Performance Parameters

Limiters used in optical receivers must satisfy stringent requirements. Consider the typical arrangement illustrated in Fig. 5.1. Here, the limiter amplifies the TIA output and delivers large voltage swings to 50-Ω loads. The following aspects of the limiter are therefore important.

Input Capacitance The limiter must exhibit a sufficiently low input capacitance so that it does not reduce the TIA bandwidth significantly. At low supply voltages, it becomes increasingly difficult to interpose a buffer between the TIA and the limiter to avoid this loading effect, especially if noise is critical. Also, the first stage of the limiter suffers from direct trade-offs between its input capacitance, input noise, and voltage gain.

Bandwidth Recall from Chapter 4 that the bandwidth of TIAs is deliberately set to approximately 70% of the data rate so as to minimize the noise. Limiters, on the other

[1] For some reason, the optical communication literature uses the term "limiting amplifier" and the RF design community the term "limiter" to refer to the same circuit. We use these terms interchangeably here.

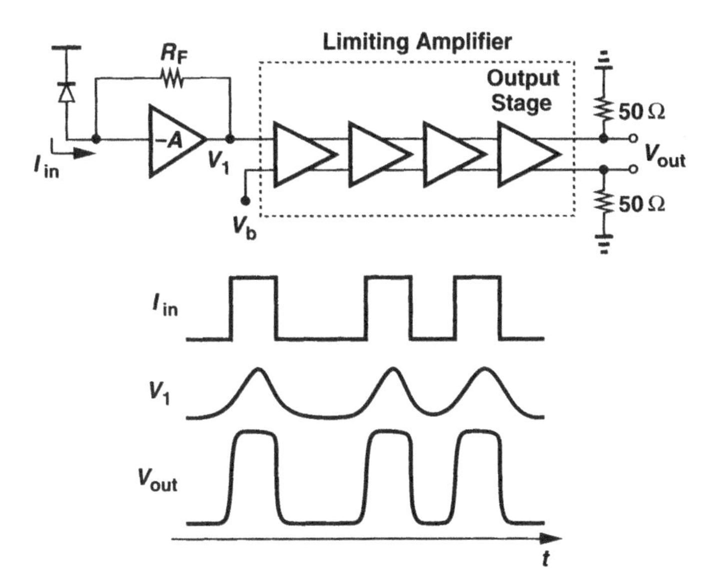

Figure 5.1 Role of a limiting amplifier in a receiver front end.

hand, are designed for a greater bandwidth—usually equal to the data rate. This is because, if two stages with equal bandwidths are cascaded, then the overall small-signal bandwidth is quite narrower. As shown in Fig. 5.1, the data transitions at the output of the TIA are relatively slow, but the limiter must amplify and clip the signal such that V_{out} exhibits a high slew rate and short rise and fall times.

Noise The input-referred noise of limiters is critical for two reasons: (1) their large bandwidth yields a greater integrated noise; and (2) the design of TIAs with a high trans-impedance gain becomes increasingly more difficult at high speeds, making the contribution of the limiter noise significant. Denoting the noise bandwidths of the TIA and the limiter by B_T and B_L, respectively, we express the overall input-referred noise current of the TIA/limiter cascade as

$$\overline{I_{n,in,tot}^2} = \overline{I_{n,T}^2} B_T + \frac{\overline{V_{n,L}^2} B_L}{R_T^2}, \tag{5.1}$$

where $I_{n,T}$ is the TIA input-referred noise current, $V_{n,L}$ is the limiter input-referred noise voltage, and R_T is the TIA gain. It is desirable to choose $V_{n,L}$ so small that the second term in (5.1) becomes negligible.

Gain The first stage of a limiting amplifier must provide enough gain so as to minimize the noise contributed by the following stages. While it is possible to design each of the subsequent stages for a low gain, the power dissipation and the total noise may still prove prohibitively high. For this reason, limiters typically employ only three to four stages.

Output Drive Capability In the arrangement of Fig. 5.1, the output stage must deliver large currents to the 50-Ω loads. For example, for a 1-V differential output swing, a current of 10 mA must be steered between the two resistors. The very large transistors required in the output driver both exhibit a large input capacitance and introduce a long time constant at the output nodes. Thus, the stage preceding the driver must tolerate a heavy capacitive load while achieving a wide band.

Jitter Limiting amplifiers may introduce jitter in the signal. It is desirable to maintain the limiter jitter below a few percent of the bit period.

Offset Voltage The differential stages shown in Fig. 5.1 suffer from a finite offset voltage, which, after experiencing a large gain in the limiter, may saturate the latter stages. For this reason, limiters usually incorporate offset cancellation.

5.1.2 Cascaded Gain Stages

As our first step toward understanding the properties of limiters, we study the small-signal and large-signal bandwidths of cascaded gain stages. Consider the circuit depicted in Fig. 5.2, where each stage is represented by an ideal voltage amplifier A_0, an output resistance

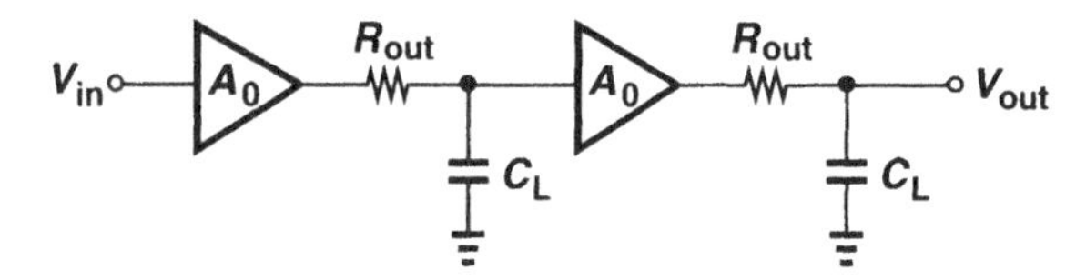

Figure 5.2 Cascade of small-signal amplifiers.

R_{out}, and a load capacitance C_L. The overall transfer function is given by

$$H(s) = \left(\frac{A_0}{1+\dfrac{s}{\omega_0}}\right)^2, \tag{5.2}$$

where $\omega_0 = (R_{out}C_L)^{-1}$ is the -3-dB bandwidth of each stage. To determine the maximum bit rate that can be amplified by the cascade with reasonable ISI, we employ the -3-dB bandwidth of the overall circuit as a measure of its speed. We must therefore assume $s = j\omega_{-3dB}$ and calculate the value of ω_{-3dB} such that

$$\left[\frac{A_0}{\sqrt{1+(\dfrac{\omega_{-3dB}}{\omega_0})^2}}\right]^2 = \left(\frac{A_0}{\sqrt{2}}\right)^2. \tag{5.3}$$

Thus,

$$\omega_{-3dB} = \omega_0\sqrt{\sqrt{2}-1} \tag{5.4}$$

$$\approx 0.644\omega_0, \tag{5.5}$$

revealing that cascading two identical stages decreases the bandwidth by about 36%. Similarly, for N identical stages, the -3-dB bandwidth is given by:

$$\omega_{-3dB} = \omega_0 \sqrt{\sqrt[N]{2} - 1}. \tag{5.6}$$

For $N \geq 2$, this expression can be approximated as:

$$\omega_{-3dB} \approx \omega_0 \frac{0.9}{\sqrt{N}}. \tag{5.7}$$

The foregoing observation suggests that each stage in a limiting amplifier must achieve a very wide band such that the overall bandwidth given by Eq. (5.7) is still on the order of R_b, where R_b denotes the bit rate. For example, if $N = 4$ and we choose $\omega_{-3dB} = 2\pi \times R_b$, then ω_0 must be at least equal to $\omega_0 \approx 2.3 \times 2\pi \times R_b$. In other words, the bandwidth of each stage must exceed $2.3R_b$.

From these calculations, we arrive at a gain-bandwidth trade-off: if each stage provides a small gain and a wide band, then N must be large so as to achieve the required gain and vice versa. Consequently, for a given overall gain, A_{tot}, an optimum value of N exists that yields the maximum overall bandwidth. In order to obtain this value, we assume the transfer function of each stage is expressed as

$$H(s) = \frac{\dfrac{B}{\omega_0}}{1 + \dfrac{s}{\omega_0}}, \tag{5.8}$$

where B denotes the gain-bandwidth product and is assumed to be constant for a given power dissipation in a given technology. Equation (5.8) embodies the gain-speed trade-off of each stage: if a higher ω_0 is chosen, then the gain, B/ω_0, is reduced proportionally.

For N identical stages, $A_{tot} = (B/\omega_0)^N$ and hence $\omega_0 = B/\sqrt[N]{A_{tot}}$. Substituting this value in Eq. (5.6), we have

$$\omega_{-3dB} = B \frac{\sqrt{\sqrt[N]{2} - 1}}{\sqrt[N]{A_{tot}}}, \tag{5.9}$$

which, with the approximation given by (5.7), reduces to

$$\omega_{-3dB} \approx B \frac{0.9}{\sqrt{N}\sqrt[N]{A_{tot}}}. \tag{5.10}$$

To minimize the denominator, $D = \sqrt{N}\sqrt[N]{A_{tot}}$, we take its natural logarithm: $\ln D = (1/2)\ln N + (1/N)\ln A_{tot}$, and differentiate with respect to N:

$$\frac{1}{D}\frac{\partial D}{\partial N} = \frac{1}{2N} - \frac{1}{N^2}\ln A_{tot}. \tag{5.11}$$

The derivative vanishes at

$$N_{opt} = 2 \ln A_{tot}. \tag{5.12}$$

Interestingly, N_{opt} is independent of the gain-bandwidth product, B. The corresponding bandwidth is thus equal to

$$\omega_{max} = \frac{0.9B}{\sqrt{2 \ln A_{tot}}} A_{tot}^{-\frac{1}{2 \ln A_{tot}}} \tag{5.13}$$

$$\approx \frac{0.636B}{\sqrt{\ln A_{tot}}} A_{tot}^{-\frac{1}{2 \ln A_{tot}}}. \tag{5.14}$$

It is instructive to study the behavior of ω_{-3dB} given by Eq. (5.10) as a function of N for a given A_{tot}. For example, if a limiting amplifier requires a gain of $A_{tot} = 100$ (40 dB), then the behavior shown in Fig. 5.3 is obtained. As predicted by (5.12), the bandwidth

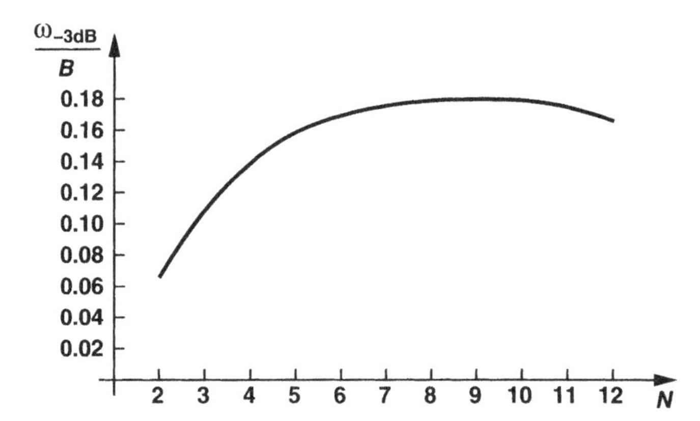

Figure 5.3 Normalized bandwidth as a function of N for $A_{tot} = 100$.

reaches a maximum at $N \approx 10$. However, the plot also reveals only an incremental change in ω_{-3dB}/B for $N > 5$. In fact, as N goes from 5 to 10, the bandwidth increases by less than 15%. Furthermore, with $N = 10$, the gain per stage is small, making the noise contributed by all of the stages significant. For these reasons, typical high-gain limiters employ no more than five gain stages.

Another important result of the above study is that the maximum bandwidth for a given gain of 100 is only $0.18B$. While this may imply that B must be on the order of $2\pi(5 \times R_b)$ so as to minimize ISI, in practice, the small-signal bandwidth need not be this large. This is because the second or third stage typically senses sufficiently large input swings, thereby experiencing switching and operating in the large-signal mode. Illustrated in Fig. 5.4 for a cascade of three differential pairs, this effect manifests itself in the drain currents of M_3-M_4 and, more visibly, M_5-M_6. Owing to relatively complete switching of the tail current

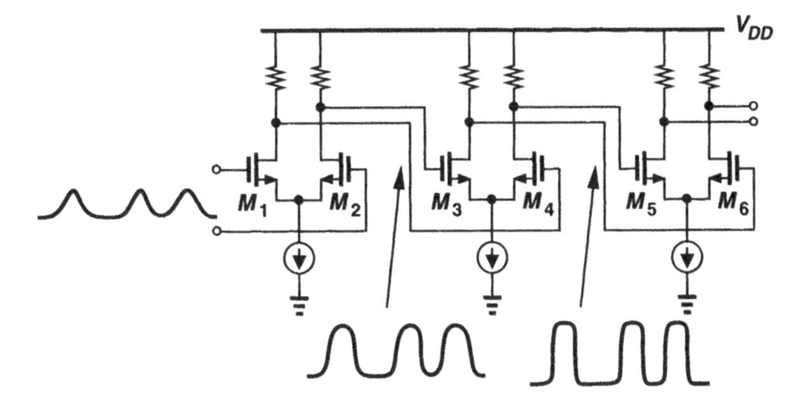

Figure 5.4 Effect of limiting on bandwidth.

and the finite delay of each stage, the last two differential pairs process the zero crossings of the data at different times, violating the assumptions behind Eq. (5.2).

To better understand why the small-signal bandwidth of limiting stages provides a conservative measure of the large-signal speed, let us consider the differential pair in Fig. 5.5, assuming that M_1 and M_2 sense a relatively large input swing and exhibit a high small-

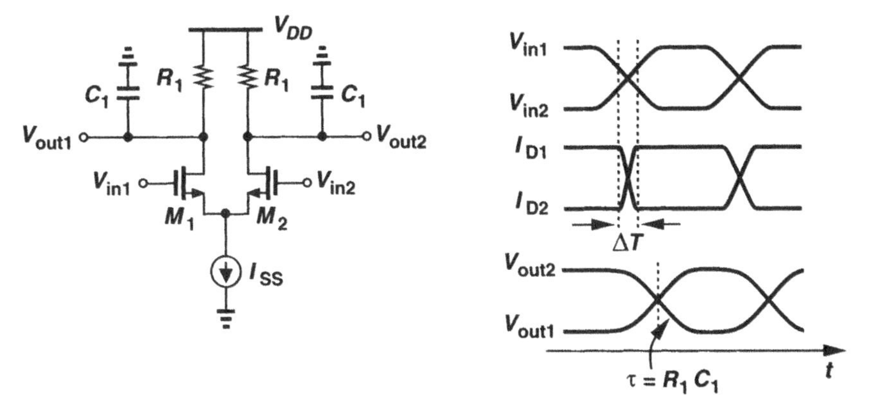

Figure 5.5 Delay through a differential pair.

signal transconductance in equilibrium.[2] Even though the input signals suffer from long transition times, the high transconductance enables M_1 and M_2 to steer I_{SS} during ΔT, sharpening the signal edges considerably. Upon flowing through the loads consisting of R_1 and C_1, I_{D1} and I_{D2} generate output voltage waveforms with a time constant of R_1C_1—as if the input signal had nearly zero transition times. We conclude that, in a cascade of identical stages operating with sufficiently large signals, the speed is limited by only that of one stage. This is, of course, a common effect in cascaded digital gates.

[2]The term "equilibrium" refers to the condition where the differential input voltage is around zero.

If an amplifier is to deliver signals to 50-Ω loads, then it must employ high currents and large output transistors in the last stage. The large input capacitance of this stage in turn requires that the preceding circuit provide a relatively low output impedance. Thus, as conceptually illustrated in Fig. 5.6, the cascade must be "tapered" in device dimensions and

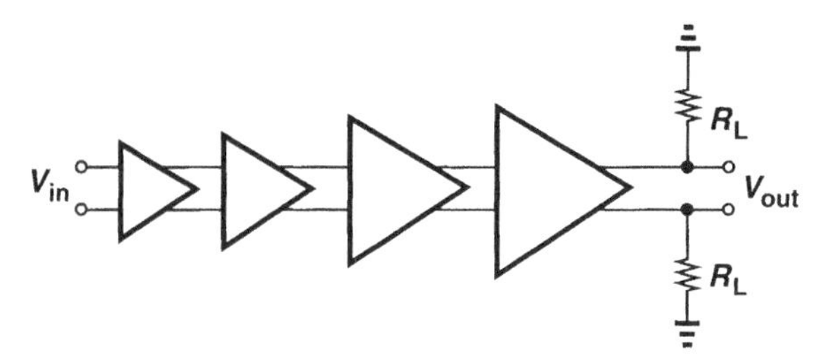

Figure 5.6 Tapered output buffer.

bias currents from the first stage to the last so as to maintain a wide band while delivering high output currents.

5.1.3 AM/PM Conversion

A nonlinear, dynamic system can potentially convert signal amplitude variations to phase disturbance. For example, a nonlinear, low-pass circuit converts amplitude modulation (AM) to phase modulation (PM).

To understand the origin of AM/PM conversion, let us consider the differential pair of Fig. 5.7, where $R_1 = R_2$ $(= R)$ and $C_1 = C_2$ $(= C)$ determine the bandwidth. For

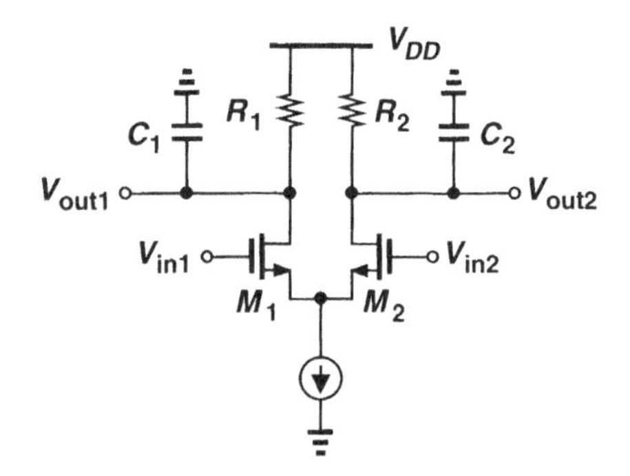

Figure 5.7 AM/PM conversion in a differential pair.

moderate input swings, the voltage-to-current characteristic of the differential pair can be approximated as $I_{out} \approx \alpha_1 V_{in} + \alpha_3 V_{in}^3$, where I_{out} and V_{in} are differential quantities. Now suppose $V_{in} = V_m \sin \omega t$ and hence

$$I_{out} = \alpha_1 V_m \sin \omega t + \alpha_3 V_m^3 \sin^3 \omega t \tag{5.15}$$

$$= (\alpha_1 V_m + \frac{3}{4}\alpha_3 V_m^3) \sin \omega t - \frac{3}{4}\alpha_3 V_m^3 \sin 3\omega t. \tag{5.16}$$

Despite harmonic distortion, I_{out} still crosses zero at $t = n\pi/\omega$, where n is an integer. In other words, variation of V_m (amplitude modulation) does not affect the phase of I_{out}. However, upon flowing through the frequency-dependent load, the first and third harmonics experience different phase shifts:

$$V_{out}(t) = A_1(\alpha_1 V_m + \frac{3}{4}\alpha_3 V_m^3)\sin(\omega t + \theta_1) - A_3\frac{3}{4}\alpha_3 V_m^3 \sin(3\omega t + \theta_2), \tag{5.17}$$

where $A_1 = R/\sqrt{1 + R^2C^2\omega^2}$, $\theta_1 = -\tan^{-1}(RC\omega)$, $A_3 = R/\sqrt{1 + 9R^2C^2\omega^2}$, and $\theta_2 = -\tan^{-1}(3RC\omega)$. The key point here is that, if V_m varies, the output zero-crossing points shift in time. Thus, if the input signal contains random noise on its amplitude, the output signal suffers from random shift in its zero crossings, i.e., jitter.

Before quantifying AM/PM conversion, we should note that the signal applied to limiters in an optical receiver may exhibit amplitude variations resulting from various sources. In fibers carrying several channels on different wavelengths, crosstalk among the signals introduces amplitude modulation. Furthermore, nonidealities such as package parasitics and limited bandwidth create ISI and hence amplitude variation.

To estimate the jitter resulting from AM/PM conversion, we obtain the dependence of signal delay upon the input amplitude. If the output zero crossings deviate from their ideal points in time by Δt, then we can substitute $t = n\pi/\omega + \Delta t$ in Eq. (5.17) and, assuming a small Δt, obtain

$$\begin{aligned} V_{out}(\frac{n\pi}{\omega} + \Delta t) &= A_1(\alpha_1 V_m + \frac{3}{4}\alpha_3 V_m^3)(\sin\theta_1 + \omega\cdot\Delta t\cos\theta_1) \\ &\quad - A_3\frac{3}{4}\alpha_3 V_m^3(\sin\theta_2 + 3\omega\cdot\Delta t\cos\theta_2). \end{aligned} \tag{5.18}$$

Setting the right-hand side to zero, we have:

$$\Delta t = \frac{-A_1(\alpha_1 + \frac{3}{4}\alpha_3 V_m^2)\sin\theta_1 + A_3\frac{3}{4}\alpha_3 V_m^2\sin\theta_2}{A_1(\alpha_1 + \frac{3}{4}\alpha_3 V_m^2)\omega\cdot\cos\theta_1 - A_3\frac{3}{4}\alpha_3 V_m^2(3\omega)\cos\theta_2}. \tag{5.19}$$

To arrive at a simpler result, let us assume that the second term in the denominator of (5.19) is much less than the first. Thus,

$$\Delta t \approx -\frac{\tan\theta_1}{\omega} + \frac{1}{\omega}\frac{A_3}{A_1}\frac{\frac{3}{4}\alpha_3 V_m^2}{\alpha_1 + \frac{3}{4}\alpha_3 V_m^2}\frac{\sin\theta_2}{\cos\theta_1}, \tag{5.20}$$

or, if $\alpha_1 \gg (3/4)\alpha_3 V_m^2$,

$$\Delta t \approx -\frac{\tan\theta_1}{\omega} + \frac{3}{4}\frac{1}{\omega}\frac{A_3}{A_1}\frac{\alpha_3}{\alpha_1}V_m^2\frac{\sin\theta_2}{\cos\theta_1}. \tag{5.21}$$

Note that the first term in (5.21) corresponds to the RC delay in the absence of nonlinearity, i.e., if α_3 or V_m are very small.[3] The second term reveals the dependence of Δt upon nonlinearity and the signal amplitude. For example, if the delay variation must remain less than 5% of the period, then we have

$$\frac{A_3}{A_1}\frac{\alpha_3}{\alpha_1}V_m^2\frac{\sin\theta_2}{\cos\theta_1} < 0.419. \tag{5.22}$$

In practice, the delay variation is measured for the minimum and maximum specified values of V_m. The result is quite conservative because the actual signal experiences much less amplitude noise.

5.2 Broadband Techniques

A cascade of simple differential pairs with resistive loads, e.g., as shown in Fig. 5.4, often proves inadequate as a broadband amplifier, especially if the input amplitude is small and the first several stages operate linearly. Many broadband techniques have been invented to allow higher speeds in a given IC technology.

5.2.1 Inductive Peaking

As described in Chapter 4, inductive peaking can substantially increase the bandwidth of gain stages. If chip area is critical, inductive peaking can also be realized by means of *active* devices. For example, consider the source follower shown in Fig. 5.8(a), where resistor R_S is placed in series with the gate of M_1. Proper choice of values can yield an

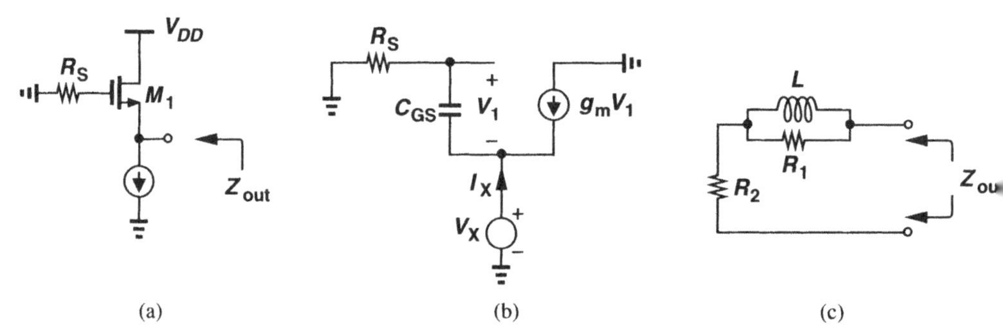

Figure 5.8 (a) Source follower providing an inductive output, (b) equivalent circuit, (c) simplified network.

inductive output impedance. Neglecting the gate-drain overlap capacitance, the source-bulk capacitance, channel-length modulation, and body effect, we construct the small-signal equivalent circuit depicted in Fig. 5.8(b), writing

$$V_1 C_{GS} s + g_m V_1 = -I_X. \tag{5.23}$$

[3]The actual delay in this case is given by $-\theta_1/\omega$. The slight discrepancy is left for the reader to resolve.

Since $V_1 C_{GS} s R_S + V_1 = -V_X$, we have

$$Z_{out} = \frac{R_S C_{GS} s + 1}{g_m + C_{GS} s}. \tag{5.24}$$

Note that $|Z_{out}(s = 0)| = 1/g_m$ and $|Z_{out}(s = \infty)| = R_S$. Thus, if $R_S \gg 1/g_m$, then $|Z_{out}|$ *increases* with frequency, exhibiting an inductive behavior. It can be shown [1] that under this condition the output impedance is modeled as illustrated in Fig. 5.8(c), where $R_1 = R_S - 1/g_m$, $R_2 = 1/g_m$, and

$$L = \frac{C_{GS}}{g_m}(R_S - \frac{1}{g_m}). \tag{5.25}$$

In order to obtain a high-quality inductor, we must maximize R_1 and minimize R_2. However, since the Q of the parallel combination is given by $R_1/(L\omega)$ and since $R_1/L = g_m/C_{GS}$, the value of R_S does not impact the Q significantly. Thus, the transistor transconductance plays the principal role in the circuit.

The primary drawback of the active inductor of Fig. 5.8(a) is the large voltage headroom consumption. To maximize the transconductance without introducing substantial source-bulk and gate-drain overlap capacitance, the bias current of the device must be maximized, thereby leading to a large gate-source voltage. At supply voltages below 1.5 V, it may simply be impossible to utilize this circuit as the load of a differential pair.

Figure 5.9(a) shows another example, where the load exhibits an impedance of

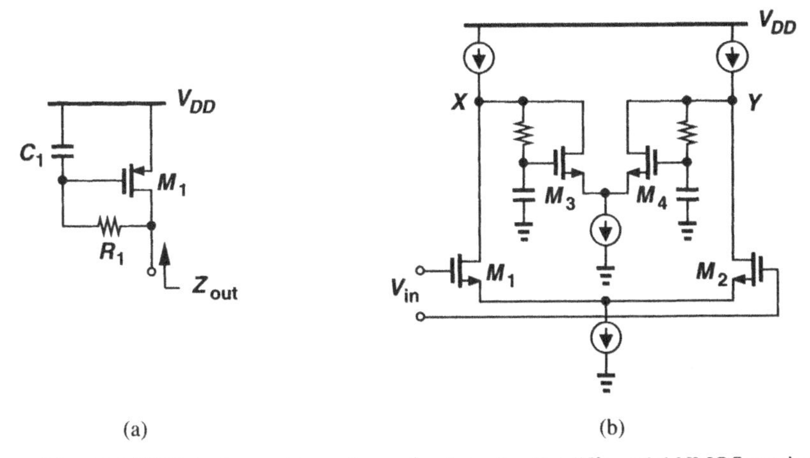

Figure 5.9 (a) PMOS device configured as active inductor, (b) differential NMOS version of (a).

$(1/g_{m1})||r_{O1}$ at low frequencies and $R_G||r_{O1}$ at high frequencies, providing inductive behavior if $1/g_{m1} < R_1$. As with the topology in Fig. 5.8(a), this type of load severely limits the voltage headroom.

Figure 5.9(b) depicts a modification where the active inductors are realized by NMOS devices to allow low-voltage operation. In practice, however, the large capacitances introduced at nodes X and Y by the load current sources and M_3-$M4$ limit the bandwidth to well below that achievable with passive inductors.

5.2.2 Capacitive Degeneration

In order to create a broadband response, it is possible to degenerate the transistors in a differential pair such that their effective transconductance *increases* at high frequencies, thereby compensating for the gain roll-off resulting from the pole at the output node. Shown in Fig. 5.10(a), such an arrangement employs both capacitive and resistive degen-

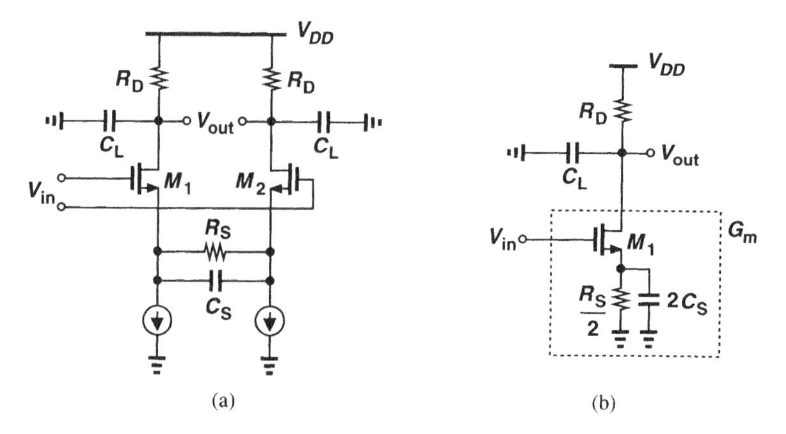

Figure 5.10 (a) Differential pair with capacitive degeneration, (b) half-circuit equivalent .

eration. Using the half circuit of Fig. 5.10(b), we express the equivalent transconductance as

$$G_m = \frac{g_m}{1 + g_m(\frac{R_S}{2} || \frac{1}{2C_S s})} \tag{5.26}$$

$$= \frac{g_m(R_S C_S s + 1)}{R_S C_S s + 1 + g_m R_S/2}. \tag{5.27}$$

The transconductance therefore contains a zero at $1/(R_S C_S)$ and a pole at $(1 + g_m R_S/2)/(R_S C_S)$ [Fig. 5.11(a)]. If the zero cancels the pole at the drain, i.e., if $R_S C_S =$

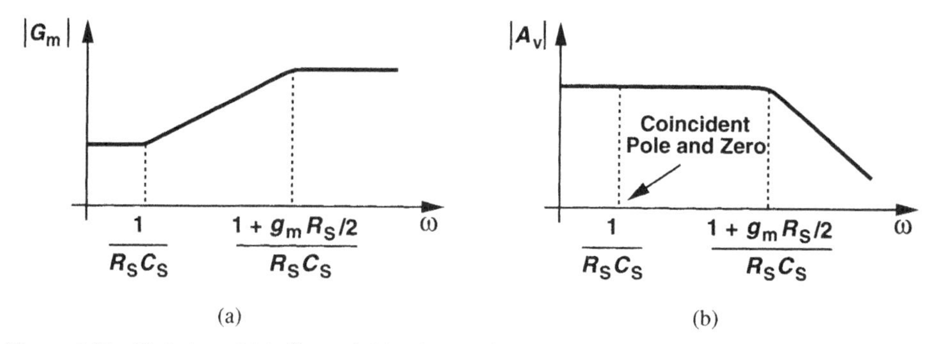

Figure 5.11 Variation of (a) G_m and (b) voltage gain with frequency.

$R_D C_L$, then the overall amplifier's bandwidth is extended to $(1 + g_m R_S/2)/(R_S C_S) = (1 + g_m R_S/2)/(R_D C_L)$. In other words, as illustrated in Fig. 5.11(b), the speed is increased by a factor of $1 + g_m R_S/2$. This is, of course, obtained at the cost of a proportional reduction in the gain because $A_v = g_m R_D/(1 + g_m R_S/2)$ at low frequencies. The thermal noise of R_S may also pose difficulties.

The reader may wonder if the above degeneration technique offers any advantage over simply reducing the load resistance by the same factor and thus increasing the bandwidth proportionally. An important benefit of capacitive/resistive degeneration stems from the change in the *input* impedance of the amplifier and hence the load seen by the preceding stage. To understand this point, we analyze the circuit of Fig. 5.10(b) while including the output resistance of the previous stage as R_G in series with the gate (Fig. 5.12). Since $g_m V_1 = I_{out}$, the current through C_{GS} is given by $(I_{out}/g_m)C_{GS}s$ and that through the

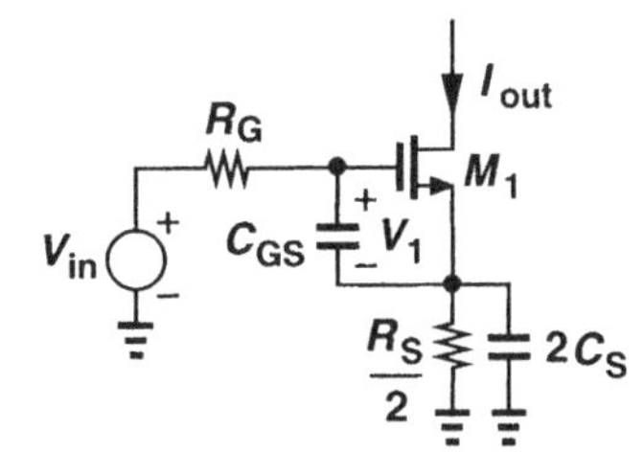

Figure 5.12 Input capacitance of a capacitively-degenerated stage.

parallel combination of $R_S/2$ and $2C_S$ by $(I_{out}/g_m)C_{GS}s + I_{out}$. Summing the voltage drops across R_G, C_{GS} and $(R_S/2)||(2C_S s)^{-1}$ and equating the result to V_{in}, we have

$$\frac{I_{out}}{g_m}C_{GS}sR_G + \frac{I_{out}}{g_m}C_{GS}s + (\frac{I_{out}}{g_m}C_{GS}s + I_{out})\frac{R_S/2}{R_S C_S s + 1} = V_{in}. \tag{5.28}$$

It follows that

$$\frac{I_{out}}{V_{in}} = \frac{g_m(R_S C_S s + 1)}{R_G C_{GS} R_S C_S s^2 + (R_G C_{GS} + R_S C_S + R_S C_{GS}/2)s + 1 + g_m R_S/2}. \tag{5.29}$$

Note that the zero still appears at $1/(R_S C_S)$. To estimate the poles, we write

$$(\frac{s}{\omega_{p1}} + 1)(\frac{s}{\omega_{p2}} + 1) = \frac{s^2}{\omega_{p1}\omega_{p2}} + (\frac{1}{\omega_{p1}} + \frac{1}{\omega_{p2}})s + 1 \tag{5.30}$$

and assume $\omega_{p1} \ll \omega_{p2}$, obtaining

$$(\frac{s}{\omega_{p1}} + 1)(\frac{s}{\omega_{p2}} + 1) \approx \frac{s^2}{\omega_{p1}\omega_{p2}} + \frac{1}{\omega_{p1}}s + 1. \tag{5.31}$$

From (5.29) and (5.31), we have

$$\omega_{p1} \approx \frac{1 + g_m R_S/2}{R_G C_{GS} + R_S C_S + R_S C_{GS}/2} \tag{5.32}$$

$$\omega_{p2} \approx \frac{1}{R_S C_S} + \frac{1}{R_G C_{GS}} + \frac{1}{2 R_G C_S}. \tag{5.33}$$

Since $g_m R_S$ is usually below 5 (to avoid reducing the gain of the stages excessively), the assumption $\omega_{p1} \ll \omega_{p2}$ is reasonable: ω_{p1} is inversely proportional to the sum of three *time constants* whereas ω_{p2} is equal to the sum of three *pole magnitudes*.

The key observation here is that, if $R_G C_{GS} \gg R_S(C_S + C_{GS}/2)$, then

$$\omega_{p1} \approx \frac{1 + g_m R_S/2}{R_G C_{GS}}; \tag{5.34}$$

that is, the input pole magnitude is increased by a factor of $1 + g_m R_S/2$. Figure 5.13 summarizes the performance of three stages with and without degeneration, assuming a

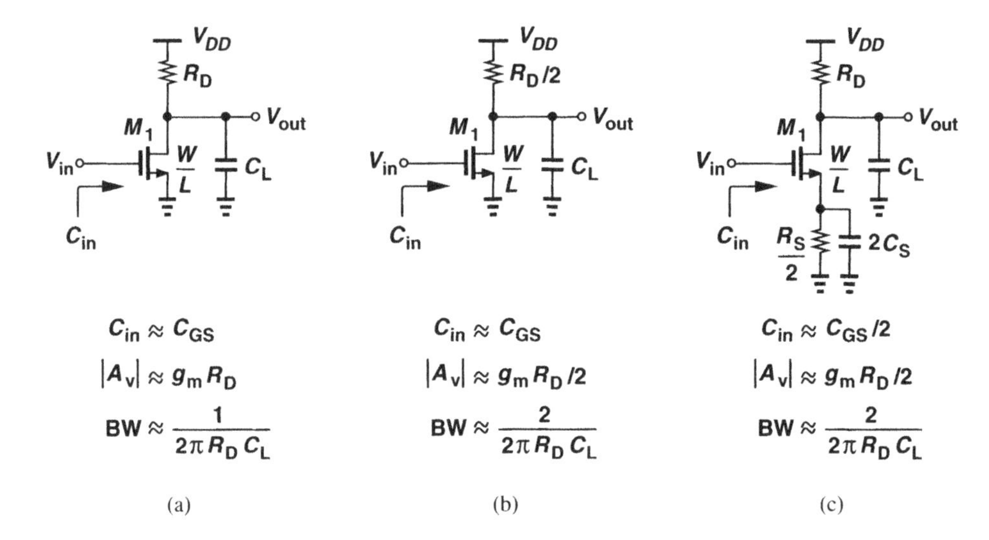

Figure 5.13 Three stages with and without capacitive degeneration.

$g_m R_S$ of 2, equal bias currents, and negligible Miller effect.

In a cascade of differential pairs, it is possible to allow the zero resulting from capacitive degeneration to cancel the pole due to the gate-source capacitance of the transistors themselves. That is, ω_{p1} is made equal to $1/(R_S C_S)$:

$$\frac{1 + g_m R_S/2}{R_G C_{GS} + R_S C_S + R_S C_{GS}/2} = \frac{1}{R_S C_S}. \tag{5.35}$$

It follows that

$$\frac{g_m R_S^2}{2} C_S = R_G C_{GS} + R_S \frac{C_{GS}}{2}, \tag{5.36}$$

and hence

$$\frac{C_S}{C_{GS}} = \frac{1}{g_m R_S}\left(\frac{2R_G}{R_S} + 1\right). \tag{5.37}$$

Under this condition, ω_{p2} and the drain-substrate capacitance limit the bandwidth.

5.2.3 Cherry-Hooper Amplifier

The Cherry-Hooper amplifier incorporates local feedback in the drain (or collector) network to improve the speed. To arrive at this topology, let us first consider the cascade of common-source stages shown in Fig. 5.14(a), where C_X denotes the total capacitance seen

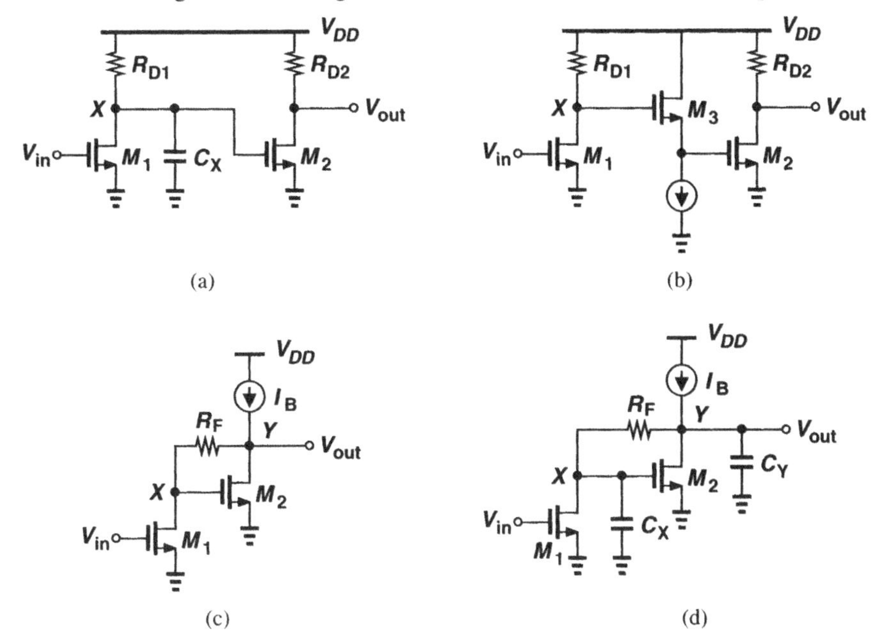

Figure 5.14 (a) Cascade of two CS stages, (b) use of source follower as buffer, (c) two stages with feedback, (d) circuit of (c) with node capacitances.

from node X to ground. The voltage gain and bandwidth of the first stage directly trade with each other: $|A_v| = g_{m1}R_{D1}$ and $\omega_{p,X} = (R_{D1}C_X)^{-1}$. Since M_2 must be wide enough to allow a reasonable gain in the second stage, C_{GS2} and the Miller effect of C_{GD2} may severely limit the bandwidth at node X.

It is possible to interpose a source follower between the two stages, as shown in Fig. 5.14(b), thereby isolating node X from the input capacitance of the second CS stage. However, the follower consumes substantial voltage headroom, limiting the allowable bias voltage across R_{D1} and hence the voltage gain. The follower may also attenuate the signal by as much as a factor of two if body effect and channel-length modulation are significant.

Now consider the topology depicted in Fig. 5.14(c). Here, R_F establishes feedback around M_2 by sensing the output voltage and returning a proportional current to X. Note that the circuit contains two paths to the output: one through M_2 and another through R_F. It is desirable to make the signal flowing through R_F negligible because it opposes that generated by M_2. Let us first calculate the low-frequency voltage gain of the circuit, assuming I_B is ideal and channel-length modulation is negligible. Given by $g_{m1}V_{in}$, the small-signal drain current of M_1 flows through R_F, creating a voltage drop across it equal to $g_{m1}V_{in}R_F$. Thus,

$$V_{out} - g_{m1}V_{in}R_F = V_X. \tag{5.38}$$

The small-signal current produced by M_2 $(= g_{m2}V_X)$ must flow through R_F:

$$g_{m2}(V_{out} - g_{m1}V_{in}R_F) = -g_{m1}V_{in}. \tag{5.39}$$

It follows that

$$\frac{V_{out}}{V_{in}} = g_{m1}R_F - \frac{g_{m1}}{g_{m2}}. \tag{5.40}$$

If $R_F \gg g_{m2}^{-1}$, then the gain is equal to that of a simple CS stage having a load resistance of R_F.

The key advantage of the circuit of Fig. 5.14(c) over those in Figs. 5.14(a) and (b) lies in the small-signal resistance seen at nodes X and Y. The reader can prove that this resistance is equal to g_{m2}^{-1} for both nodes, a value typically much less than R_F. Thus, if the capacitances at these nodes are modeled as shown in Fig. 5.14(d), then the pole frequencies are on the order of $\omega_{p,X} \approx g_{m2}/C_X$ and $\omega_{p,Y} \approx g_{m2}/C_Y$, much higher than those of the circuits in Figs. 5.14(a) and (b). For example, if we assume $C_X \approx C_{GS2}$, then $\omega_{p,X} \approx g_{m2}/C_{GS2} \approx 2\pi f_{T2}$, where f_{T2} denotes the unity current gain frequency ("transit" frequency) of M_2.

In summary, the circuit of Fig. 5.14(c) provides a voltage gain of approximately $g_{m1}R_F$ but with low resistance levels at nodes X and Y, yielding only high-frequency poles. Called the Cherry-Hooper amplifier [2], this topology has been extensively used in broadband applications.[4]

While intuitively appealing, the approximations $\omega_{p,X} \approx g_{m2}/C_X$ and $\omega_{p,Y} \approx g_{m2}/C_Y$ are inaccurate. This is because, at high frequencies, C_Y shunts the output node, thereby lowering the loop gain and *increasing* the impedance seen at the gate of M_2. Similarly, C_X shunts node X, yielding a higher output impedance. We must therefore analyze the circuit more carefully. Using the equivalent circuit shown in Fig. 5.15, where $I_{in} = g_{m1}V_{in}$, we note that the parallel combination of R_F and C_{GD2} carries a current equal to $I_{in} + V_X C_X s$, sustaining a voltage drop of $(I_{in} + V_X C_X s)R_F/(R_F C_{GD2}s + 1)$. Adding this drop to V_X and equating the result to V_{out}, we have

$$(I_{in} + V_X C_X s)\frac{R_F}{R_F C_{GD2}s + 1} + V_X = V_{out}. \tag{5.41}$$

[4]In the original Cherry-Hooper amplifier, the load of the second stage is realized as a resistor rather than a current source.

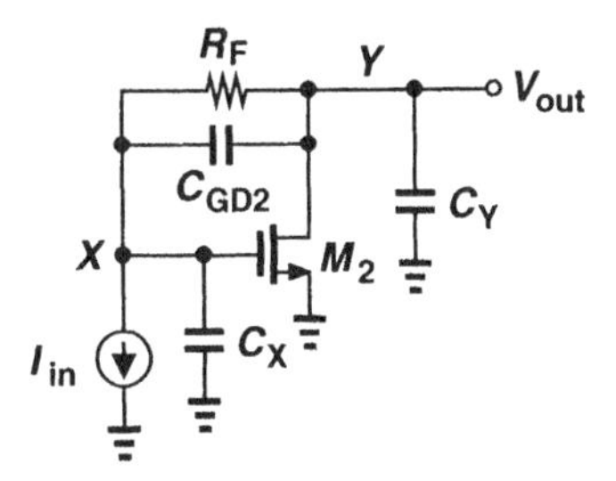

Figure 5.15 Equivalent circuit of Cherry-Hooper amplifier.

Also, summing the currents at the output node yields

$$-V_{out}C_Y s - g_{m2}V_X = I_{in} + V_X C_X s, \tag{5.42}$$

and hence

$$V_X = \frac{-V_{out}C_Y s - I_{in}}{g_{m2} + C_X s}. \tag{5.43}$$

Substituting for V_X in (5.41), we obtain

$$\frac{V_{out}}{I_{in}} = \frac{R_F^2 C_X C_{GD2} s^2 + (g_{m2}R_F C_{GD2} + C_X - C_{GD2})R_F s + g_{m2}R_F - 1}{R_F(C_X C_Y + C_{GD2}C_Y + C_{GD2}C_X)s^2 + (C_Y + g_{m2}R_F C_{GD2} + C_X)s + g_{m2}}. \tag{5.44}$$

This equation does not offer much insight but, as a special case, suppose the two poles are *equal*. The reader can then prove that

$$\omega_{p1} = \omega_{p2} = \frac{2g_{m2}}{C_X + C_Y + g_{m2}R_F C_{GD2}}. \tag{5.45}$$

If the third term in the denominator is negligible, the pole frequency is given by g_{m2} and the *average* value of C_X and C_Y. Thus, ω_{p1} and ω_{p2} are still much higher than those without feedback, e.g., $(R_F C_X)^{-1}$ or $(R_F C_Y)^{-1}$.

Used in limiters, the differential version of the Cherry-Hooper amplifier assumes the configurations shown in Fig. 5.16(a).[5] Since PMOS or *pnp* current sources typically contribute substantial capacitance to the output nodes, I_1 and I_2 are often replaced by resistors [Fig. 5.16(b)].

While allowing high speeds, the Cherry-Hooper amplifier faces difficulties at low supply voltages. In Fig. 5.16(b), for example, I_{SS1} must flow through the feedback resistors and $I_{SS1}+I_{SS2}$ through the load resistors. The minimum supply voltage is therefore equal to

$$V_{DD,min} = \frac{(I_{SS1} + I_{SS2})}{2}R_D + \frac{I_{SS1}}{2}R_F + V_{GS3,4} + V_{ISS2}, \tag{5.46}$$

[5]This circuit requires common-mode feedback.

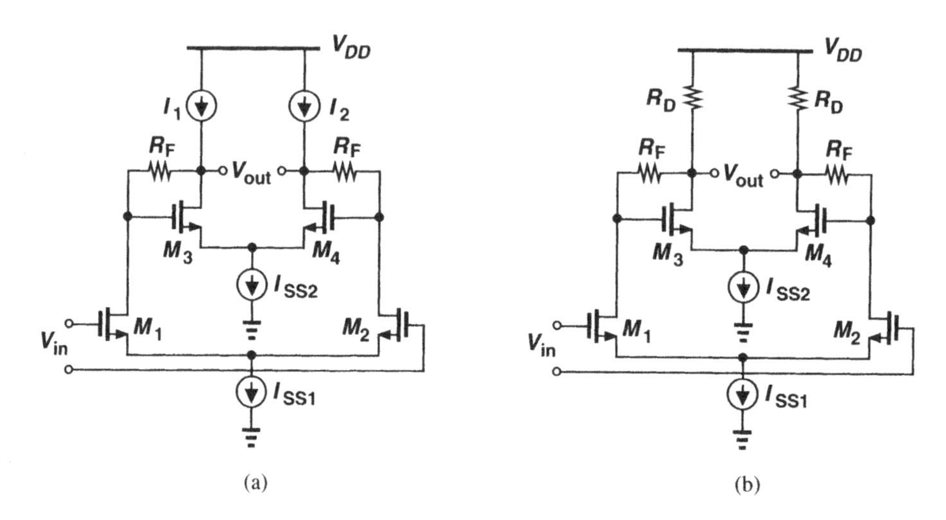

Figure 5.16 Differential Cherry-Hooper amplifier with (a) current-source loads and (b) resistive loads.

where V_{ISS2} denotes the minimum voltage required across I_{SS2}.[6] This constraint limits the voltage gain of the circuit.

To alleviate the gain-headroom trade-off, the circuit can be modified as shown in Fig. 5.17, where resistors or current sources provide part or all of the bias current of the input

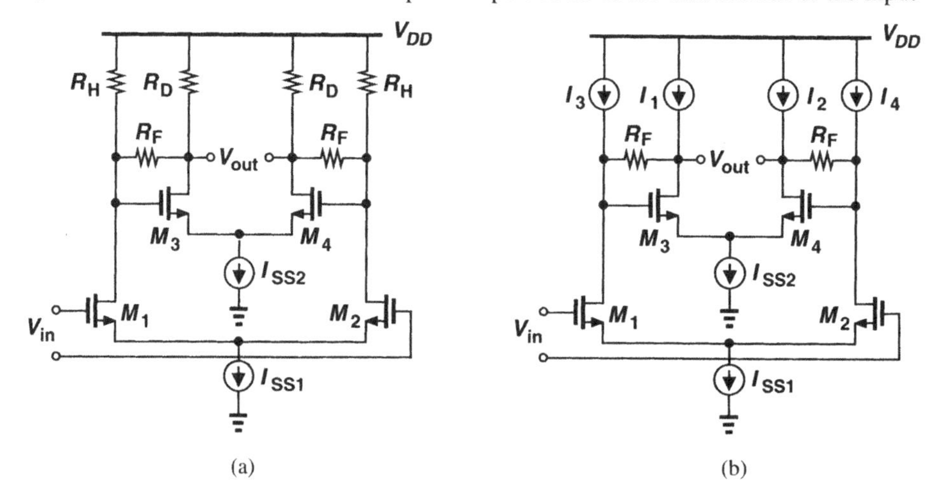

Figure 5.17 Modified Cherry-Hooper amplifier with (a) resistive loads and (b) current-source loads.

differential pair. In Fig. 5.17(a), R_H must be much greater than the input resistance of the second stage ($\approx g_{m2}^{-1}$ if R_D is large) to avoid degrading the gain. In Fig. 5.17(b), the current

[6] We assume $V_{GS3,4} > V_{GS1,2} - V_{TH1,2}$.

sources connected to the drains of M_1-M_4 may introduce substantial capacitance at these nodes.[7] This is because, with their low mobility, PMOS devices require a large width to carry the required currents while consuming a reasonable voltage headroom.

Figure 5.18 shows a variant of the Cherry-Hooper amplifier. Here, emitter followers are

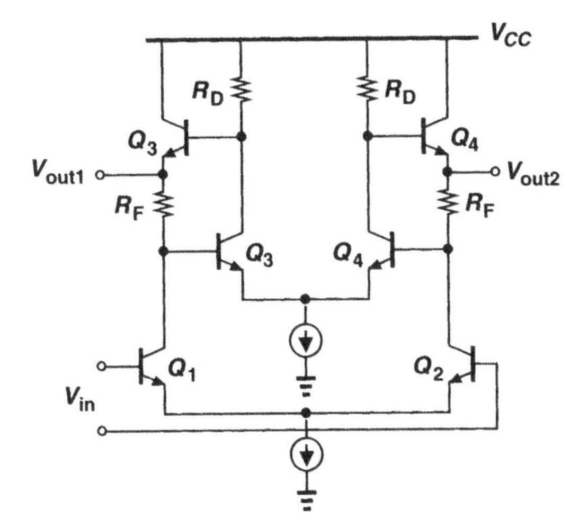

Figure 5.18 Cherry-Hooper amplifier with emitter followers in feedback loop.

inserted in the feedback path so as to drive the load capacitance. In addition to isolating the collectors of $Q3$ and $Q4$ from the load capacitance, the other advantage of this circuit is that feedback lowers the output impedance of the followers. However, the voltage headroom consumed by the followers prohibits the use of this configuration at low supply voltages.

It is also worth noting that the benefits of feedback in a Cherry-Hooper amplifier begin to diminish as the signal amplitude increases and the differential pair experiences large-signal operation. This means that the technique proves more useful in the first few stages of a limiting amplifier than in the last few.

5.2.4 f_T Doublers

The input capacitance of differential pairs in a cascade of stages is the principal limiting factor in achieving a broad bandwidth. As explained in Section 5.3, this issue becomes even more serious in output buffers that must deliver large currents to off-chip loads.

A method of reducing the input capacitance of differential pairs while maintaining the same gain is realized by "f_T doublers." Consider the circuit shown in Fig. 5.19(a), where the device dimensions and bias currents are chosen according to gain and voltage headroom requirements. We wish to modify the circuit such that the input capacitance decreases while the voltage gain remains unchanged. The small-signal behavior of the circuit is expressed as

$$V_{out} = g_m(V_{in1} - V_{in2})R_D, \tag{5.47}$$

[7] The circuit requires common-mode feedback.

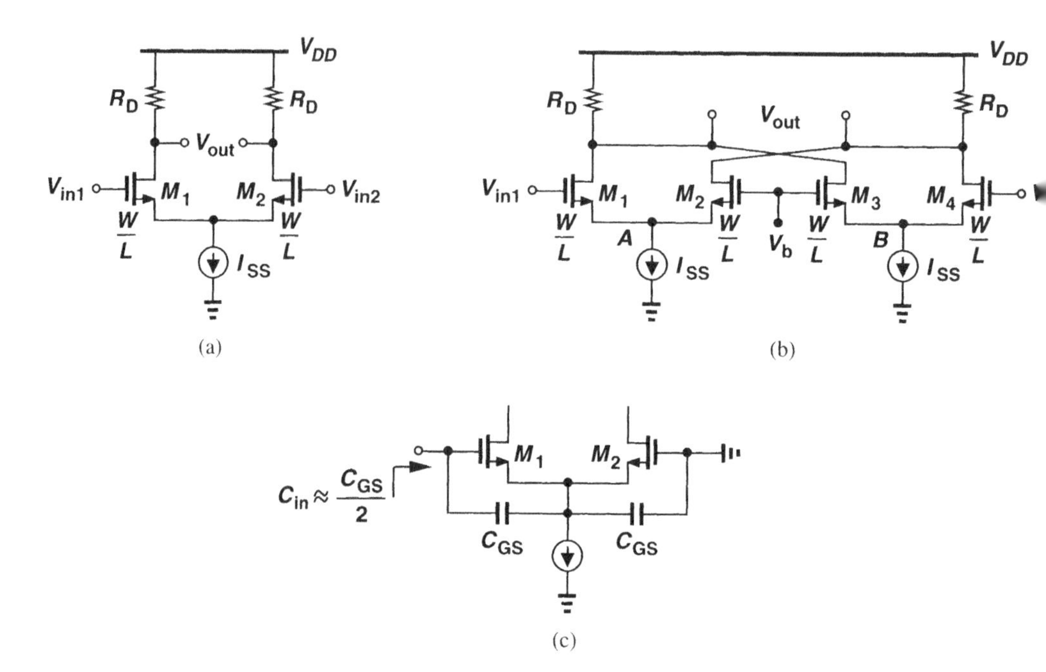

Figure 5.19 (a) Simple differential stage, (b) f_T doubler, (c) input capacitance of f_T doubler.

where g_m denotes the transconductance of each transistor. Now suppose two such differential pairs are configured as shown in Fig. 5.19(b), where the input ports are placed in *series* while the output ports are connected in parallel. (The load resistors are still equal to R_D.) The bias voltage V_b is chosen equal to the common-mode level of V_{in1} and V_{in2}, allowing the differential pairs to operate with zero systematic offset. Using superposition to calculate V_{out} in terms of V_{in1} and V_{in2}, we have

$$V_{out} = g_m(V_{in1} - V_{in2})R_D. \tag{5.48}$$

The circuit thus provides the same voltage gain but with a lower input capacitance. In fact, if the parasitic capacitance at nodes A and B is negligible, then the input capacitance seen by V_{in1} or V_{in2} is roughly equal to $C_{GS}/2$ [Fig. 5.19(c)]. Since this circuit halves the input capacitance while maintaining the same overall transconductance, it is called an f_T doubler.

The f_T doubler of Fig. 5.19(b) suffers from several drawbacks. First, the power dissipation is doubled.[8] Second, the total current flowing through the load resistors is doubled, possibly driving the transistors into the triode region. Third, the total capacitance contributed by the transistors to the *output* nodes is doubled, lowering the output pole. Fourth, if the source-bulk junction capacitance of the transistors and the capacitance introduced

[8] In bipolar implementations, it is easier to ensure that the two differential pairs experience full switching. The total bias current of such an f_T doubler need not be higher than that of a single differential pair.

by the tail current sources is not negligible, the input capacitance is higher than $C_{GS}/2$. Despite these issues, f_T doublers still prove useful in broadband output buffers (Section 5.3).

5.3 Output Buffers

Various subsystems of an optical fiber transceiver must drive off-chip loads at high speeds. For example, the SERDES output must travel on a printed-circuit board (PCB) to reach the laser driver (Fig. 5.20). To minimize reflections (which introduce ISI in random data),

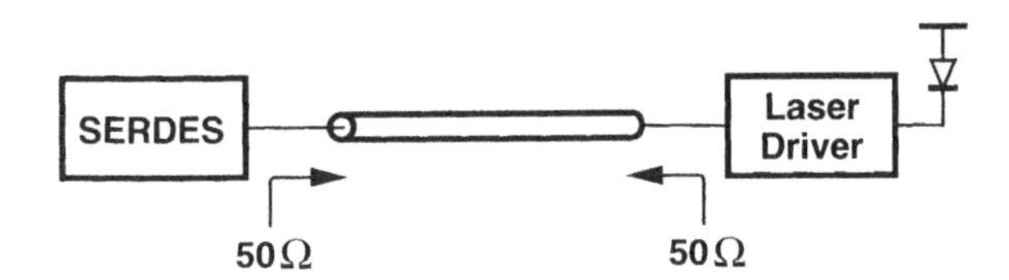

Figure 5.20 Connection of a SERDES and a laser driver through a transmission line.

the characteristic impedance of the trace carrying the signal is matched to that of the laser driver input (across the entire signal band). With typical impedance levels of 50 Ω, the SERDES must provide a large output current so as to produce reasonable voltage swings.

An output buffer topology commonly used in today's high-speed systems is the "open-drain" or "open-collector" differential pair. Illustrated in Fig. 5.21, the circuit produces a differential current that flows through differential transmission lines and is absorbed by two termination resistors at the far end. For example, a step of I_{SS} in I_{D1} and I_{D2} sees an impedance of $Z_0 = R_L$ at each of nodes A and B, immediately creating a voltage swing of $I_{SS}Z_0$. The voltage steps then travel to the far end, reaching the termination resistors after some delay ΔT. If $Z_0 = R_L$, then voltage and current equations are satisfied at the far end, producing no reflection (Chapter 2).

It is important to note that the differential pair exhibits a *high* output impedance, failing to provide impedance matching at the near-end interface. Nonetheless, if the input impedance seen at the far end is matched to that of the lines, the signal experiences no reflection and driving the line by open-drain transistors does not pose serious issues. We now study various aspects of this topology.

5.3.1 Differential Signaling

While requiring two pads and two package pins for every signal, differential transmission and reception offers many important benefits.

Noise Coupling The large transients generated by a differential circuit and coupled to sensitive nodes cancel each other to the first order. For example, the waveforms V_A and V_B in Fig. 5.21 may switch at very high speeds with a single-ended amplitude of 0.5 V, thereby coupling electrically and magnetically through bond wires to other signals. With differential operation and reasonable physical symmetry, the noise produced by "aggressors" is minimized.

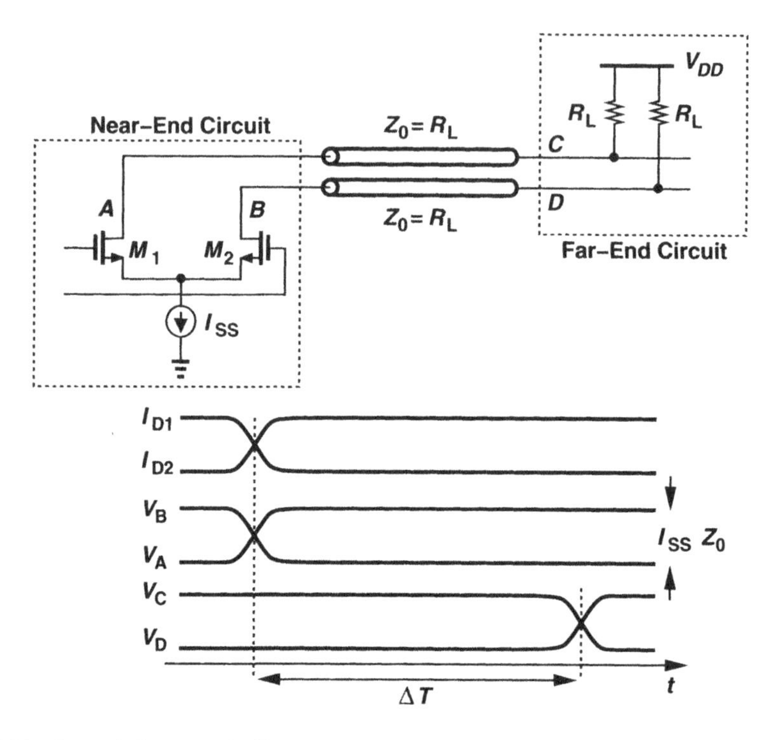

Figure 5.21 Open-drain output buffer.

The second advantage of differential operation stems from the noise coupling experienced by "victims." If processed and distributed differentially, sensitive signals receive the noise generated by aggressors as a *common-mode* disturbance, rejecting it by virtue of symmetry. The two properties are illustrated in Fig. 5.22.

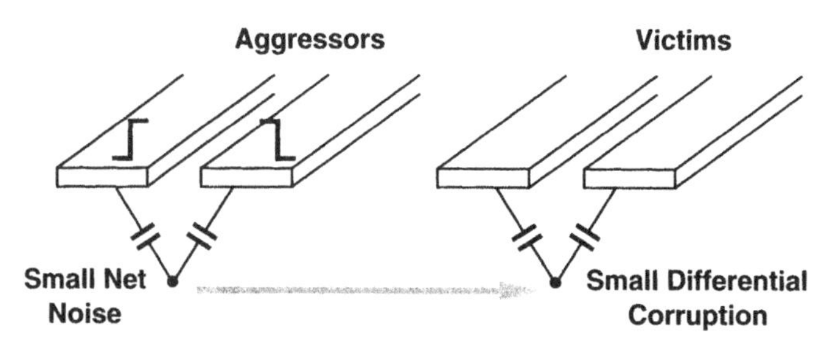

Figure 5.22 Noise generation and pickup by differential circuits.

In reality, mismatches between devices and connections do limit the above benefits. Nevertheless, with mismatches of, say, 1%, the differential noise produced by an aggressor is suppressed by 40 dB and that sensed by a victim is rejected by the same amount. In other words, ±0.5-V differential swings of an aggressor are reduced to ±0.05 mV when they appear on a victim signal.

Supply Noise Another important property of differential signaling relates to the transient currents drawn from the supply voltage. In Fig. 5.21, for example, the total current flowing through the termination resistors is approximately equal to I_{SS} at any point in time. Thus, even if V_{DD} suffers from a finite output impedance due to bond wire and package inductance, it experiences negligible transient changes. By contrast, if the signal were single-ended, e.g., if, as depicted in Fig. 5.23, only the drain current of M_1 or M_2 were

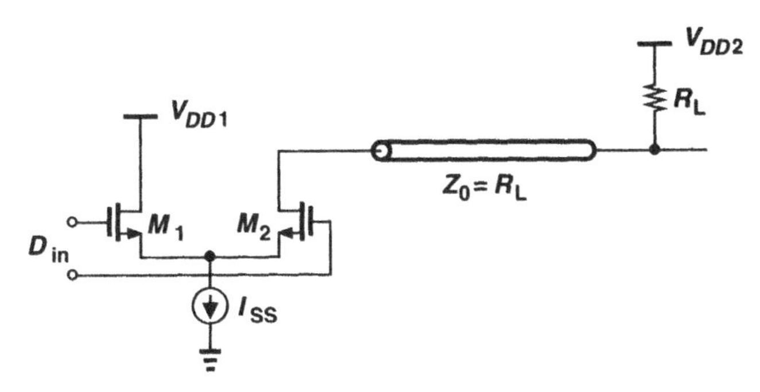

Figure 5.23 Open-drain buffer with single-ended output.

to drive the transmission line, each data pulse would create a large current change in both V_{DD1} and V_{DD2}, requiring extremely small parasitic inductance.

It is worth noting that even the differential topology of Fig. 5.21 introduces some transient currents in the supply. Illustrated in Fig. 5.24, this effect occurs as a result of the finite

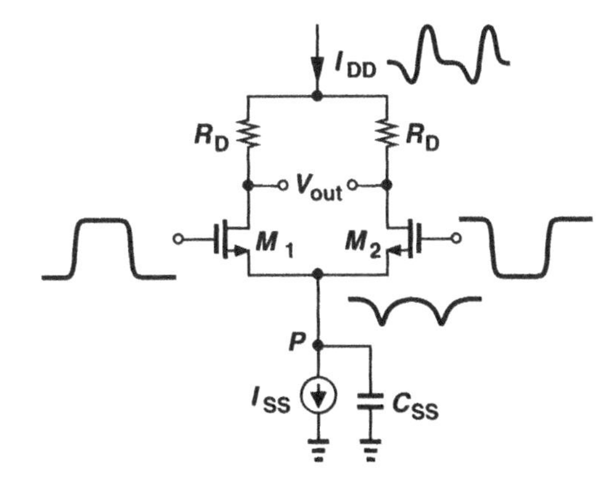

Figure 5.24 Supply current transients in a differential stage.

capacitance, C_{SS}, seen at the common-source node to ground. During switching, M_1 and M_2 draw a current equal to I_{SS} plus the displacement current required to charge C_{SS}. As a result, the total current drawn from the supply is equal to $I_{SS} + C_{SS}dV_P/dt$. This issue is alleviated through the use of on-chip bypass capacitors between V_{DD} and ground.

Package Parasitics Differential operation also relaxes package parasitic requirements for signal lines. Consider the arrangement shown in Fig. 5.25, where two transmission

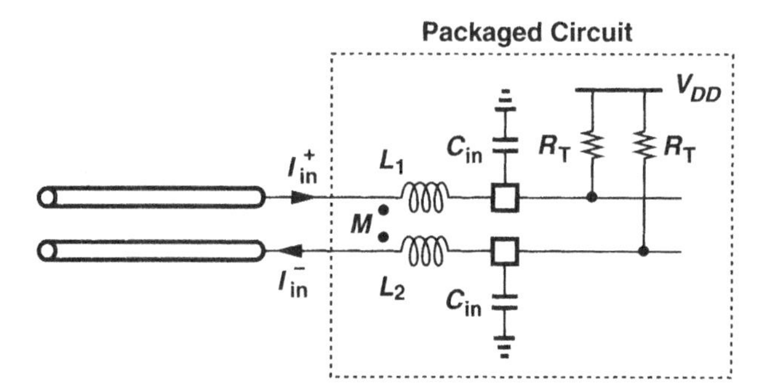

Figure 5.25 Effect of package parasitics.

lines carry differential signals to a packaged circuit. The package and bond wire parasitics are represented by L_1 and L_2, and various capacitances are lumped into C_{in}. The mutual coupling between the inductors, M, yields the following voltage drops across each:

$$V_{L1} = I_{in}^{+} L_1 s - I_{in}^{-} M s \tag{5.49}$$

$$V_{L2} = -I_{in}^{-} L_2 s + I_{in}^{+} M s. \tag{5.50}$$

With $L_1 = L_2 = L_P$ and $I_{in}^{+} = I_{in}^{-}$, we have

$$V_{L1} = I_{in}^{+}(L_P - M)s \tag{5.51}$$

$$V_{L2} = I_{in}^{-}(-L_P + M)s. \tag{5.52}$$

The key observation here is that mutual coupling *reduces* the transient drop across each inductor, thus lowering the effective inductance that appears in each signal line. With careful design of the chip pad frame and the package, M may approach $0.5L_P$ to $0.75L_P$, yielding a small equivalent inductance in (5.51) and (5.52). For single-ended signals, on the other hand, this property cannot be exploited.

Stability In the presence of package parasitics, a single-ended multistage amplifier may become unstable. Figure 5.26 illustrates two examples of this effect. In Fig. 5.26(a), the parasitic inductance in series with V_{DD} provides a finite impedance at node X, thereby allowing the transient current in the last stage to develop a voltage at this node. This voltage then feeds back to node P through R_{D1}, possibly driving the last two common-source stages into oscillation. In Fig. 5.26(b), the ground inductance creates a positive-feedback path from M_2 to M_1, potentially leading to instability.

We should recognize that, even with a small L_D, the circuits of Fig. 5.26 may oscillate because the transistors exhibit a high cut-off frequency, allowing a high-frequency noise component to experience sufficient gain around the loop. By contrast, the differential counterparts, shown in Fig. 5.27, are much more stable. In Fig. 5.27(a), the last stage draws a small transient current from the supply and, more importantly, the feedback through the supply line appears as a *common-mode* disturbance at nodes P and Q. In Fig. 5.27(b), the

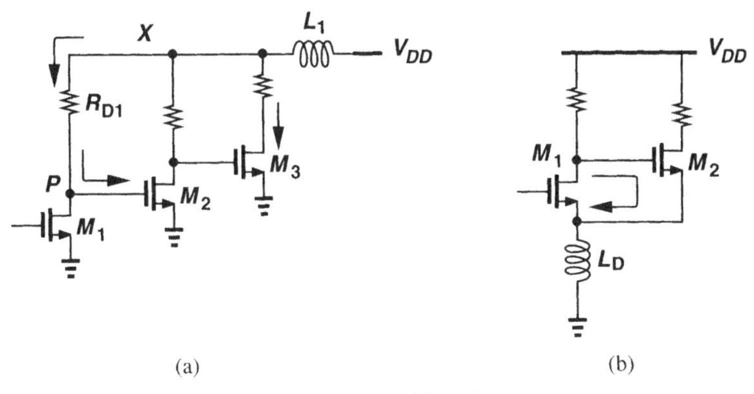

Figure 5.26 Examples of instability due to parasitic inductance.

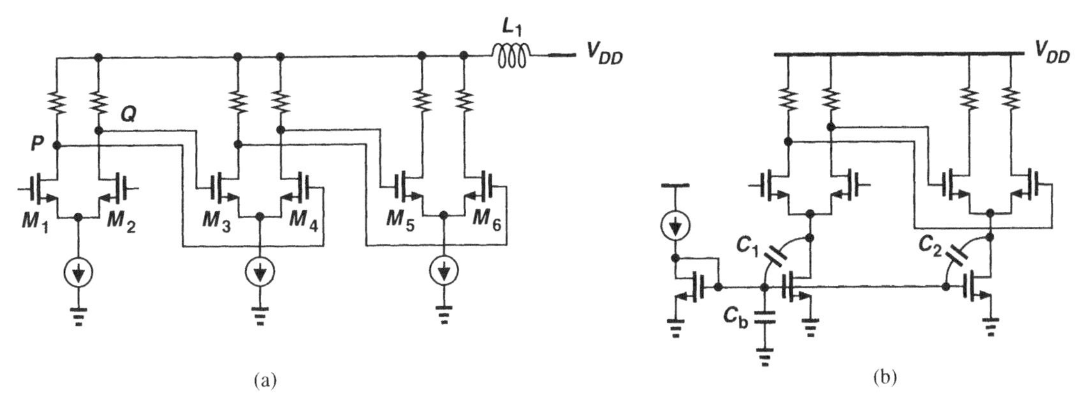

Figure 5.27 Feedback paths in cascaded differential stages.

ground inductance does not affect the primary feedback path, i.e., the path consisting of the overlap capacitances C_1 and C_2. Capacitor C_b suppresses this feedback.[9]

5.3.2 Double Termination

In the arrangement of Fig. 5.21, the buffer exhibits a relatively high output impedance, a benign effect if the far end is perfectly matched. Indeed, this method is used for frequencies as high as a few gigahertz. However, the package parasitics and the input capacitance of the circuit at the far end (e.g., as depicted in Fig. 5.25) inevitably introduce impedance mismatches. At higher speeds, such mismatches create significant reflections that travel back to the buffer, see a heavy impedance mismatch at the near end, are reflected again, and reach the far end with some delay with respect to the original signal. As a result, the

[9]With the parasitic capacitances that appear from the common-source node of each differential pair to ground, the circuit may experience *common-mode* oscillation. Careful simulations are therefore necessary to ensure this does not happen.

reception at the far end may experience substantial ISI.

To better understand the above phenomenon, let us consider the simple case illustrated in Fig. 5.28, where only the input capacitance of the far-end circuit is taken into account

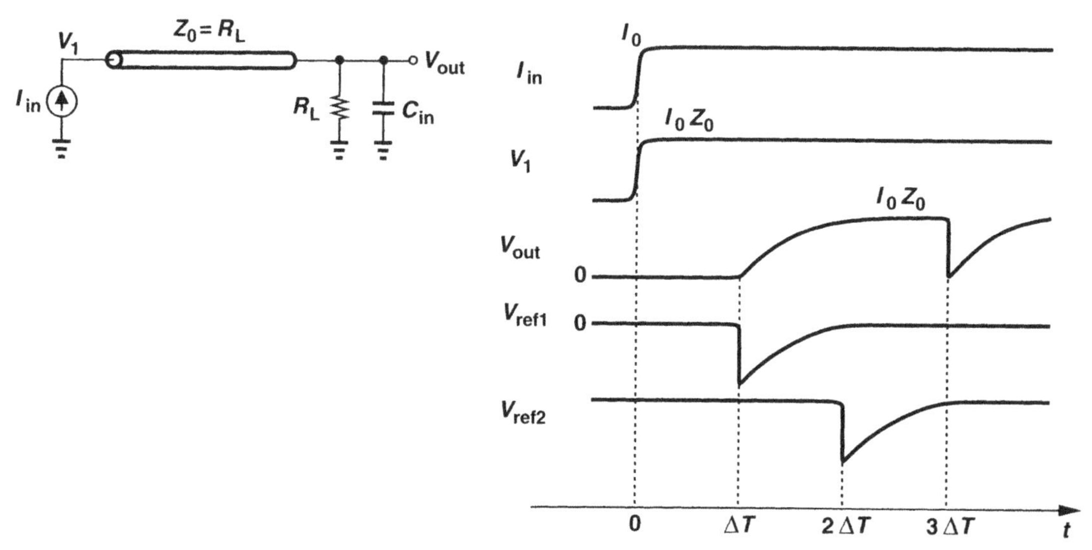

Figure 5.28 Reflections in a T line.

and the load resistance matches the line characteristic impedance. Assuming I_{in} steps from 0 to I_0 at $t = 0$, we note that the current initially sees an impedance equal to Z_0, forcing V_1 to jump to $I_0 Z_0$. The resulting voltage and current pulses travel along the line, reaching the far end after a delay, ΔT. The current pulse flows through the load impedance, generating a voltage waveform given by

$$V_{out} = I_0 R_L[1 - \exp\frac{-(t - \Delta T)}{\tau}]u(t - \Delta T), \tag{5.53}$$

where $\tau = R_L C_{in} = Z_0 C_{in}$ and $u(t)$ denotes the unit step. Since this voltage waveform differs from the voltage pulse launched by I_{in}, a voltage equal to the difference between the two must be reflected from the far end:

$$V_{ref1}(t) = I_0 Z_0[1 - \exp\frac{-(t - \Delta T)}{\tau}]u(t - \Delta T) - I_0 Z_0 u(t - \Delta T) \tag{5.54}$$

$$= -I_0 Z_0 \exp\frac{-(t - \Delta T)}{\tau}u(t - \Delta T). \tag{5.55}$$

The reflected component reaches the near end at $t = 2\Delta T$, sees a high impedance, and is reflected in its entirety:

$$V_{ref2}(t) = -I_0 Z_0 \exp\frac{-(t - 2\Delta T)}{\tau}u(t - 2\Delta T). \tag{5.56}$$

This component travels to the far end, changing V_{out} by approximately the same value:

$$V_{out}(t) = I_0 R_L \left[1 - \exp\frac{-(t-\Delta T)}{\tau}\right] u(t-\Delta T) - I_0 Z_0 \exp\frac{-(t-3\Delta T)}{\tau} u(t-3\Delta T). \tag{5.57}$$

The impedance mismatch at the far end leads to more reflections but the effect is typically negligible. The second term in this equation exhibits an amplitude proportional to I_0 and a delay proportional to the line length, potentially corrupting subsequent bits in a random sequence.

The foregoing analysis reveals that the high output impedance of the buffer may introduce significant ISI in the presence of mismatches at the far end. For this reason, buffers operating at speeds higher than a few gigahertz incorporate "double termination." Illustrated in Fig. 5.29, the idea is to add on-chip resistors $R_1 = R_2 = Z_0$ so that the output

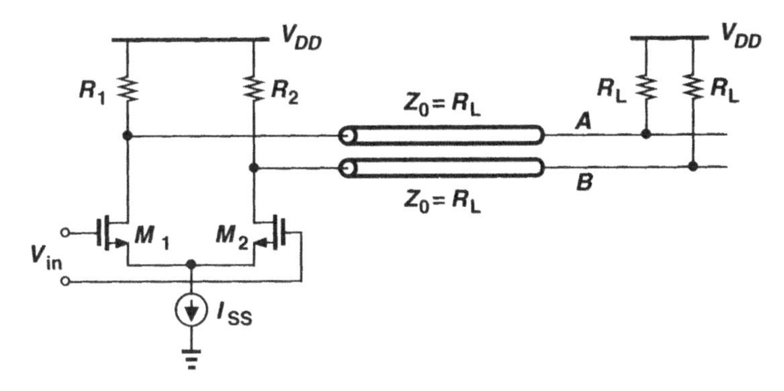

Figure 5.29 Doubly-terminated output buffer.

impedance of the buffer approaches the characteristic impedance of the transmission line, thereby minimizing reflections from the near end.

The principal difficulty with double termination is the inevitable doubling of the power dissipation. To deliver a given voltage swing to the far end, the buffer tail current must be twice that of the high-impedance topology of Fig. 5.21. For example, with $Z_0 = 50\ \Omega$, a tail current of 20 mA is required to generate a differential swing of 1 V_{pp} at nodes A and B.

The actual value of the on-chip termination resistors in Fig. 5.29 is sometimes chosen to be *higher* than the line impedance so as to accommodate package parasitics [3]. Consider, for example, the simple package model shown in Fig. 5.30(a), where R_{out} must match Z_0 to minimize reflections from the near end. Let us first analyze the section consisting of R_1, C_1, and L_1 [Fig. 5.30(b)]. For a narrow frequency band, the parallel combination of R_1 and C_1 can be modeled by a series network [Fig. 5.30(c)]:

$$\frac{R_1}{R_1 C_1 s + 1} = \frac{R_S C_S s + 1}{C_S s}. \tag{5.58}$$

For $s = j\omega$, we have $R_1 C_1 = (R_S C_S \omega^2)^{-1}$ and $R_1 C_1 + R_S C_S - R_1 C_S = 0$. If $R_1 \gg R_S$

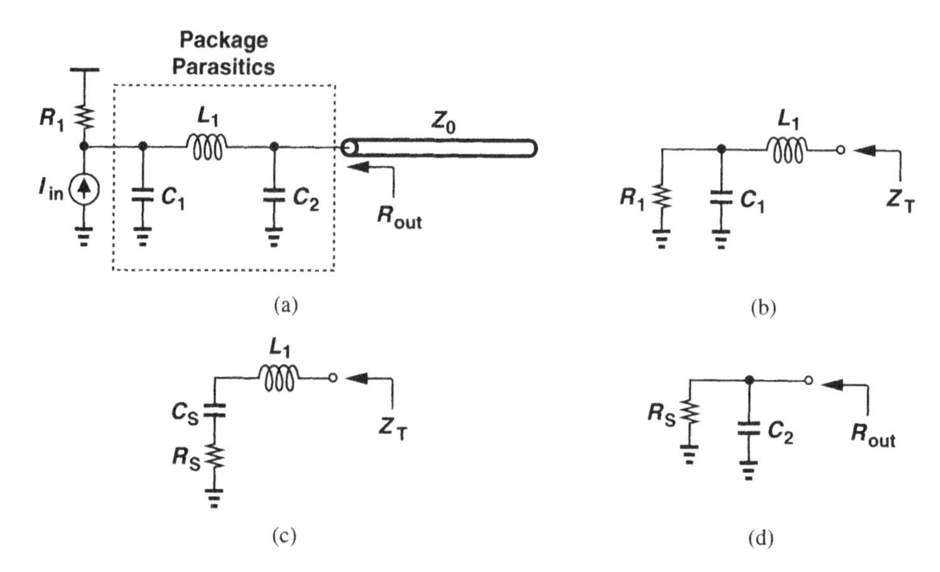

Figure 5.30 Effect of package parasitics on output impedance.

(i.e., the capacitor has a high Q), then $C_S \approx C_1$ and

$$R_S = \frac{1}{R_1(C_1\omega^2)} \tag{5.59}$$

$$= \frac{R_1}{Q^2}, \tag{5.60}$$

where $Q = R_1C_1\omega$. If L_1 and C_1 resonate, the circuit of Fig. 5.30(c) reduces to a single resistance equal to R_S that is much *smaller* than R_1 if Q exceeds 3. That is, the network consisting of L_1 and C_1 transforms R_1 to a lower value. Consequently, in the vicinity of the resonance frequency of L_1 and C_1, the overall circuit of Fig. 5.30(a) can be simplified as shown in Fig. 5.30(d), revealing that the output impedance is equivalent to the parallel combination of R_S and C_2 in a narrow band.

While assuming a narrow range around the resonance frequency of L_1 and C_1, the above analysis still yields the important conclusion that the output impedance may be lowered by package parasitics, mandating a value for R_1 greater than Z_0. Accurate package modeling and careful simulations can determine the value of R_1 that yields minimum ISI. A value of 75 Ω to 100 Ω typically satisfies this condition, saving substantial power as well.

5.3.3 Predriver Design

In order to steer the large tail current to the two outputs, differential buffers must employ wide transistors, thus exhibiting a high input capacitance. A "predriver" stage is therefore necessary to drive the output stage with large, fast transitions. Various trade-offs require that the two stages be designed as a single circuit. As an example, consider the cascade shown in Fig. 5.31. Suppose I_2 is large enough to generate the required output voltage

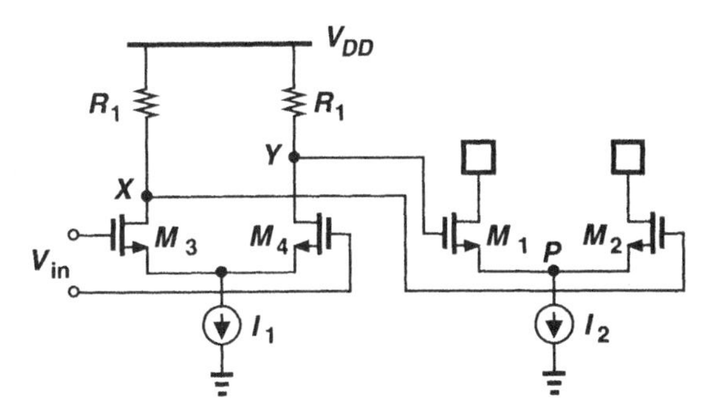

Figure 5.31 Cascade of a differential pair and output buffer.

swings, and M_1 and M_2 are wide enough to steer 90% of I_2 to one side if they sense a differential swing of 1 V_{pp} at nodes X and Y. This means that $I_1 R_1$ must be equal to 0.5 V if M_3 and M_4 experience complete switching. What happens if the width of the output transistors $W_{1,2}$ is doubled? Then, the differential swing at X and Y required to steer 90% of I_2 decreases (but not by as much as a factor of 2), allowing the use of a *smaller* value for R_1. Unfortunately, since R_1 cannot be halved, the time constant at X and Y still increases. Conversely, if the width of M_1 and M_2 is halved, R_1 must increase (but by a factor of less than 2), *decreasing* the time constant at X and Y.

The above observation implies that reducing the width of the output transistors and increasing the value of R_1 can maintain a constant output swing while enhancing the predriver speed. However, as $W_{1,2}$ decreases and R_1 increases, the minimum voltage at node P falls, eventually driving I_2 into the triode region. In addition, M_3 and M_4 also enter the triode region if V_X and V_Y swing below the input common-mode level by more than one threshold voltage. These issues become increasingly more difficult as the supply voltage scales down.

The trade-off between speed and voltage headroom can be alleviated by inductive peaking. As shown in Fig. 5.32, inductors placed in series with the predriver load resistors resonate with the total capacitance at nodes X and Y, increasing the bandwidth and hence

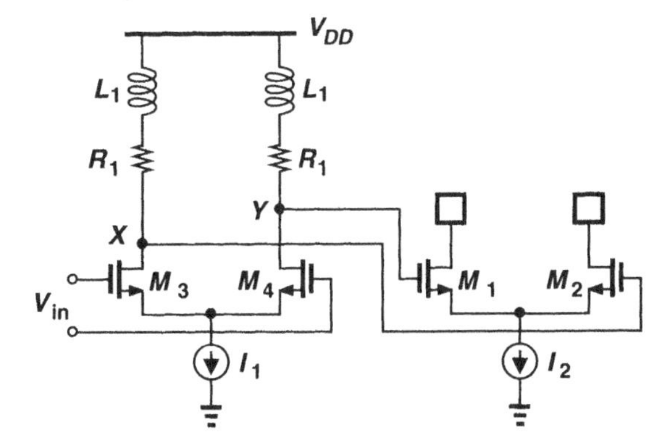

Figure 5.32 Inductive peaking in a predriver.

allowing the use of wide output transistors. Of course, as the width of M_1 and M_2 increases, their drain junction and overlap capacitance eventually limits the bandwidth at the *output* nodes, necessitating inductive peaking at the output as well.

To achieve even higher speeds, the output stage can be configured as an f_T doubler. Depicted in Fig. 5.33, the overall buffer may also incorporate inductive peaking in the predriver.

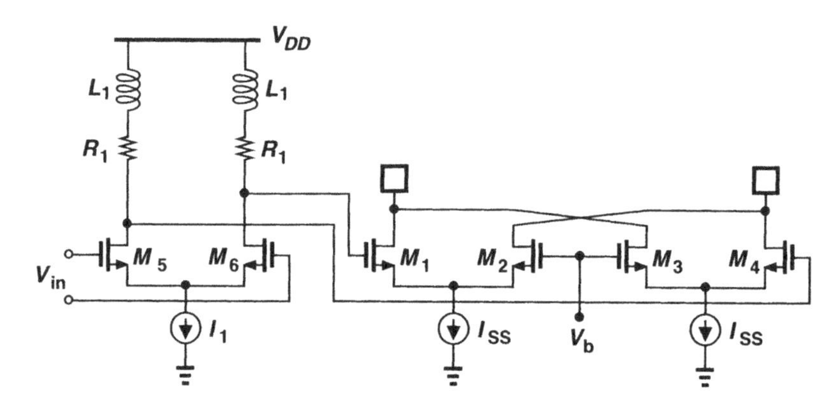

Figure 5.33 Inductive peaking prior to an f_T doubler.

Figure 5.34(a) shows another example employing high-speed techniques to deliver a differential voltage swing of approximately 340 mV to 75-Ω on-chip termination resistors and 50-Ω off-chip loads. The output stage also uses inductive peaking to achieve faster transitions [3]. The simulated eye diagram for a data rate of 40 Gb/s in 0.13-μm technology is shown in Fig. 5.34(b).

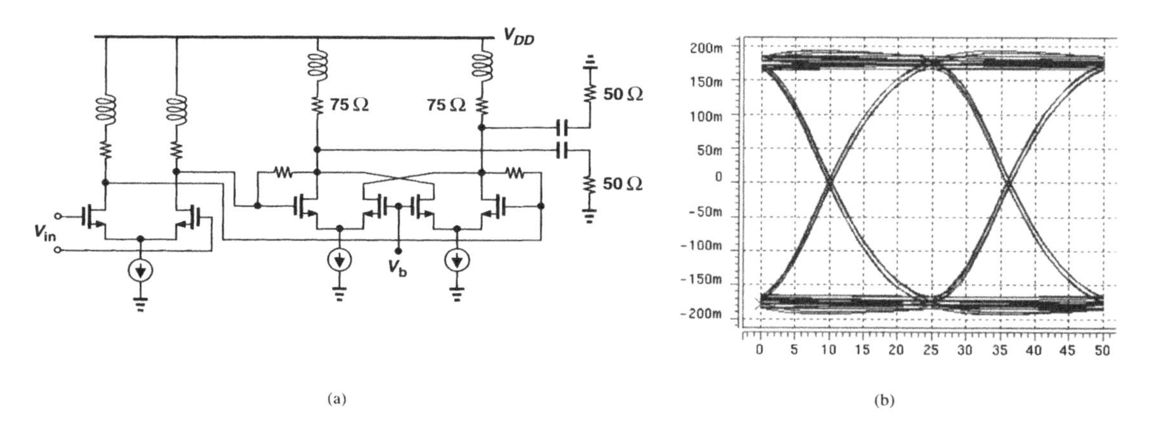

Figure 5.34 (a) High-speed output buffer, (b) simulated output eye diagram.

5.4 Distributed Amplification

Our study of broadband amplification thus far indicates fundamental gain-bandwidth trade-offs, especially if the circuit must employ large transistors to deliver high currents to termination resistors. A method of overcoming such trade-offs is based on "distributed amplification," originally introduced by Ginzton in 1948 [4]. DAs are particularly suited to driving low-impedance loads but they also present a relatively low impedance to the preceding stage. Before describing the principle of distributed amplification, we briefly review monolithic transmission lines.

5.4.1 Monolithic Transmission Lines

The integration of T lines in monolithic technologies is a relatively new development, dating back only a few decades. The performance and utility of T lines depend on three parameters: inductance, capacitance, and loss, with the latter two trading with each other heavily in monolithic implementations. For this reason, only IC technologies that provide low-resistance and/or low-capacitance interconnects can accommodate high-quality T lines.

To identify T lines that lend themselves to fabrication in VLSI technologies, we first review the back end of a typical CMOS process. As depicted in Fig. 5.35 for a 0.13-μm generation, the process provides a silicided polysilicon layer and eight metal layers.

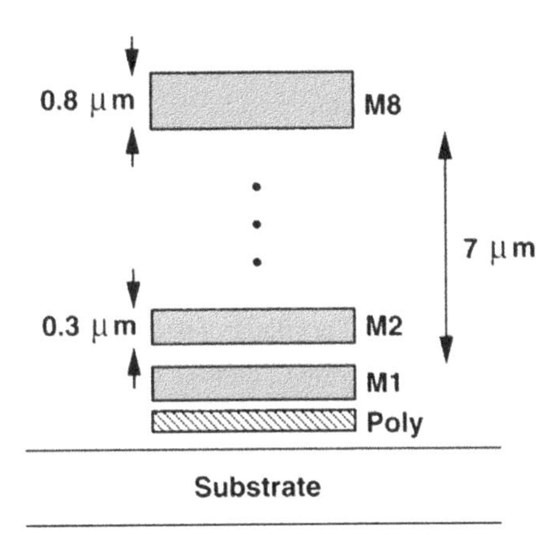

Figure 5.35 Back end of a typical CMOS process.

Due to its high resistivity (about 5 to 10 $\Omega/\square$), the polysilicon layer is ill-suited to T line fabrication. Each of the first seven metal layers has a thickness of about 0.3 μm and a sheet resistance, R_{sh}, of 70 m$\Omega/\square$. The top layer typically has a thickness of 0.8 μm and $R_{sh} = 25$ m$\Omega/\square$. The metals are separated by two types of dielectrics: a 0.7-μm layer with $\epsilon_r \approx 3.5$ and a 0.1-μm layer with $\epsilon_r \approx 7$.

The above back end example suggests three T line structures as candidates for monolithic integration. Illustrated in Fig. 5.36(a), the first is the microstrip topology, where a signal strip lies on top of a ground plane. As explained later, it is desirable to minimize the capacitance per unit length of the line, C_0, hence the use of M8 and M1 for signal

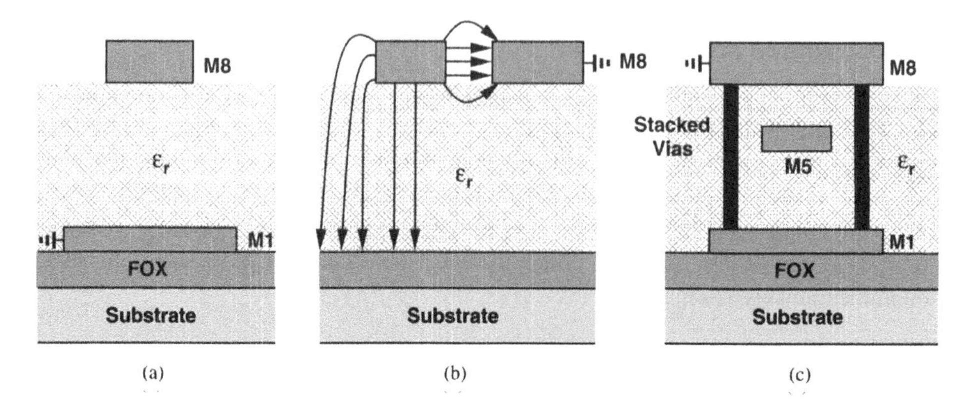

Figure 5.36 Monolithic T line geometries: (a) microstrip, (b) coplanar line, and (c) stripline.

and ground, respectively. Shown in Fig. 5.36(b), the second structure, a coplanar T line, employs widely-spaced signal and ground strips in M8 so as to reduce C_0. However, the signal line still exhibits substantial capacitance to the substrate.

Depicted in Fig. 5.36(c), the third structure is the stripline configuration, where the signal line is realized in an intermediate metal layer and it is shielded by ground planes in M1 and M8. Stacked vias provide vertical walls, connecting the two ground planes while confining the fields within the shield.

In circuit design, several properties of T lines become critical: characteristic impedance, loss, wave velocity, and field confinement.

Characteristic Impedance For a lossless T line, the characteristic impedance, Z_0, is given by $Z_0 = \sqrt{L_0/C_0}$, where L_0 and C_0 denote the inductance and capacitance per unit length, respectively. Most circuit applications of T lines require that C_0 be minimized so as to accommodate larger transistor capacitances while still achieving a high value for Z_0. From this viewpoint, the microstrip and coplanar structures of Fig. 5.36 are advantageous over the stripline. For most T lines, the exact value of Z_0 must be obtained through the use of electromagnetic field simulators, but to develop intuition, closed-form expressions prove useful. In particular, for the microstrip geometry shown in Fig. 5.37, the characteristic impedance is given by [5]:

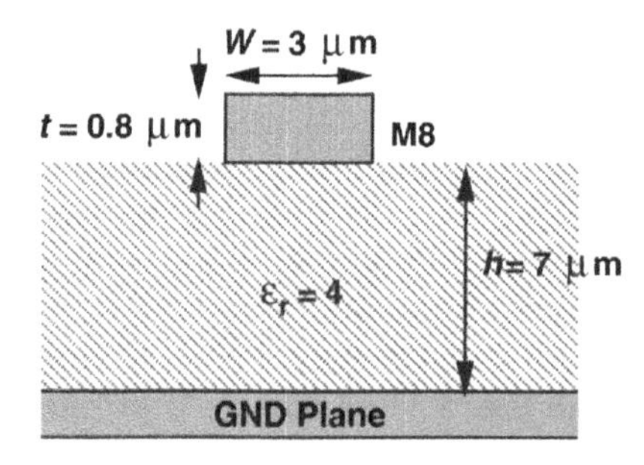

Figure 5.37 Example of microstrip dimensions.

$$Z_0 = \frac{377}{\sqrt{\epsilon_r}} \frac{h}{W_e} \frac{1}{1 + 1.735\epsilon_r^{-0.0724}(\frac{W_e}{h})^{-0.836}}, \tag{5.61}$$

where

$$W_e = W + \frac{t}{\pi}(1 + \ln\frac{2h}{t}). \tag{5.62}$$

In the back-end example of Fig. 5.35, $h \approx 7\ \mu$m, $t \approx 0.8\ \mu$m, and $\epsilon_r \approx 4$. If $W = 3$ μm, then $W_e = 3.98\ \mu$m and $Z_0 \approx 94.3\ \Omega$. For $W = 4\ \mu$m, Z_0 falls to 85.9 Ω.

Loss The loss of microstrips arises from three sources: low-frequency resistance of interconnects, skin effect in interconnects, and dielectric loss.[10] The loss due to skin effect in the microstrip of Fig. 5.37 is expressed as [6]:

$$\alpha_{skin} = \frac{4.34}{\pi} \frac{R_s}{hZ_0} \left[1 - \left(\frac{W'}{4h}\right)^2\right] \left\{1 + \frac{h}{W'} + \frac{h}{\pi W'}\left[\ln\left(\frac{2h}{t} + 1\right) - \frac{1 + t/h}{1 + t/(2h)}\right]\right\} \tag{5.63}$$

for $2 \geq W/h \geq 1/(2\pi)$. In this equation,

$$W' = W + \frac{t}{\pi} \ln\left(1 + \frac{2h}{t}\right), \tag{5.64}$$

R_s denotes the sheet resistance due to skin effect:

$$R_s = \sqrt{\pi f \mu_0 \rho}, \tag{5.65}$$

and α_{skin} is expressed in dB/cm. For aluminum interconnects and $W = 3\ \mu$m,

$$\alpha_{skin} = 2.156 \times 10^{-5}\sqrt{f}\ \mathrm{dB/cm}. \tag{5.66}$$

For example, at $f = 40$ GHz, $\alpha_{skin} = 0.431$ dB/mm.

It is instructive to compare the skin loss of a microstrip at $f = 40$ GHz with that due to its low-frequency resistance. This is accomplished by obtaining an equivalent resistance per unit length, R_{skin}, due to skin effect. Recall from Chapter 2 that the voltage wave at point x along a line is expressed as

$$V(x) = V_0 e^{-\alpha x} e^{-j\beta x}, \tag{5.67}$$

where α denotes the loss and $\beta = 2\pi/\lambda$. The value of α_{skin} given by (5.63) is related to α as

$$\alpha_{skin} = -20\log(e^{-\alpha x}). \tag{5.68}$$

[10] If the skin depth in the ground plane is greater than the metal thickness, the substrate may introduce additional loss.

Thus, $\alpha = 0.0496 \text{ mm}^{-1}$. For a T line with negligible dielectric loss,

$$\alpha \approx \frac{R_{skin}}{2Z_0}, \tag{5.69}$$

which, for $Z_0 = 94.3\ \Omega$, yields $R_{skin} = 9.35\ \Omega$/mm. On the other hand, the low-frequency resistance of the metal 8 wire is given by $(25\ \text{m}\Omega/\square) \times (1000\ \mu\text{m}/3\ \mu\text{m}) \approx 8.33\ \Omega$/mm. The apparent resistance of the line therefore increases by about 12% as the frequency goes from zero to 40 GHz.

In coplanar lines, the loss through the substrate may also become significant [7]. Resulting from the electric field lines that terminate on the substrate (Fig. 5.38), this loss is much more significant in silicon technologies than in III–V processes.

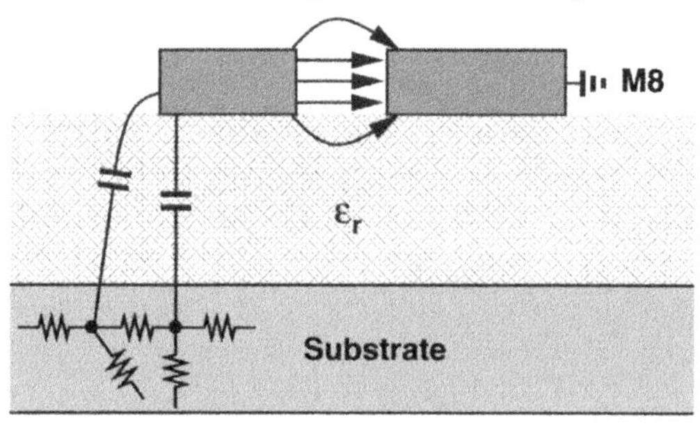

Figure 5.38 Substrate loss in a coplanar line.

The trade-off between the characteristic impedance and the loss limits the performance of circuits using T lines. The resistivity of interconnects in standard silicon processes further exacerbates this issue.

Wave Velocity The propagation velocity, v, in T lines determines the end-to-end delay that a signal experiences, playing a critical role in distributed oscillators. For a lossless line,

$$v = \frac{1}{\sqrt{L_0 C_0}}. \tag{5.70}$$

An empirical expression for v in microstrips appears as [5]:

$$v = \frac{3 \times 10^8}{\sqrt{1 + 0.6(\epsilon_r - 1)(\frac{W_e}{h})^{0.0297}}} \quad \text{if } \frac{W}{h} < 0.6$$

$$= \frac{3 \times 10^8}{\sqrt{1 + 0.63(\epsilon_r - 1)(\frac{W_e}{h})^{0.1255}}} \quad \text{if } \frac{W}{h} > 0.6,$$

where W_e is given by (5.62). For the above example with $W = 3\ \mu\text{m}$, $v = 1.803 \times 10^8$ m/s, and hence $L_0 = 523$ pH/mm and $C_0 = 58.8$ fF/mm.

Field Confinement The leakage of fields from a T line to its environment can increase the loss whereas the leakage in the opposite direction can lead to noise coupling. In the coplanar geometry of Fig. 5.38, for example, noise in the substrate couples through the capacitances to the signal line. In this respect, the microstrip and particularly the stripline exhibit a higher performance. It therefore appears that the microstrip provides a reasonable compromise among Z_0, loss, and field confinement.

5.4.2 Distributed Amplifiers

As described earlier, lossless uniform transmission lines exhibit two important properties: (1) infinite bandwidth and (2) a constant delay and hence a linear phase response. Even though practical T lines suffer from some loss, we can still exploit these properties to construct broadband amplifiers.

Ideal Amplifier As the first step toward understanding distributed amplification, consider a simple common-source stage that is driven by a finite source impedance, R_S (Fig. 5.39). Invoking the Miller effect for C_{GD}, we can approximate the input and output poles

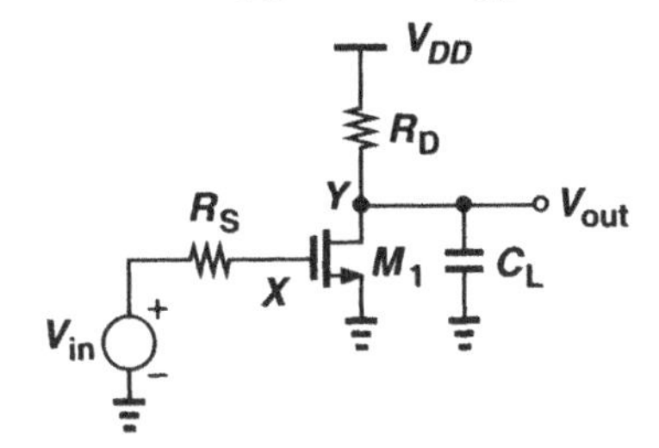

Figure 5.39 Simple common-source stage.

as

$$|\omega_{in}| \approx \frac{1}{R_S[C_{GS} + (1 + g_{m1}R_D)C_{GD}]} \quad (5.71)$$

$$|\omega_{out}| \approx \frac{1}{R_D C_L}, \quad (5.72)$$

where C_L includes the drain junction and the gate-drain overlap capacitance of M_1 and the load capacitance (resulting from the subsequent stage). If the width of M_1 is increased to achieve greater voltage gain, both ω_{in} and ω_{out} decrease in magnitude. Thus, the circuit suffers from gain-bandwidth trade-offs.

Let us now decompose the large transistor into smaller equal units and uniformly *distribute* them along a transmission line. Illustrated in Fig. 5.40(a), such an arrangement employs *two* T lines, one for the input and another for the output. The capacitances introduced by each transistor decrease both the line characteristic impedance and the wave velocity. We denote the loaded values by Z_{0L} and v_{0L}, respectively, assuming for now that the loaded input and output lines exhibit roughly equal parameters. For proper operation, both lines must be terminated as shown in Fig. 5.40(b). (The bias network for the transistors is not shown.)

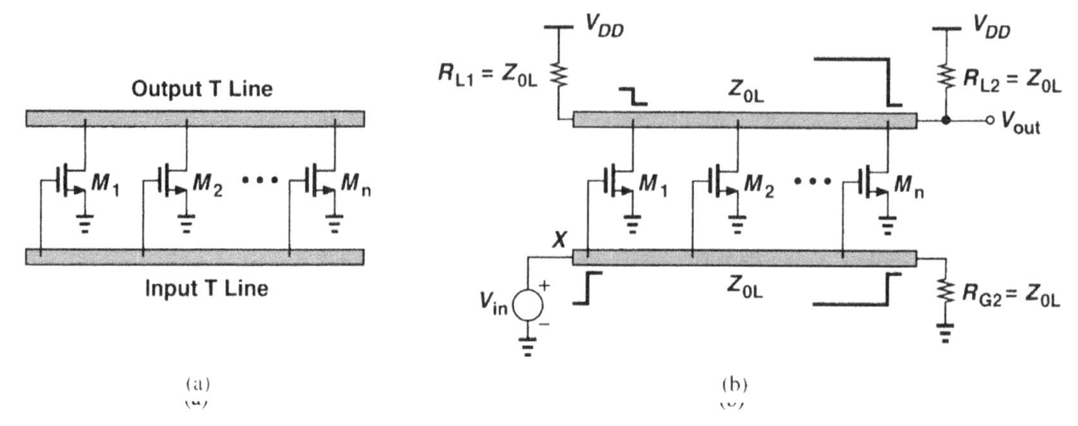

Figure 5.40 (a) Basic distributed amplifier, (b) addition of termination resistors.

We first analyze the circuit of Fig. 5.40(b) qualitatively. Suppose V_{in} jumps from 0 to V_0 at $t = 0$. The resulting step begins to travel to the right, reaching the gates of the transistors after some delay. Upon sensing the step, each transistor draws a current equal to $g_m V_0$ from the output line. This current sees an instantaneous impedance of Z_{0L} to the left and to the right and hence splits equally, thereby creating two voltage steps of height $(g_m V_0)(Z_{0L}/2)$, one traveling toward R_{L1} and the other toward R_{L2}. If the wave velocities of the input and output lines are equal, then each transistor senses the input step at exactly the same time as the output step produced by the preceding transistor(s) arrives at its drain. Since the waves reaching the ends of both lines are absorbed by R_{G2}, R_{L1}, and R_{L2}, no reflections corrupt the signal.

The key observation here is that, in a distributed amplifier, the transistor capacitances do *not* translate to a time constant. Rather, they simply lower the characteristic impedance of the input and output T lines. In other words, with lossless lines, the DA of Fig. 5.40(b) can in principle provide infinite gain with infinite bandwidth! Each common-source stage along with an additional length of T lines increases the gain by $g_m Z_{0L}/2$.[11] (The DA therefore trades delay for gain.)

Let us now quantify the behavior of a DA using lossless T lines. Suppose the amplifier incorporates n common-source stages uniformly distributed over a length l. Neglecting Miller effect, we express the loaded capacitance per unit length of the input line as $C_0 + nC_{GS}/l$. We also assume that the output line is loaded such that it exhibits the same characteristics as the input line. Thus,

$$Z_{0L} = \sqrt{\frac{L_0}{C_0 + \dfrac{nC_{GS}}{l}}}. \tag{5.73}$$

[11] In contrast to *cascaded* stages, here the gains of the CS stages are added rather than multiplied.

The total voltage gain is equal to

$$A_v = \frac{ng_m Z_{0L}}{2} \tag{5.74}$$

$$= \frac{ng_m}{2}\sqrt{\frac{L_0}{C_0 + \dfrac{nC_{GS}}{l}}}. \tag{5.75}$$

For simplicity, we assume a constant overdrive voltage, $V_{GS} - V_{TH}$, for the MOSFETs, writing

$$g_m = \mu_n C_{OX}\frac{W}{L}(V_{GS} - V_{TH}) \tag{5.76}$$

$$= \frac{2I_D}{V_{GS} - V_{TH}}. \tag{5.77}$$

This means both W/L and I_D must be scaled proportionally to scale g_m.[12] Equation (5.75) is then rewritten as

$$A_v = \frac{n}{2}\mu_n C_{OX}\frac{W}{L}(V_{GS} - V_{TH})\sqrt{\frac{L_0}{C_0 + \dfrac{nC_{GS}}{l}}}. \tag{5.78}$$

In the ideal case, the capacitance of the line itself, C_0, is negligible with respect to nC_{GS}/l. We also assume $C_{GS} \approx (2/3)C_{OX} \cdot WL$. Thus,

$$A_v \approx \frac{3}{4}n\mu_n\frac{C_{GS}}{L}(V_{GS} - V_{TH})\sqrt{\frac{L_0}{\dfrac{nC_{GS}}{l}}} \tag{5.79}$$

$$\approx \frac{3}{4}\frac{\mu_n}{L^2}(V_{GS} - V_{TH})\sqrt{\frac{nC_{GS}}{l}L_0 \cdot l^2} \tag{5.80}$$

$$\approx \frac{3}{4}\frac{\mu_n}{L^2}(V_{GS} - V_{TH})\frac{l}{v}. \tag{5.81}$$

The voltage gain is therefore proportional to the end-to-end delay, l/v.

Another form of Eq. (5.81) can be derived as follows. For MOSFETs, the transconductance and the gate-source capacitance yield the transit frequency as:

$$2\pi f_T = \frac{g_m}{C_{GS}}, \tag{5.82}$$

[12]For any active device, scaling the device width and bias current by k scales the transconductance by the same factor.

if the gate-drain overlap capacitance is neglected. Consequently, (5.75) reduces to

$$A_v = \frac{n}{2} \cdot 2\pi f_T C_{GS} \sqrt{\frac{L_0}{C_0 + \dfrac{nC_{GS}}{l}}}, \tag{5.83}$$

which, with $C_0 \ll nC_{GS}/l$, gives

$$A_v \approx \pi f_T \frac{l}{v}. \tag{5.84}$$

For example, the voltage gain approaches a value of π for a delay equal to $1/f_T$.

The derivations leading to Eq. (5.81) have assumed that n can be arbitrarily large: the longer the line is, the more gain stages are required. In practice, however, the supply voltage limits n or the bias current of each stage.

If a distributed amplifier is driven by an on-chip circuit, e.g., a predriver, then its input impedance loads the preceding stage, possibly degrading the overall gain of the cascade. Modeling the cascade as shown in Fig. 5.41, we note that the output voltage of the DA is equal to $I_1 Z_{0L} A_v$, i.e., it is the *product* of Z_{0L} and A_v that determines the gain of the

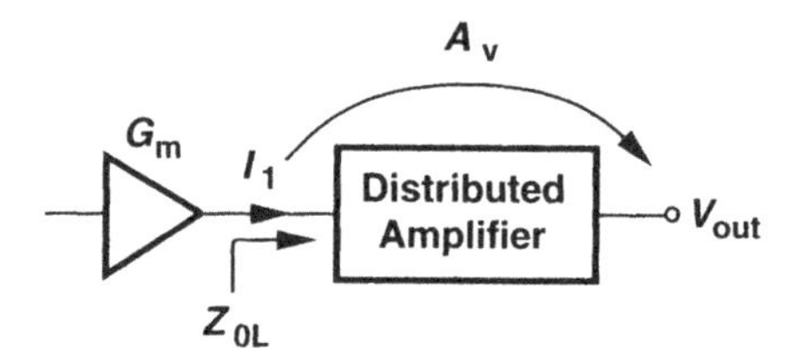

Figure 5.41 Cascade of a predriver and a distributed amplifier.

overall circuit. From this perspective, we have

$$A_v Z_{0L} \approx \pi f_T \frac{l}{v} Z_{0L} \tag{5.85}$$

$$= \pi f_T l L_0. \tag{5.86}$$

Interestingly, for a given f_T, the product is independent of the transistor gate-source capacitance, suggesting that L_0 must be maximized.

Nonidealities In reality, DAs provide neither infinite gain nor infinite bandwidth. The reasons are as follows. (1) The supply voltage limits the number of common-source stages and hence the maximum achievable gain. (2) As depicted in Fig. 5.42, the resistive loss in the input line gradually attenuates the signal applied to the gates of the transistors, leaving little amplitude for the stages near the far end and making them useless. (3) Resistive losses in the output line also limit the bandwidth. (4) The output resistance of the transistors appears as the G conductance in Fig. 2.35, thus raising the loss in the output T line. (5) The Miller multiplication of each transistor's gate-drain overlap capacitance becomes more

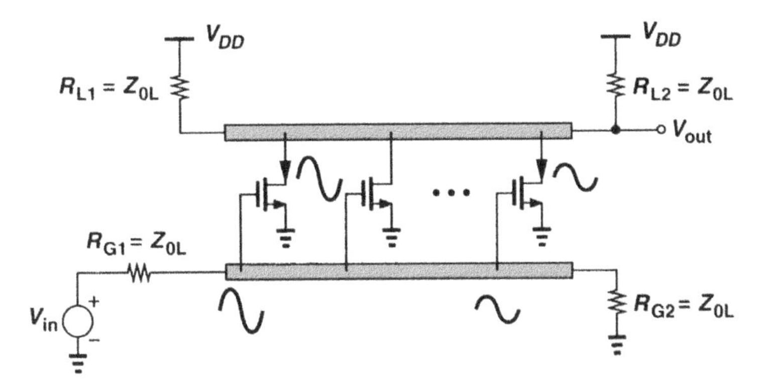

Figure 5.42 Loss in a distributed amplifier.

significant as the wave travels toward the far end. We now study some of these issues in more detail.

The loss of T lines limits the performance of distributed amplifiers. Intuitively, we note that, even though every section increases the gain monotonically, the input line attenuation yields smaller signal levels at the end of each section. Thus, beyond a certain point, adding sections yields negligible increase in the gain while consuming power and voltage headroom.

Let us now determine how the number of sections must be chosen in the presence of loss in the input line. Consider the circuit of Fig. 5.43, where each section incorporates a

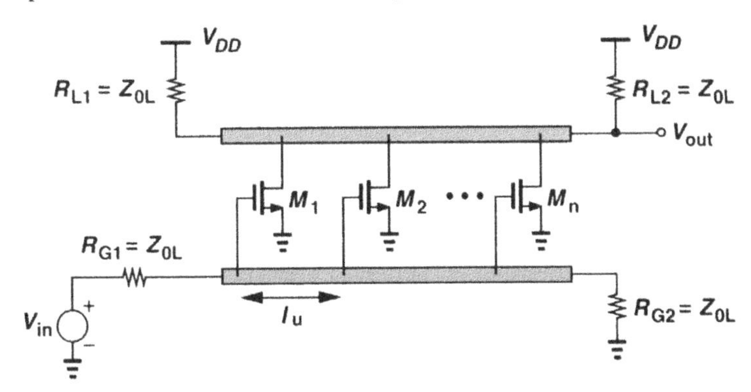

Figure 5.43 Example of DA for loss calculations.

transistor with transconductance g_m and a line of length l_u. Recall from Chapter 2 that a wave propagating on a lossy line can be expressed as

$$V(x) = V_0 e^{-\alpha x} e^{-j\beta x}. \tag{5.87}$$

Since the input signal amplitude drops by a factor of $\exp(-kl_u)$ after k sections, we express the total gain resulting from n sections as[13]

$$A_v = \sum_{k=0}^{n-1} e^{-\alpha k l_u} \frac{g_m Z_{0L}}{2} \tag{5.88}$$

$$= \frac{1 - e^{-\alpha n l_u}}{1 - e^{-\alpha l_u}} \frac{g_m Z_{0L}}{2}. \tag{5.89}$$

In a well-designed DA, it is reasonable to assume $\alpha n l_u < 0.5$, allowing the approximation $\exp \epsilon \approx 1 + \epsilon + \epsilon^2/2$ for both the numerator and the denominator of the first fraction in (5.89):

$$A_v \approx n \frac{1 - \dfrac{\alpha n l_u}{2}}{1 - \dfrac{\alpha l_u}{2}} \frac{g_m Z_{0L}}{2}. \tag{5.90}$$

As expected, if $\alpha \to 0$, then A_v approaches $n g_m Z_{0L}/2$. Since $\alpha l_u \ll 1$, we assume $1/(1 - \epsilon) \approx 1 + \epsilon$ and rewrite (5.90) as

$$A_v \approx n(1 - \alpha l_u \frac{n-1}{2}) \frac{g_m Z_{0L}}{2}, \tag{5.91}$$

where the second-order term is neglected.

Equation (5.91) reveals that as n increases, the term in the parentheses slows down the growth of the gain. Of course, with the assumption $\alpha n l_u < 0.5$ made earlier, the factor $1 - \alpha l_u(n - 1)/2$ does not fall below 0.75. Now suppose we choose n such that this term is equal to 0.8:

$$1 - \alpha l_u \frac{n-1}{2} = 0.8, \tag{5.92}$$

obtaining the number of sections as:

$$n = \frac{0.4}{\alpha l_u} + 1. \tag{5.93}$$

For example, if $\alpha l_u = 0.1$, then $n = 5$ and $n[1 - \alpha l_u(n - 1)/2] = 4$. With this value of αl_u, if n is raised to 6, then $n[1 - \alpha l_u(n - 1)/2] = 4.5$, a relatively small increase.

The output resistance of short-channel MOSFETs biased at high currents can increase the loss of the *output* T line in a distributed amplifier. Appearing from each drain node to ground, such resistance can be included as the G conductance in Eq. (2.43):

$$\alpha = \frac{1}{2}(\frac{R_S}{Z_{0L}} + \frac{m}{r_O} Z_{0L}). \tag{5.94}$$

[13]The loss in the output line is neglected here.

where it is assumed m transistors, each having an output resistance of r_O, are used per unit length of the T line. As an example, suppose $R_S = 10\ \Omega$/mm and $Z_{0L} = 80\ \Omega$. With $r_O = \infty$, the attenuation factor $\alpha \approx 0.0625\ \text{mm}^{-1}$ and hence $|20\log[\exp(-0.0625)]| = 0.542$ dB/mm. If a DA employs ten transistors per millimeter and if the transistor output resistances must raise α by no more than 20%, then

$$\frac{1}{2}\left(\frac{R_S}{Z_{0L}} + \frac{10}{r_O}Z_{0L}\right) \le 0.0625\ \text{mm}^{-1} \times 1.2. \tag{5.95}$$

It follows that

$$r_O \ge 32\ \text{k}\Omega. \tag{5.96}$$

With typical current levels used in DAs, it is difficult to achieve such a high output resistance in deep-submicron technologies.

Fig. 5.44(a) illustrates the effect of Miller multiplication: the apparent voltage gain seen from the gate to drain of transistor number m is m times that seen by transistor number

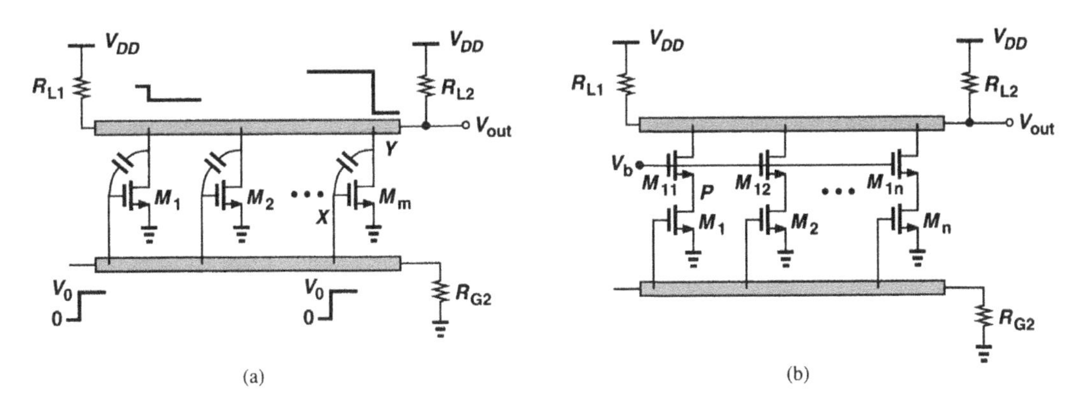

Figure 5.44 (a) Miller effect in a DA, (b) reduction of Miller effect by cascoding.

1. From another point of view, the voltage on the input T line rises by V_0 whereas that on the output T line falls by $mg_mZ_{0L}/2$, thereby opposing the change at node X. As a result, the effective capacitance per unit length of the input T line increases from left to right, lowering both the characteristic impedance and the wave velocity.

The effect of Miller multiplication can be reduced through the use of cascode stages [Fig. 5.44(b)] [8]. In this case, however, the capacitance associated with the cascode junction (node P) cannot be absorbed by a transmission line, thus limiting the bandwidth. This capacitance arises from the drain junction and gate-drain overlap of M_1, source junction of M_{11}, and the gate oxide of M_{11}.

Cascaded DAs The issues described above can be alleviated by cascading two or more distributed amplifiers. Illustrated in Fig. 5.45, such a cascade allows the output current of the first DA to flow through the input T line of the next. Since each DA incorporates

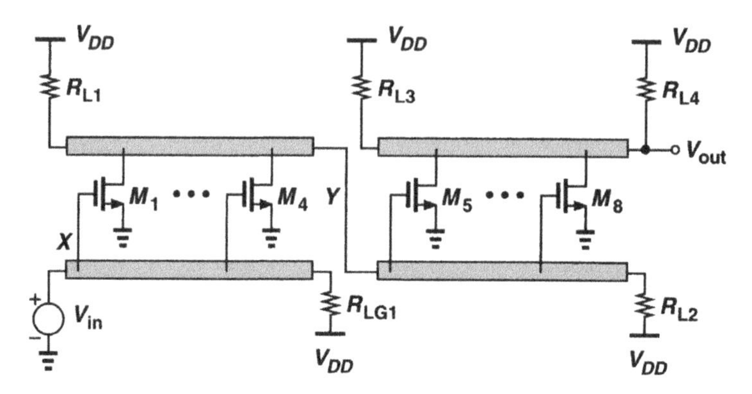

Figure 5.45 Cascaded distributed amplifiers.

only a few sections, the above effects do not accumulate through the chain. Unfortunately, the drain bias voltage of M_1-M_4 must be high enough to provide the gate-source voltage required for M_5-M_8, further limiting the voltage drop across $R_{L1}||R_{L2}$ and hence the gain. Without M_5-M_8, the first DA could allow V_Y to be lower than V_X by one MOS threshold voltage, a substantial difference in voltage headroom.

5.4.3 Distributed Amplifiers with Lumped Devices

The use of actual transmission lines in DAs has posed difficulties in VLSI technologies. The tight trade-off between the capacitance and loss of integrated T lines has made such implementations more amenable to GaAs technologies because their semi-insulating substrates and thick interconnect layers relax these trade-offs considerably. Fortunately, modern VLSI technologies offer a large number of metal layers (e.g., eight in the 0.13-μm generation) that can form transmission lines with reasonable performance. Nevertheless, it is sometimes desirable to construct DAs without T lines. For example, in the frequency range below 10 GHz, the lines would be prohibitively long (from both layout and loss points of view).

Since it is possible to approximate T lines by a series of lumped LC sections, we postulate that a DA can simply incorporate a chain of inductors and capacitors in both the input and output networks. Depicted in Fig. 5.46, such an arrangement may not require explicit capacitors as the transistor and inductor capacitances can be used to define the characteristic impedance and wave velocity.

The circuit of Fig. 5.46 becomes an attractive choice at low frequencies because, for a given series dc resistance, spiral inductors provide much more inductance than transmission lines do. Consequently, the wave velocity, $v = 1/\sqrt{L_0 C_0}$, can be *reduced* substantially, allowing a higher gain with a reasonable "length":

$$A_v = \pi f_T \frac{l}{v}. \tag{5.97}$$

Note that v can also be lowered by increasing C_0, but the resulting reduction of the characteristic impedance makes the design of the circuit driving the DA more difficult.

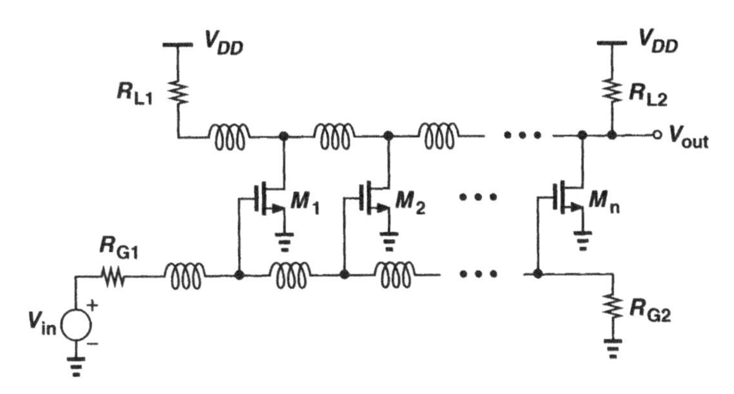

Figure 5.46 DA using lumped elements.

The choice between lumped inductors and T lines depends on both the operation frequency range and the environment. The transition from the former to the latter falls in the vicinity of 20 to 40 GHz.

5.5 Other Broadband Techniques

The problem of broadband design becomes particularly difficult in limiting amplifiers and laser drivers. The former require a multitude of gain stages, each of which limits the bandwidth, and the latter must deliver high currents, thus presenting a large input capacitance to the preceding stage. In this section, we study a number of circuit techniques that prove useful in the design of such amplifiers.

It is important to appreciate the speed limitations posed by a cascade of stages. For n identical gain cells, each having a bandwidth of f_0, the overall bandwidth is given by

$$f_{tot} = f_0 \sqrt[m]{2^{1/n} - 1}, \tag{5.98}$$

where m is equal to 2 for first-order stages and 4 for second-order stages [9]. For example, if a total bandwidth of 10 GHz is desired, then a five-stage amplifier requires an f_0 of 26 GHz for first-order cells and 16 GHz for second-order cells. Note that actual second-order gain stages suffer from other nonidealities, such as Miller effect, that exacerbate this issue [10].

5.5.1 T-Coil Peaking

As seen in Chapter 4, shunt inductive peaking increases the bandwidth by allowing less signal current through the load resistor and hence forcing a greater signal current through the load capacitance [Fig. 5.47(a)]. "T-coils" provide an alternative means of increasing the bandwidth.

Illustrated in Fig. 5.47(b), a T-coil consists of two coupled inductors and a bridge capacitor. The circuit has three terminals: E for receiving a current (from a transistor), F for connecting to the power supply through a resistor, and G for driving the load capacitance

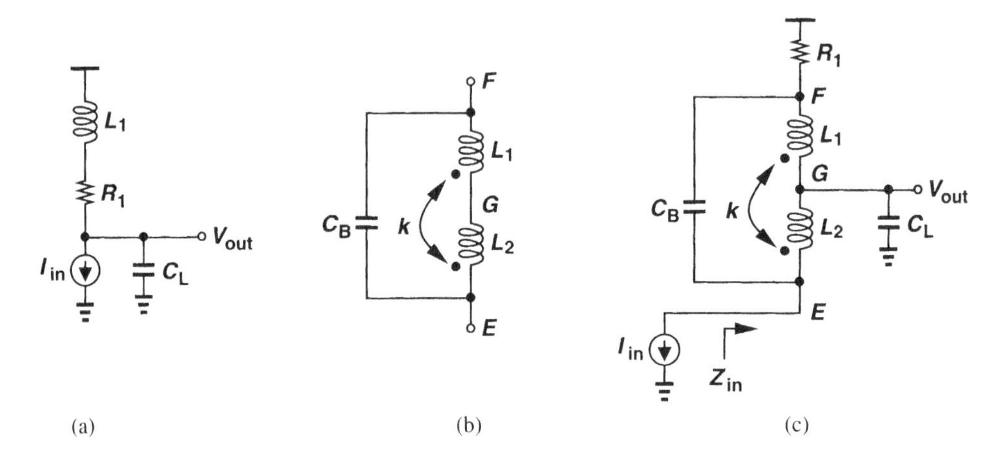

Figure 5.47 (a) Shunt peaking, (b) basic T-coil, (c) input impedance of T-coil with a capacitive load.

and hence the next stage. We then connect the T-coil to the surrounding components as shown in Fig. 5.47(c). This arrangement offers two interesting attributes [11], which we now study.

The first attribute is that the T-coil components can be chosen so that the input impedance, Z_{in}, in Fig. 5.47(c) remains resistive and equal to R_L at *all* frequencies and for *any* value of C_L. We can intuitively explain this property at the two extremes: at very low frequencies, L_1 and L_2 short E and F and, therefore, $Z_{in} = R_1$; at very high frequencies, L_1 and L_2 are open, C_B is a short, and again $Z_{in} = R_1$. It can be proved that $Z_{in} = R_1$ at all frequencies if the following conditions hold:

$$L_1 = L_2 = \frac{C_L R_1^2}{4}\left(1 + \frac{1}{4\zeta^2}\right) \tag{5.99}$$

$$C_B = \frac{C_L}{16\zeta^2} \tag{5.100}$$

$$k = \frac{4\zeta^2 - 1}{4\zeta^2 + 1}, \tag{5.101}$$

where ζ denotes the damping factor of the transfer function V_{out}/I_{in} [11, 12]. For a well-behaved frequency response (uniform group delay), we have $\zeta = \sqrt{3}/2$ and $k = 1/2$. The above conditions thus reduce to

$$L_1 = L_2 = \frac{C_L R_1^2}{3} \tag{5.102}$$

$$C_B = \frac{C_L}{12} \tag{5.103}$$

$$k = \frac{1}{2}. \tag{5.104}$$

As explained below, these equations readily allow us to design the T-coil components.

The second attribute of T-coil peaking is that it provides a greater bandwidth enhancement than shunt or series peaking does. If the conditions given by Eqs. (5.99)-(5.101) hold, then in Fig. 5.47(c),

$$\frac{V_{out}}{I_{in}} = \frac{R_1}{\frac{1}{4}(\frac{1-k}{1+k})R_1^2C_L^2s^2 + \frac{1}{2}R_1C_Ls + 1}. \tag{5.105}$$

With $k = 1/2$, the bandwidth is broadened by factor of 2.72, which is a 70% increase over that provided by inductive peaking with a similar response.

It is interesting to note that the circuit of Fig. 5.47(c) does not contain a zero even though C_B appears to provide a feedforward path. The absence of a zero can be explained as follows. Suppose at a complex frequency $s = s_1$, the output, V_{out}, is equal to zero. Then, C_L carries no current and I_{in} flows entirely through R_1, generating $V_F = -I_{in}R_1$. Since V_E is also equal to $-I_{in}R_1$, we conclude that the current through C_B is equal to zero and $V_{out} = V_E + (L_2 + M)s_1I_{in}$, where $M = k\sqrt{L_1L_2}$. This output voltage cannot be zero unless $I_{in} = 0$, contradicting our assumption that I_{in} is a finite excitation.

T-Coil Realization With $L_1 = L_2 = L$ in Fig. 5.47(c), we can design the T-coil as a symmetric spiral inductor as shown in Fig. 5.48(a) [11]. The center tap serves as the

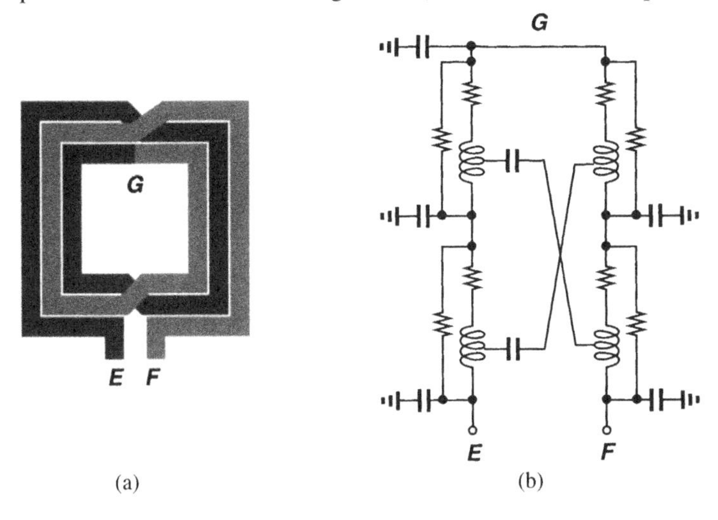

Figure 5.48 (a) Spiral T-coil, (b) equivalent circuit.

output node and the line spacing is chosen to yield $k = 1/2$. The design proceeds as follows. First, we note that the total inductance between E and F in Fig. 5.47(c) is equal to $L_1 + L_2 + 2M = 2L + 2kL = 3L$. Also, Eq. (5.102) requires that $3L = C_LR_T^2$. Next, based on the values of C_L and R_T, we make an initial guess for the number of turns and the outer dimension of the symmetric inductor to reach $L_{EF} = 3L = C_LR_T^2$. Finally, using a field solver such as ASITIC, we adjust the line spacing to obtain $k = 1/2$ [between L_{EG} and L_{GF} in Fig. 5.48(a)] and the outer dimension to restore the inductance value to

$C_L R_T^2$. This means that L_{EG} and L_{GF} must be simulated as two separate but interwound spirals so that their mutual coupling can be measured.

For broadband design, T-coils must be modeled such that their response remains accurate for at least the last decade of the band of interest. (It is assumed that the circuit plays a negligible role at lower frequencies.) Figure 5.48(b) depicts a distributed model that serves this purpose [11]. Here, the spiral is decomposed into a number of sections, each including an inductance, series and parallel resistances, and a parasitic capacitance to the substrate. The resistances are chosen to yield the estimated Q in the middle of the last decade of the frequency band. The fringe (interwinding) capacitance is also included. Since this capacitance equivalently appears between nodes E and F, it should be subtracted from the value of the bridge capacitance, C_B, in Fig. 5.47(c).

Differential Implementation Differential circuits employing T-coils must incorporate two symmetric spirals. Shown in Fig. 5.49 is an example, where a differential pair drives

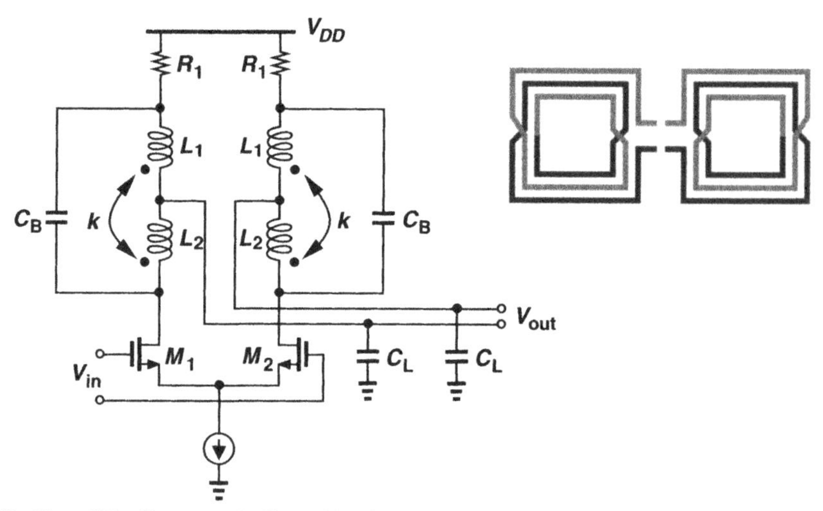

Figure 5.49 Use of T-coil to cancel effect of load capacitance.

the load capacitance through a T-coil. In this case, C_L represents the input capacitance of the next stage plus a fraction of the capacitance of each spiral to the substrate. The T-coils are placed symmetrically in the circuit's layout.

5.5.2 Negative Capacitance

In order to reduce the effect of capacitance at a given node, a network presenting a negative capacitance can be added to the node. Differential signal paths lend themselves to this method particularly well.

Negative capacitance can be realized through the use of Miller effect with positive feedback. As illustrated in Fig. 5.50(a), two capacitors are placed between the inputs and their corresponding noninverting outputs. The single-ended input capacitance resulting from C_F

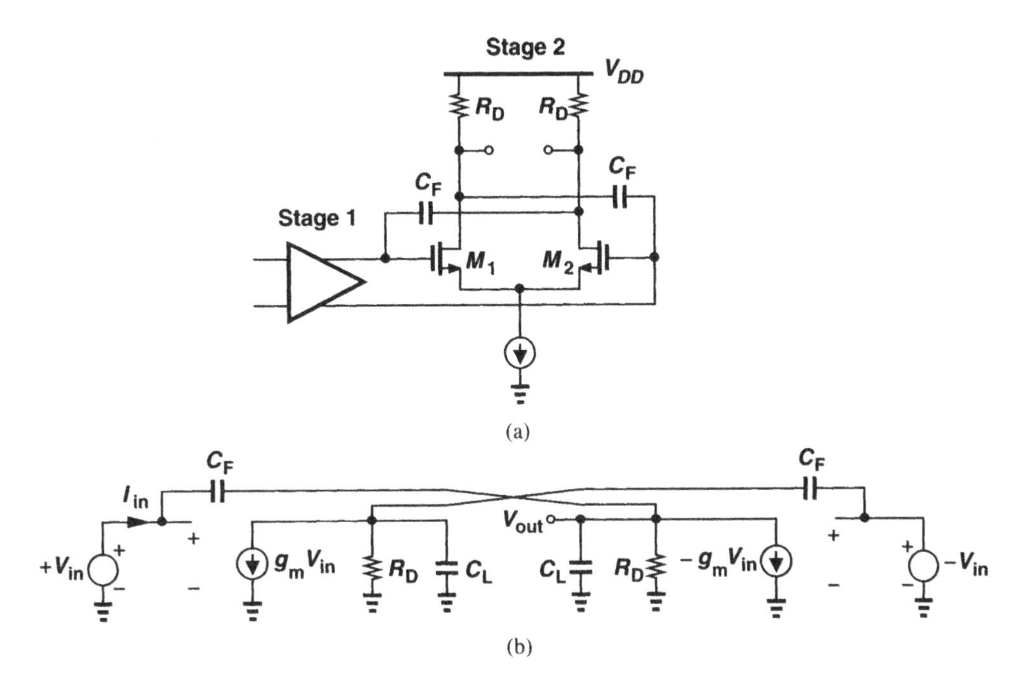

Figure 5.50 (a) Miller negative capacitance, (b) equivalent circuit.

is equal to

$$C_{in} = (1 - A_v)C_F \tag{5.106}$$

$$= (1 - g_mR_D)C_F, \tag{5.107}$$

where channel-length modulation is neglected. Thus, the value of C_F can be chosen so as to partially cancel the positive input capacitance. For example, if $C_F = C_{GD}$, then the effect of C_{GD} is removed, and if $C_F > C_{GD}$, then a fraction of C_{GS} is also cancelled.

The topology of Fig. 5.50(a) does have a bandwidth limitation: the feedback capacitors load the *output* nodes and hence cannot be very large. In other words, as C_F increases from zero, the bandwidth at the output of Stage 1 increases but, eventually, the bandwidth at the output of Stage 2 decreases. To analyze this trend, consider the small-signal equivalent circuit shown in Fig. 5.50(b), where the two input voltages are assumed to be differential. Neglecting C_L for now, we wish to determine the input impedance, V_{in}/I_{in}, and the transfer function, V_{out}/V_{in}. The current I_{in} flows through C_F and joins $+g_mV_{in}$, generating a single-ended output voltage equal to

$$V_{out} = (I_{in} + g_mV_{in})R_D. \tag{5.108}$$

Adding the voltage drop across C_F to this value and equating the result to V_{in}, we have

$$V_{in} = \frac{I_{in}}{C_F s} + (I_{in} + g_m V_{in})R_D. \tag{5.109}$$

It follows that

$$\frac{V_{in}}{I_{in}} = \frac{R_D}{1 - g_m R_D} + \frac{1}{(1 - g_m R_D)C_F s}. \tag{5.110}$$

The single-ended input impedance therefore consists of a resistance in series with a capacitance, both of which are negative if $g_m R_D \geq 1$. In practice, the first term is overwhelmed by the second.

To compute the transfer function, we find I_{in} from (5.110) and substitute in (5.108):

$$\frac{V_{out}}{V_{in}}(s) = \frac{g_m + C_F s}{R_D C_F s + 1} R_D. \tag{5.111}$$

As predicted, C_F creates a pole at the output node, possibly limiting the bandwidth.

The behavior of the above circuit becomes more complex if the load capacitance seen by Stage 2 is taken into account. To this end, we repeat the foregoing derivations but with R_D replaced by $R_D||(C_L s)^{-1} = R_D/(R_D C_L s + 1)$. It follows that

$$\frac{V_{in}}{I_{in}} = \frac{R_D(C_F + C_L)s + 1}{R_D C_L s + 1 - g_m R_D} \cdot \frac{1}{C_F s}. \tag{5.112}$$

The real and imaginary parts of this impedance are expressed as follows:

$$Re\{\frac{V_{in}}{I_{in}}\} = \frac{g_m C_L - (1 - g_m R_D)C_F}{R_D^2 C_L^2 \omega^2 - (1 - g_m R_D)^2} \cdot \frac{R_D}{C_F} \tag{5.113}$$

$$Im\{\frac{V_{in}}{I_{in}}\} = \frac{R_D^2(C_L + C_F)\omega^2 + 1 - g_m R_D}{R_D^2 C_L^2 \omega^2 - (1 - g_m R_D)^2} \cdot \frac{-1}{C_F \omega}. \tag{5.114}$$

As expected, at low frequencies, the ω^2 terms in (5.114) are negligible, yielding an input capacitance of $(1 - g_m R_D)C_F$. But as the frequency increases, the denominator of (5.114) reaches zero, producing zero Miller multiplication at a frequency of

$$\omega_1 = \frac{1}{R_D C_L} - \frac{g_m}{C_L}. \tag{5.115}$$

This value is *lower* than the output pole of Stage 2 before C_F is added, limiting the benefit of the negative Miller capacitance technique.

The feedback capacitors in Fig. 5.50(a) must track the transistor capacitances in the presence of process variations. If realized as a regular MOS device, C_F exhibits a small value because the difference between the input and output common levels of the differential pair is typically less than the threshold voltage, failing to create an inversion layer within the device. A preferred solution is to implement C_F as MOS varactors; as explained in Chapter 7, these devices provide a reasonable capacitance even with $V_{GS} = 0$.

Negative capacitance can also be created by means of a "negative impedance converter" (NIC). Shown in Fig. 5.51(a) is an example [10], where a cross-coupled pair, M_1-M_2,

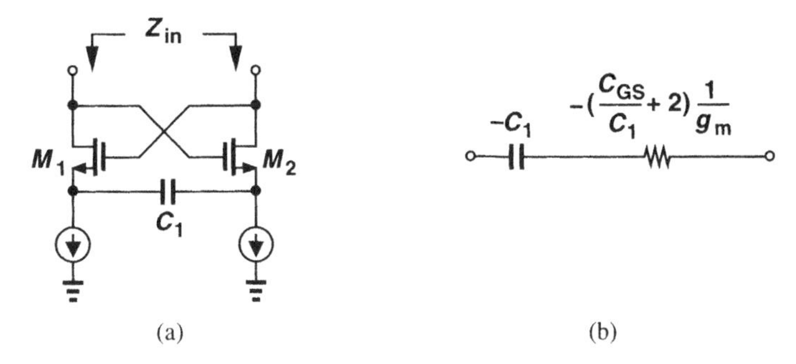

Figure 5.51 (a) Use of negative impedance converter to produce negative capacitance, (b) equivalent circuit.

transforms C_1 to a negative capacitance. The reader can show that

$$Z_{in} = \frac{1}{C_1 s}\frac{(C_{GS} + 2C_1)s + g_m}{-C_{GS}s + g_m} \tag{5.116}$$

As expected, for $C_1 = \infty$, Z_{in} reduces to a negative resistance of $2/g_m$ in parallel with a positive capacitance of $C_{GS}/2$. Let us now write

$$\frac{1}{Z_{in}} = \frac{-1 + C_{GS}/g_m}{\frac{1}{C_1 s} + (\frac{C_{GS}}{C_1} + 2)\frac{1}{g_m}}. \tag{5.117}$$

At frequencies well below the f_T of the transistors, the second term in the numerator is negligible, yielding an impedance consisting of a negative capacitance equal to $-C_1$ and a negative resistance equal to $-(C_{GS}/C_1 + 2)/g_m$ [Fig. 5.51(b)] [10].

Figure 5.52 shows a typical arrangement employing this technique to cancel the capacitance at nodes X and Y (primarily, the input capacitance of M_3-M_4). For complete cancellation, $2C_1 = C_X = C_Y$. However, if, due to mismatches, $2C_1 > C_X$, then the cross-coupled pair may turn into a relaxation oscillator. It can be shown that the condition of oscillation is given by [13]

$$g_{m1,2}R_D \geq \frac{C_1}{C_1 - C_X/2}. \tag{5.118}$$

For example, if $C_1 = 1.2(C_X/2)$ then $g_m R_D$ must remain less than 6 to avoid oscillation. Of course, even without actual oscillation, the circuit may still exhibit significant ringing and hence ISI. For this reason, the cross-coupled pair can only partially cancel C_X and C_Y if a small ISI is desired.

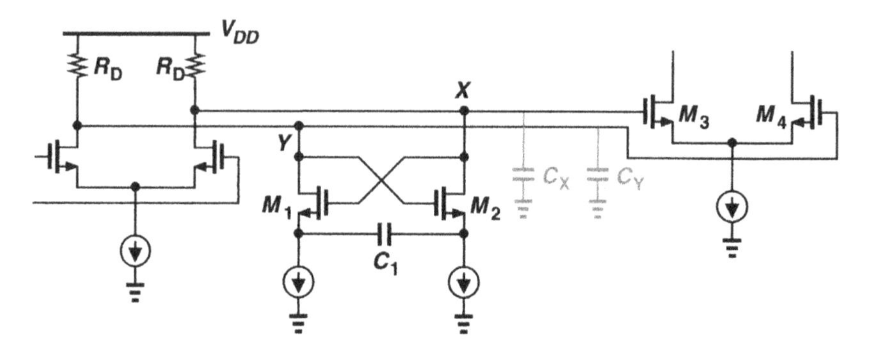

Figure 5.52 Use of negative capacitance to broaden bandwidth.

The negative resistance presented by the above cross-coupled pair can also lead to ISI. This occurs if inductors are added in series with the load resistors so as to further increase the bandwidth. If the magnitude of the negative resistance becomes comparable with R_D, then a topology resembling an LC oscillator is formed, producing ringing in the step response of the overall circuit. In practice, proper choice of C_1 cancels a considerable fraction of C_X and C_Y, and inductive peaking broadens the bandwidth with the remaining capacitance [10].

5.5.3 Active Feedback

A cascade of gain stages can incorporate active feedback to achieve a broader bandwidth. Consider the arrangement shown in Fig. 5.53, where each G_m stage can be realized as a

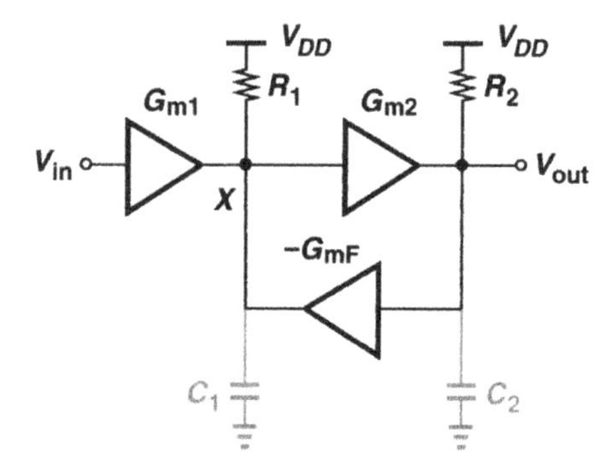

Figure 5.53 Active feedback.

simple differential pair [10]. The feedback circuit, $-G_{mF}$, senses the output voltage and returns a proportional current to node X. The transfer function is obtained by summing the currents drawn by G_{m1} and $-G_{mF}$ from R_1, multiplying the result by $R_1||(C_1s)^{-1}$ and G_{m2}, and equating this current to $V_{out}/[R_2||(C_2s)^{-1}]$:

$$G_{m2}\left[G_{m1}V_{in}R_1 - G_{mF}V_{out}\left(R_1||\frac{1}{C_1s}\right)\right] = V_{out}\left(\frac{R_2C_2s+1}{R_2}\right). \tag{5.119}$$

It follows that [10]

$$\frac{V_{out}}{V_{in}}(s) = \frac{A_0\omega_n^2}{s^2 + 2\zeta\omega_n s + \omega_n^2}, \tag{5.120}$$

where

$$A_0 = \frac{G_{m1}G_{m2}R_1R_2}{1 + G_{m2}G_{mF}R_1R_2} \tag{5.121}$$

$$\zeta = \frac{1}{2}\frac{R_1C_1 + R_2C_2}{\sqrt{R_1R_2C_1C_2(1 + G_{m2}G_{mF}R_1R_2}} \tag{5.122}$$

$$\omega_n^2 = \frac{1 + G_{m2}G_{mF}R_1R_2}{R_1R_2C_1C_2}. \tag{5.123}$$

For a maximally-flat Butterworth response, we choose $\zeta = \sqrt{2}/2$, obtaining a -3-dB bandwidth of $\omega_{-3dB} = \omega_n$. Multiplying (5.121) and (5.123) gives

$$A_0\omega_{-3dB}^2 = \frac{G_{m1}G_{m2}}{C_1C_2} \tag{5.124}$$

and hence

$$A_0\omega_{-3dB} = \frac{G_{m1}G_{m2}}{C_1C_2}\frac{1}{\omega_{-3dB}}. \tag{5.125}$$

Since G_{m1}/C_1 and G_{m2}/C_2 are dominated by C_{GS} and on the order of $2\pi f_T$, we have

$$A_0\omega_{-3dB} = f_T\frac{f_T}{f_{-3dB}}, \tag{5.126}$$

where $f_{-3dB} = \omega_{-3dB}/(2\pi)$. In other words, active feedback increases the gain-bandwidth product beyond the f_T of the technology by a factor equal to the ratio of f_T and the cell bandwidth [10].

How much does active feedback improve the bandwidth for a given voltage gain? To arrive at some intuitive results, let us assume $R_1 = R_2 = R$, $C_1 = C_2 = C$, and $G_{m1} = G_{m2} = G_m$ in Fig. 5.53. Setting ζ to $\sqrt{2}/2$ in Eq. (5.122), we obtain

$$1 + G_mG_{mF}R^2 = 2 \tag{5.127}$$

and

$$A_0 = \frac{G_m^2R^2}{2} \tag{5.128}$$

$$\omega_n = \frac{\sqrt{2}}{RC}. \tag{5.129}$$

We now eliminate the active feedback circuit (the G_{mF} stage) and lower the load resistance so as to obtain the same gain as with feedback. From Eq. (5.128), this requires that the new load resistance, R', be equal to $R/\sqrt{2}$. The bandwidth of the simple cascade is given by Eq. (5.98):

$$\omega_{-3dB} = \frac{1}{R'C}\sqrt[2]{2^{1/2}-1} \tag{5.130}$$

$$\approx \frac{0.644}{R'C} \tag{5.131}$$

$$\approx \frac{0.644\sqrt{2}}{RC}. \tag{5.132}$$

By comparison, Eq. (5.129) reveals a bandwidth of $\sqrt{2}/(RC)$ with active feedback, a factor of 1.55 higher.

It is possible to combine broadband techniques in amplifiers to achieve greater speeds. For example, [14] employs active feedback and T-coil peaking in a Cherry-Hooper stage.

5.5.4 Triple-Resonance Peaking

The bandwidth of cascaded stages can be increased by a factor of $2\sqrt{3} \approx 3.5$ using "triple-resonance peaking" (TRP) [15]. We study this technique here.

Triple-Resonance Topology To arrive at the concept of TRP, first consider the inductively-peaked cascade of two stages shown in Fig. 5.54(a), where it is assumed M_1 and M_2 contribute approximately equal capacitances ($C/2$) to node X. As the frequency approaches $\omega_1 = 1/\sqrt{L_1 C}$, the impedance of L_1 rises, allowing a greater fraction of I_{D1} to flow through $C_1 + C_2$ and hence extend the bandwidth. This is the conventional shunt-peaking operation.

Let us now insert an inductor L_2 in series with C_2 [Fig. 5.54(b)] such that L_2 and C_2 resonate at ω_1, thus acting as a *short* and absorbing all of I_{D1}. In this case, I_{D1} flows through L_2 and C_2 rather than $C_1 + C_2$, producing a more gradual gain roll-off. For L_2 and C_2 to resonate at ω_1, we have

$$L_2 = 2L_1. \tag{5.133}$$

(If C_1 and C_2 are not equal, L_1/L_2 can be slightly adjusted.)

To minimize peaking, the output voltage at this frequency, $I_{D1}/(C_2\omega_1)$, must be equal to that at low frequencies, $I_{D1}R_1$, yielding

$$R_1 = 2\sqrt{\frac{L_1}{C}}. \tag{5.134}$$

This amplifier exhibits the frequency response shown in Fig. 5.54(c), revealing *three* distinct resonance frequencies. To understand the operation of the circuit and derive the required relationships among different components, we examine the frequency response around each of the resonance frequencies.

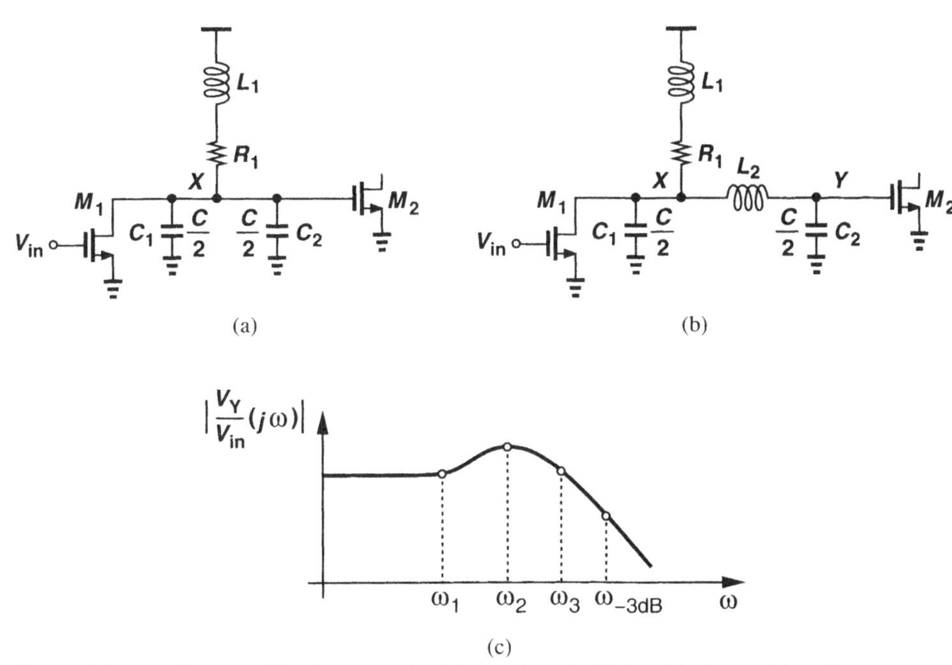

Figure 5.54 (a) Shunt peaking in a cascade, (b) insertion of additional inductor, (c) triple resonance behavior.

Frequency Response Let us begin with the simplified model shown in Fig. 5.55(a). It is helpful to first compute the input impedance of the π network consisting of C_1, L_2 and C_2:

$$Z_\pi(s) = \frac{L_2C_2s^2 + 1}{L_2C_1C_2s^2 + C_1 + C_2} \cdot \frac{1}{s}. \tag{5.135}$$

For $s = j\omega$,

$$Z_\pi(j\omega) = \frac{1 - L_2C_2\omega^2}{C_1 + C_2 - L_2C_1C_2\omega^2} \cdot \frac{1}{j\omega}. \tag{5.136}$$

We thus observe a series resonance between L_2 and C_2 and a parallel resonance between L_2 and $C_1C_2/(C_1 + C_2)$. At $\omega = \omega_1 = 1/\sqrt{L_2C_2}$, the series resonance of L_2 and C_2 forces all of I_{in} to flow through C_2. Moreover, this resonance also changes the *sign* of the network's input impedance, Z_X, because around $\omega = \omega_1$, we have $Z_X(j\omega) \approx Z_\pi(j\omega)$ (why?). That is, V_X becomes *negative* as ω exceeds ω_1. Illustrated in Fig. 5.55(b), this behavior requires that currents I_1 and I_2 flow from ground to node X. Thus, the total current flowing through C_2 reaches $I_1 + I_2 + I_{in}$. We now observe that (a) the capacitive current, I_2, generates a relatively frequency-independent output voltage, and (b) the inductive current, I_1, introduces a roll-up in the frequency response.

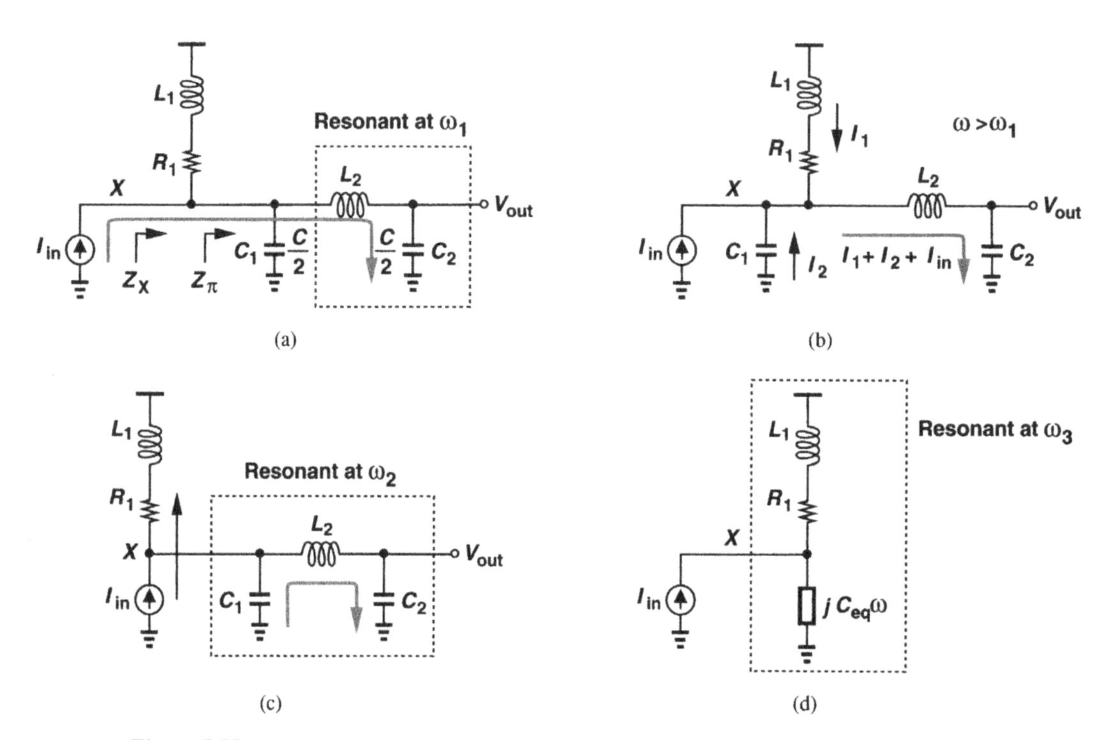

Figure 5.55 Behavior of triple-resonance network at different frequencies.

Due to the roll-up, $|V_{out}/I_{in}|$ continues to rise with frequency until the π network begins to resonate [Fig. 5.55(c)]. This resonance frequency is given by

$$\omega_2 = \frac{1}{\sqrt{L_2 \dfrac{C_1 C_2}{C_1 + C_2}}} \tag{5.137}$$

$$= \sqrt{2}\omega_1. \tag{5.138}$$

How does the π network behave at this frequency? First, it presents an *infinite* impedance to node X because $L_2 j\omega_2 + (C_2 j\omega_2)^{-1} = -(C_1 j\omega_2)^{-1}$, i.e., the two parallel branches (C_1, L_2 in series with C_2) exhibit equal and opposite impedances at $\omega = \omega_2$. As a result, all of I_{in} flows through R_1 and L_1, yielding $V_X = I_{in}(R_1 + L_1 s)$. Second, since C_1 and C_2 carry equal and opposite currents at this frequency,

$$|V_{out}| = |V_X| \tag{5.139}$$

$$= |I_{in}|\sqrt{R_1^2 + L_1^2\omega_2^2} \tag{5.140}$$

$$= |I_{in}|\sqrt{\frac{3}{2}}R_1. \tag{5.141}$$

(Recall that R_1 was chosen equal to $2\sqrt{L_1/C}$.) It follows that the frequency response of the triple-resonance network exhibits a peaking of $\sqrt{3/2} \approx 1.8$ dB.

For $\omega > \omega_2$, the input impedance of the π network becomes capacitive [Eq. (5.136)] and $|V_{out}/I_{in}|$ begins to fall. As shown in Fig. 5.55(d), this capacitive impedance eventually resonates with L_1 at a third frequency, ω_3, which is obtained by finding $(R_1 + L_1 j\omega)||Z_\pi(j\omega)$ and setting its phase to zero. Since $L_2 = 2L_1$, $C_1 = C_2 = C$, and $R_1 = 2\sqrt{L_1/C}$, we have

$$\omega_3 = \sqrt[4]{6}\omega_1. \tag{5.142}$$

The -3-dB bandwidth exceeds this value and is approximately equal to

$$\omega_3 \approx \sqrt{3}\omega_1 \tag{5.143}$$

$$\approx \frac{2\sqrt{3}}{R_1 C}. \tag{5.144}$$

That is, triple-resonance peaking improves the bandwidth by a factor of $2\sqrt{3}$ in the ideal case. With inductor parasitics, especially those of L_2, the factor is somewhat less. Also, this technique requires twice as many inductors as simple inductive peaking or T-coil peaking.

An alternative topology called the "reversed triple-resonance network" (RTRN) has been described in [16]. Illustrated in Fig. 5.56, this method places the series RL branch

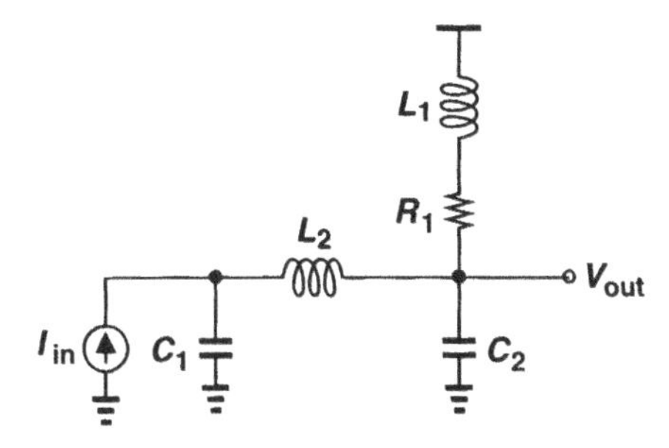

Figure 5.56 Alternative triple-resonance network.

at the output node rather than at the input node. If C_1 and C_2 are unequal, i.e., if $C_1 = (1 - \alpha)C/2$ and $C_2 = (1 + \alpha)C/2$, then L_1 and L_2 can be chosen so as to obtain a flat frequency response [16]:

$$L_1 = (1 - \alpha)^2 C R_1^2/4 \tag{5.145}$$

$$L_2 = (1 - \alpha) C R_1^2/2. \tag{5.146}$$

References

1. B. Razavi, *Design of Analog CMOS Integrated Circuits*, Chicago: McGraw-Hill, 2001.

2. E. M. Cherry and D. E. Hooper, "The Design of Wideband Transistor Feedback Amplifiers," *Proc. IEE*, vol. 110, pp. 375–389, Feb. 1963.
3. H.-M. Rein and M. Moller, "Design Considerations for Very High Speed Si Bipolar ICs Operating up to 50 Gb/s," *IEEE Journal of Solid-State Circuits*, vol. 31, pp. 1076–1090, August 1996.
4. E. L. Ginzton et al, "Distributed Amplification," *Proc. IRE*, vol. 36, pp. 956–969, Aug. 1948.
5. R. S. Carson, *High-Frequency Amplifiers*, 2nd Ed., Wiley, 1982.
6. R. A. Pucel, D. J. Masse, and C. P. Hartwig, "Losses in Microstrip," *IEEE Trans. Microwave Theory and Techniques*, vol. MTT-16, pp. 342–350, June 1968.
7. V. Milanovic et al., "Micromachined Microwave Transmission Lines in CMOS Technology," *IEEE Trans. Microwave Theory and Techniques*, vol. 45, pp. 630–635, May 1997.
8. H. Shigematsu et al., "A 49-GHz Preamplifier with a Transconductance Gain of 52 dBΩ Using InP HEMTs," *IEEE Journal of Solid-State Circuits*, vol. 36, pp. 1309–1313, Sept. 2001.
9. R. P. Jindal, "Gigahertz-band high-gain low-noise AGC amplifiers in fine-line NMOS," *IEEE J. Solid-State Circuits*, vol. 22, pp. 512.521, Aug. 1987.
10. S. Galal and B. Razavi, "10-Gb/s limiting amplifier and laser/modulator driver in 0.18-um CMOS techno logy," *IEEE Journal of Solid-State Circuits*, vol. 38, pp. 2138-2146, Dec. 2003.
11. S. Galal and B. Razavi, "Broadband ESD protection circuits in CMOS technology," *IEEE Journal of Solid-State Circuits*, vol. 38, pp. 2334-2340, Dec. 2003.
12. D. L. Feucht, *Handbook of Analog Circuit Design*, San Diego, CA: Academic, 1990.
13. B. Razavi, "A study of phase noise in CMOS oscillators," *IEEE Journal of Solid-State Circuits*, vol. 31, pp. 331-343, March 1996.
14. C. Lee and S.-I. Liu, "A 35-Gb/s limiting amplifier in 0.13-um CMOS technology," *Dig. VLSI Symp. Circuits*, pp. 152-153, 2006.
15. S. Galal and B. Razavi, "40-Gb/s amplifier and ESD protection circuit in 0.18-um CMOS technology," *IEEE Journal of Solid-State Circuits*, vol. 39, pp. 2389-2396, Dec. 2004.
16. C.-F. Liao and S.-I. Liu, "40 Gb/s transimpedance-AGC amplifier and CDR circuit for b roadband data receivers in 90 nm CMOS," *IEEE Journal of Solid-State Circuits*, vol. 43, pp. 642-655, March 2008.

Chapter 6

Oscillator Fundamentals

Oscillators are an integral part of many electronic systems. Applications range from clock generation in microprocessors to carrier synthesis in cellular telephones, requiring vastly different oscillator topologies and performance parameters. Robust, high-performance oscillator design in CMOS technology continues to pose interesting challenges. As described in Chapter 8, oscillators are usually embedded in a phase-locked system.

This chapter deals with the analysis and design of CMOS oscillators, more specifically, voltage-controlled oscillators (VCOs). Beginning with a general study of oscillation in feedback systems, we introduce ring oscillators and LC oscillators along with methods of varying the frequency of oscillation. We then describe a mathematical model of VCOs that will be used in the analysis of PLLs in Chapter 8.

6.1 General Considerations

A simple oscillator produces a periodic output, usually in the form of voltage. As such, the circuit has no input while sustaining the output indefinitely. How can a circuit oscillate? Negative feedback systems may oscillate, i.e., an oscillator is a badly-designed feedback amplifier![1] Consider the unity-gain negative feedback circuit shown in Fig. 6.1, where

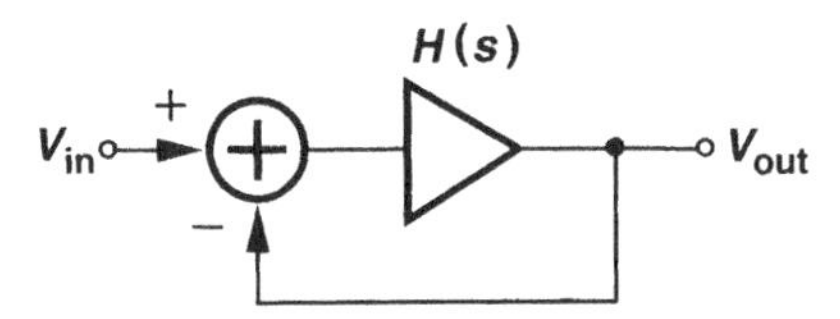

Figure 6.1 Feedback system.

$$\frac{V_{out}}{V_{in}}(s) = \frac{H(s)}{1 + H(s)}. \tag{6.1}$$

[1] It is said, "In the high-frequency world, amplifiers oscillate and oscillators don't."

If the amplifier itself experiences so much phase shift at high frequencies that the overall feedback becomes positive, then oscillation may occur. More accurately, if for $s = j\omega_0$, $H(j\omega_0) = -1$, then the closed-loop gain approaches infinity at ω_0. Under this condition, the circuit amplifies its own noise components at ω_0 indefinitely. In fact, as conceptually illustrated in Fig. 6.2, a noise component at ω_0 experiences a total gain of unity and a phase shift of 180°, returning to the subtractor as a negative replica of the input. Upon

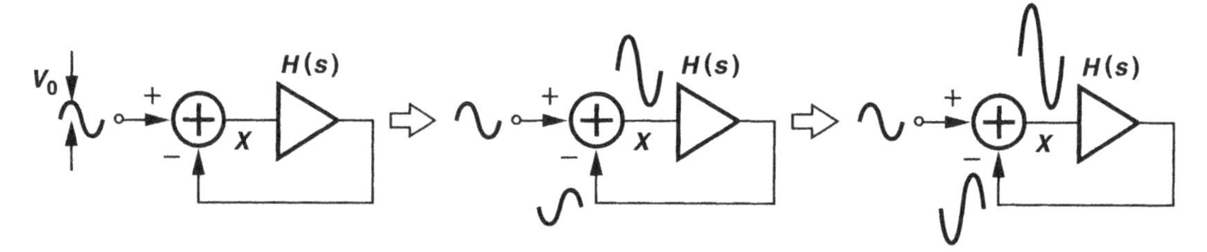

Figure 6.2 Evolution of oscillatory system with time.

subtraction, the input and the feedback signals give a larger difference. Thus, the circuit continues to "regenerate," allowing the component at ω_0 to grow.

For the oscillation to begin, a loop gain of unity or greater is necessary. This can be seen by following the signal around the loop over many cycles and expressing the amplitude of the subtractor's output in Fig. 6.2 as a geometric series (if $\angle H(j\omega_0) = 180°$):

$$V_X = V_0 + |H(j\omega_0)|V_0 + |H(j\omega_0)|^2 V_0 + |H(j\omega_0)|^3 V_0 + \cdots. \tag{6.2}$$

If $|H(j\omega_0)| > 1$, the above summation diverges whereas if $|H(j\omega_0)| < 1$, then

$$V_X = \frac{V_0}{1 - |H(j\omega_0)|} < \infty. \tag{6.3}$$

In summary, if a negative-feedback circuit has a loop gain that satisfies two conditions:

$$|H(j\omega_0)| \geq 1 \tag{6.4}$$

$$\angle H(j\omega_0) = 180°, \tag{6.5}$$

then the circuit may oscillate at ω_0. Called "Barkhausen criteria," these conditions are necessary but not sufficient [1]. To ensure oscillation in the presence of temperature and process variations, we typically choose the loop gain to be at least twice or three times the required value.

We may state the second Barkhausen criterion as $\angle H(j\omega) = 180°$ or a *total* phase shift of 360°. This should not be confusing: if the system is designed to have a low-frequency negative feedback, it already produces 180° of phase shift in the signal traveling around the loop (as represented by the subtractor in Fig. 6.1), and $\angle H(j\omega) = 180°$ denotes an additional *frequency-dependent* phase shift that, as illustrated in Fig. 6.2, ensures the feedback signal *enhances* the original signal. Thus, the three cases illustrated in Fig. 6.3 are

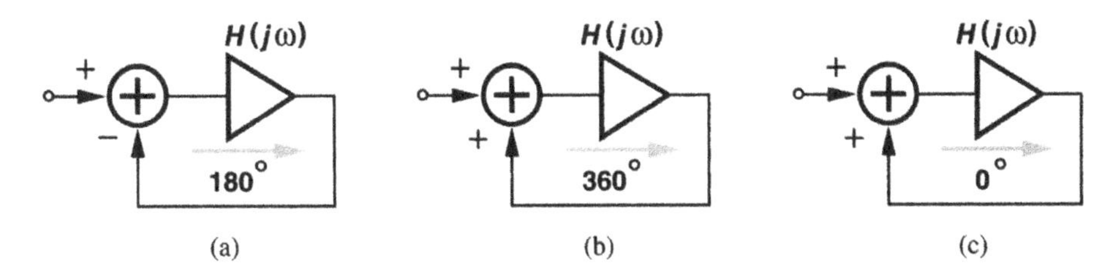

Figure 6.3 Various views of oscillatory feedback system.

equivalent in terms of the second criterion. We say the system of Fig. 6.3(a) exhibits a frequency-dependent phase shift of 180° (denoted by the arrow) and a dc phase shift of 180°. The difference between Figs. 6.3(b) and (c) is that the open-loop amplifier in the former contains enough stages with proper polarities to provide a total phase shift of 360° at ω_0 whereas that in the latter produces *no* phase shift at ω_0. Examples of these topologies are presented later in this chapter.

CMOS oscillators in today's technology are typically implemented as "ring oscillators" or "LC oscillators." We study each type in the following sections.

6.2 Ring Oscillators

A ring oscillator consists of a number of gain stages in a loop. To arrive at the actual implementation, we begin by attempting to make a single-stage feedback circuit oscillate.

As an example, let us explain why a single common-source stage does not oscillate if it is placed in a unity-gain loop. From Fig. 6.4, it is seen that the open-loop circuit contains only one pole, thereby providing a maximum frequency-dependent phase shift of 90° (at

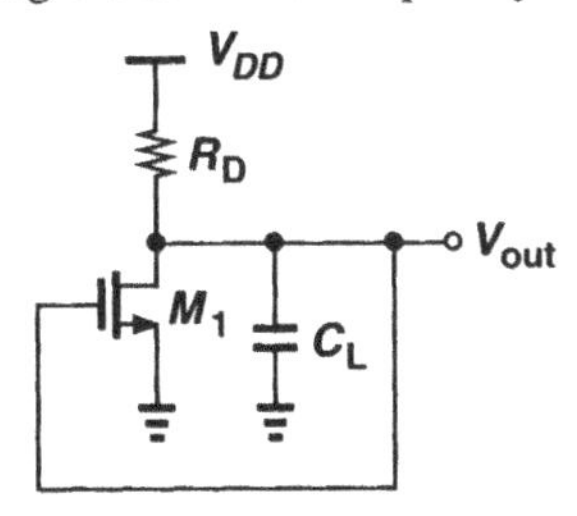

Figure 6.4 CS stage with feedback.

a frequency of infinity). Since the common-source stage exhibits a dc phase shift of 180° due to the signal inversion from the gate to the drain, the maximum total phase shift is 270°. The loop therefore fails to sustain oscillation growth.

The above example suggests that oscillation may occur if the circuit contains multiple stages and hence multiple poles. Indeed, such a topology is considered *undesirable* in amplifier design because it leads to inadequate phase margin in op amps. We therefore surmise that if the circuit of Fig. 6.4 is modified as shown in Fig. 6.5, then two significant poles appear in the signal path, allowing the frequency-dependent phase shift to approach 180°. Unfortunately, this circuit exhibits *positive* feedback near zero frequency due to the

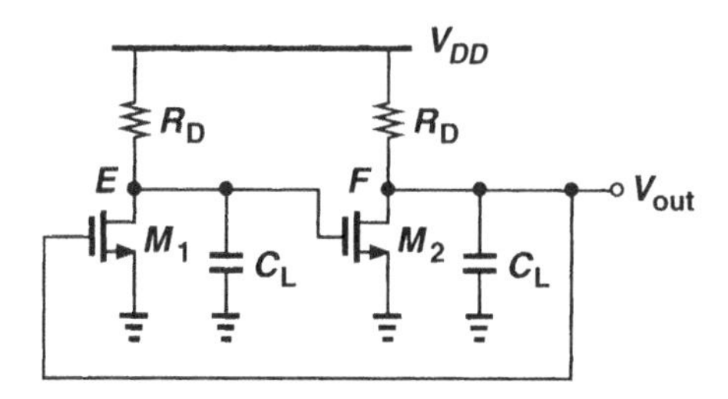

Figure 6.5 Two-pole feedback system.

signal inversion through each common-source stage. As a result, it simply "latches up" rather than oscillates. That is, if V_E rises, V_F falls, thereby turning M_1 off and allowing V_E to rise further. This may continue until V_E reaches V_{DD} and V_F drops to near zero, a state that will remain indefinitely.

To gain more insight into the oscillation conditions, let us assume an ideal inverting stage (with zero phase shift at all frequencies) is inserted in the loop of Fig. 6.5, providing *negative* feedback near zero frequency and eliminating the problem of latch-up (Fig. 6.6). Does this circuit oscillate? We note that the loop contains only two poles: one at E and

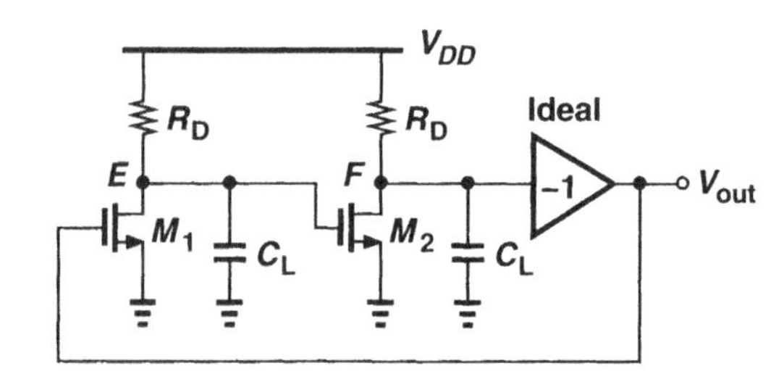

Figure 6.6 Two-pole feedback system with additional signal inversion.

another at F. The frequency-dependent phase shift can therefore reach 180°, but at a frequency of infinity. Since the loop gain vanishes at very high frequencies, we observe that the circuit does not satisfy both of Barkhausen's criteria at the same frequency (Fig. 6.7), failing to oscillate.

The foregoing discussion points to the need for greater phase shift around the loop, suggesting the possibility of oscillation if the third inverting stage in Fig. 6.6 contains a pole that contributes significant phase. We then arrive at the topology depicted in Fig. 6.8. If the three stages are identical, the total phase shift around the loop, ϕ, reaches $-135°$ at $\omega = \omega_{p,E} (= \omega_{p,F} = \omega_{p,G})$ and $-270°$ at $\omega = \infty$. Consequently, ϕ equals $-180°$ at $\omega < \infty$, where the loop gain can be still greater than or equal to unity. This circuit indeed oscillates if the loop gain is sufficient, and it is an example of a ring oscillator.

It is instructive to calculate the minimum voltage gain per stage in Fig. 6.8 that is necessary for oscillation. Neglecting the effect of the gate-drain overlap capacitance and de-

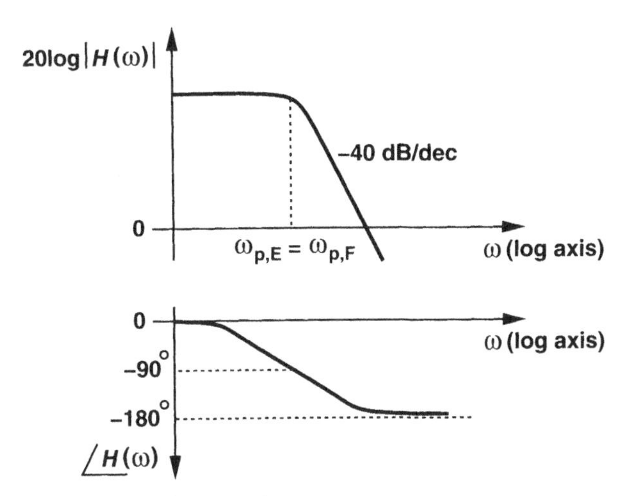

Figure 6.7 Loop gain characteristics of a two-pole system.

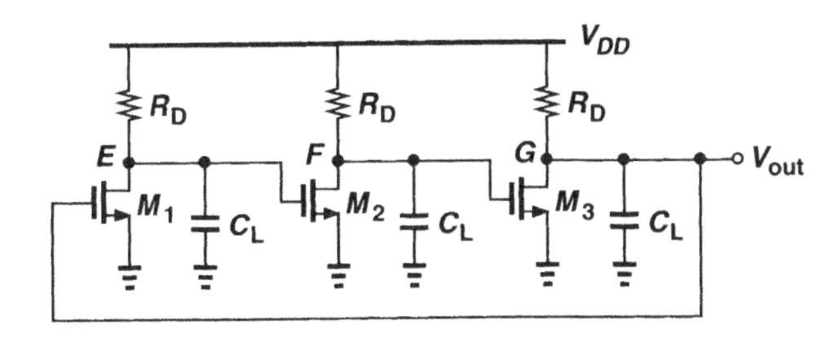

Figure 6.8 Three-stage ring oscillator.

noting the transfer function of each stage by $-A_0/(1+s/\omega_0)$, we have for the loop gain:

$$H(s) = -\frac{A_0^3}{(1+\dfrac{s}{\omega_0})^3}. \tag{6.6}$$

The circuit oscillates only if the frequency-dependent phase shift equals 180°, i.e., if each stage contributes 60°. The frequency at which this occurs is given by

$$\tan^{-1}\frac{\omega_{osc}}{\omega_0} = 60° \tag{6.7}$$

and hence

$$\omega_{osc} = \sqrt{3}\omega_0. \tag{6.8}$$

The minimum voltage gain per stage must be such that the magnitude of the loop gain at ω_{osc} is equal to unity:

$$\frac{A_0^3}{\left[\sqrt{1+(\frac{\omega_{osc}}{\omega_0})^2}\right]^3} = 1. \tag{6.9}$$

It follows from (6.8) and (6.9) that

$$A_0 = 2. \tag{6.10}$$

In summary, a three-stage ring oscillator requires a low-frequency gain of 2 per stage, and it oscillates at a frequency of $\sqrt{3}\omega_0$, where ω_0 is the 3-dB bandwidth of each stage.

Let us now examine the waveforms at the three nodes of the oscillator of Fig. 6.8. Since each stage contributes a frequency-dependent phase shift of 60° as well as a low-frequency signal inversion, the waveform at each node is 240° (or 120°) out of phase with respect to its neighboring nodes (Fig. 6.9). The ability of generating multiple phases is a very useful

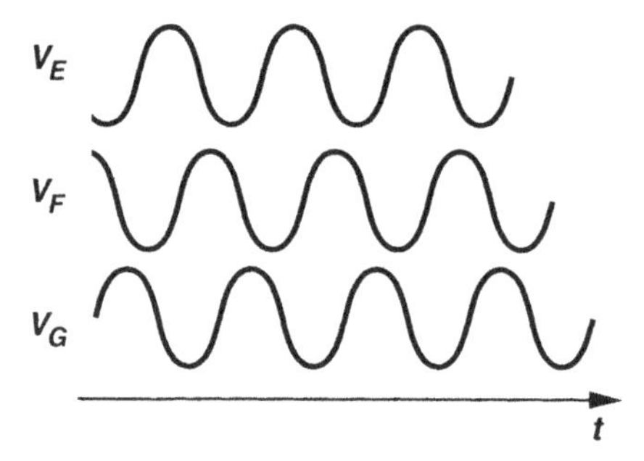

Figure 6.9 Waveforms of a three-stage ring oscillator.

property of ring oscillators.

Amplitude Limiting The natural question at this point is: what happens if in the three-stage ring of Fig. 6.8, $A_0 \neq 2$? We know from Barkhausen's criteria that if $A_0 < 2$, the circuit fails to oscillate, but what if $A_0 > 2$? To answer this question, we first model the oscillator by a linear feedback system, as depicted in Fig. 6.10. Note that the feedback is positive (i.e., V_{out} is *added* to V_{in}) because $H(s)$ in Eq. (6.6) already includes the negative polarity resulting from three inversions in the signal path. The closed-loop transfer function is:

$$\frac{V_{out}(s)}{V_{in}(s)} = \frac{\frac{-A_0^3}{(1+s/\omega_0)^3}}{1+\frac{A_0^3}{(1+s/\omega_0)^3}} \tag{6.11}$$

$$= \frac{-A_0^3}{(1+s/\omega_0)^3 + A_0^3}. \tag{6.12}$$

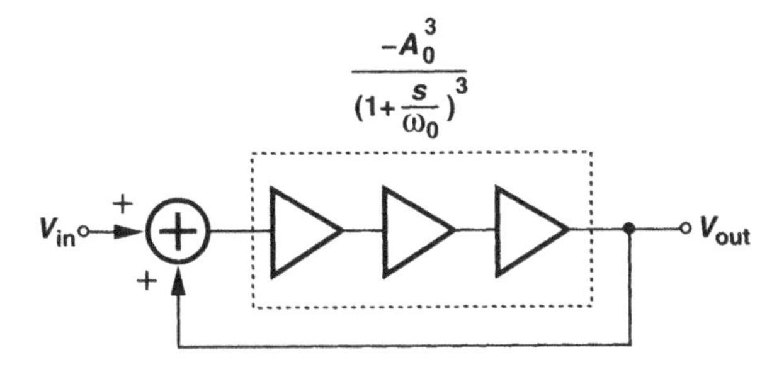

Figure 6.10 Linear model of three-stage ring oscillator.

The denominator of (6.12) can be expanded as:

$$(1+\frac{s}{\omega_0})^3 + A_0^3 = (1+\frac{s}{\omega_0}+A_0)\left[(1+\frac{s}{\omega_0})^2 - (1+\frac{s}{\omega_0})A_0 + A_0^2\right]. \tag{6.13}$$

Thus, the closed-loop system exhibits three poles:

$$s_1 = (-A_0 - 1)\omega_0 \tag{6.14}$$

$$s_{2,3} = [\frac{A_0(1 \pm j\sqrt{3})}{2} - 1]\omega_0. \tag{6.15}$$

Since A_0 itself is positive, the first pole leads to a decaying exponential term, $\exp[(-A_0 - 1)\omega_0 t]$, which can be neglected in the steady state. Figure 6.11 illustrates the locations of the poles for different values of A_0, revealing that for $A_0 > 2$, the two

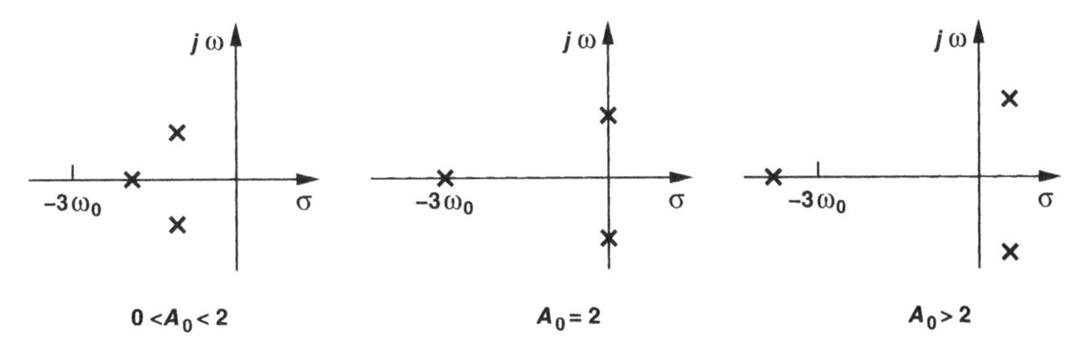

Figure 6.11 Poles of three-stage ring oscillator for various values of gain.

complex poles exhibit a positive real part and hence give rise to a growing sinusoid. Neglecting the effect of s_1, we express the output waveform as

$$V_{out}(t) = a\exp(\frac{A_0 - 2}{2}\omega_0 t)\cos(\frac{A_0\sqrt{3}}{2}\omega_0 t). \tag{6.16}$$

Thus, if $A_0 > 2$, the exponential envelope grows to infinity.

In practice, as the oscillation amplitude increases, the stages in the signal path experience nonlinearity and eventually "saturation," limiting the maximum amplitude. We may say the poles begin in the right half plane and eventually move to the imaginary axis to stop the growth. If the small-signal loop gain is greater than unity, the circuit must spend enough time in saturation so that the "average" loop gain is still equal to unity.[2]

As an example, consider Fig. 6.12, which shows a differential implementation of the oscillator of Fig. 6.8. What is the maximum voltage swing of each stage? If the gain per

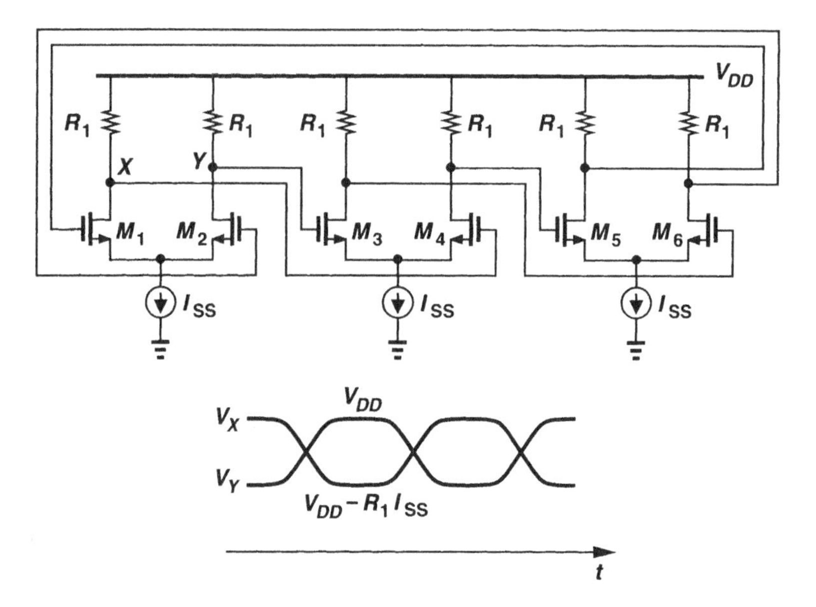

Figure 6.12 Differential oscillator.

stage is well above 2, then the amplitude grows until each differential pair experiences complete switching, that is, until I_{SS} is completely steered to one side every half cycle. As a result, the swing at each node is equal to $I_{SS}R_1$. From the waveforms shown in Fig. 6.12, we also observe that each stage is in its high-gain region for only a fraction of the period, (e.g., when $|V_X - V_Y|$ is small).

A simple implementation of ring oscillators that does not require resistors is depicted in Fig. 6.13. Suppose the circuit is released with an initial voltage at each node equal to the trip point of the inverters, V_{trip}.[3] With identical stages and no noise in the devices, the circuit would remain in this state indefinitely,[4] but noise components disturb each node voltage, yielding a growing waveform. The signal eventually exhibits rail-to-rail swings.

---

[2] While intuitive, these statements are not rigorous. The concepts of transfer function, poles, and loop gain are difficult to apply to a nonlinear circuit.

[3] The trip point of an inverter is the input voltage that results in an equal output voltage.

[4] This is indeed how SPICE predicts the circuit's behavior. To start the oscillation in SPICE, one of the nodes must be initialized at a different voltage.

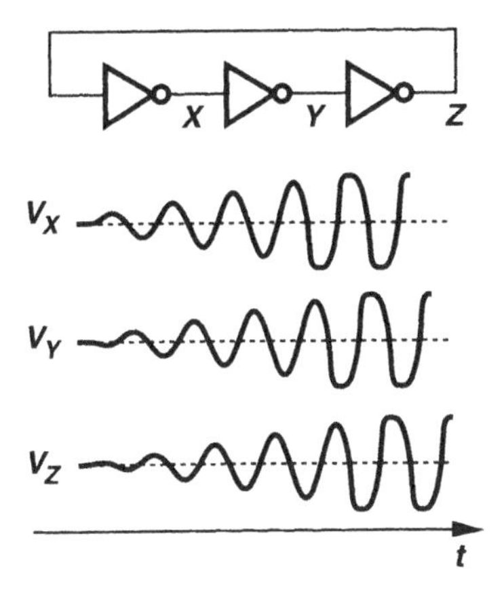

Figure 6.13 Ring oscillator using CMOS inverters.

Let us now assume the circuit of Fig. 6.13 begin with $V_X = V_{DD}$ (Fig. 6.14). Under this condition, $V_Y = 0$ and $V_Z = V_{DD}$. Thus, when the circuit is released, V_X begins to fall

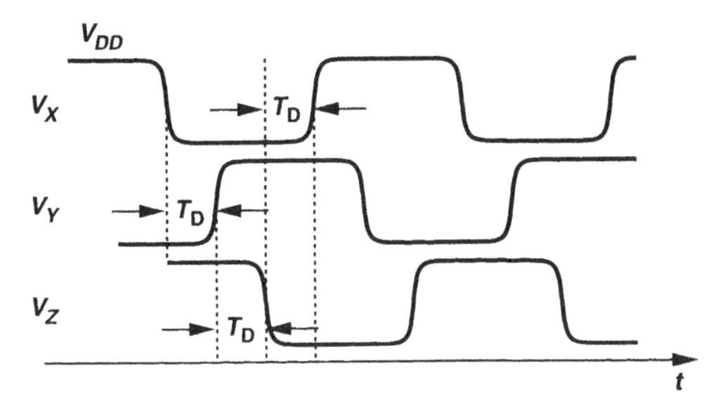

Figure 6.14 Waveforms of ring oscillator when one node is initialized at V_{DD}.

to zero (because the first inverter senses a high input), forcing V_Y to rise to V_{DD} after one inverter delay, T_D, and V_Z to fall to zero after another inverter delay. The circuit therefore oscillates with a delay of T_D between consecutive node voltages, yielding a period of $6T_D$.

The above small-signal and large-signal analyses raise an interesting question. While the small-signal oscillation frequency is given by $A_0\sqrt{3}\omega_0/2$ [from Eq. (6.16)], the large-signal value is $1/(6T_D)$. Are these two values equal? Not necessarily. After all, ω_0 is determined by the small-signal output resistance and capacitance of each inverter near the trip point whereas T_D results from the large-signal, nonlinear current drive and capacitances of each stage. In other words, when the circuit is released with all inverters at their trip point, the oscillation begins with a frequency of $\sqrt{3}A_0\omega_0/2$ but, as the amplitude grows and the circuit becomes nonlinear, the frequency shifts to $1/(6T_D)$ (which is a lower value).

Ring oscillators employing more than three stages are also feasible. The total number of inversions in the loop must be odd so that the circuit does not latch up. For example, as shown in Fig. 6.15(a), a ring can incorporate five inverters, providing a frequency of

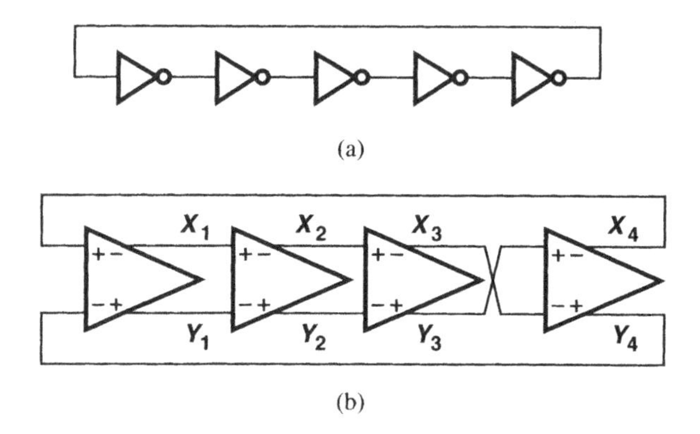

Figure 6.15 (a) Five-stage single-ended ring oscillator, (b) four-stage differential ring oscillator.

$1/(10T_D)$. On the other hand, the differential implementation can utilize an *even* number of stages by simply configuring one stage such that it does not invert. Illustrated in Fig. 6.15(b), this flexibility demonstrates another advantage of differential circuits over their single-ended counterparts.

What is the minimum required voltage gain per stage in the four-stage oscillator of Fig. 6.15(b)? How many signal phases are provided by the circuit? Using a notation similar to that for Fig. 6.8, we have:

$$H(s) = -\frac{A_0^4}{(1+\dfrac{s}{\omega_0})^4}. \tag{6.17}$$

For the circuit to oscillate, each stage must contribute a frequency-dependent phase shift of $180°/4 = 45°$. The frequency at which this occurs is given by $\tan^{-1}\omega_{osc}/\omega_0 = 45°$ and hence $\omega_{osc} = \omega_0$. The minimum voltage gain is therefore derived as

$$\frac{A_0}{\sqrt{1+(\dfrac{\omega_{osc}}{\omega_0})^2}} = 1. \tag{6.18}$$

That is, $A_0 = \sqrt{2}$. As expected, this value is lower than that required in a three-stage ring.

With 45° of phase shift per stage, the oscillator provides four phases and their complements. This is illustrated in Fig. 6.16.

The number of stages in a ring oscillator is determined by various requirements, including speed, power dissipation, noise immunity, etc. In most applications, three to five stages provide optimum performance (for differential implementations).

As an example, let us determine the maximum voltage swings and the minimum supply voltage of a ring oscillator incorporating differential pairs with resistive loads (e.g., as in Fig. 6.12) if no transistor must enter the triode region. We assume each stage experiences

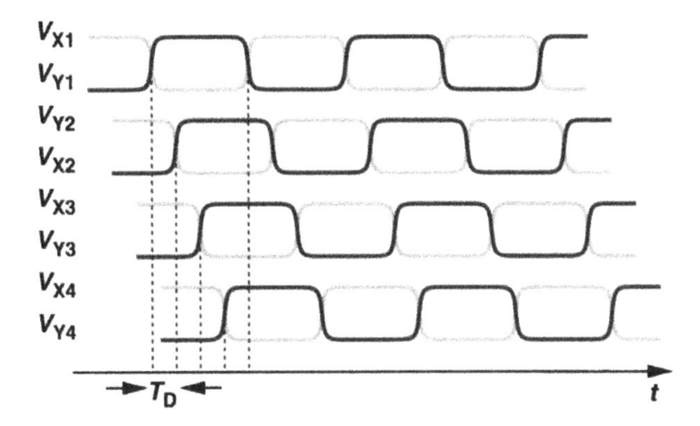

Figure 6.16 Oscillator output phases.

complete switching. Figure 6.17(a) shows two stages in cascade. If each stage experiences complete switching, then each drain voltage, e.g., V_X or V_Y, varies between V_{DD} and

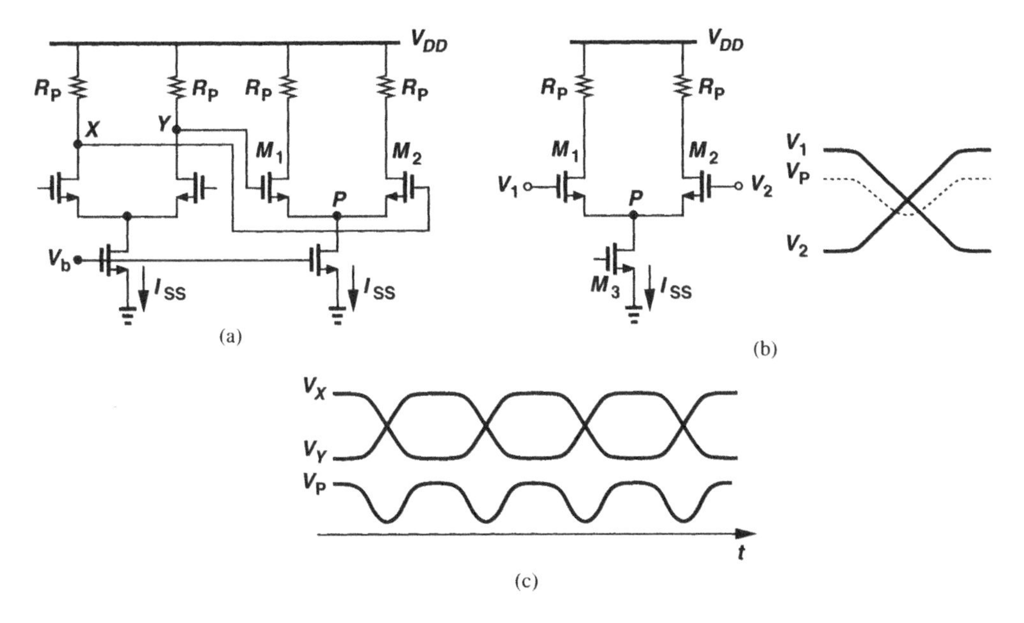

Figure 6.17 Stages in a ring oscillator.

$V_{DD} - I_{SS}R_P$. Thus, when M_1 is fully on, its gate and drain voltages are equal to V_{DD} and $V_{DD} - I_{SS}R_P$, respectively. For this transistor to remain in saturation, we have $I_{SS}R_P \leq V_{TH}$; i.e., the peak-to-peak swing at each drain must not exceed V_{TH}.

How is the minimum supply voltage determined? If V_{DD} is lowered, the voltage at the common-source node of each differential pair, e.g., V_P in Fig. 6.17(a), falls, eventually driving the tail transistor into the triode region. We must therefore calculate V_P for the worst case, noting that V_P does vary with time because M_1 and M_2 carry unequal currents when the input difference becomes large.

Now consider the stand-alone circuit of Fig. 6.17(b), assuming the inputs vary between V_{DD} and $V_{DD} - I_{SS}R_P$. How does V_P vary? When the gate voltage of M_1, V_1, is equal to V_{DD} and M_1 carries all of I_{SS},

$$V_P = V_{DD} - \sqrt{\frac{2I_{SS}}{\mu_n C_{ox}(W/L)_{1,2}}} - V_{TH}. \tag{6.19}$$

As V_1 falls and V_2 rises, so does V_P because, so long as M_2 is off, M_1 operates as a source follower. When the difference between V_1 and V_2 reaches $\sqrt{2}(V_{GS,eq} - V_{TH})$, where $V_{GS,eq}$ denotes the equilibrium overdrive of each transistor, M_2 turns on. To calculate V_P after this point, we note that $I_{D1} + I_{D2} = I_{SS}$, $V_{GS1} = V_1 - V_P$, and $V_{GS2} = V_2 - V_P$. Thus,

$$\frac{1}{2}\mu_n C_{ox}(\frac{W}{L})_{1,2}(V_1 - V_P - V_{TH})^2 + \frac{1}{2}\mu_n C_{ox}(\frac{W}{L})_{1,2}(V_2 - V_P - V_{TH})^2 = I_{SS}. \tag{6.20}$$

Expanding the quadratic terms and rearranging the result, we have

$$2V_P^2 - 2(V_1 - V_{TH} + V_2 - V_{TH})V_P + (V_1 - V_{TH})^2 + (V_2 - V_{TH})^2 - \frac{2I_{SS}}{\mu_n C_{ox}(W/L)_{1,2}} = 0. \tag{6.21}$$

It follows that

$$V_P = \frac{1}{2}[V_1 + V_2 - 2V_{TH} \pm \sqrt{-(V_1 - V_2)^2 + \frac{4I_{SS}}{\mu_n C_{ox}(W/L)_{1,2}}}]. \tag{6.22}$$

If V_1 and V_2 vary differentially, they can be expressed as $V_1 = V_{CM} + \Delta V$ and $V_2 = V_{CM} - \Delta V$, where $V_{CM} = V_{DD} - I_{SS}R_P/2$, yielding

$$V_P = V_{CM} - V_{TH} \pm \frac{1}{2}\sqrt{-(2\Delta V)^2 + \frac{4I_{SS}}{\mu_n C_{ox}(W/L)_{1,2}}}. \tag{6.23}$$

This expression reveals why node P is considered a virtual ground in small-signal operation: if $|\Delta V|$ is much less than the maximum overdrive voltage, then V_P is relatively constant. Since the term under the square root reaches a maximum for $\Delta V = 0$ (equilibrium condition),

$$V_{P,min} = V_{CM} - V_{TH} - \sqrt{\frac{I_{SS}}{\mu_n C_{ox}(W/L)_{1,2}}}. \tag{6.24}$$

As expected, the last term in (6.24) represents the overdrive voltage of each transistor in equilibrium (where $I_{D1} = I_{D2} = I_{SS}/2$).

Figure 6.17(c) shows typical waveforms in the oscillator. Note that V_P varies at twice the oscillation frequency. This property is sometimes exploited in "frequency doublers."

To determine the minimum supply voltage, we write $V_{P,min} \geq V_{ISS}$, where V_{ISS} denotes the minimum required voltage across I_{SS}. Thus,

$$V_{DD} - \frac{R_P I_{SS}}{2} - V_{TH} - \sqrt{\frac{I_{SS}}{\mu_n C_{ox}(W/L)_{1,2}}} \geq V_{ISS}, \tag{6.25}$$

and

$$V_{DD} \geq V_{ISS} + V_{TH} + \sqrt{\frac{I_{SS}}{\mu_n C_{ox}(W/L)_{1,2}}} + \frac{R_P I_{SS}}{2}. \tag{6.26}$$

The terms on the right are: the voltage headroom consumed by a current source, one threshold voltage, the equilibrium overdrive, and half of the swing at each node.

In CMOS technologies lacking high-quality resistors, the implementation of Fig. 6.17(a) must be modified. While a PMOS transistor operating in the deep triode region can serve as the load [Fig. 6.18(a)], the gate voltage must be set so as to define the on-

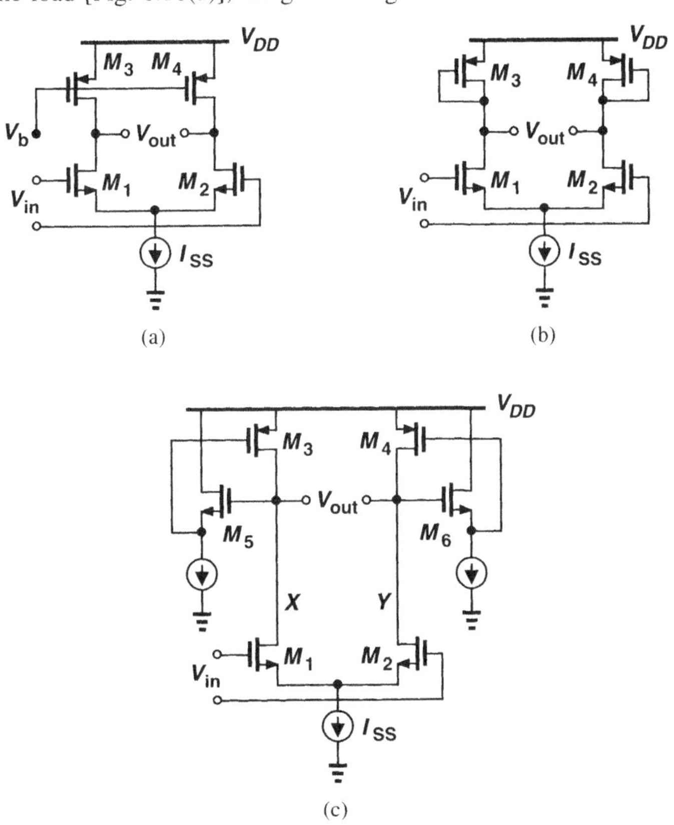

Figure 6.18 Differential stages using PMOS loads.

resistance accurately. Alternatively, a diode-connected load can be utilized [Fig. 6.18(b)] but at the cost of one threshold voltage in the headroom. Figure 6.18(c) shows a more efficient load where an NMOS source follower is inserted between the drain and gate of each PMOS transistor. With the output sensed at nodes X and Y, M_3 and M_4 consume only a voltage headroom equal to $|V_{DS3,4}|$. If $V_{GS5} \approx V_{TH3}$, then M_3 operates at the edge of the triode region and the small-signal resistance of the load is roughly equal to $1/g_{m3}$ (with the assumption $\lambda = \gamma = 0$).

The load of Fig. 6.18(c) exhibits another interesting property as well. Since the gate-source capacitance of M_3 is driven by the source follower, the time constant associated with the load is smaller than that of a diode-connected transistor. Also, the finite output resistance of the follower may yield an inductive behavior for the load.

6.3 LC Oscillators

Monolithic inductors have gradually appeared in bipolar and CMOS technologies in the past 10 years, making it possible to design oscillators based on passive resonant circuits. Before delving into such oscillators, it is instructive to review basic properties of RLC circuits.

As shown in Fig. 6.19(a), an inductor L_1 placed in parallel with a capacitor C_1 resonates at a frequency $\omega_{res} = 1/\sqrt{L_1 C_1}$. At this frequency, the impedances of the induc-

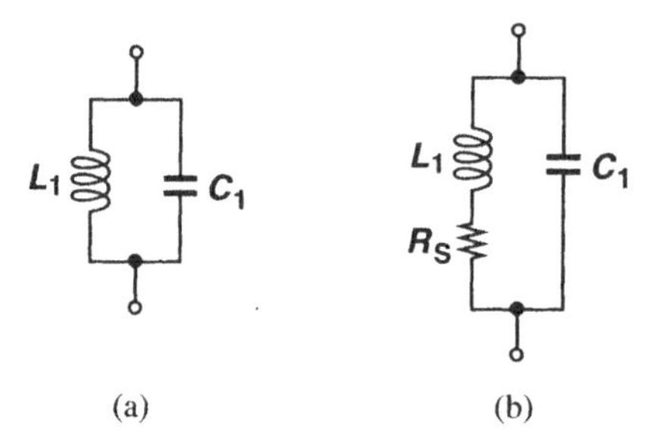

Figure 6.19 (a) Ideal and (b) realistic LC tanks.

tor, $jL_1\omega_{res}$, and the capacitor, $1/(jC_1\omega_{res})$, are equal and opposite, thereby yielding an infinite impedance. We say the circuit has an infinite quality factor, Q. In practice, inductors (and capacitors) suffer from resistive components. For example, the series resistance of the metal wire used in the inductor can be modeled as shown in Fig. 6.19(b). We define the Q of the inductor as $L_1\omega/R_S$. For this circuit, the reader can show that the equivalent impedance is given by

$$Z_{eq}(s) = \frac{R_S + L_1 s}{1 + L_1 C_1 s^2 + R_S C_1 s}, \tag{6.27}$$

and hence,

$$|Z_{eq}(s = j\omega)|^2 = \frac{R_S^2 + L_1^2\omega^2}{(1 - L_1 C_1 \omega^2)^2 + R_S^2 C_1^2 \omega^2}. \tag{6.28}$$

That is, the impedance does not go to infinity at any $s = j\omega$. We say the circuit has a finite Q. The magnitude of Z_{eq} in (6.28) reaches a peak in the vicinity of $\omega = 1/\sqrt{L_1C_1}$, but the actual resonance frequency has some dependency on R_S.

The circuit of Fig. 6.19(b) can be transformed to an equivalent topology that more easily lends itself to analysis and design. To this end, we first consider the series combination shown in Fig. 6.20(a). For a narrow frequency range, it is possible to convert the circuit to

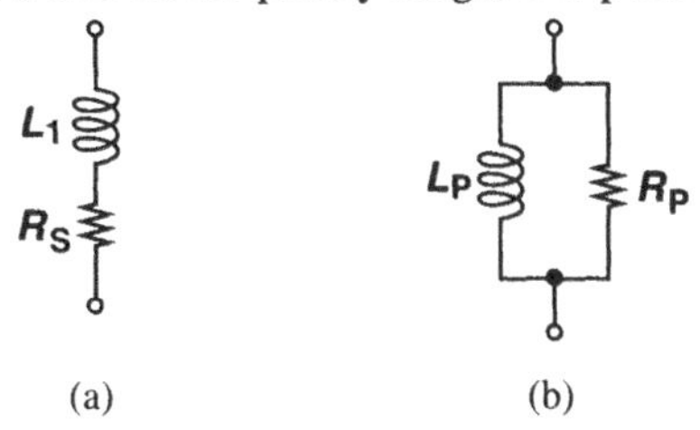

Figure 6.20 Conversion of a series combination to a parallel combination.

the parallel configuration of Fig. 6.20(b). For the two impedances to be equivalent:

$$L_1s + R_S = \frac{R_PL_Ps}{R_P + L_Ps}. \tag{6.29}$$

Considering only the steady-state response, we assume $s = j\omega$ and rewrite (6.29) as

$$(L_1R_P + L_PR_S)j\omega + R_SR_P - L_1L_P\omega^2 = R_PL_Pj\omega. \tag{6.30}$$

This relationship must hold for all values of ω (in a narrow range), mandating that

$$L_1R_P + L_PR_S = R_PL_P \tag{6.31}$$

$$R_SR_P - L_1L_P\omega^2 = 0. \tag{6.32}$$

Calculating R_P from the latter and substituting in the former, we have

$$L_P = L_1(1 + \frac{R_S^2}{L_1^2\omega^2}). \tag{6.33}$$

Recall that $L_1\omega/R_S = Q$, a value typically greater than 3 for monolithic inductors. Thus,

$$L_P \approx L_1 \tag{6.34}$$

and

$$R_P \approx \frac{L_1^2\omega^2}{R_S} \tag{6.35}$$

$$\approx Q^2R_S. \tag{6.36}$$

In other words, the parallel network has the same reactance but a resistance Q^2 times the series resistance. This concept holds valid for a first-order RC network as well if the Q of the series combination is defined as $1/(C\omega)/R_S$.

The above transformation allows the conversion illustrated in Fig. 6.21, where $C_P = C_1$. The equivalence of course breaks down as ω departs susbtantially from the resonance

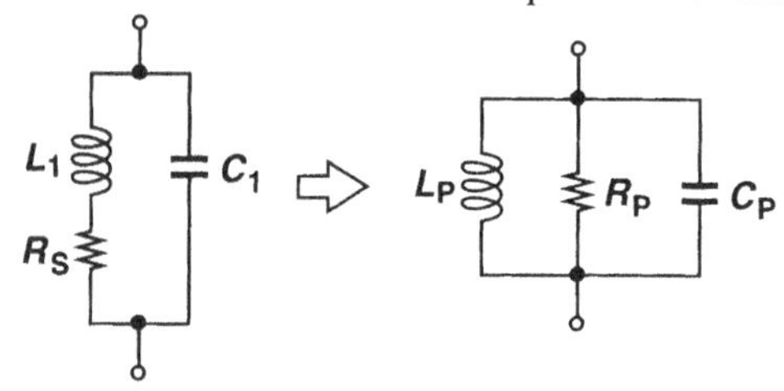

Figure 6.21 Conversion of a tank to three parallel components.

frequency. The insight gained from the parallel combination is that at $\omega_1 = 1/\sqrt{L_p C_p}$, the tank reduces to a simple resistor; i.e., the phase difference between the voltage and current of the tank drops to zero. Plotting the magnitude of the tank impedance versus frequency [Fig. 6.22(a)], we note that the behavior is inductive for $\omega < \omega_1$ and capacitive for $\omega > \omega_1$.

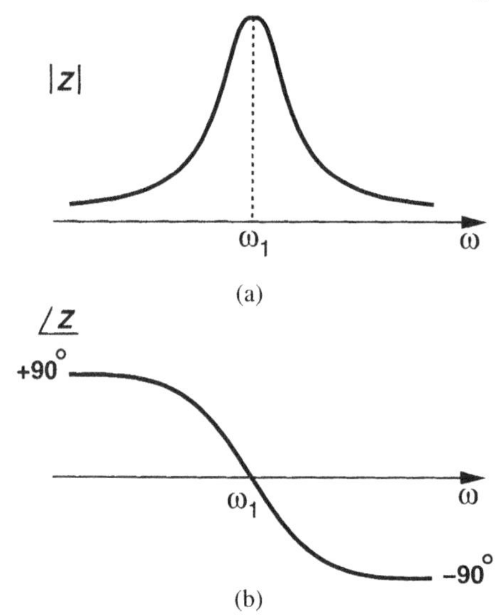

Figure 6.22 (a) Magnitude and (b) phase of the impedance of an LC tank as a function of frequency.

We then surmise that the phase of the impedance is positive for $\omega < \omega_1$ and negative for $\omega > \omega_1$ [Fig. 6.22(b)]. These observations prove useful in studying LC oscillators. (Why do we expect the phase shift to approach $+90°$ at very low frequencies and $-90°$ at very high frequencies?)

Let us now consider the "tuned" stage of Fig. 6.23(a), where an LC tank operates as the load. At resonance, $jL_p\omega = 1/(jC_p\omega)$ and the voltage gain equals $-g_{m1}R_P$. (Note that

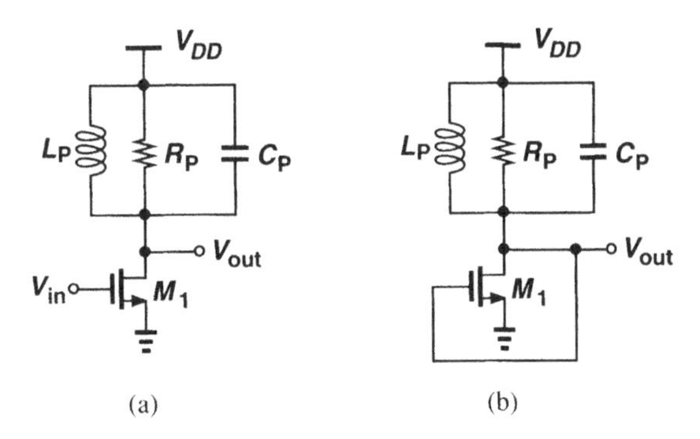

Figure 6.23 (a) Tuned gain stage, (b) stage of (a) in feedback.

the gain of the circuit is very small at frequencies near zero.) Does this circuit oscillate if the output is connected to the input [Fig. 6.23(b)]? At resonance, the total phase shift around the loop is equal to 180° (rather than 360°). Also, from Fig. 6.22(b), the frequency-dependent phase shift of the tank never reaches 180°. Thus, the circuit does not oscillate.

Before modifying the circuit for oscillatory behavior, let us observe another interesting property of the gain stage of Fig. 6.23(a) that distinguishes it from a common-source topology using a resistive load. Suppose, as shown in Fig. 6.24, the stage is biased at a drain current I_1. If the series resistance of L_p is small, the dc level of V_{out} is close to V_{DD}. How

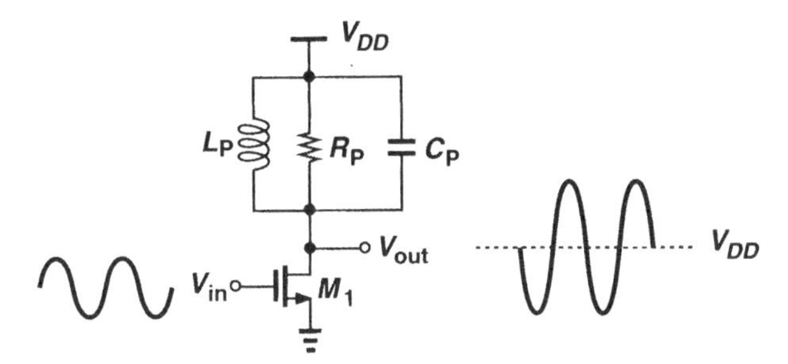

Figure 6.24 Output signal levels in a tuned stage.

does V_{out} vary if a small sinusoidal voltage at the resonance frequency is applied to the input? We expect V_{out} to be an inverted sinusoid with an average value near V_{DD} because the inductor cannot sustain a large dc drop. In other words, if the average value of V_{out} deviates significantly from V_{DD}, then the inductor series resistance must carry an average current greater than I_1. Thus, the peak output level in fact *exceeds* the supply voltage, an important and often useful attribute of the LC load. For example, with proper design, the output peak-to-peak swing can be larger than V_{DD}.

We now study two types of LC oscillators.

6.3.1 Crossed-Coupled Oscillator

Suppose we place two stages of Fig. 6.23(a) in a cascade, as depicted in Fig. 6.25. While similar to the topology of Fig. 6.5, this configuration does not latch up because its low-

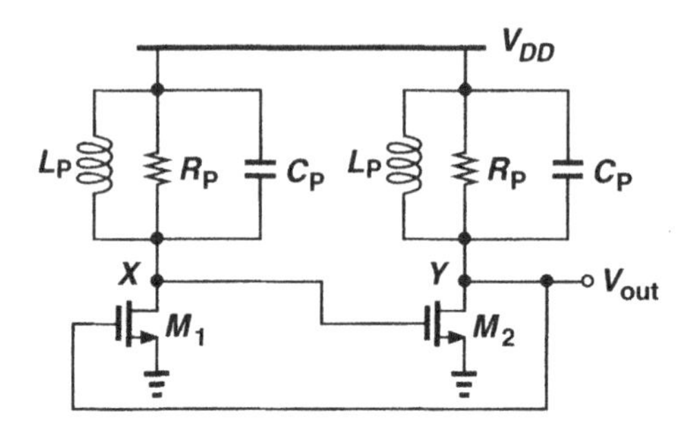

Figure 6.25 Two tuned stages in a feedback loop.

frequency gain is very small. Furthermore, at resonance, the total phase shift around the loop is zero because each stage contributes a zero frequency-dependent phase shift. That is, if $g_{m1}R_Pg_{m2}R_P \geq 1$, then the loop oscillates. Note that V_X and V_Y are differential waveforms. (Why?)

We now sketch the open-loop voltage gain and phase of the circuit shown in Fig. 6.25, neglecting transistor capacitances. The magnitude of the transfer function has a shape similar to that in Fig. 6.22(a) but with sharper rise and fall because it results from the *product* of those of the two stages. The total phase at low frequencies is given by signal inversion by each common-source stage plus a 90° phase shift due to each tank. A similar behavior occurs at high frequencies. The gain and phase are sketched in Fig. 6.26. From these plots, the reader can prove that the circuit cannot oscillate at any other frequency.

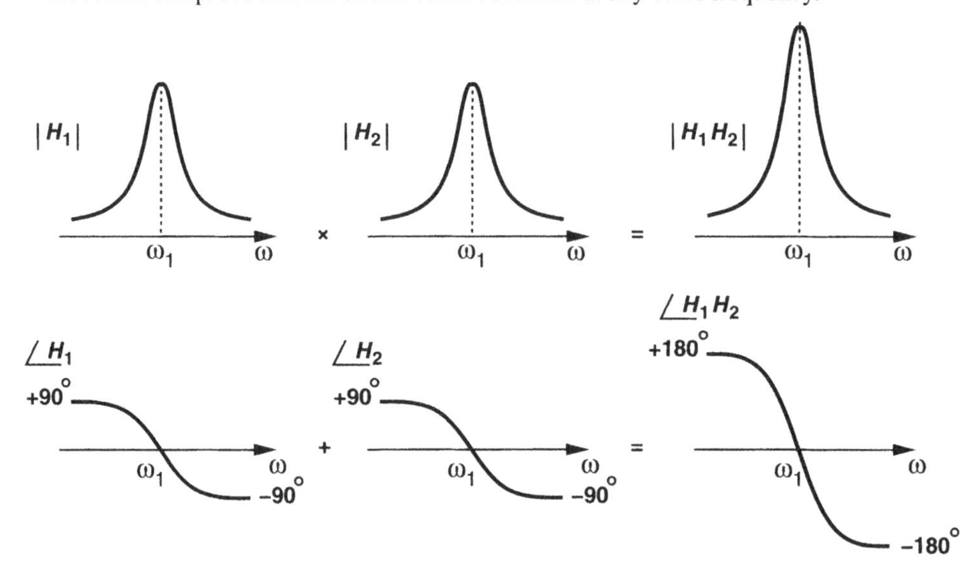

Figure 6.26 Loop gain characteristics of the circuit shown in Fig. 6.25.

The circuit of Fig. 6.25 serves as the core of many LC oscillators and is sometimes drawn as in Fig. 6.27(a) or (b). However, the drain currents of M_1 and M_2 and hence the

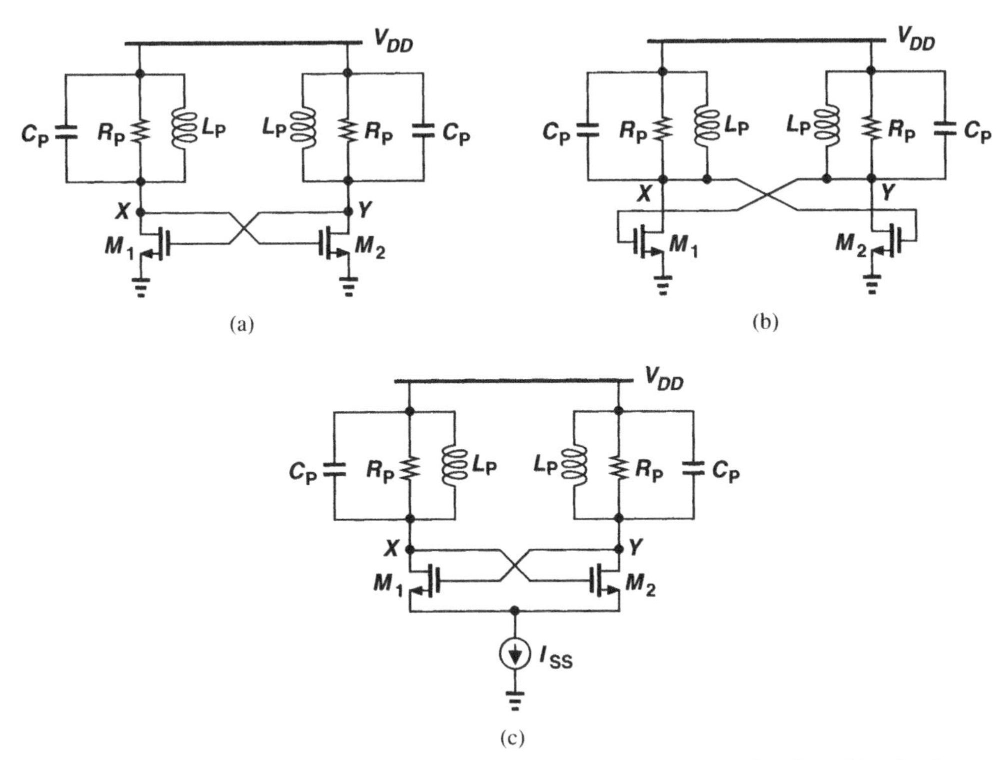

Figure 6.27 (a) Redrawing of the oscillator shown in Fig. 6.25, (b) another redrawing of the circuit, (c) addition of tail current source to lower supply sensitivity.

output swings heavily depend on the supply voltage. Since the waveforms at X and Y are differential, the drawing in Fig. 6.27(b) suggests that M_1 and M_2 can be converted to a differential pair as depicted in Fig. 6.27(c), where the total bias current is defined by I_{SS}.

For the circuit of Fig. 6.27(c), we plot V_X and V_Y and I_{D1} and I_{D2} as the oscillation begins. If the circuit begins with zero difference between V_X and V_Y, then $V_X = V_Y \approx V_{DD}$. The two transistors share the tail current equally. If $(g_{m1,2}R_P)^2 \geq 1$, where R_P is the equivalent parallel resistance of the tank at resonance, then noise components at the resonance frequency are continually amplified by M_1 and M_2, allowing the oscillation to grow. The drain currents of M_1 and M_2 vary according to the instantaneous value of $V_X - V_Y$ (as in a differential pair).

As shown in Fig. 6.28, the oscillation amplitude grows until the loop gain drops at the peaks. In fact, if $g_{m1,2}R_P$ is large enough, the difference between $V_X - V_Y$ reaches a level that steers the entire tail current to one transistor, turning the other off. Thus, in the steady state, I_{D1} and I_{D2} vary between zero and I_{SS}.

The oscillator of Fig. 6.27(c) is constructed in fully differential form. The supply sensitivity of the circuit, however, is nonzero even with perfect symmetry. This is because the drain junction capacitances of M_1 and M_2 vary with the supply voltage.

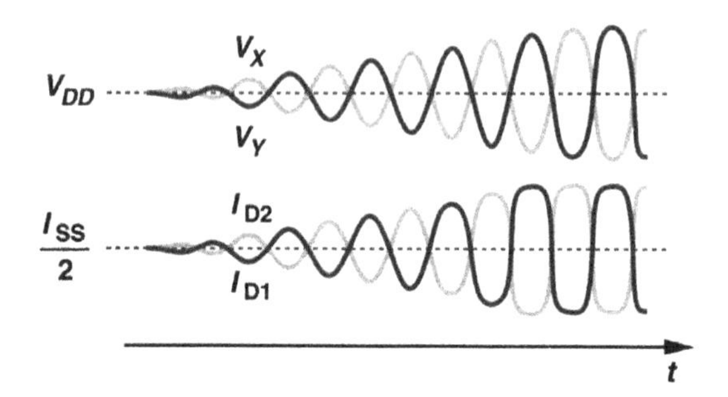

Figure 6.28 Oscillation behavior.

6.3.2 Colpitts Oscillator

An LC oscillator may be realized with only one transistor in the signal path. Consider the gain stage of Fig. 6.23(a) again and recall that the drain voltage cannot be applied to the gate because the overall phase shift at resonance equals 180° rather than 360°. Also, recall that in a common-gate stage, the phase shift from the source to the drain is zero. We then surmise that if, as shown in Fig. 6.29(a), the drain voltage is returned to the source rather than the gate, the circuit may oscillate. The coupling must incorporate a capacitor to avoid

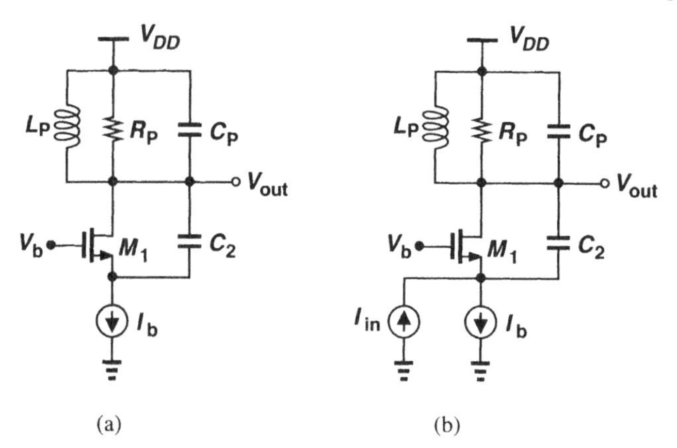

Figure 6.29 (a) Tuned stage with feedback applied from drain to source, (b) addition of input current to calculate closed-loop gain.

disturbing the bias point of M_1.

Unfortunately, owing to insufficient loop gain, the circuit of Fig. 6.29(a) does not oscillate. To prove this point, we invoke the view of Fig. 6.1, where an oscillator is considered a feedback system with infinite closed-loop gain. Applying an input current as depicted in Fig. 6.29(b) and neglecting transistor parasitics, we obtain the closed-loop gain as:

$$\frac{V_{out}}{I_{in}} = L_P s || \frac{1}{C_P s} || R_P \tag{6.37}$$

because M_1 and C_2 directly conduct the input current to the tank. Since the closed-loop gain cannot be equal to infinity at any frequency, the circuit fails to oscillate.

The reader may wonder why the input to the feedback system is realized as a current source applied to the source of the transistor rather than a voltage source applied to its gate. Let us perform the analysis with the latter stimulus. From Fig. 6.30, we note that with a finite variation of V_{in}, the change in I_b is still zero if the bias current source is

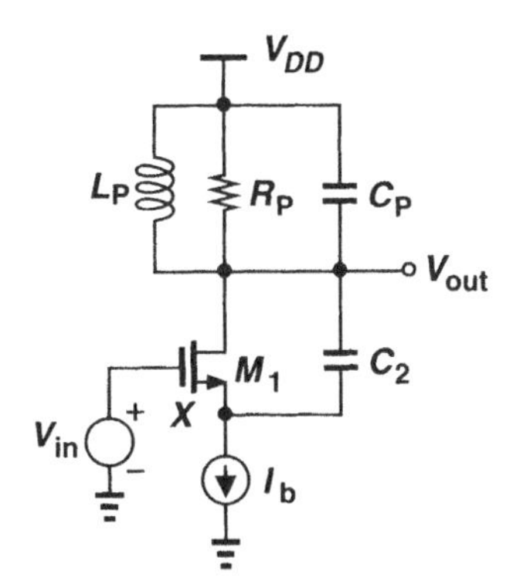

Figure 6.30 Oscillator with voltage stimulus.

ideal. Thus, if the source-bulk junction capacitance of M_1 is neglected, the change in the tank current is zero, yielding $V_{out}/V_{in} = 0$. Interestingly, V_X does vary with V_{in}, but M_1 generates a small-signal current that cancels that through C_2. The reader can prove that $V_X/V_{in} = g_m/(g_m + C_2 s)$.

The above example reveals two important points. First, to excite a circuit into oscillation, the stimulus can be applied at different points. (That is, the noise of any device in the loop can initiate the oscillation.[5]) Second, in Fig. 6.30, V_{out}/V_{in} is zero because the impedance connected between the source of M_1 and ground is infinity. We then add a capacitor from this node to ground as shown in Fig. 6.31(a), seeking conditions of oscillation. Note that the capacitor in parallel with L_P is removed. The reason will become clear later.

Approximating M_1 by a single voltage-dependent current source, we construct the equivalent circuit of Fig. 6.31(b). Since the current through the parallel combination of L_P and R_P is given by $V_{out}/(L_P s) + V_{out}/R_P$, the total current through C_1 is equal to $I_{in} - V_{out}/(L_P s) - V_{out}/R_P$, yielding

$$V_1 = -(I_{in} - \frac{V_{out}}{L_P s} - \frac{V_{out}}{R_P})\frac{1}{C_1 s}. \tag{6.38}$$

Writing the current through C_2 as $(V_{out} + V_1)C_2 s$, we sum all of the currents at the output

[5]This is because the natural frequencies of a linear (observable) system do not depend on the location of the stimulus. Of course, the type of stimulus (voltage or current) must be chosen such that when it is set to zero, the circuit returns to its original topology. For example, driving the gate of M_1 in Fig. 6.30 by a current changes the natural frequencies of the circuit.

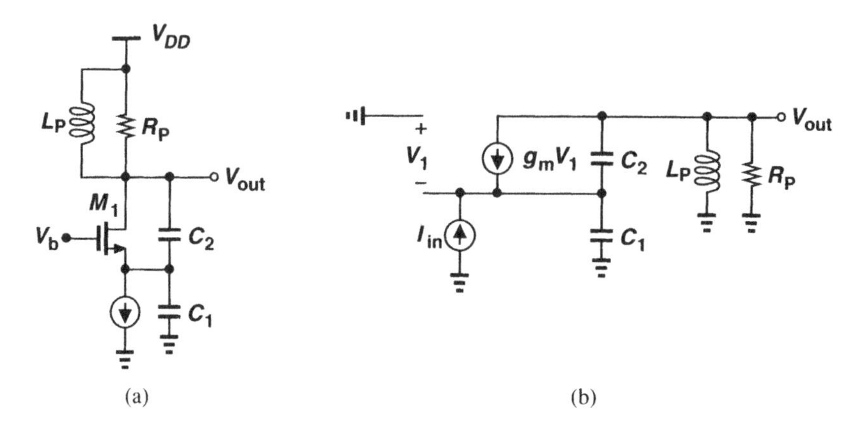

Figure 6.31 (a) Colpitts oscillator, (b) equivalent circuit of (a) with input stimulus.

node:

$$-g_m(I_{in} - \frac{V_{out}}{L_P s} - \frac{V_{out}}{R_P})\frac{1}{C_1 s} + [V_{out} - (I_{in} - \frac{V_{out}}{L_P s} - \frac{V_{out}}{R_P})\frac{1}{C_1 s}]C_2 s + \frac{V_{out}}{L_P s} + \frac{V_{out}}{R_P} = 0. \tag{6.39}$$

It follows that

$$\frac{V_{out}}{I_{in}} = \frac{R_P L_P s(g_m + C_2 s)}{R_P C_1 C_2 L_P s^3 + (C_1 + C_2)L_P s^2 + [g_m L_P + R_P(C_1 + C_2)]s + g_m R_P}. \tag{6.40}$$

Note that, as expected, (6.40) reduces to $(L_P s || R_P)$ if $C_1 = 0$. The circuit oscillates if the closed-loop transfer function goes to infinity at an imaginary value of s, $s_R = j\omega_R$. Consequently, both the real and imaginary parts of the denominator must drop to zero at this frequency:

$$-R_P C_1 C_2 L_P \omega_R^3 + [g_m L_P + R_P(C_1 + C_2)]\omega_R = 0 \tag{6.41}$$

$$-(C_1 + C_2)L_P \omega_R^2 + g_m R_P = 0. \tag{6.42}$$

Since with typical values, $g_m L_P \ll R_P(C_1 + C_2)$, Eq. (6.41) yields:

$$\omega_R^2 = \frac{1}{L_P \dfrac{C_1 C_2}{C_1 + C_2}}, \tag{6.43}$$

and Eq. (6.42) results in

$$g_m R_P = \frac{(C_1 + C_2)^2}{C_1 C_2} \tag{6.44}$$

$$= \frac{C_1}{C_2}(1 + \frac{C_2}{C_1})^2. \tag{6.45}$$

Recognizing that $g_m R_P$ is the voltage gain from the source of M_1 to the output (if $g_{mb} = 0$), we determine the ratio C_1/C_2 for minimum required gain. The reader can prove that the minimum occurs for $C_1/C_2 = 1$, requiring

$$g_m R_P \geq 4. \tag{6.46}$$

Equation (6.46) demonstrates an important disadvantage of the Colpitts oscillator with respect to the cross-coupled topology of Fig. 6.27(c). The former demands a voltage gain of at least 4 at resonance and the latter, only unity. This issue is critical if the inductor suffers from a low Q and hence a small R_P, a common situation in CMOS technologies. As a consequence, the cross-coupled scheme is used more widely.

The foregoing analysis neglected the capacitance that appears in parallel with the inductor. If this capacitance, C_P, is included in the equivalent circuit, Eq. (6.43) is modified as:

$$\omega_R^2 = \frac{1}{L_P(C_P + \frac{C_1 C_2}{C_1 + C_2})}, \tag{6.47}$$

whereas (6.46) remains unchanged. Thus, C_P is simply included in parallel with the series combination of C_1 and C_2.

6.3.3 One-Port Oscillators

Our development of oscillators thus far has been based on feedback systems. An alternative view that provides more insight into the oscillation phenomenon employs the concept of "negative resistance." To arrive at this view, let us first consider a simple tank that is stimulated by a current impulse [Fig. 6.32(a)]. The tank responds with a decaying oscillatory behavior because, in every cycle, some of the energy that reciprocates between the capacitor and the inductor is lost in the form of heat in the resistor. Now suppose a resistor equal to $-R_P$ is placed in parallel with R_P and the experiment is repeated [Fig. 6.32(b)]. Since $R_P||(-R_P) = \infty$, the tank oscillates indefinitely. Thus, if a one-port circuit exhibiting a negative resistance is placed in parallel with a tank [Fig. 6.32(c)], the combination may oscillate. Such a topology is called a one-port oscillator.

How can a circuit provide a negative resistance? Recall that feedback multiplies or divides the input and output impedances of circuits by a factor equal to one plus the loop gain. Thus, if the loop gain is sufficiently *negative* (i.e., the feedback is sufficiently positive), a negative resistance is achieved. As a simple example, let us apply positive feedback around a source follower. The follower introduces no signal inversion and neither must the feedback network. As depicted in Fig. 6.33(a), we implement the feedback by a common-gate stage and add the current source I_b to provide the bias current of M_2. From the equivalent circuit in Fig. 6.33(b) (where channel-length modulation and body effect are neglected), we have

$$I_X = g_{m2} V_2 = -g_{m1} V_1 \tag{6.48}$$

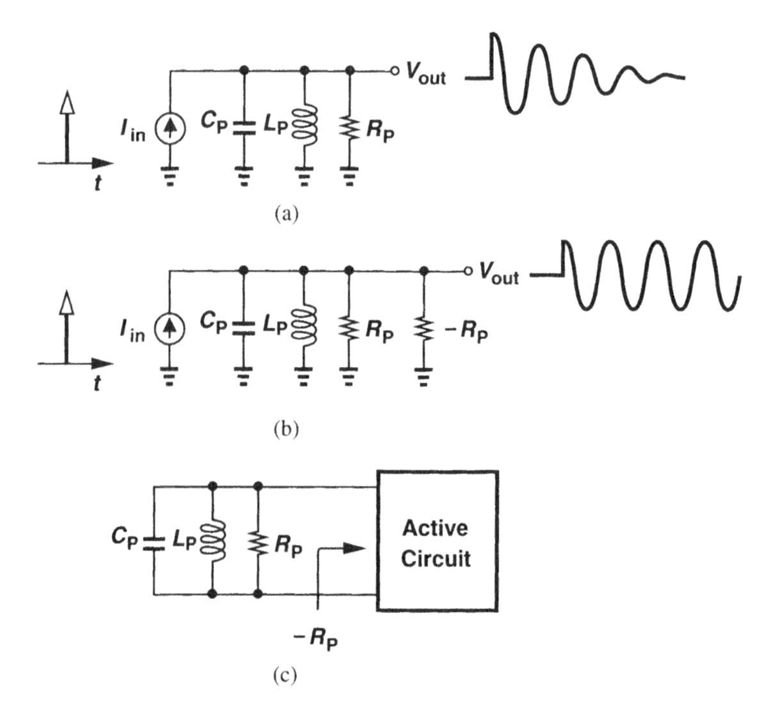

Figure 6.32 (a) Decaying impulse response of a tank, (b) addition of negative resistance to cancel loss in R_P, (c) use of an active circuit to provide negative resistance.

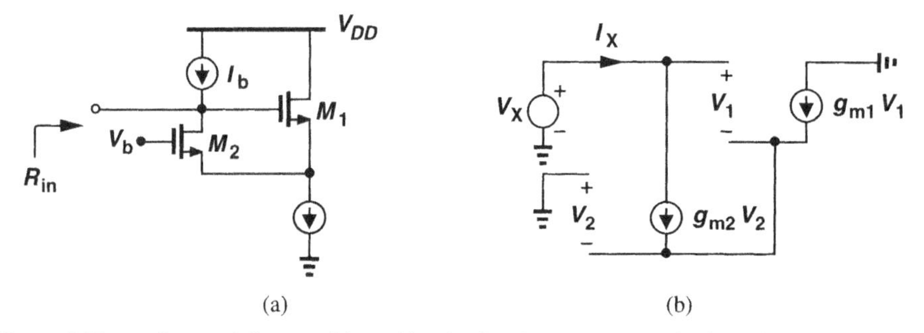

Figure 6.33 (a) Source follower with positive feedback to create negative impedance, (b) equivalent circuit of (a) to calculate the input impedance.

and

$$V_X = V_1 - V_2 \tag{6.49}$$

$$= -\frac{I_X}{g_{m1}} - \frac{I_X}{g_{m2}}. \tag{6.50}$$

Thus,

$$\frac{V_X}{I_X} = -\left(\frac{1}{g_{m1}} + \frac{1}{g_{m2}}\right), \tag{6.51}$$

and, if $g_{m1} = g_{m2} = g_m$, then

$$\frac{V_X}{I_X} = \frac{-2}{g_m}. \tag{6.52}$$

Negative resistance becomes more intuitive if we bear in mind that it is an *incremental* quantity; that is, negative resistance indicates that if the applied voltage *increases*, the current drawn by the circuit *decreases*. In Fig. 6.33(a), for example, if the input voltage increases, so does the source voltage of M_1, decreasing the drain current of M_2 and allowing part of I_b to flow to the input source.

With a negative resistance available, we can now construct an oscillator as illustrated in Fig. 6.34. Here, R_P denotes the equivalent parallel resistance of the tank and, for os-

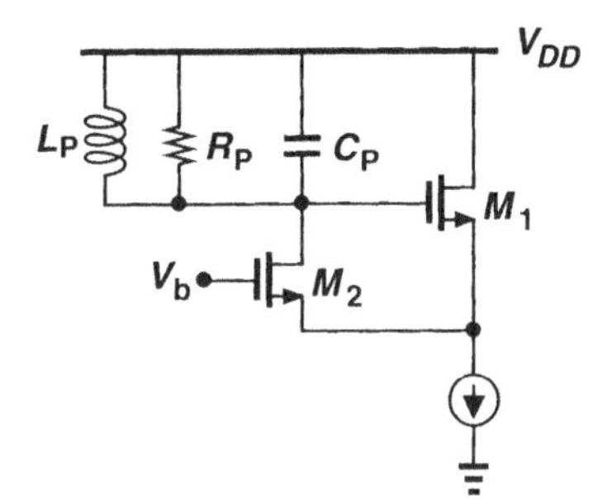

Figure 6.34 Oscillator using negative input resistance of a source follower with positive feedback.

cillation build-up, $R_P - 2/g_m \geq 0$. Note that the inductor provides the bias current of M_2, obviating the need for a current source. If the small-signal resistance presented by M_1 and M_2 to the tank is less negative than $-R_P$, then the circuit experiences large swings such that each transistor is nearly off for part of the period, thereby yielding an "average" resistance of $-R_P$.

The circuit of Fig. 6.34 is similar to the stage of Fig. 6.29(a) but with the feedback capacitor replaced by a source follower. More interestingly, the circuit can be redrawn as in Fig. 6.35(a), bearing a resemblance to Fig. 6.27(c). In fact, if the drain current of M_1 flows through a tank and the resulting voltage is applied to the gate of M_2, the topology of Fig. 6.35(b) is obtained. Ignoring bias paths and merging the two tanks into one (Fig. 6.36), we note that the cross-coupled pair must provide a negative resistance of $-R_P$ between nodes X and Y to enable oscillation. The reader can prove that this resistance is equal to $-2/g_m$ and hence it is necessary that $R_P \geq 1/g_m$. Thus, the circuit can be viewed as either a feedback system or a negative resistance in parallel with a lossy tank. This topology is also called a "negative-G_m oscillator."

As another method of creating negative resistance, consider the topology depicted in Fig. 6.37(a), where none of the nodes is grounded and channel-length modulation, body

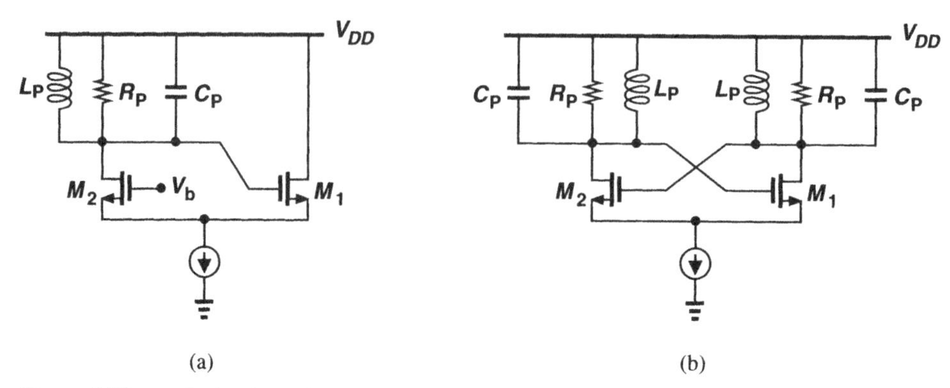

Figure 6.35 (a) Redrawing of the topology shown in Fig. 6.34, (b) differential version of (a).

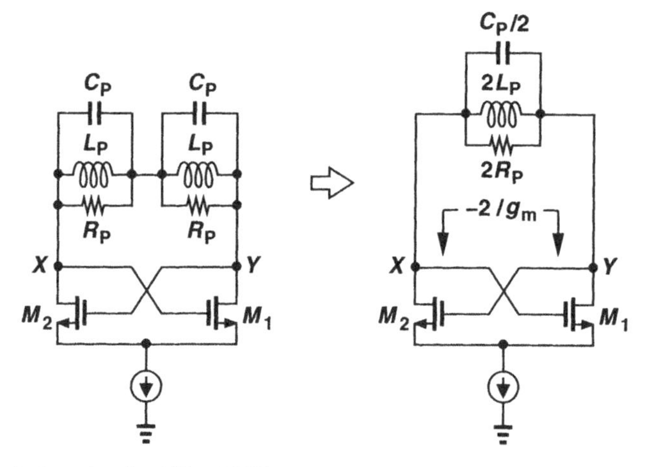

Figure 6.36 Equivalent circuit of Fig. 6.35(b).

effect, and transistor capacitances are neglected. Since the drain current of M_1 is equal to $(-I_X/C_1s)g_m$, we have

$$V_X = (I_X - \frac{-I_X}{C_1 s} g_m)\frac{1}{C_2 s} + \frac{I_X}{C_1 s} \tag{6.53}$$

and hence

$$\frac{V_X}{I_X} = \frac{g_m}{C_1 C_2 s^2} + \frac{1}{C_2 s} + \frac{1}{C_1 s}. \tag{6.54}$$

For $s = j\omega$, this impedance consists of a negative resistance equal to $-g_m/(C_1C_2\omega^2)$ in series with the series combination of C_1 and C_2 [Fig. 6.37(b)]. Thus, as shown in Fig. 6.37(c), if an inductor is placed between the gate and drain of M_1, the circuit may oscillate. Of the three nodes in the circuit, one can be an ac ground, resulting in the three different topologies illustrated in Fig. 6.38. The circuit of Fig. 6.38(a) is in fact based on a source

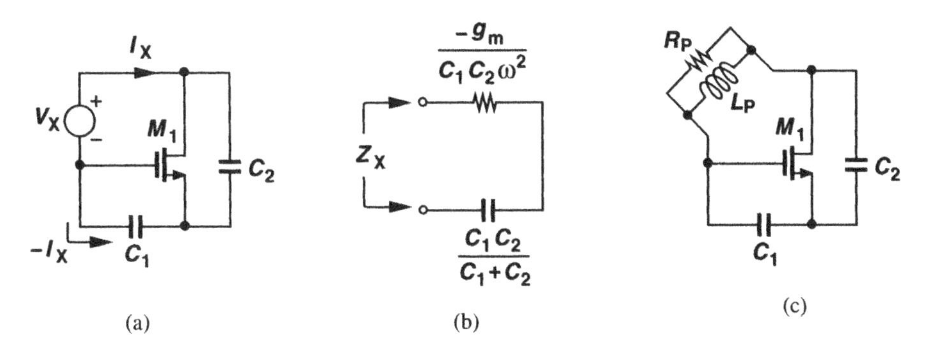

Figure 6.37 (a) Circuit topology providing negative resistance, (b) equivalent circuit of (a), (c) oscillator using (a).

follower, whose input impedance contains a negative real part. The configuration of Fig.

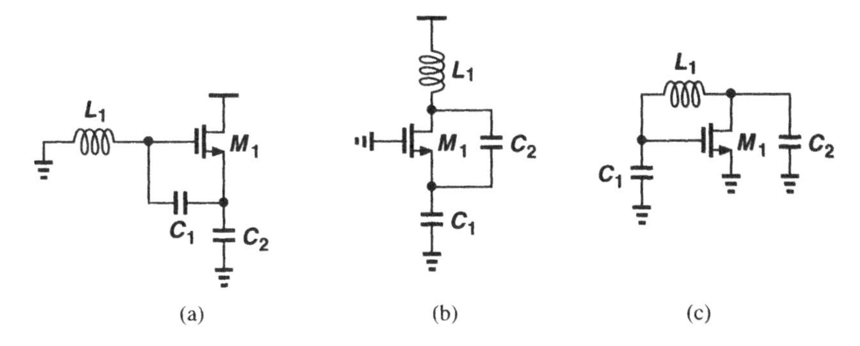

Figure 6.38 Oscillator topologies derived from the circuit of Fig. 6.37(c).

6.38(b) is a Colpitts oscillator.

It is instructive to redraw the circuits of Fig. 6.38 with proper biasing. The circuits are redrawn in Fig. 6.39.

6.4 Voltage-Controlled Oscillators

Most applications require that oscillators be "tunable," i.e., their output frequency be a function of a control input, usually a voltage. An ideal voltage-controlled oscillator is a circuit whose output frequency is a linear function of its control voltage (Fig. 6.40):

$$\omega_{out} = \omega_0 + K_{VCO} V_{cont}. \tag{6.55}$$

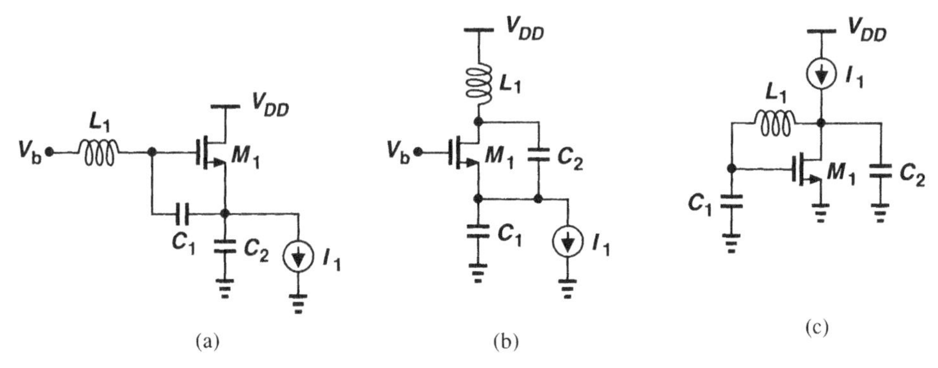

Figure 6.39 Oscillator circuits including biasing.

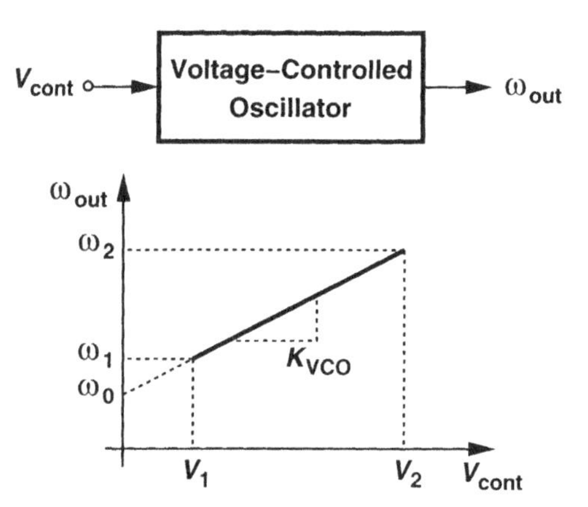

Figure 6.40 Definition of a VCO.

Here, ω_0 represents the intercept corresponding to $V_{cont} = 0$ and K_{VCO} denotes the "gain" or "sensitivity" of the circuit (expressed in rad/s/V).[6] The achievable range, $\omega_2 - \omega_1$, is called the "tuning range."

In the negative-G_m oscillator of Fig. 6.27(c), we assume $C_P = 0$, consider only the drain junction capacitance, C_{DB}, of M_1 and M_2, and explain why V_{DD} can be viewed as the control voltage. We also determine the gain of the VCO. Since C_{DB} varies with the drain-bulk voltage, if V_{DD} changes, so does the resonance frequency of the tank. Noting that the average voltage across C_{DB} is approximately equal to V_{DD}, we write

$$C_{DB} = \frac{C_{DB0}}{(1 + \frac{V_{DD}}{\phi_B})^m}, \tag{6.56}$$

[6] A more familiar unit is Hz/V but one must be careful with the dimension of K_{VCO} in the context of phase-locked loops.

and

$$K_{VCO} = \frac{\partial \omega_{out}}{\partial V_{DD}} \tag{6.57}$$

$$= \frac{\partial \omega_{out}}{\partial C_{DB}} \cdot \frac{\partial C_{DB}}{\partial V_{DD}}. \tag{6.58}$$

With $\omega_{out} = 1/\sqrt{L_P C_{DB}}$, we have

$$K_{VCO} = \frac{-1}{2\sqrt{L_P C_{DB}} C_{DB}} \cdot \frac{-mC_{DB}}{\phi_B(1 + \frac{V_{DD}}{\phi_B})} \tag{6.59}$$

$$= \frac{m}{2\phi_B(1 + \frac{V_{DD}}{\phi_B})} \cdot \omega_{out}. \tag{6.60}$$

Note that the relationship between ω_{out} and V_{cont} is nonlinear because K_{VCO} varies with V_{DD} and ω_{out}.

Before modifying the oscillators studied in the previous sections for tunability, we summarize the important performance parameters of VCOs.

Center Frequency The center frequency (i.e., the midrange value in Fig. 6.40) is determined by the environment in which the VCO is used. For example, in the clock generation network of a microprocessor, the VCO may be required to run at the clock rate or even twice that. Today's CMOS VCOs achieve center frequencies as high as 10 GHz.

Tuning Range The required tuning range is dictated by two parameters: (1) the variation of the VCO center frequency with process and temperature and (2) the frequency range necessary for the application. The center frequency of some CMOS oscillators may vary by a factor of two at the extremes of process and temperature, thus mandating a sufficiently wide ($\geq 2\times$) tuning range to guarantee that the VCO output frequency can be driven to the desired value. Also, some applications incorporate clock frequencies that must vary by one to two orders of magnitude depending on the mode of operation, demanding a proportionally wide tuning range.

An important concern in the design of VCOs is the variation of the output phase and frequency as a result of noise on the control line. For a given noise amplitude, the noise in the output frequency is proportional to K_{VCO} because $\omega_{out} = \omega_0 + K_{VCO}V_{cont}$. Thus, to minimize the effect of noise in V_{cont}, the VCO gain must be *minimized*, a constraint in direct conflict with the required tuning range. In fact, if, as shown in Fig. 6.40, the allowable range of V_{cont} is from V_1 to V_2 (e.g., from 0 to V_{DD}) and the tuning range must span at least ω_1 to ω_2, then K_{VCO} must satisfy the following requirement:

$$K_{VCO} \geq \frac{\omega_2 - \omega_1}{V_2 - V_1}. \tag{6.61}$$

Note that, for a given tuning range, K_{VCO} increases as the supply voltage decreases, making the oscillator more sensitive to noise on the control line.

Tuning Linearity As exemplified by Eq. (6.60), the tuning characteristics of VCOs exhibit nonlinearity, i.e., their gain, K_{VCO}, is not constant. As explained in Chapter 8, such nonlinearity degrades the settling behavior of phase-locked loops. For this reason, it is desirable to minimize the variation of K_{VCO} across the tuning range.

Actual oscillator characteristics typically exhibit a high gain region in the middle of the range and a low gain at the two extremes (Fig. 6.41). Compared to a linear characteristic (the gray line), the actual behavior displays a maximum gain *greater* than that predicted

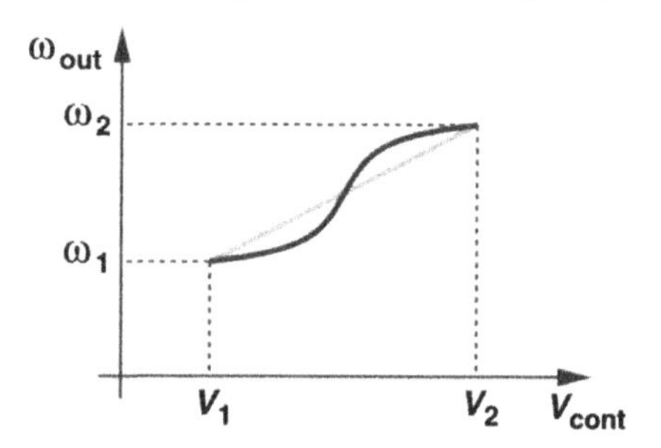

Figure 6.41 Nonlinear VCO characteristic.

by (6.61), implying that, for a given tuning range, nonlinearity inevitably leads to higher sensitivity for some region of the characteristic.

Output Amplitude It is desirable to achieve a large output oscillation amplitude, thus making the waveform less sensitive to noise. The amplitude trades with power dissipation, supply voltage, and (as explained in Section 6.4.2) even the tuning range. Also, the amplitude may vary across the tuning range, an undesirable effect.

Power Dissipation As with other analog circuits, oscillators suffer from trade-offs between speed, power dissipation, and noise. Typical oscillators drain 1 to 10 mW of power.

Supply and Common-Mode Rejection Oscillators are quite sensitive to noise, especially if they are realized in single-ended form. As seen in Eq. (6.60), even differential oscillators exhibit supply sensitivity. The design of oscillators for high noise immunity is a difficult challenge. Note that noise may be coupled to the control line of a VCO as well. For these reasons, it is preferable (but not always possible) to employ differential paths for both the oscillation signal and the control line.

Output Signal Purity Even with a constant control voltage, the output waveform of a VCO is not perfectly periodic. The electronic noise of the devices in the oscillator and supply noise lead to noise in the output phase and frequency. These effects are quantified by "jitter" and "phase noise" and determined by the requirements of each application.

6.4.1 Tuning in Ring Oscillators

Recall from Section 6.2 that the oscillation frequency, f_{osc}, of an N-stage ring equals $(2NT_D)^{-1}$, where T_D denotes the large-signal delay of each stage. Thus, to vary the frequency, T_D can be adjusted.

As a simple example, consider the differential pair of Fig. 6.42 as one stage of a ring oscillator. Here, M_3 and M_4 operate in the triode region, each acting as a variable resistor

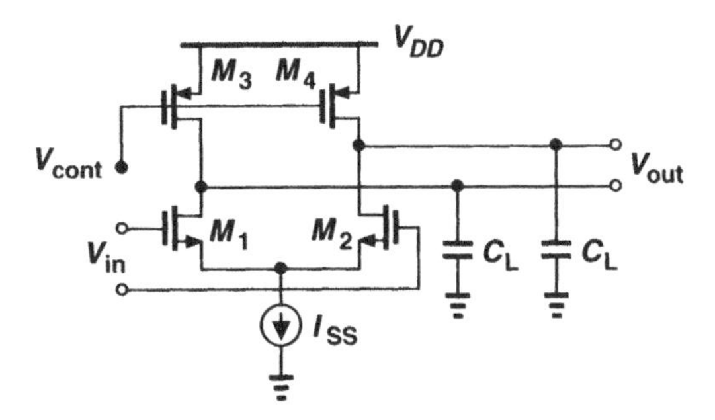

Figure 6.42 Differential pair with variable output time constant.

controlled by V_{cont}. As V_{cont} becomes more positive, the on-resistance of M_3 and M_4 increases, thus raising the time constant at the output, τ_1, and lowering f_{osc}. If M_3 and M_4 remain in deep triode region,

$$\tau_1 = R_{on3,4}C_L \tag{6.62}$$

$$= \frac{C_L}{\mu_p C_{ox}(\frac{W}{L})_{3,4}(V_{DD} - V_{cont} - |V_{THP}|)}. \tag{6.63}$$

In the above equation, C_L denotes the total capacitance seen at each output to ground (including the input capacitance of the following stage). The delay of the circuit is roughly proportional to τ_1, yielding

$$f_{osc} \propto \frac{1}{T_D} \tag{6.64}$$

$$\propto \frac{\mu_p C_{ox}(\frac{W}{L})_{3,4}(V_{DD} - V_{cont} - |V_{THP}|)}{C_L}. \tag{6.65}$$

Interestingly, f_{osc} is linearly proportional to V_{cont}.

As an example, for given device dimensions and bias currents in Fig. 6.42, we compute the maximum allowable value of V_{cont} and determine what happens if M_3 and M_4 enter saturation. Let us assume (somewhat arbitrarily) that M_3 and M_4 remain in deep triode region if $|V_{DS3,4}| \leq 0.2 \times 2|V_{GS3,4} - V_{THP}|$. If each stage in the ring experiences complete switching, then the maximum drain current of M_3 and M_4 is equal to I_{SS}. To satisfy the above condition, we must have $I_{SS}R_{on3,4} \leq 0.4(V_{DD} - V_{cont} - |V_{THP}|)$, and hence

$$\frac{I_{SS}}{\mu_p C_{ox}(\frac{W}{L})_{3,4}(V_{DD} - V_{cont} - |V_{THP}|)} \leq 0.4(V_{DD} - V_{cont} - |V_{THP}|). \tag{6.66}$$

It follows that

$$V_{cont} \le V_{DD} - |V_{THP}| - \sqrt{\frac{I_{SS}}{0.4\mu_p C_{ox}(\frac{W}{L})_{3,4}}}. \tag{6.67}$$

If V_{cont} exceeds this level by a large margin, M_3 and M_4 eventually enter saturation. Each stage then requires common-mode feedback to produce the output swings around a well-defined CM level.

The differential pair of Fig. 6.42 suffers from a critical drawback: the output swing of the circuit varies considerably across the tuning range. With complete switching, each stage provides a differential output swing of $2I_{SS}R_{on3,4}$. Thus, a tuning range of, say, two to one translates to a twofold variation in the swing.

In order to minimize the swing variation, the tail current can be adjusted by V_{cont} as well such that, as V_{cont} becomes more positive, I_{SS} decreases. The circuit nonetheless requires a means of maintaining $I_{SS}R_{on3,4}$ relatively constant. To this end, let us consider the circuit in Fig. 6.43(a), where M_5 operates in the deep triode region and amplifier A_1 applies

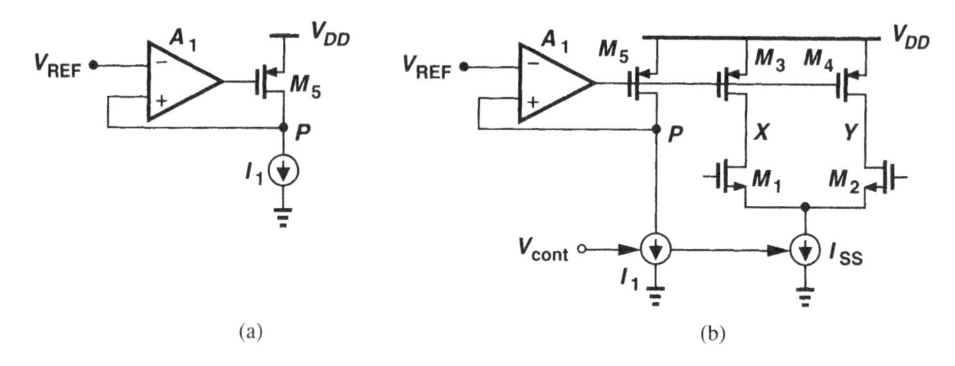

Figure 6.43 (a) Simple feedback circuit defining V_P, (b) replica biasing to define voltage swings in a ring oscillator.

negative feedback to the gate of M_5. If the loop gain is sufficiently large, the differential input voltage of A_1 must be small, giving $V_P \approx V_{REF}$ and $|V_{DS5}| \approx V_{DD} - V_{REF}$. Thus, the feedback ensures a relatively constant drain-source voltage even if I_1 varies. In fact, as I_1, say, decreases, A_1 raises the gate voltage of M_5 such that $R_{on5}I_1 \approx V_{DD} - V_{REF}$.

The topology of Fig. 6.43(a) can serve as a "replica circuit" for the stages of a ring oscillator, thereby defining the oscillation amplitude. Illustrated in Fig. 6.43(b), the idea is to "servo" the on-resistance of M_3 and M_4 to that of M_5 and vary the frequency by adjusting I_1 and I_{SS} simultaneously [2]. If M_3 and M_4 are identical to M_5 and I_{SS} to I_1, then V_X and V_Y vary from V_{DD} to $V_{DD} - V_{REF}$ as M_1 and M_2 steer the tail current to one side or the other. Thus, if process and temperature variations, say, reduce I_1 and I_{SS}, then A_1 increases the on-resistance of M_3-M_5, forcing V_P and hence V_X and V_Y (when M_1 or M_2 is fully on) equal to V_{REF}.

The bandwidth of the op amp A_1 in Fig. 6.43(b) is of some concern. If a change in V_{cont} takes a long time to change ω_{out}, then the settling speed of a PLL using this VCO degrades significantly (Chapter 8).

How does the oscillation frequency depend on I_{SS} for a VCO incorporating the stage of Fig. 6.43(b)? Noting that $R_{on3,4}I_{SS} \approx V_{DD} - V_{REF}$, we have $R_{on3,4} \approx (V_{DD} - V_{REF})/I_{SS}$ and hence

$$f_{osc} \propto \frac{1}{R_{on3,4}C_L} \tag{6.68}$$

$$\propto \frac{I_{SS}}{(V_{DD} - V_{REF})C_L}. \tag{6.69}$$

Thus, the characteristic is relatively linear.

Delay Variation by Positive Feedback To arrive at another tuning technique, recall that a cross-coupled transistor pair such as that of Fig. 6.36 exhibits a negative resistance of $-2/g_m$, a value that can be controlled by the bias current. A negative resistance $-R_N$ placed in parallel with a positive resistance $+R_P$ gives an equivalent value $+R_N R_P/(R_N - R_P)$, which is more positive if $|-R_N| > |+R_P|$. This idea can be applied to each stage of a ring oscillator as illustrated in Fig. 6.44(a). Here, the load of the differential pair consists of resistors R_1 and R_2 ($R_1 = R_2 = R_P$) and the cross-

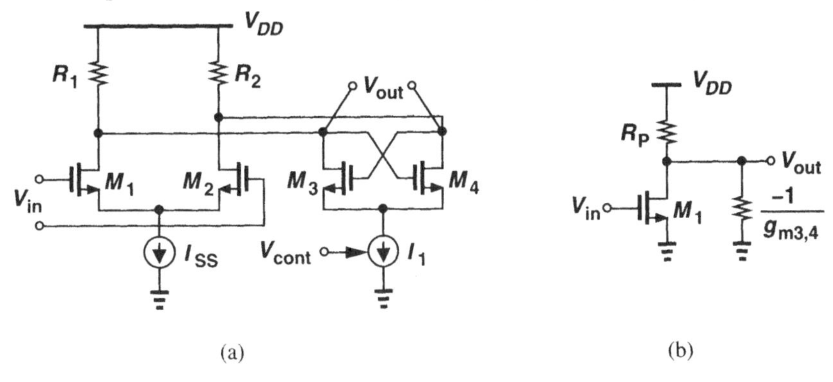

Figure 6.44 (a) Differential stage with variable negative-resistance load, (b) half-circuit equivalent of (a).

coupled pair M_3-M_4. As I_1 increases, the small-signal differential resistance $-2/g_{m3,4}$ becomes less negative and, from the half circuit of Fig. 6.44(b), the equivalent resistance $R_P||(-1/g_{m3,4}) = R_P/(1 - g_{m3,4}R_P)$ increases, thereby lowering the frequency of oscillation.

An important issue in the circuit of Fig. 6.44(a) is that as I_1 varies, so do the currents steered by M_3 and M_4 to R_1 and R_2. Thus, the output voltage swing is not constant across the tuning range. To minimize this effect, I_{SS} can be varied in the *opposite* direction such that the total current steered between R_1 and R_2 remains constant. In other words, it is desirable to vary I_1 and I_{SS} *differentially* while their sum is fixed, a characteristic provided by a differential pair. Illustrated in Fig. 6.45, the idea is to employ a differential pair M_5-M_6 to steer I_T to M_1-M_2 or M_3-M_4 so that $I_{SS} + I_1 = I_T$. Since I_T must flow through

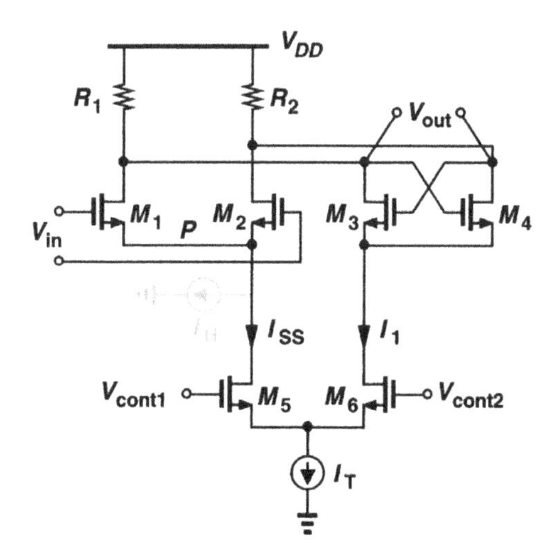

Figure 6.45 Use of a differential pair to steer current between M_1-M_2 and M_3-M_4.

R_1 and R_2, if M_1-M_4 experience complete switching in each cycle of oscillation, then I_T is steered to R_1 (through M_1 and M_3) in half a period and to R_2 (through M_2 and M_4) in the other half, giving a differential swing of $2R_PI_T$.

In the circuit of Fig. 6.45, V_{cont1} and V_{cont2} can be viewed as differential control lines if they vary by equal and opposite amounts. Such a topology provides higher noise immunity for the control input than if V_{cont} is single-ended. Now, note that as V_{cont1} decreases and V_{cont2} increases, the cross-coupled pair exhibits a greater transconductance, thereby raising the time constant at the output nodes. But what happens if all of I_T is steered by M_6 to M_3 and M_4? Since M_1 and M_2 carry no current, the gain of the stage falls to zero, prohibiting oscillation. To avoid this effect, a small constant current source, I_H, can be connected from node P to ground, thereby ensuring M_1 and M_2 always remain on. With typical values, this ring oscillator provides a two-to-one tuning range and reasonable linearity.

As an example, we calculate the minimum value of I_H in Fig. 6.45 to guarantee a low-frequency gain of 2 when all of I_T is steered to the cross-coupled pair. The small-signal voltage gain of the circuit equals $g_{m1,2}R_P/(1 - g_{m3,4}R_P)$. Assuming square-law devices, we have

$$\sqrt{\mu_n C_{ox}(\frac{W}{L})_{1,2} I_H} \frac{R_P}{1 - \sqrt{\mu_n C_{ox}(\frac{W}{L})_{3,4} I_T} R_P} \geq 2. \tag{6.70}$$

That is,

$$I_H \geq \frac{4\left[1-\sqrt{\mu_n C_{ox}(\frac{W}{L})_{3,4} I_T R_P}\right]^2}{\mu_n C_{ox}(\frac{W}{L})_{1,2} R_P^2}. \tag{6.71}$$

An important drawback of using the differential pair M_5-M_6 in the circuit of Fig. 6.45 is the additional voltage headroom that it consumes. As depicted in Fig. 6.46, for M_5 to

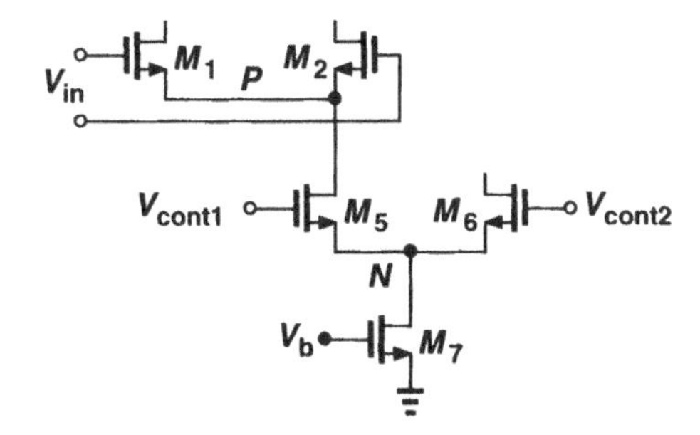

Figure 6.46 Headroom calculation for a current-steering topology.

remain in saturation, V_P must be sufficiently higher than V_N. When $V_{cont1} = V_{cont2}$, the minimum allowable drain-source voltage of M_5 is equal to its equilibrium overdrive voltage, implying that, compared to that calculated for Fig. 6.17, the supply voltage must be higher by this value. Note also that if V_{cont1} or V_{cont2} is allowed to vary above its equilibrium value by more than V_{TH}, then M_5 or M_6 enters the triode region.

The above observation reveals a trade-off between voltage headroom and the *sensitivity* of the VCO. To minimize the sensitivity with a given tuning range, the transconductance of M_5-M_6 must be *minimized*. (That is, to steer all of the tail current, the differential pair must require a *large* $V_{cont1}-V_{cont2}$.) However, for a given tail current, $g_m = 2I_D/(V_{GS}-V_{TH})$, indicating a large equilibrium overdrive for M_5-M_6 and a correspondingly higher value for the minimum required supply voltage.

We should mention that the pair M_5-M_6 need not remain in complete saturation. If the drain voltages are low enough to drive these transistors into the triode region, then the equivalent transconductance of the differential pair drops, thus demanding a greater $V_{cont1} - V_{cont2}$ to steer the tail current. This phenomenon in fact translates to a *lower* VCO sensitivity. In practice, careful simulations are required to ensure the VCO characteristic remains relatively linear across the range of interest.[7]

At low supply voltages, it is desirable to avoid the voltage headroom consumed by M_5-M_6 in Fig. 6.45. The issue can be resolved by means of "current folding." Suppose, as illustrated in Fig. 6.47(a), a differential pair drives two current mirrors, generating I_{out1}

[7]If both M_5 and M_6 are in the triode region and $V_{cont1} \neq V_{cont2}$, then supply voltage variations affect the current steered between the two transistors, introducing noise in the frequency of oscillation.

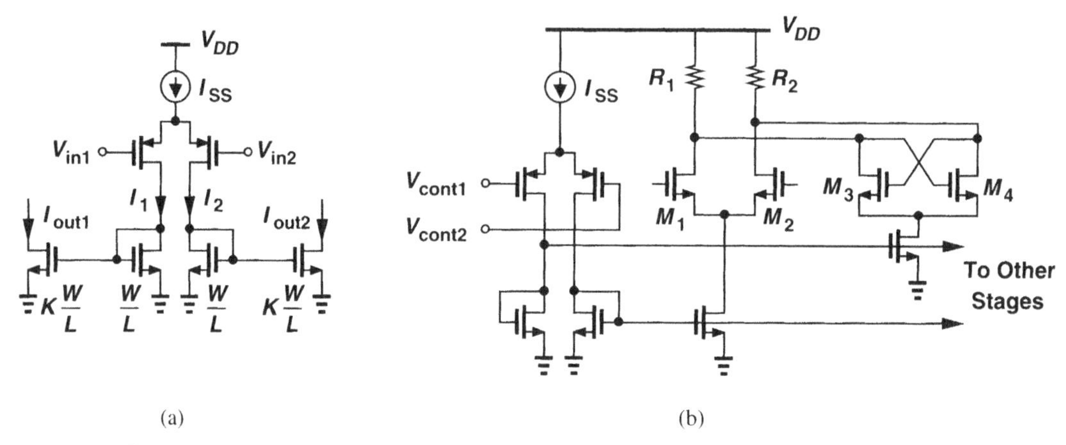

Figure 6.47 (a) Current-folding topology, (b) application of current folding to current steering.

and I_{out2}. Since $I_1 + I_2 = I_{SS}, I_{out1} = KI_1$, and $I_{out2} = KI_2$, we have $I_{out1} + I_{out2} = KI_{SS}$. Thus, as $V_{in1} - V_{in2}$ goes from a very negative value to a very positive value, I_{out1} varies from KI_{SS} to zero and I_{out2} from zero to KI_{SS} while their sum remains constant—a behavior similar to that of a differential pair.

We now utilize the topology of Fig. 6.47(a) in the gain stage of Fig. 6.44(a). Shown in Fig. 6.47(b), the resulting circuit operates from a low supply voltage.

Delay Variation by Interpolation Another approach to tuning ring oscillators is based on "interpolation" [3, 4]. As illustrated in Fig. 6.48(a), each stage consists of a fast path and a slow path whose outputs are summed and whose gains are adjusted by V_{cont} in opposite

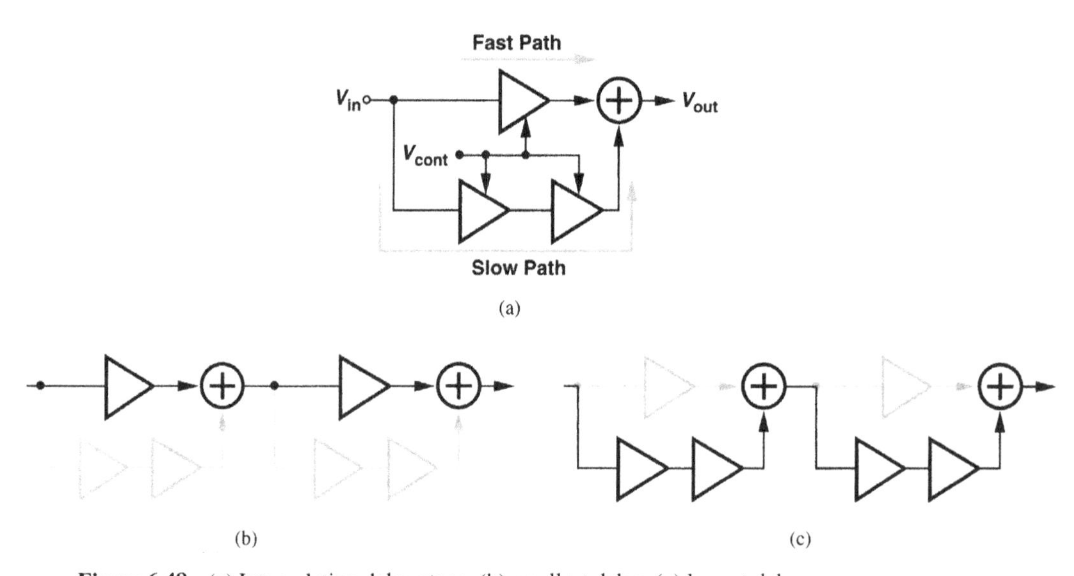

Figure 6.48 (a) Interpolating delay stage, (b) smallest delay, (c) largest delay.

directions. At one extreme of the control voltage, only the fast path is on and the slow path is disabled, yielding the maximum oscillation frequency [Fig. 6.48(b)]. Conversely, at the other extreme, only the slow path is on and the fast path is off, providing the minimum oscillation frequency [Fig. 6.48(c)]. If V_{cont} lies between the two extremes, each path is partially on and the total delay is a weighted sum of their delays.

To better understand the concept of interpolation, let us implement the topology of Fig. 6.48(a) at the transistor level. Each stage can be simply realized as a differential pair whose gain is controlled by its tail current. But how are the two outputs summed? Since the two transistors in a differential pair provide output *currents*, the outputs of the two pairs can be added in the current domain. As depicted in Fig. 6.49(a), simply shorting the outputs of two

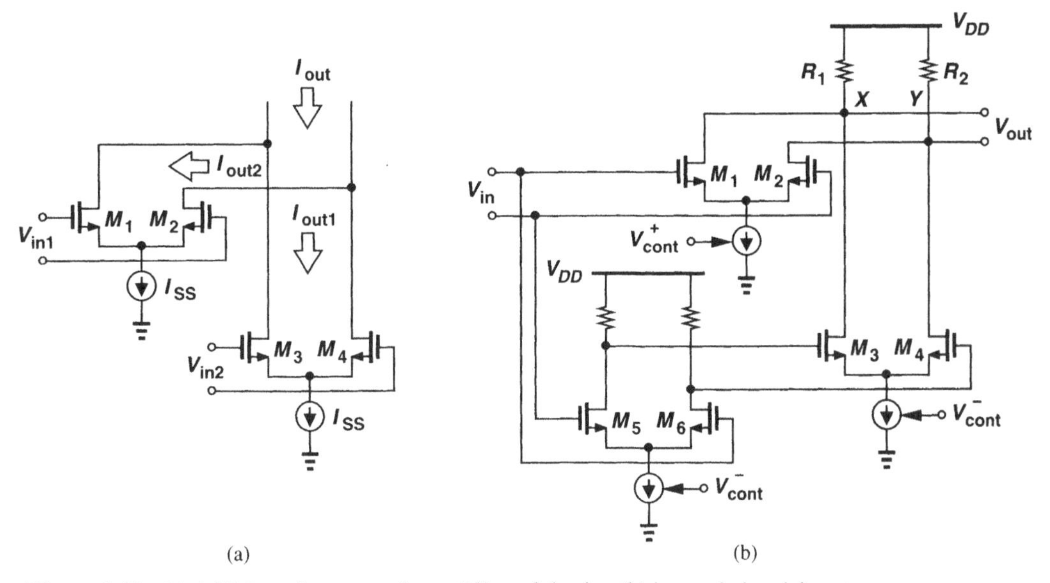

Figure 6.49 (a) Addition of currents of two differential pairs, (b) interpolating delay stage.

pairs performs the current addition, e.g., for small signals, $I_{out} = g_{m1,2}V_{in1} + g_{m3,4}V_{in2}$. The overall interpolating stage therefore assumes the configuration shown in Fig. 6.49(b), where V_{cont}^{+} and V_{cont}^{-} denote voltages that vary in opposite directions (so that when one path turns on, the other turns off). The output currents of M_1-M_2 and M_3-M_4 are summed at X and Y and flow through R_1 and R_2, producing V_{out}.

In the circuit of Fig. 6.49(b), the gain of each stage is varied by the tail current to achieve interpolation. But it is desirable to maintain constant voltage swings. We also recognize that the gain of the differential pair M_5-M_6 need not be varied because even if only the gain of M_3-M_4 drops to zero, the slow path is fully disabled. We then surmise that if the tail currents of M_1-M_2 and M_3-M_4 vary in opposite directions such that their sum remains constant, we achieve both interpolation between the two paths and constant output swings. Illustrated in Fig. 6.50, the resulting circuit employs the differential pair M_7-M_8 to steer I_{SS} between M_1-M_2 and M_3-M_4. If V_{cont} is very negative, M_8 is off and only the fast path amplifies the input. Conversely, if V_{cont} is very positive, M_7 is off and only the slow

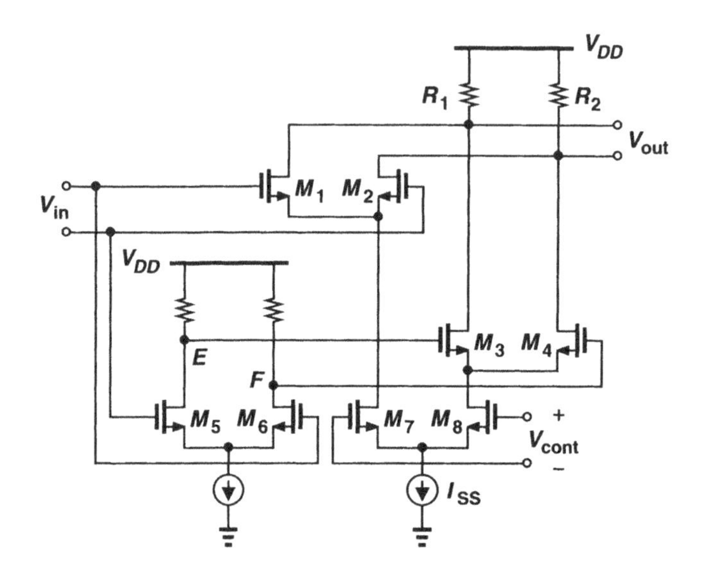

Figure 6.50 Interpolating delay stage with current steering.

path is enabled. Since the slow path in this case employs one more stage than the fast path, the VCO achieves a tuning range of roughly two to one. For operation with low supply voltages, the control pair M_7-M_8 can be replaced by the current-folding topology of Fig. 6.47(a).

As an example, we combine the tuning techniques of Figs. 6.45 and 6.50 to achieve a wider tuning range. We begin with the interpolating stage of Fig. 6.50 and add a cross-coupled pair to the output nodes [Fig. 6.51(a)]. However, to obtain constant voltage swings, the total current through the load resistors must remain constant. This is accomplished by replacing the control differential pair with the current-folding circuit of Fig. 6.47(a). Depicted in Fig. 6.51(b), the resulting configuration steers the current to M_1-M_2 to speed up the circuit and to M_3-M_4 and M_{10}-M_{11} to slow down the circuit. The tail current source dimensions are chosen such that $I_{SS1} = I_{SS2} + I_{SS3}$.

Wide-Range Tuning Except for the circuit of Fig. 6.43(b), the ring oscillator tuning techniques presented thus far achieve a tuning range of typically no more than three to one. In applications where the frequency must be varied by orders of magntitude, the topology shown in Fig. 6.52 can be used. Driven by the input, the additional PMOS transistors M_5 and M_6 pull each output node to V_{DD}, creating a relatively constant output swing even with large variations in I_{SS}. The oscillation frequency of a ring incorporating this stage can be varied by more than four orders of magnitude with less than a twofold variation in the amplitude.

6.4.2 Tuning in LC Oscillators

The oscillation frequency of LC topologies is equal to $f_{osc} = 1/(2\pi\sqrt{LC})$, suggesting that only the inductor and capacitor values can be varied to tune the frequency and other

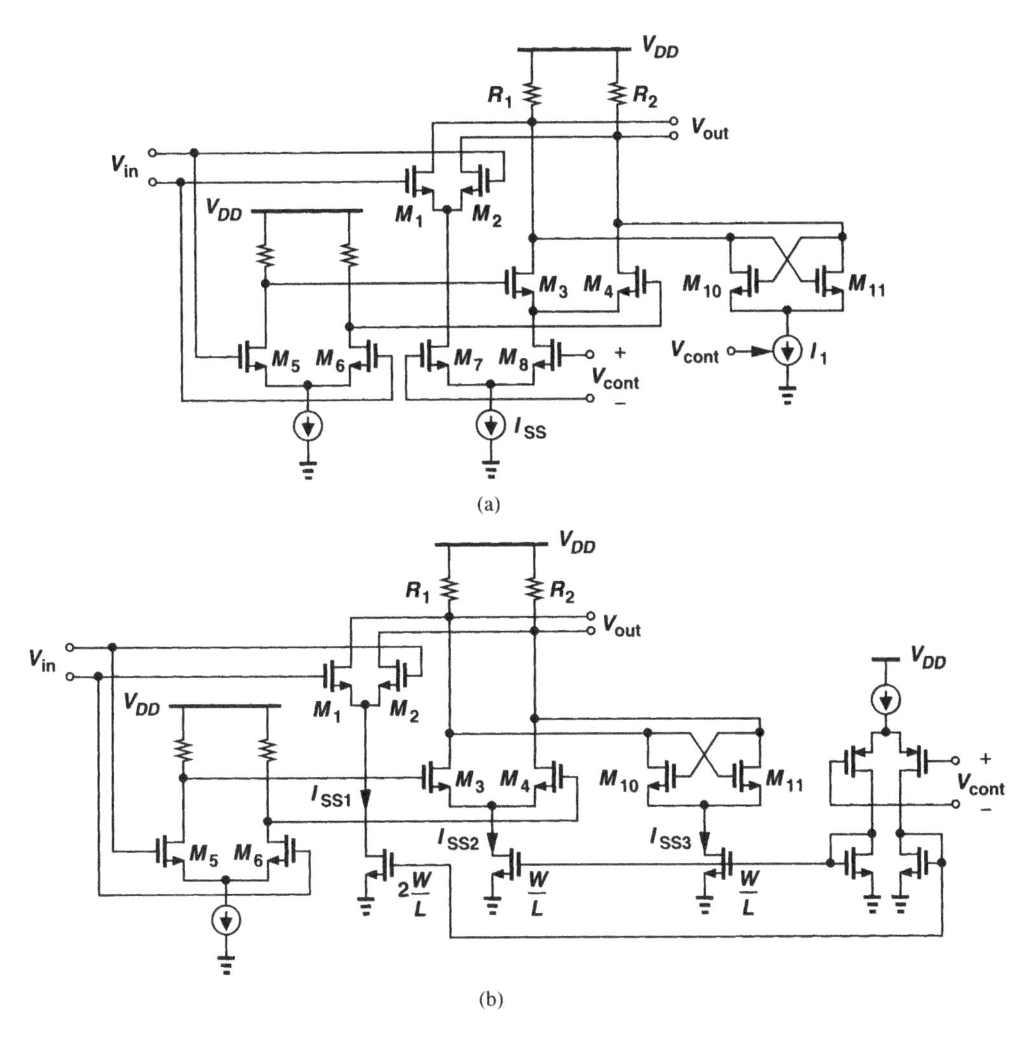

Figure 6.51 VCO using two different tuning techniques.

parameters such as bias currents and transistor transconductances affect f_{osc} negligibly. Since it is difficult to vary the value of monolithic inductors, we simply change the tank capacitance to tune the oscillator. Voltage-dependent capacitors are called "varactors."[8]

A reverse-biased *pn* junction can serve as a varactor. The voltage dependence is

[8]The term "varicap" is also used.

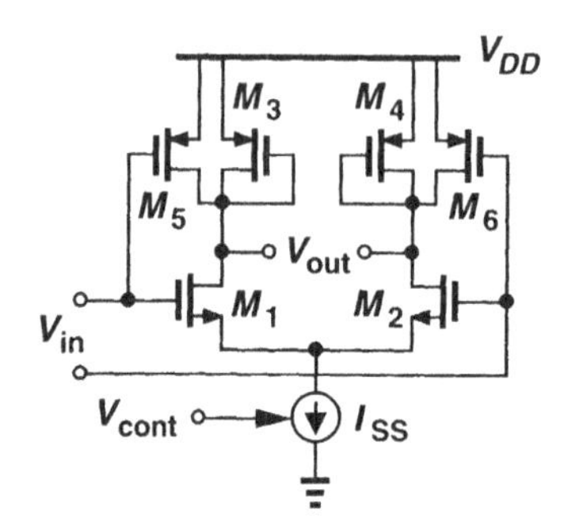

Figure 6.52 Differential stage with wide tuning range.

expressed as

$$C_{var} = \frac{C_0}{(1 + \frac{V_R}{\phi_B})^m}, \tag{6.72}$$

where C_0 is the zero-bias value, V_R the reverse-bias voltage, ϕ_B the built-in potential of the junction, and m a value typically between 0.3 and 0.4.[9] Equation (6.72) reveals an important drawback of LC oscillators: at low supply voltages V_R has a very limited range, yielding a small range for C_{var} and hence for f_{osc}. We also note that to maximize the tuning range, constant capacitances in the tank must be *minimized*.

As an example, suppose in Eq. (6.72), $\phi_B = 0.7$ V, $m = 0.35$, and V_R can vary from zero to 2 V. How much tuning range can be achieved? For $V_R = 0, C_j = C_0$ and $f_{osc,min} = 1/(2\pi\sqrt{LC_0})$. For $V_R = 2$ V, $C_j \approx 0.62C_0$ and $f_{osc,max} = 1/(2\pi\sqrt{L \times 0.62C_0}) \approx 1.27 f_{osc,min}$. Thus, the tuning range is approximately equal to 27%. As explained later, the parasitic capacitances of the inductor and the transistor(s) further limit this range because they cannot be varied by the control voltage.

Let us now add varactor diodes to a cross-coupled LC oscillator (Fig. 6.53). To avoid forward-biasing D_1 and D_2 significantly, V_{cont} must not exceed V_X or V_Y by more than a

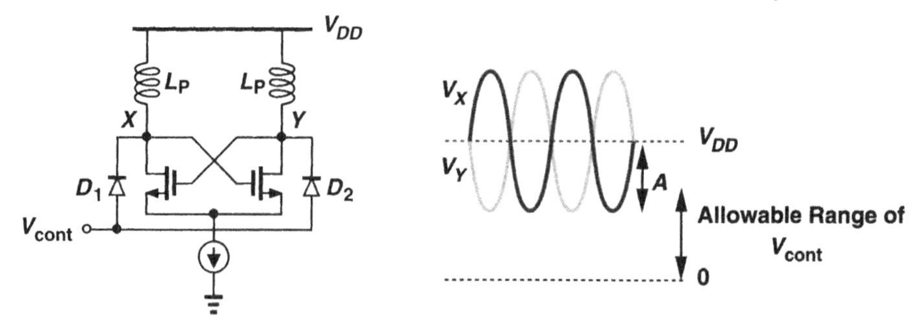

Figure 6.53 LC oscillator using varactor diodes.

[9]Note that $m = 0.5$ for an abrupt junction, but pn junctions in CMOS technology are not abrupt.

few hundred millivolts. Thus, if the peak amplitude at each node is A, then $0 < V_{cont} < V_{DD} - A + 300$ mV, where it is assumed a forward bias of 300 mV creates negligible current. Interestingly, the circuit suffers from a trade-off between the output swing and the tuning range. This effect appears in most LC oscillators.

Note that, since the swings at X and Y are typically large (e.g., 1 V_{pp} at each node), the capacitance of D_1 and D_2 *varies* with time. Nonetheless, the "average" value of the capacitance is still a function of V_{cont}, providing the tuning range.

How are varactor diodes realized in CMOS technology? Illustrated in Fig. 6.54 are two types of pn junctions. In Fig. 6.54(a), the anode is inevitably grounded whereas in Fig. 6.54(b), both terminals are floating. For the circuit of Fig. 6.53, only the floating diode can be used. To increase the capacitance of the junction, the p^+ and n^+ areas (and hence the n-well) are enlarged.

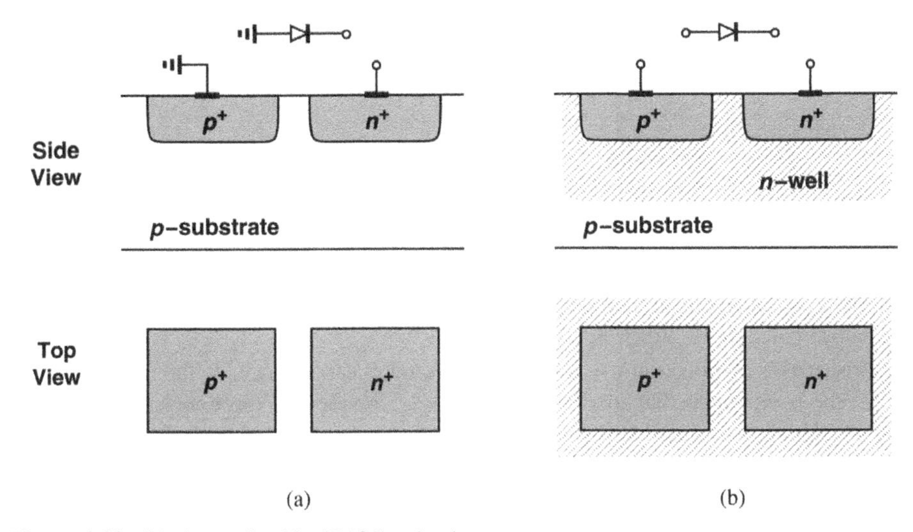

Figure 6.54 Diodes realized in CMOS technology.

Upon closer examination, the structure of Fig. 6.54(b) suffers from a number of drawbacks. First, the n-well material has a high resistivity, creating a resistance in series with the reverse-biased diode and lowering the quality factor of the capacitance. Second, the n-well displays substantial capacitance to the substrate, contributing a constant capacitance to the tank and limiting the tuning range. The diode is therefore represented as shown in Fig. 6.55, where C_n represents the (voltage-dependent) capacitance between the n-well

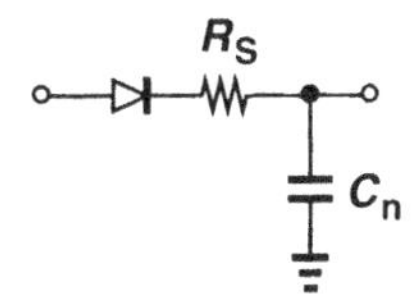

Figure 6.55 Circuit model of the varactor shown in Fig. 6.54(b).

and the substrate.[10]

In order to decrease the series resistance of the structure shown in Fig. 6.54(b), the p^+ region can be surrounded by an n^+ ring so that the displacement current flowing through the junction capacitance sees a low resistance in all four directions [Fig. 6.56(a)]. Since a single minimum-size p^+ area has a small capacitance, many of these units can be placed in

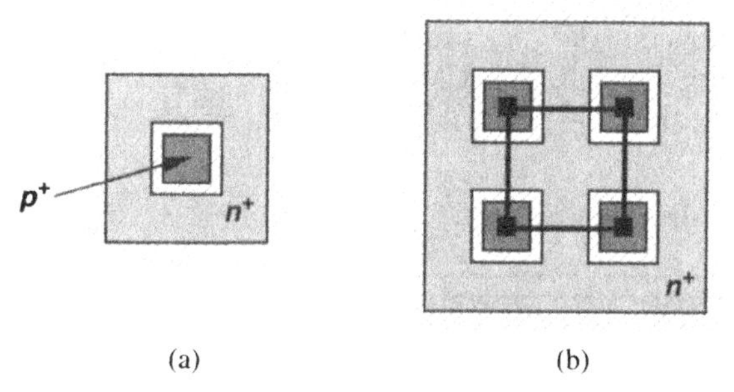

Figure 6.56 (a) Reduction of series resistance by surrounding the p^+ region by an n^+ ring, (b) several diodes in parallel.

parallel [Fig. 6.56(b)]. The n-well, however, must accommodate the entire set, exhibiting a large capacitance to the substrate.

It is instructive at this point to examine the unwanted capacitances in the circuit of Fig. 6.53, i.e., the components that are not varied by V_{cont}. We identify three such capacitances: (1) the capacitance between the n-well and the substrate associated with D_1 and D_2; (2) the capacitances contributed by the transistors to each node, i.e., $C_{GD}, 2C_{GD}$ (the factor of 2 arising from Miller effect[11]), and C_{DB}; and (3) the parasitic capacitance of the inductor itself. Monolithic inductors are typically implemented as metal spiral structures (Fig. 6.57) having relatively large dimensions ($S \approx$ 100–200 μm). Their capacitance to the substrate

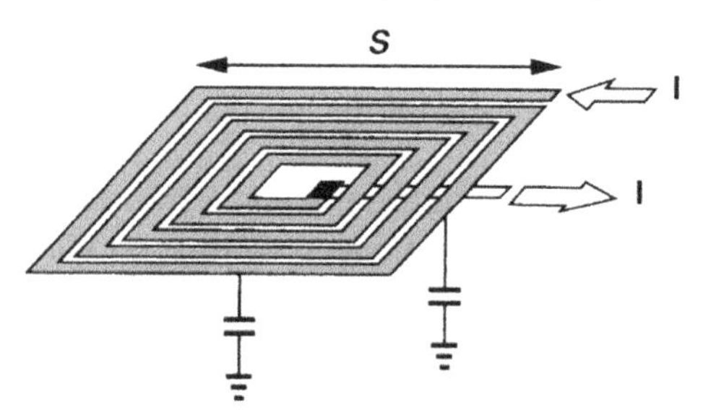

Figure 6.57 Spiral inductor structure.

is therefore quite large.

[10] In circuit simulations, C_n is replaced by a diode having proper junction capacitance.

[11] If the gate and drain voltages vary by equal and opposite amounts, the Miller multiplication factor is equal to 2 regardless of the small-signal gain.

In Fig. 6.53, it is desirable to connect the anode of the diodes to nodes X and Y, thereby eliminating the parasitic n-well capacitances from the tank. Shown in Fig. 6.58 is a topology allowing such a modification. Here, the cross-coupled pair incorporates PMOS devices, providing swings around the ground potential. However, owing to their lower mobility, the PMOS transistors must be wider than their NMOS counterparts so as to exhibit the same transconductance. This increases the second component mentioned above.

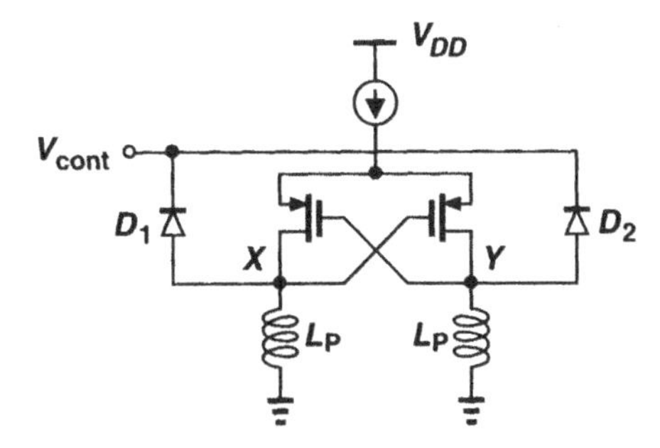

Figure 6.58 Negative-G_m oscillator using PMOS devices to eliminate n-well capacitance from the tanks.

The design of low-noise CMOS LC oscillators with acceptable tuning range is still a topic of active research. Issues such as phase noise and inductor and varactor design continue to intrigue researchers. MOS varactors have also been investigated as an alternative to pn junctions.

6.5 Mathematical Model of VCOs

The definition of the voltage-controlled oscillator given by Eq. (6.55) specifies the relationship between the control voltage and the output frequency. The dependence is "memoryless" because a change in V_{cont} immediately results in a change in ω_{out}. But how is the output signal of the VCO expressed as a function of time? To answer this question, we must review the concepts of phase and frequency.

Consider the waveform $V_0(t) = V_m \sin \omega_0 t$. The argument of the sinusoid is called the "total phase" of the signal. In this example, the phase varies linearly with time, exhibiting a slope equal to ω_0. Note that, as depicted in Fig. 6.59, every time $\omega_0 t$ crosses an integer multiple of π, $V_0(t)$ crosses zero.

Now consider two waveforms $V_1(t) = V_m \sin[\phi_1(t)]$ and $V_2(t) = V_m \sin[\phi_2(t)]$, where $\phi_1(t) = \omega_1 t, \phi_2(t) = \omega_2 t$, and $\omega_1 < \omega_2$. As illustrated in Fig. 6.60, $\phi_2(t)$ crosses integer multiples of π faster than $\phi_1(t)$ does, yielding faster variations in $V_2(t)$. We say $V_2(t)$ accumulates phase faster.

The above study reveals that the faster the phase of a waveform varies, the higher the frequency of the waveform, suggesting that the frequency[12] can be defined as the derivative

[12]The quantity $\omega = 2\pi f$ is called the "radian frequency" (and expressed in rad/s) to distinguish it from f (expressed in Hz). In this book, we call both the frequency, but use ω more often to avoid the factor 2π.

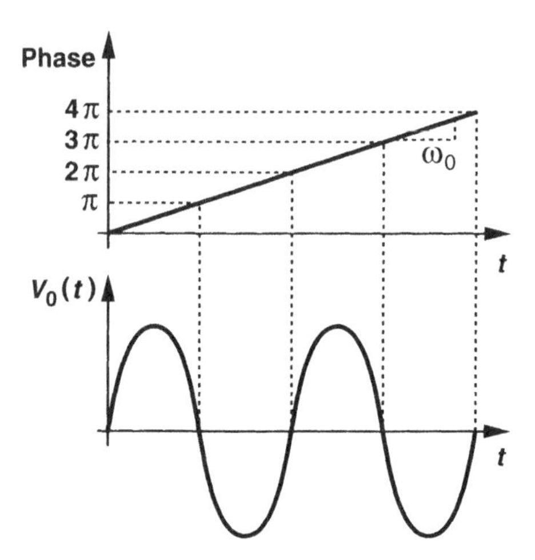

Figure 6.59 Illustration of phase of a signal.

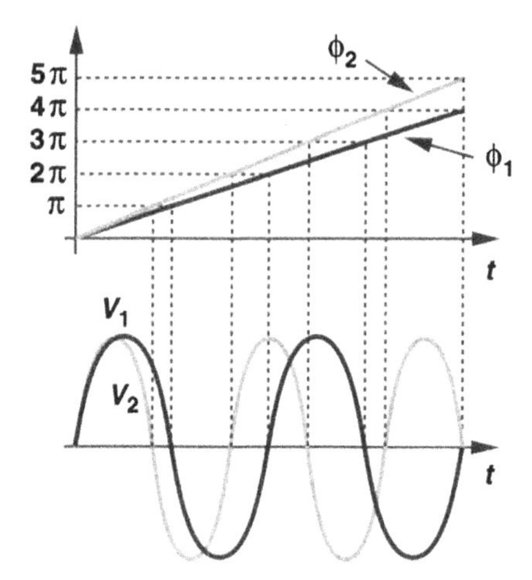

Figure 6.60 Variation of phase for two signals.

of the phase with respect to time:

$$\omega = \frac{d\phi}{dt}. \tag{6.73}$$

Figure 6.61(a) shows the phase of a sinusoidal waveform with constant amplitude as a function of time. Let us plot the waveform in the time domain. Taking the time derivative of $\phi(t)$, we obtain the behavior illustrated in Fig. 6.61(b). The frequency therefore periodically toggles between ω_1 and ω_2, yielding the waveform shown in Fig. 6.61(c). (This is a simple example of binary frequency modulation, called "frequency shift keying" and utilized in wireless pagers and many other communication systems.)

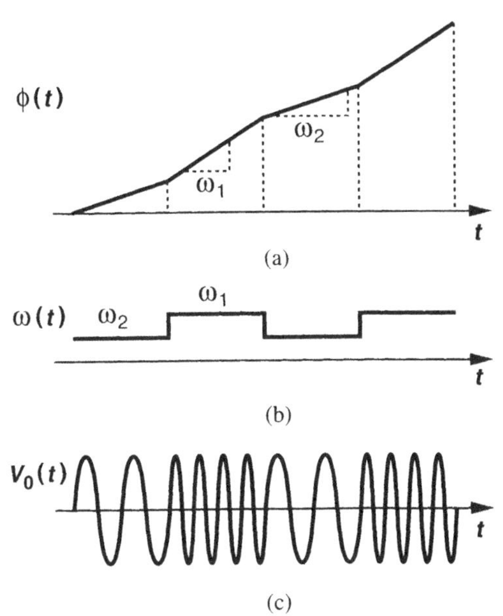

Figure 6.61 Output phase of an oscillator.

Equation (6.73) indicates that, if the frequency of a waveform is known as a function of time, then the phase can be computed as

$$\phi = \int \omega dt + \phi_0. \tag{6.74}$$

In particular, since for a VCO, $\omega_{out} = \omega_0 + K_{VCO}V_{cont}$, we have

$$V_{out}(t) = V_m \cos(\int \omega_{out} dt + \phi_0) \tag{6.75}$$

$$= V_m \cos(\omega_0 t + K_{VCO} \int V_{cont} dt + \phi_0). \tag{6.76}$$

Equation (6.76) proves essential in the analysis of VCOs and PLLs.[13] The initial phase ϕ_0 is usually unimportant and is assumed zero hereafter.

The control line of a VCO senses a rectangular signal toggling between V_1 and V_2 at a period T_m. Let us plot the frequency, phase, and output waveform as a function of time. Since $\omega_{out} = \omega_0 + K_{VCO}V_{cont}$, the output frequency toggles between $\omega_1 = \omega_0 + K_{VCO}V_1$ and $\omega_2 = \omega_0 + K_{VCO}V_2$ (Fig. 6.62). The phase is equal to the time integral of this result, rising linearly with time at a slope of ω_1 for half the input period and ω_2 for the other half. The output waveform of the VCO is similar to that shown in Fig. 6.61. Thus, a VCO can operate as a frequency modulator.

[13]Note that K_{VCO} cannot be brought out of the integral if the characteristic is nonlinear.

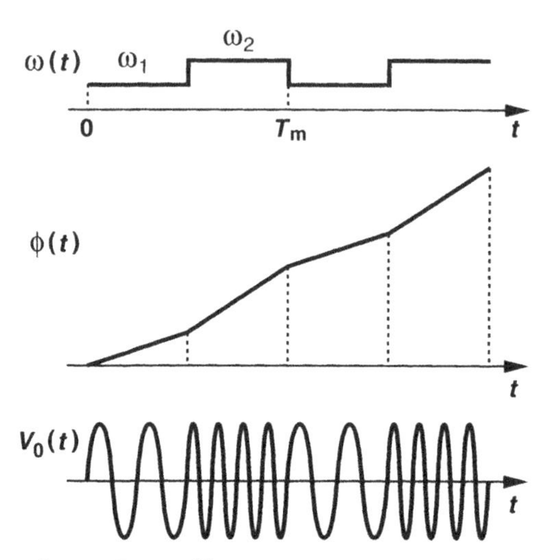

Figure 6.62 Output waveforms of an oscillator.

As explained in Chapter 8, if a VCO is placed in a phase-locked loop, then only the second term of the total phase in Eq. (6.76) is of interest. This term, $K_{VCO}\int V_{cont}dt$, is called the "excess phase," ϕ_{ex}. In fact, in the analysis of PLLs, we view the VCO as a system whose input and output are the control voltage and the excess phase, respectively:

$$\phi_{ex} = K_{VCO}\int V_{cont}dt. \tag{6.77}$$

That is, the VCO operates as an *ideal* integrator, providing a transfer function:

$$\frac{\Phi_{ex}}{V_{cont}}(s) = \frac{K_{VCO}}{s}. \tag{6.78}$$

As an example, suppose a VCO senses a small sinusoidal control voltage $V_{cont} = V_m\cos\omega_m t$. We determine the output waveform and its spectrum. The output is expressed as

$$V_{out}(t) = V_0\cos(\omega_0 t + K_{VCO}\int V_{cont}dt) \tag{6.79}$$

$$= V_0\cos(\omega_0 t + K_{VCO}\frac{V_m}{\omega_m}\sin\omega_m t) \tag{6.80}$$

$$= V_0\cos\omega_0 t\cos(K_{VCO}\frac{V_m}{\omega_m}\sin\omega_m t) \tag{6.81}$$

$$-V_0\sin\omega_0 t\sin(K_{VCO}\frac{V_m}{\omega_m}\sin\omega_m t).$$

If V_m is small enough that $K_{VCO}V_m/\omega_m \ll 1$ rad, then

$$V_{out}(t) \approx V_0 \cos\omega_0 t - V_0(\sin\omega_0 t)(K_{VCO}\frac{V_m}{\omega_m}\sin\omega_m t) \tag{6.82}$$

$$= V_0\cos\omega_0 t - \frac{K_{VCO}V_m V_0}{2\omega_m}[\cos(\omega_0-\omega_m)t - \cos(\omega_0+\omega_m)t]. \tag{6.83}$$

The output therefore consists of three sinusoids having frequencies of $\omega_0, \omega_0 - \omega_m$, and $\omega_0 + \omega_m$. The spectrum is shown in Fig. 6.63. The components at $\omega_0 \pm \omega_m$ are called "sidebands."

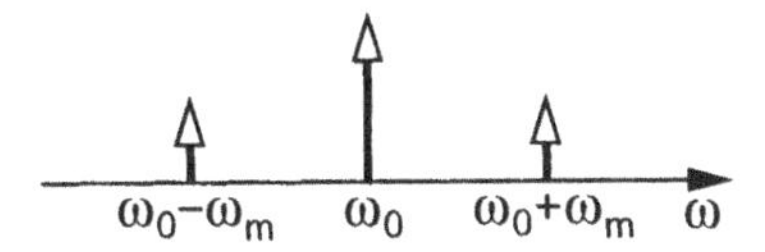

Figure 6.63 Appearance of sidebands at VCO output.

The above example reveals that variation of the control voltage with time may create unwanted components at the output. Indeed, when a VCO operates in the steady state, the control voltage must experience very little variation.[14] This issue is studied in Chapter 8.

A common mistake in expressing the phase of signals arises from the familiar form $V_m \cos\omega_0 t$. Here, the phase is equal to the product of frequency and time, creating the impression that such equality holds in all conditions. We may even deduce that, since the output frequency of a VCO is given by $\omega_0 + K_{VCO}V_{cont}$, the output waveform can be written as $V_m\cos[(\omega_0 + K_{VCO}V_{cont})t]$. To understand why this is incorrect, let us compute the frequency as the derivative of the phase:

$$\omega = \frac{d}{dt}[(\omega_0 + K_{VCO}V_{cont})t] \tag{6.84}$$

$$= K_{VCO}\frac{dV_{cont}}{dt}t + \omega_0 + K_{VCO}V_{cont}. \tag{6.85}$$

The first term in this expression is redundant, vanishing only if $dV_{cont}/dt = 0$. Thus, in the general case, the phase cannot be written as the product of time and frequency.

Our study of VCOs in this section has assumed sinusoidal output waveforms. In practice, depending on the type and speed of the oscillator, the output may contain significant harmonics, even approaching a rectangular waveform. How should Eq. (6.76) be modified in this case? We expect that $V_{out}(t)$ can be expressed as a Fourier series:

$$V_{out}(t) = V_1\cos(\omega_0 t + \phi_1) + V_2\cos(2\omega_0 t + \phi_2) + \cdots. \tag{6.86}$$

We also note that if the (fundamental) frequency of a rectagular waveform is changed by Δf, the frequency of its second harmonic must change by $2\Delta f$, etc. Thus, if V_{cont} varies

[14]Except when the VCO senses a signal to perform frequency modulation.

by ΔV, then the frequency of the first harmonic varies by $K_{VCO}\Delta V$, the frequency of the second harmonic by $2K_{VCO}\Delta V$, etc. That is,

$$V_{out}(t) = V_1 \cos(\omega_0 t + K_{VCO} \int V_{cont} dt + \theta_1) + V_2 \cos(2\omega_0 t + 2K_{VCO} \int V_{cont} dt + \theta_2) + \cdots, \tag{6.87}$$

where $\theta_1, \theta_2, \cdots$ are constant phases necessary for the representation of each harmonic in the Fourier series expansion.

Equation (6.87) suggests that the harmonics of an oscillator output can be readily taken into account. For this reason, we often limit our calculations to the first harmonic even though we may draw the waveforms in rectangular shape rather than sinusoidal shape.

References

1. N. M. Nguyen and R. G. Meyer, "Start-up and Frequency Stability in High-Frequency Oscillators," *IEEE Journal of Solid-State Circuits,* vol. 27, pp. 810–820, May 1992.
2. I. A. Young, J. K. Greason, and K. L. Wong, "A PLL Clock Generator with 5 to 110 MHz of Lock Range for Microprocessors," *IEEE Journal of Solid-State Circuits,* vol. SC-27, pp.1599–1607, Nov. 1992.
3. B. Lai and R. C. Walker, "A Monolithic 622 Mb/sec Clock Extraction and Data Retiming Circuit," *ISSCC Dig. Tech. Papers,* pp. 144–145, Feb. 1991.
4. S. K. Enam and A. A. Abidi, "NMOS ICs for Clock and Data Regeneration in Gigabit-per-Second Optical-Fiber Receivers," *IEEE Journal of Solid-State Circuits,* vol. SC-27, pp. 1763–1774, Dec. 1992.

Chapter 7

LC Oscillators

Having developed oscillator fundamentals in Chapter 6, we focus in this chapter on one type that has become increasingly more popular, namely, LC oscillators. Despite their limited tuning range, such oscillators continue to penetrate various applications because (a) their behavior has been gradually demystified through extensive research in RF design; (b) they provide large output swings (i.e., high slew rates) at high frequencies; (c) they can potentially operate from lower supply voltages than ring oscillators can; and (d) they exhibit substantially less phase noise than ring oscillators do.

This chapter deals with the design of various LC oscillator topologies amenable to integration in BiCMOS and CMOS technologies. We begin with an overview of monolithic inductors and varactors. Next, we study the design of differential LC VCOs. Finally, we analyze multiphase and distributed oscillators.

7.1 Monolithic Inductors

The performance of LC oscillators heavily depends on the quality of inductors and varactors. We are fortunate to inherit the vast body of knowledge developed on this subject in RF design.

As mentioned in Chapter 6, monolithic inductors are typically realized as spiral structures. The mutual coupling between every two turns results in a relatively large inductance per unit area. For example, a straight 500-μm line exhibits an inductance of roughly 0.5 nH whereas, if converted to a spiral of the same total length, it provides an inductance of about 1 nH (Fig. 7.1). The design of the inductor seeks to maximize the Q and minimize the parasitic capacitance. The area consumed by the inductor is also of concern in some applications.

The calculation of the inductance of a spiral is not straightforward. Various accurate [1] and approximate [2, 3] methods have been proposed. A physics-based expression that exhibits less than a few percent of error is given in [4] and can be reduced to the following

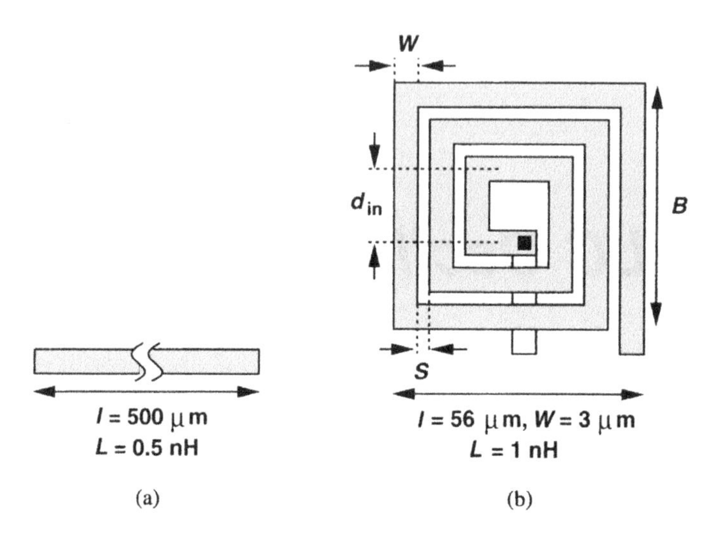

Figure 7.1 Inductance of a wire as a straight or spiral structure.

for a square spiral:

$$L = \frac{\mu_0 l}{2\pi}\left[\ln\frac{l}{N(W+t)} - 0.2 - 0.47N + (N-1)(\ln(\sqrt{1+k^2}+k) - \sqrt{1+\frac{1}{k^2}}+\frac{1}{k})\right], \tag{7.1}$$

where, with the notation in Fig. 7.1,

$\mu_0 = 4\pi \times 10^{-7}$ H/m is the vacuum permeability,

$l = (4N+1)[d_{in} + N(W+S)]$ is the total length of the spiral wire,[1]

t is the conductor thickness,

N is the number of turns,

$k = 3l/[4N(N+1)(W+S)]$.

The value of inductance depends on primarily two parameters: the number of turns and the diameter of each turn. But the line width and spacing also affect the results because they determine how many turns and with what diameters can be accommodated in a given area. For example, in Fig. 7.2, $L_1 > L_2$ because each turn of L_1 encloses a greater amount

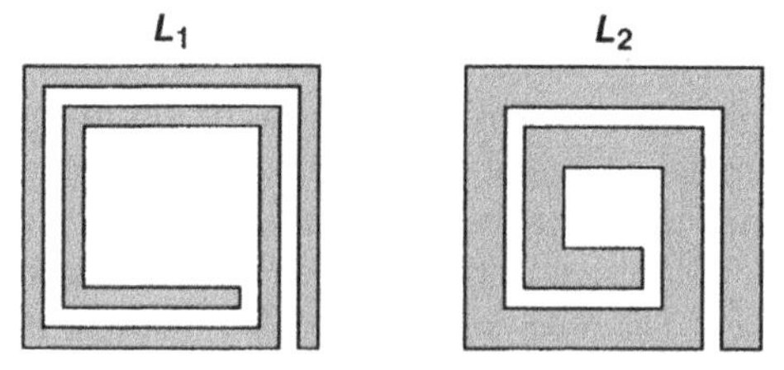

Figure 7.2 Relationship between inductance and linewidth.

[1]This equation holds for an integer number of turns. More general expressions are given in [4].

of magnetic flux than its counterpart in L_2.

7.1.1 Loss Mechanisms

Let us now consider the nonidealities that limit the Q of inductors. Recall from Chapter 6 that the finite Q of an LC tank manifests itself as a decaying impulse response because of the energy loss in every cycle. We must therefore identify sources of dissipation in an inductor and its environment.

The first loss mechanism relates to the series resistance of the metal wire comprising the spiral. This resistance arises from two components: one due to the low-frequency resistivity of the metal and another resulting from skin effect. With a sheet resistance of 30 to 70 m$\Omega/\square$, metal interconnects in VLSI technologies seriously limit the Q at low frequencies, necessitating wide lines. Figure 7.3(a) illustrates an example where a 2-nH inductor is

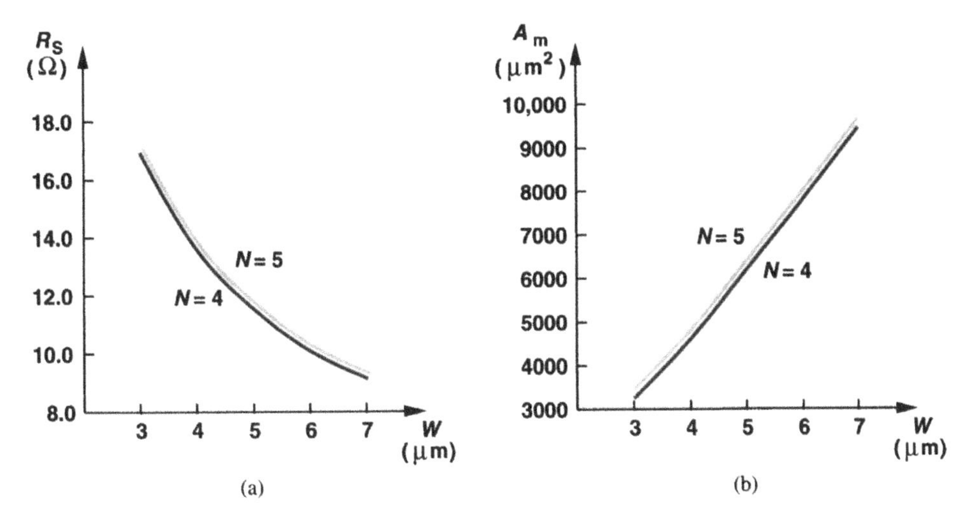

Figure 7.3 Dependence of (a) series resistance and (b) area of inductors on linewidth and number of turns.

designed with various line widths for $N = 4$ and $N = 5$. Note that, to obtain a given inductance, B (the outer dimension) is adjusted as W varies (but S is kept constant). We make two important observations. (1) For a given N, the low-frequency resistance, R_S, becomes a weaker function of W as W increases because a wider line inevitably requires a greater length if the inductance must remain unchanged. In other words, beyond $W \approx 5\ \mu$m, doubling W reduces R_S by less than twofold. (2) For a given W, the value of R_S is a very weak function of N; i.e., the total length of the spiral, l, is quite independent of N.

Increasing W yields a greater area and hence a larger capacitance for the inductor. The reader can show that the area of a square spiral (excluding the gaps and the inner hole) is given by

$$A_m = 4NW[B - W - (N-1)(S+W)]. \tag{7.2}$$

Figure 7.3(b) plots the area of the inductor for the 2-nH example, revealing a relatively linear dependence on W. Note that, for a given W, A_m remains relatively constant as N varies from 4 to 5. This is to be expected because Fig. 7.3(a) implies that the total line length is constant.

The series resistance due to skin effect manifests itself at high frequencies. The skin depth of aluminum is approximately equal to 1.4 μm at $f = 10$ GHz. For a straight wire, the resulting sheet resistance is given by

$$R_{skin} = \frac{\rho}{\delta}, \tag{7.3}$$

where ρ denotes the resistivity and δ the skin depth:

$$\delta = \sqrt{\frac{\rho}{\pi f \mu}}. \tag{7.4}$$

That is,

$$R_{skin} = \sqrt{\pi f \rho \mu}. \tag{7.5}$$

In a spiral structure, on the other hand, the proximity of adjacent turns leads to a complex current distribution. Interestingly, such current distribution also changes the *inductance* value because the area and hence the magnetic flux enclosed by each turn change. For example, as shown in Fig. 7.4, the current may concentrate at the edges of the wire even

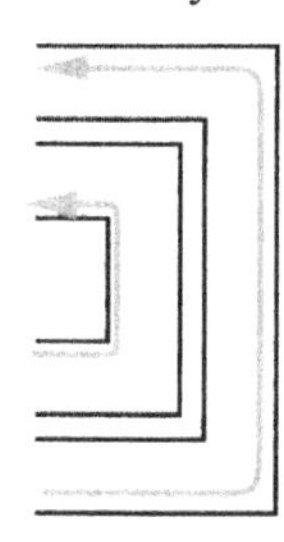

Figure 7.4 Distribution of current at high frequencies.

if the line is quite wide. Thus, accurate modeling, especially at frequencies above 10 GHz, may require electromagnetic field simulations.

The second loss mechanism arises from the capacitance between the inductor and the substrate and the substrate resistance. Illustrated in Fig. 7.5, this phenomenon lowers the Q because, as the potential at each point on the inductor varies with time, the displacement current generated in the capacitance flows through the substrate resistance. Note that if the resistance were zero or infinite, no loss would occur.

The third loss mechanism relates to magnetic coupling between the inductor and the substrate. Recall from basic electromagnetics that Lenz's law states: The current induced by a magnetic field generates another magentic field that opposes the first field. Thus, as the current through an inductor varies with time, it creates an "eddy" current in the substrate (Fig. 7.6). Again, since the substrate resistance is neither zero nor infinite, some of the

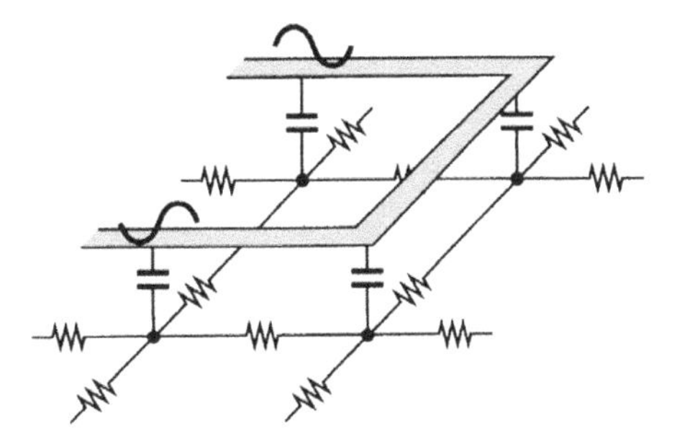

Figure 7.5 Substrate loss due to capacitive coupling.

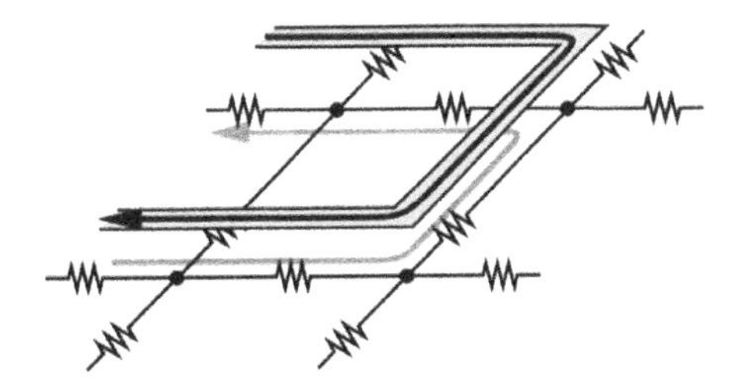

Figure 7.6 Substrate loss due to magnetic coupling.

inductor energy is lost in the form of heat.

It is possible to view the above phenomenon as transformer coupling, with the inductor and the substrate acting as the primary and the secondary, respectively [Fig. 7.7(a)]. Constructing the equivalent circuit of Fig. 7.7(b) and assuming a mutual inductance of M

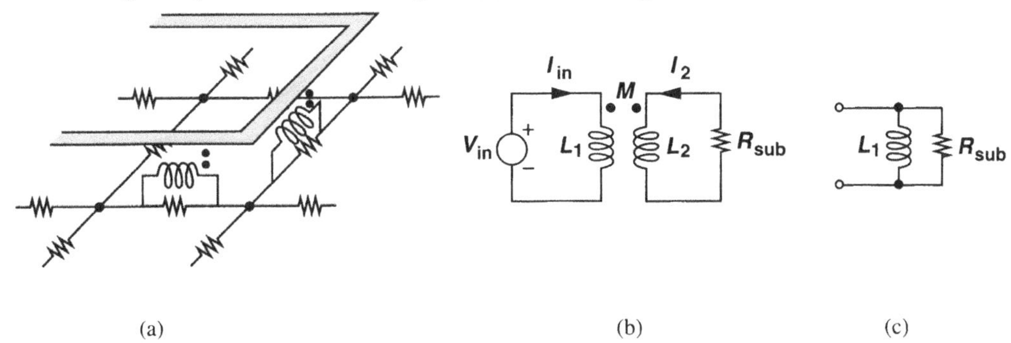

Figure 7.7 (a) Magnetic coupling viewed as transformer action, (b) equivalent circuit, (c) simplified model.

between the inductor and the substrate, we have

$$V_{in} = L_1 s I_{in} + M s I_2 \tag{7.6}$$

$$-R_{sub} I_2 = I_2 L_2 s + M s I_{in}. \tag{7.7}$$

Thus,

$$\frac{V_{in}}{I_{in}} = L_1 s - \frac{M^2 s^2}{R_{sub} + L_2 s}. \tag{7.8}$$

If $L_1 = L_2 = M$ (i.e., complete magnetic coupling to the substrate), then

$$\frac{V_{in}}{I_{in}} = L_1 s || R_{sub}, \tag{7.9}$$

yielding the equivalent circuit of Fig. 7.7(c). Thus, to the first order, the substrate resistance appears in parallel with the inductor, reducing the Q.

Equation (7.8) provides more insight if we consider an extreme case where R_{sub} is small, e.g., $R_{sub} \ll |L_2 s|$. We can then factor $L_2 s$ out and approximate the result as

$$\frac{V_{in}}{I_{in}} = (L_1 - \frac{M^2}{L_2})s + \frac{M^2}{L_2^2} R_{sub}. \tag{7.10}$$

Thus, eddy currents reduce the equivalent *inductance* as well.

All of the above three losses become more significant as the frequency rises: (a) skin effect is proportional to $\sqrt{f}$; (b) the impedance of the capacitors in Fig. 7.5 drops with f; and (c) the quality factor of the equivalent circuit in Fig. 7.7(c), $Q = R_{sub}/(L\omega)$, also falls with f. Consequently, each inductor design must be optimized for a certain frequency range.

It is instructive to study the impact of the substrate resistance on the Q. Let us begin with capacitive coupling and consider the simple circuit in Fig. 7.8(a). The reader can show that the power dissipated in R_{sub} is given by

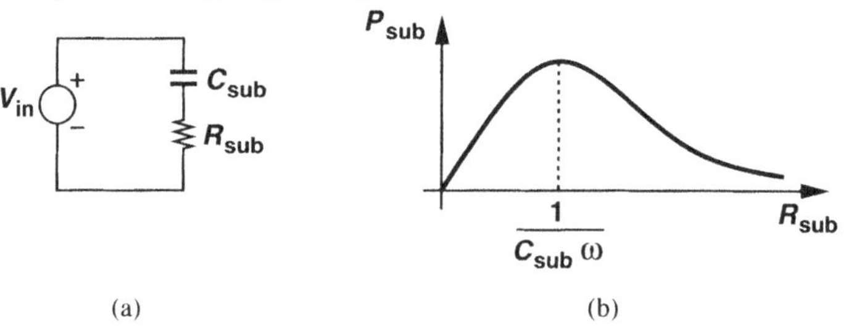

Figure 7.8 (a) Circuit to compute substrate loss, (b) loss as a function of substrate resistance.

$$P_{sub1} = \frac{R_{sub} C_{sub}^2 \omega^2}{R_{sub}^2 C_{sub}^2 \omega^2 + 1} V_{in}^2. \tag{7.11}$$

Plotted in Fig. 7.8(b), P_{sub1} reaches a maximum for $R_{sub1} = (C_{sub}\omega)^{-1}$. Thus, low loss mandates operating well below or well above R_{sub1}. Since at high frequencies, the term $(C_{sub}\omega)^{-1}$ is relatively small, it is preferable to choose a *high* value for R_{sub} so as to minimize the loss.

For magnetic coupling, the simple result expressed by Eq. (7.9) also suggests that the substrate resistance must be maximized. For this reason, lightly-doped substrates have become popular in RF and high-speed circuits utilizing inductors—even though they are more prone to latch-up than heavily-doped substrates are.

7.1.2 Inductor Modeling

With the nonidealities studied above, we can now develop a circuit model for monolithic inductors. This task poses two issues: (1) both the inductor and the substrate are three-dimensional distributed structures and can only be *approximated* by a lumped model; (2) some parameters such as the inductance and skin depth *vary* with frequency, making it difficult to fit the model in a broad bandwidth. For most oscillators, a narrowband model suffices. For broadband circuits employing inductive peaking, the model must be accurate near the upper end of the band (Chapter 5).

Let us begin with a model embodying metal losses. Shown in Fig. 7.9(a), a series resis-

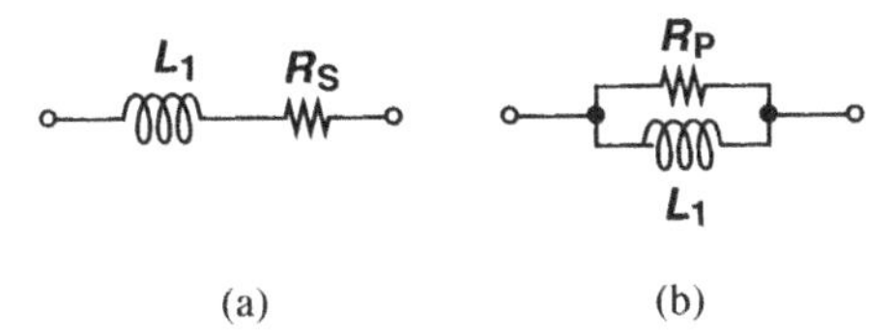

Figure 7.9 (a) Series and (b) parallel models of an inductor.

tor can represent both low-frequency and skin resistance. If R_S is independent of the frequency, then the model is valid for a limited range. The loss can alternatively be modeled by a parallel resistance [Fig. 7.9(b)] because, as explained in Chapter 6, the two networks are equivalent for a narrow frequency range.

Some broadband circuits may require that the variation of the skin resistance with frequency be included in the model. First, consider an extreme case where the model must be valid at only dc and at a high frequency. As depicted in Fig. 7.10(a), we select a series resistance, R_{S1}, equal to that due to skin effect and *shunt* the combination of R_{S1} and L_1

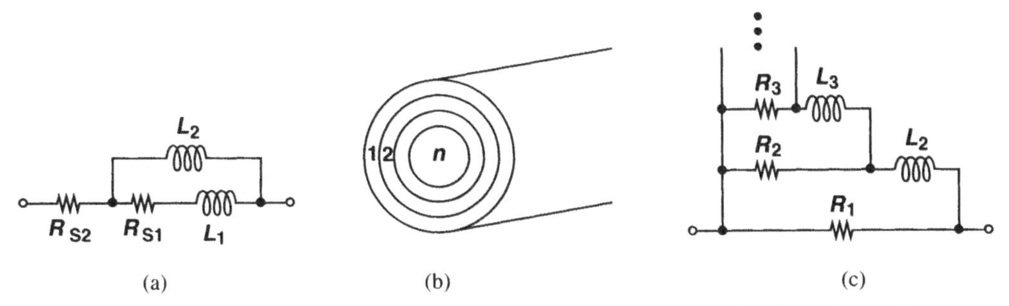

Figure 7.10 (a) Simple model including frequency-dependent resistance, (b) view of a conductor at high frequencies, (c) resulting skin effect model.

with a large inductor, L_2. We then add R_{S2} in series to model the low-frequency resistance of the wire. Thus, at high frequencies, L_2 is open and $R_{S1} + R_{S2}$ embody the overall loss, and at very low frequencies, L_2 is a short and the network reduces to R_{S2}.

The foregoing principle can be extended to broadband modeling of skin effect [5]. Depicted in Fig. 7.10(b) for a coaxial cable, this approach views the signal line as a set of concentric cylinders, each having some (low-frequency) resistance and inductance, arriving at the circuit in Fig. 7.10(c) for one section of the distributed model. Here, the branch consisting of R_j and L_j represents the impedance of cylinder number j. At low frequencies, the current is uniformly distributed through the conductor and the model is reduced to $R_1||R_2||\cdots||R_n$ [5]. As the frequency increases, the current moves away from the inner cylinders, an effect modeled by the rising impedance of the the inductors in each branch. In [5], a constant ratio R_j/R_{j+1} is maintained to simplify the model while achieving a reasonable agreement between the simulated and measured transient behavior of coaxial lines.

The key feature of this method is that it reduces the problem of skin effect to the calculation of low-frequency resistances and inductances. In this book, we continue with the simple model of Fig. 7.9(a) because, along with other effects added below, it provides a reasonable approximation for most circuits of interest.

We now add the effect of capacitive coupling. Figure 7.11(a) shows a one-dimensional

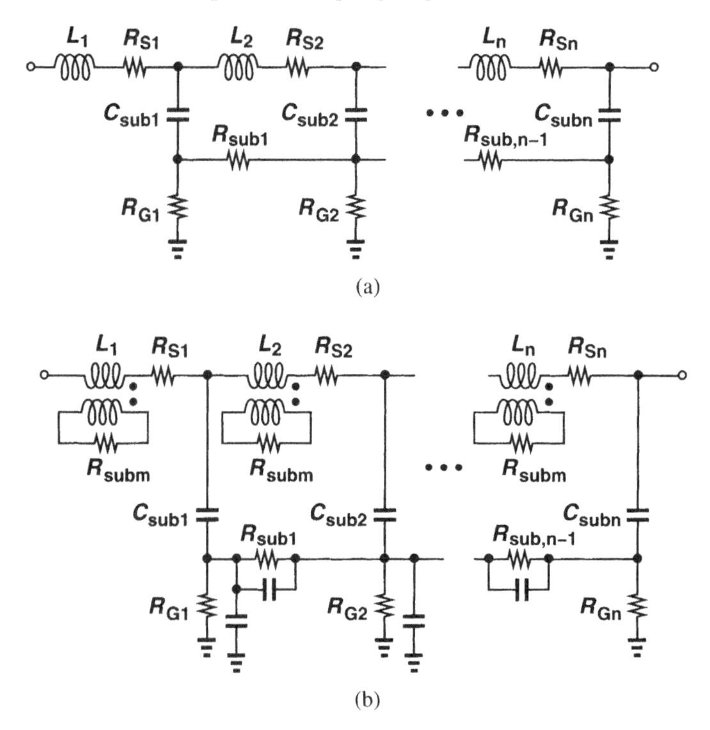

Figure 7.11 Inductor model including (a) capacitive coupling to the substrate, and (b) magnetic coupling to the substrate.

uniformly distributed model, where the total inductance and series resistance are decomposed into equal segments $L_1, \cdots, L_n$ ($L_1 + \cdots + L_n = L_{tot}$) and $R_{S1}, \cdots, R_{Sn}$ ($R_{S1} + \cdots + R_{Sn} = R_{Stot}$), respectively.[2] The nodes in the substrate are connected to each other by $R_{sub1}, \cdots, R_{subn}$ and to ground by $R_{G1}, \cdots, R_{Gn}$. The total capacitance between the spiral and the substrate is decomposed into $C_{sub1}, \cdots, C_{subn}$ ($C_{sub1} + \cdots + C_{subn} = C_{subtot}$).

Let us include the effect of magnetic losses. This is accomplished as shown in Fig. 7.11(b), where each inductor segment is coupled to the substrate through a transformer. Proper choice of mutual coupling and R_{subm} allow accurate representation of this type of loss. In practice, the capacitance between different nodes in the substrate must also be included, leading to the network depicted in Fig. 7.11(b).

The model developed thus far is relatively complete but quite complex, making it difficult to fit the values of various parameters to the data obtained from inductor measurements. For this reason, more compact models have been proposed that more easily lend themselves to parameter extraction and fitting. Figure 7.12(a) shows an example where the distributed network of Fig. 7.11(b) has been converted to a lumped circuit and the

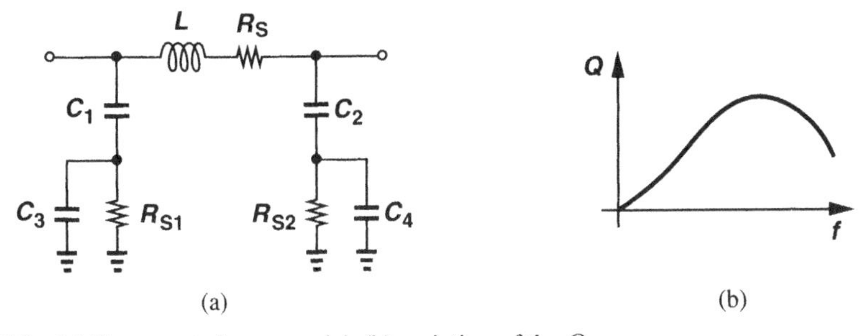

Figure 7.12 (a) Compact inductor model, (b) variation of the Q.

transformer coupling is omitted. Such a representation is justified because it fits measured inductor characteristics with reasonable accuracy. This model also predicts the Q variation shown in Fig. 7.12(b): at low frequencies, the inductor impedance is small and the Q is limited by R_S; at high frequencies, C_1 and C_2 allow R_{S1} and R_{S2} to appear in parallel with the inductor, lowering the Q.

An inaccuracy resulting from the model of Fig. 7.12(a) arises if one end of the inductor, e.g., the right end, is grounded. The model predicts that in this case the parasitic capacitance consists of C_1 and C_3, which are usually assumed to be equal to C_2 and C_4, respectively. In reality, however, if the right end is grounded, then, owing to the distributed nature of the capacitance, C_1 and C_2 must be only one-third of the overall value [7]. This point exemplifies the utility of distributed models.

We should mention that spiral inductors contain two other parasitic capacitances. Illustrated in Fig. 7.13(a), the first results from the underpass wire connecting the inner end of the inductor. The model is modified as shown in Fig. 7.13(b) to include this effect. The second capacitance component arises from the fringe field between adjacent turns [Fig.

[2] A more accurate model would include mutual coupling between every two segments as well. In that case, $L_{tot} = L_1 + \cdots + L_n + nM$.

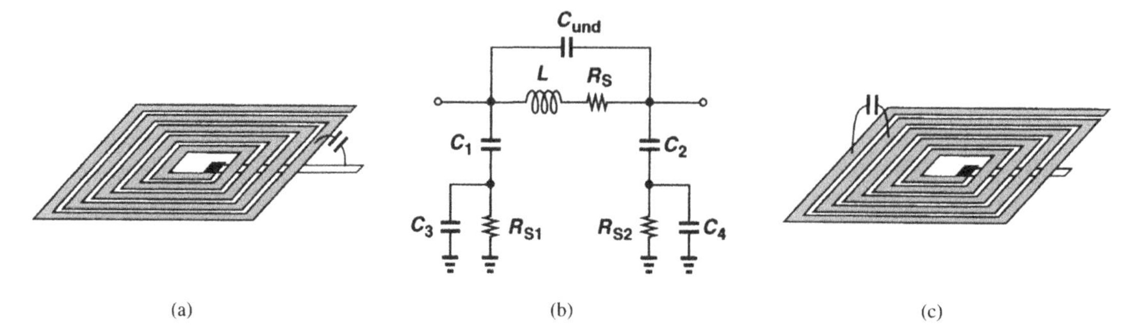

Figure 7.13 (a) Overlap capacitance associated with the underpass, (b) inductor model reflecting the overlap capacitance, (c) interwinding capacitance.

7.13(c)]. While the effect of this component can be embodied in C_{und}, it is usually negligible because the potential difference between adjacent turns is small and, therefore, so is the electric energy stored in the fringe capacitance.

7.1.3 Inductor Design Guidelines

The design of monolithic inductors is a complex task, often requiring optimization programs. Nonetheless, some general principles can be followed to arrive at a "first cut."

For a given inductance, different combinations of line width, number of turns, and outer dimension can be used, leading to a large design space. However, the dc resistance of the inductor, R_{dc}, often constrains the choice of these values. In particular, the line must be sufficiently wide so that R_{dc} does not significantly limit the Q. For example, if seeking an overall Q of 5, we may choose $L\omega/R_{dc} = 10$, assuming that other losses lower the Q to the target value. This choice in turn narrows down the design space.

The inner turns of a spiral contribute little inductance while suffering from all of the loss mechanisms. More importantly, as shown in Fig. 7.14, the currents flowing in opposite

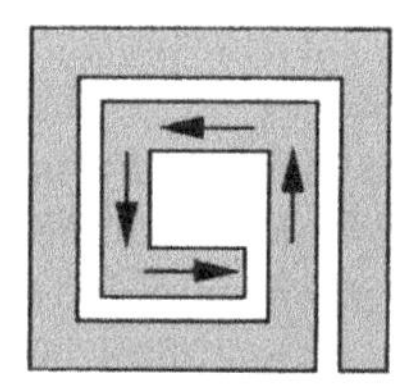

Figure 7.14 Reduction of inductance due to currents flowing in opposite directions.

directions in these turns *reduce* the inductance. For these reasons, the first four or five inner turns are usually removed. Also, the adjacent turns are typically spaced by the minimum allowed by the technology. Increasing the spacing generally *degrades* the performance.[3]

[3]Except in symmetric inductors, to be discussed later in this section.

In order to reduce the loss resulting from capacitive coupling to the substrate, a conductive "shield" can be placed under the inductor. As shown in Fig. 7.15(a), if the shield is held at a constant potential, then the electric field lines produced by the inductor terminate on

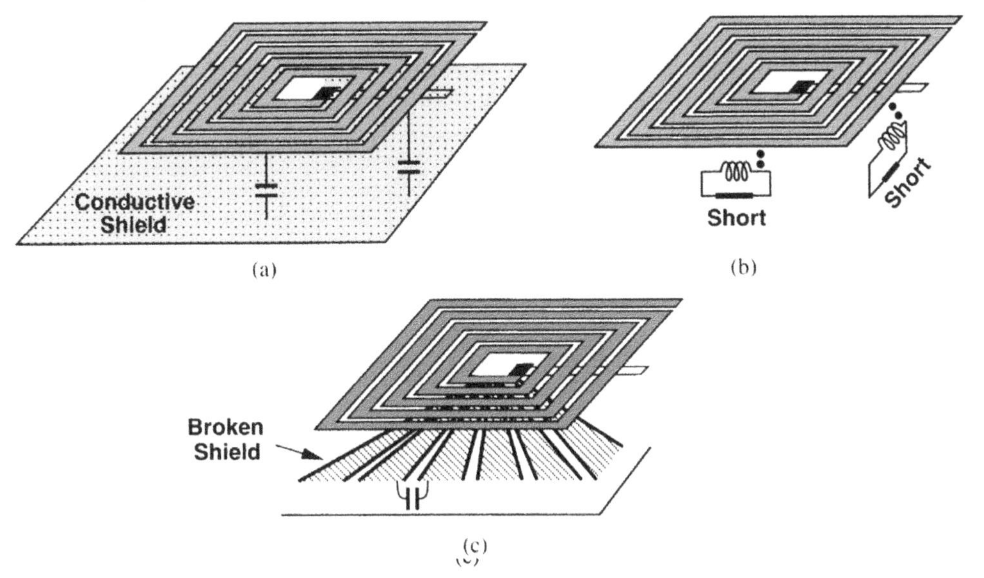

Figure 7.15 (a) Addition of conductive shield to suppress loss due to capacitive coupling to substrate, (b) effect on inductor, (c) patterned shield.

the shield and the displacement currents flow through a *small* resistance. Consequently, the resistive loss is suppressed. However, a continuous shield introduces a low resistance for eddy currents as well, acting as a *short* across the secondary of the equivalent transformer [Fig. 7.15(b)]. Setting R_{sub} to zero in Eq. (7.8), we note that

$$\frac{V_{in}}{I_{in}} = (L_1 - \frac{M^2}{L_2})s. \tag{7.12}$$

In other words, a short across the secondary *reduces* the overall inductance—even to zero if $M = L_2 = L_1$.

To resolve the above difficulty, the shield is broken periodically in the direction perpendicular to that of the current flow [Fig. 7.15(c)]. The resulting gaps stop the eddy currents while contributing some lossy capacitive coupling to the substrate. Thus, the width of the gaps is typically minimized, even at the cost of increasing the fringe capacitance between adjacent sections of the shield. Note that the fringe capacitances appear in *series* with the eddy current, thereby exhibiting a relatively high impedance and impacting the performance negligibly. Depending on the frequency of operation and the inductance value, the broken shield technique may increase the Q by 5 to 10%. Note that the shield creates little change in *magnetic* coupling; eddy currents continue to flow through the substrate.

The type of the material to be used in the shield for maximum Q has been debated. The following layers have been considered: n-well, silicided p^+ or n^+ diffusion, silicided polysilicon, and metal, with the parasitic capacitance increasing in the same order. At high frequencies, n-well or p^+ or n^+ diffusion may prove the optimum choice.

Inductor Structures We have thus far considered only square spirals. For a given value of inductance, a *circular* spiral exhibits less length and hence a smaller resistance. The circular structure is often approximated by an octagonal shape so as to facilitate generation of masks used in the fabrication process (Fig. 7.16).

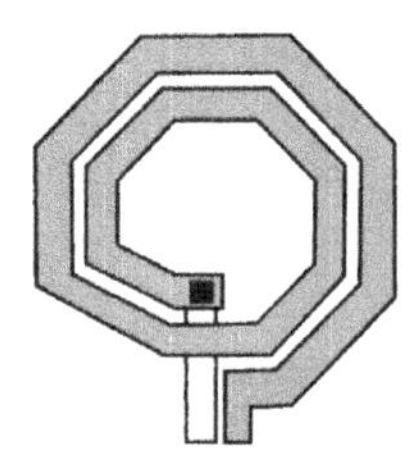

Figure 7.16 Octagonal inductor.

Another interesting modification of monolithic inductors is particularly suited to differential circuits. Shown in Fig. 7.17(a), a *symmetric* structure of value L can replace two

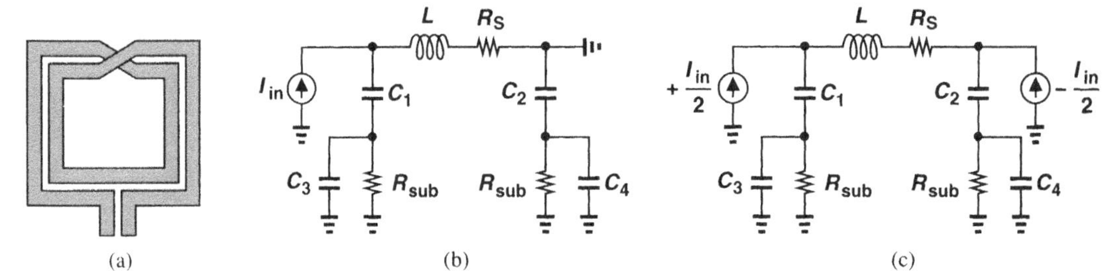

Figure 7.17 (a) Symmetric inductor, (b) inductor driven by single-ended stimulus, (c) inductor driven by differential stimuli.

asymmetric inductors of value of $L/2$, thereby occupying less area. More importantly, differential inductors generally exhibit a higher Q than their single-ended counterparts [6]. To understand this property, we employ the compact model of Fig. 7.12, driving the inductor by a single-ended or differential current source [Figs. 7.17(b) and (c)]. We note that in Fig. 7.17(b), the high-frequency resistance in parallel with the tank is equal to R_{sub} whereas in Fig. 7.17(c) it is equal to $2R_{sub}$.[4] Thus, the effect of substrate loss, as modeled by R_{sub}, diminishes considerably for a differential stimulus. For frequencies in the range of tens of gigahertz, differential structures provide Q's of around 10 and single-ended geometries around 5 to 6.

In symmetric inductors, the interwinding (fringe) capacitance does become significant. Consider the structure shown in Fig. 7.18(a), where the spiral is driven by differential voltages and viewed as four segments in series. Modeling each segment by an inductor and

[4] Equivalently, R_{sub} appears in parallel with $L/2$.

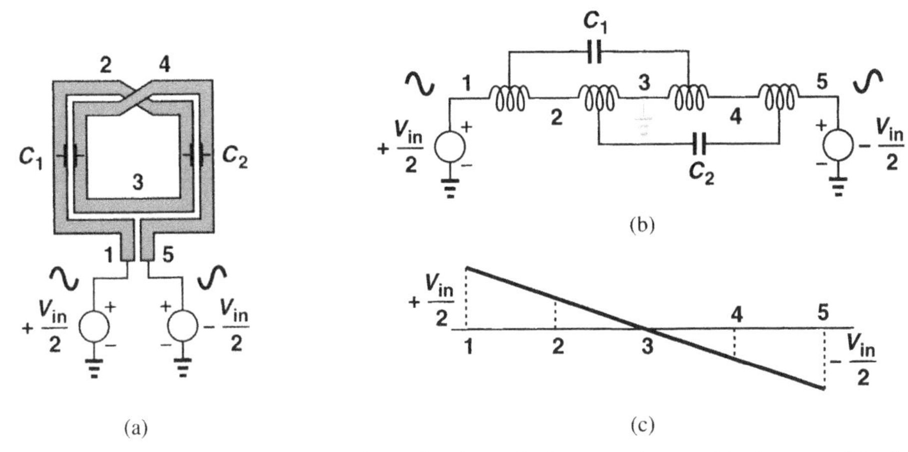

Figure 7.18 (a) Symmetric inductor including interwinding capacitance, (b) circuit model of inductor, (c) voltage profile through the inductor.

including the fringe capacitance between the segments, we arrive at the representation in Fig. 7.18(b). Note that symmetry creates a virtual ground at node 3. This model reveals that C_1 and C_2 sustain large voltages, e.g., as much as $V_{in}/2$ if we assume a linear voltage profile from node 1 to node 5 [Fig. 7.18(c)]. Consequently, the self-resonance frequency of symmetric inductors is typically limited by the interwinding capacitance rather than the capacitance to the substrate. The line-to-line spacing may therefore be chosen several times the minimum allowable value.

Circuits requiring large inductance values can incorporate "stacked" structures to save substantial area. Depicted in Fig. 7.19 is an example, where two stacked spirals are placed

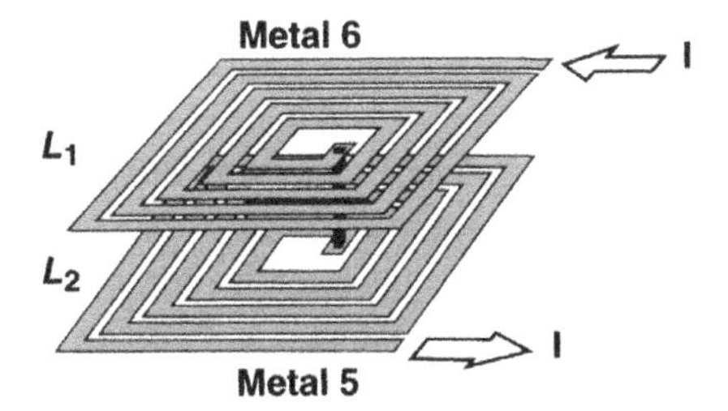

Figure 7.19 Stacked inductor.

in series, yielding a total inductance equal to

$$L_{tot} = L_1 + L_2 + 2M. \tag{7.13}$$

Since the lateral dimensions of the spirals are much greater than their vertical separation, L_1 and L_2 exhibit almost perfect magnetic coupling, i.e., $M \approx L_1 = L_2$ and $L_{tot} \approx 4L_1$. Similarly, n stacked spirals operating in series raise the total inductance by approximately a factor of n^2.

In reality, the multiplication factor falls somewhat below n^2 for square spirals. This is because the edges of one inductor that are perpendicular to the edges of the other provide no mutual coupling [7]. Measurements indicate a multiplication factor of 3.5 for $n = 2$ [7].

The stacked structure of Fig. 7.19 suffers from capacitance between L_1 and L_2 and between L_2 and the substrate [Fig. 7.20(a)]. To reduce the effect of the latter, the end point

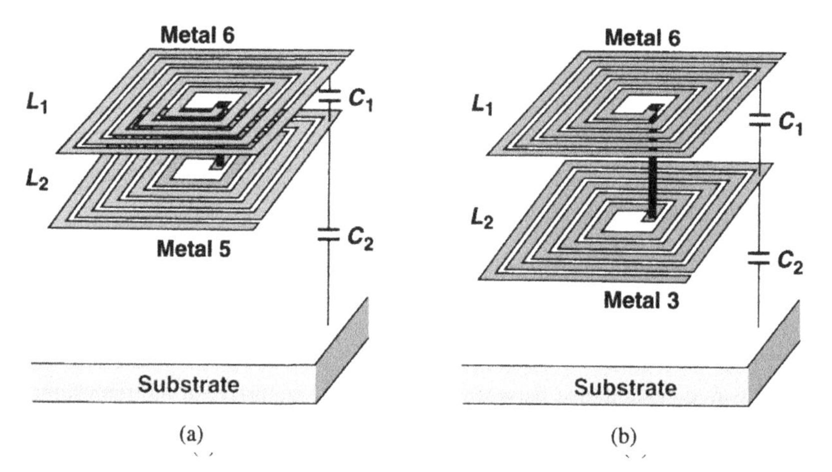

Figure 7.20 Stacked inductor consisting of (a) metal 6 and metal 5 layers, (b) metal 6 and metal 3 layers.

of L_2 is connected to ac ground. It can be proved that the equivalent capacitance of such a configuration is given by [7]

$$C_{eq} = \frac{4C_1 + C_2}{12}. \tag{7.14}$$

Interestingly, the interlayer capacitance contributes four times as much as the bottom-layer capacitance.[5] Thus, the overall capacitance can be reduced if C_1 is reduced, e.g., by moving the bottom spiral down [7]. Illustrated in Fig. 7.20(b), such a modification employs a lower layer of metal, thereby lowering C_{eq} significantly. For example, in a technology with five metal layers, moving L_2 from metal 4 to metal 2 decreases C_{eq} by a factor of 3.8 [7].

7.2 Monolithic Varactors

In Chapter 6, we introduced pn junctions as a candidate for voltage-dependent capacitors required in VCOs. While used in both bipolar and CMOS technologies, such varactors become less attractive at low supply voltages for two reasons. First, pn junctions suffer from a limited tuning range that trades with nonlinearity in the C-V characteristic. As the junction capacitance expression in Chapter 6 suggests, the capacitance varies slowly under reverse bias and sharply under forward bias, thereby introducing significant nonlinearity in

[5] Similar results hold for n stacked spirals [7].

the VCO characteristic. Second, at low supply voltages, it becomes increasingly more difficult to select the oscillator common-mode level and signal swings so as to avoid forward biasing the diodes.

An alternative varactor that does not exhibit the above shortcomings is derived from MOSFETs. With their *nonmonotonic* C-V characteristics (Fig. 7.21), regular MOSFETs

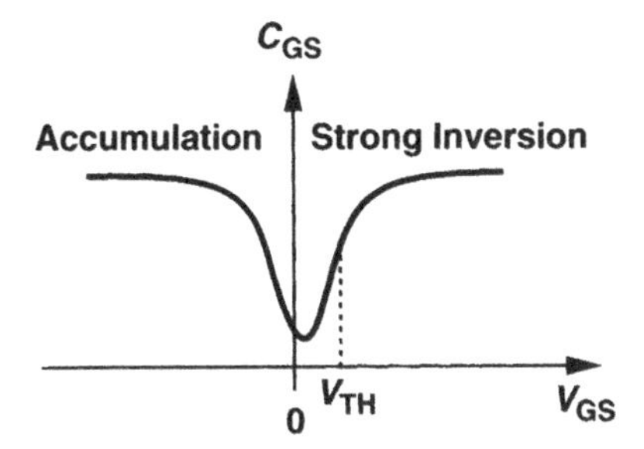

Figure 7.21 C-V characteristic of a MOSFET.

may serve as varactors if the VCO design guarantees the gate-source voltage remains in the monotonic region. However, the device suffers from a large source-drain resistance in the vicinity of minimum capacitance due to the low carrier concentration in the channel. A simple modification resolves these difficulties. Called an "accumulation-mode MOS varactor" and shown in Fig. 7.22(a), this structure resembles an NMOS transistor placed inside an n-well. If V_G is below V_S, the electrons in the n-well are repelled from the silicon/oxide interface and a depletion region is formed [Fig. 7.22(b)]. Under this condition, the equivalent capacitance is given by the series combination of the oxide and depletion capacitances. As V_G exceeds V_S, the interface attracts electrons from the n^+ source and drain terminals, creating a channel [Fig. 7.22(c)]. The overall capacitance therefore rises to that of the oxide, exhibiting the characteristic shown in Fig. 7.22(d). Since the material under the gate oxide is n-type, the concept of strong inversion does not apply here.

MOS varactors have proved quite useful in VCO design. While providing a dynamic range, C_{max}/C_{min}, of 2.5 to 3 with $-1\text{ V} \leq V_G - V_S \leq +1\text{ V}$, these devices comfortably tolerate both positive and negative voltages, allowing large VCO swings. In fact, the characteristic of Fig. 7.22(d) indicates that MOS varactors *should* operate with positive and negative biases so as to provide maximum dynamic range.

MOS varactors nevertheless suffer from a trade-off between the dynamic range and the Q. This is because, for minimum channel length, the gate-source and gate-drain overlap capacitances of the device constitute a significant fraction of the overall capacitance, thereby limiting the tuning range. For example, in a typical 0.18-μm CMOS technology with $C_{ov} \approx 0.3$ fF/μm, $C_{ox} \approx 10$ fF/μm^2, and $L_{eff} \approx 0.16$ μm, the ratio of the total overlap capacitance to the oxide capacitance reaches 2×0.3 fF/μm/(10 μm$\times$0.16 μm) $= 37.5\%$. To alleviate this issue, the channel length may be increased to about three times the minimum value, lowering the above ratio to 10 to 15%.[6] This remedy, however, increases the

[6] If the Q of the inductor is high enough to necessitate a high varactor Q as well, then a minimum length may be mandatory here.

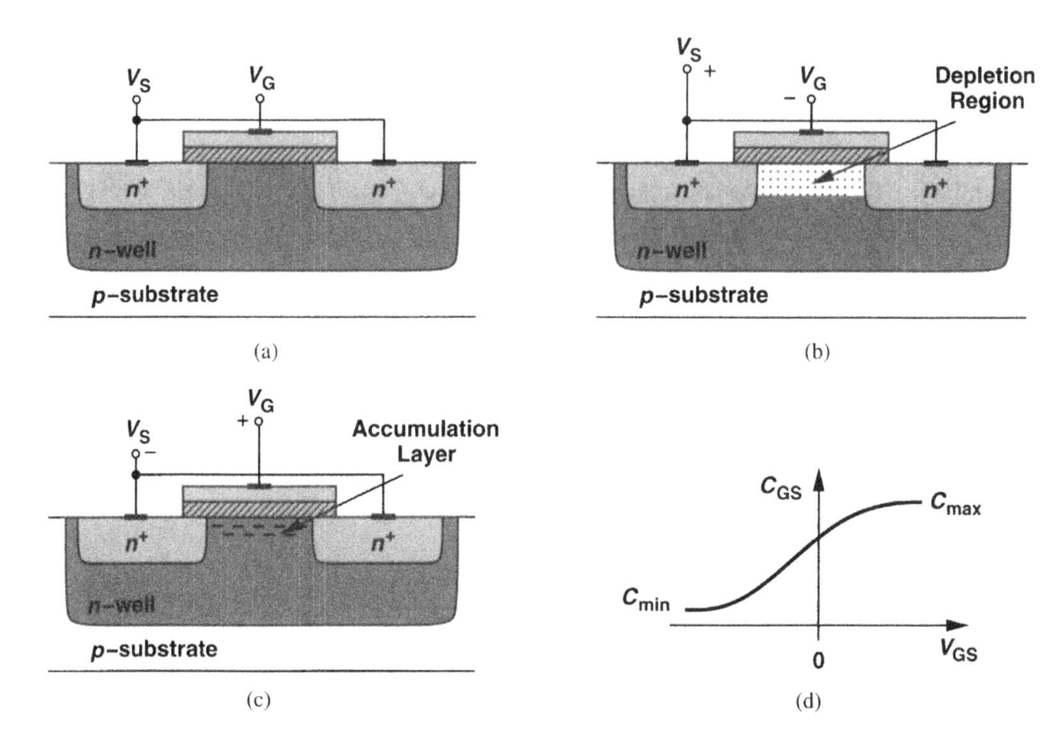

Figure 7.22 (a) MOS varactor structure, (b) varactor in depletion mode, (b) varactor in accumulation mode, (d) C-V characteristic.

resistance between the source and drain terminals (Fig. 7.23), thus degrading the Q.

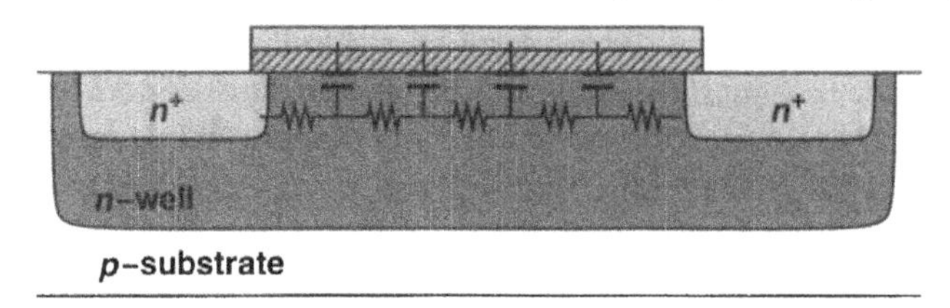

Figure 7.23 Effect of channel resistance in MOS varactor.

Despite this trade-off, MOS varactors exhibit Q's of several tens in the gigahertz range, proving efficient means of frequency control, especially at low supplies.

7.3 Basic LC Oscillators

Our treatment of LC oscillators in Chapter 6 suggests that the negative-G_m configuration is well-suited to full integration even with low-Q inductors. In this section, we study several variants of this topology.

Figure 7.24(a) shows the circuit of Chapter 6, including MOS varactors to achieve tun-

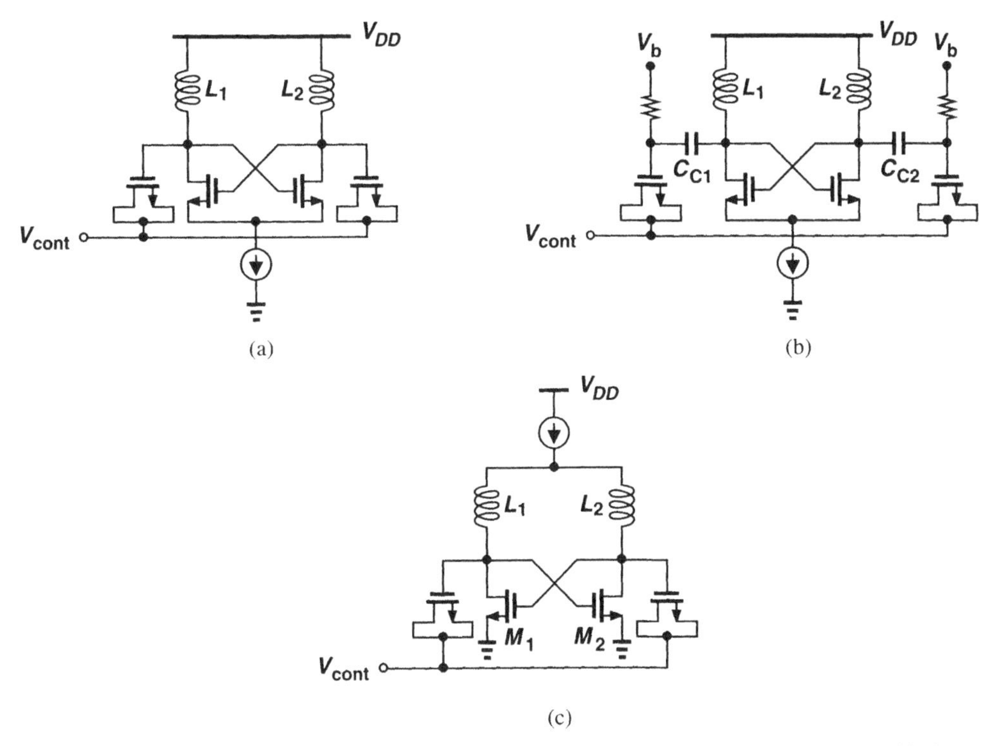

Figure 7.24 LC VCO with (a) tail biasing, (b) tail biasing and capacitive coupling, (c) top biasing.

ing. The principal drawback of this arrangement is that it does not exploit the full dynamic range of the varactors. Since the low dc drop across the inductors results in an output common-mode level near V_{DD}, the gate-source voltage of the varactors remains positive for $V_{cont} < V_{DD}$. To resolve this issue, the varactors can be capacitively coupled to the oscillator, with their gate voltages defined independently [Fig. 7.24(b)]. For example, if $V_b = V_{DD}/2$ and V_{cont} varies from zero to V_{DD}, then the average voltage across the varactors assumes values from $-V_{DD}/2$ to $+V_{DD}/2$.

In the circuit of Fig. 7.24(b), the coupling capacitors must be sufficiently large to present the entire varactor capacitance to the tank. If $C_{C1} = C_{C2} = nC_{var,max}$, where $C_{var,max}$ denotes the maximum varactor capacitance, then the equivalent capacitance of the series combination is given by

$$C_{eq} = \frac{nC_{var,max}C_{var}}{nC_{var,max} + C_{var}} \tag{7.15}$$

$$\approx C_{var}\left(1 - \frac{C_{var}}{nC_{var,max}}\right). \tag{7.16}$$

Thus, the dynamic range of C_{eq} is less than that of C_{var}. For example, if $n = 5$ and $C_{var,min} < C_{var} < C_{var,max}$, then $C_{var,min} < C_{eq} < 0.83C_{var,max}$.

While it is desirable to use large coupling capacitors, their bottom-plate parasitic loads the tank, both lowering the oscillation frequency and limiting the tuning range. Illustrated in Fig. 7.25(a), typical metal sandwich capacitors suffer from a bottom-plate parasitic of

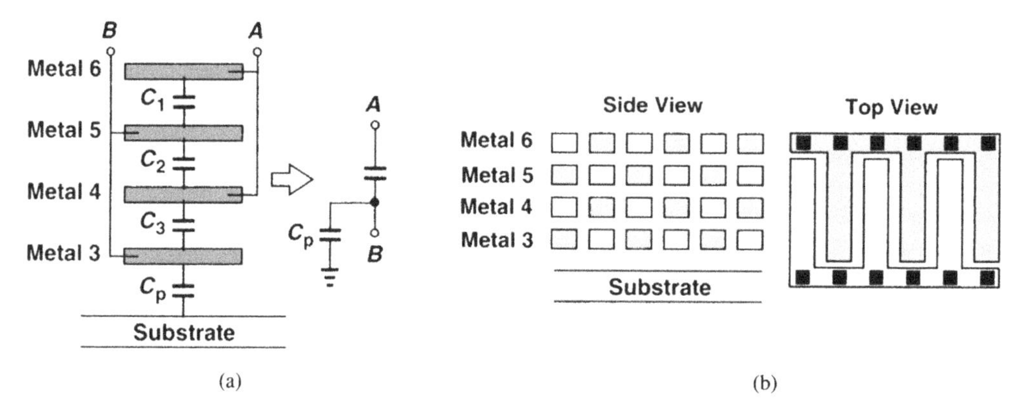

Figure 7.25 (a) Metal sandwich capacitor, (b) fringe capacitor.

about 15%. With $n = 5$, C_B may reach $0.75C_{var,max}$, contributing a large amount of fixed capacitance to the tank. In this respect, the "fringe capacitor" shown in Fig. 7.25(b) [8] proves a better choice as its bottom-plate parasitic typically falls below 5%.

A VCO topology that avoids capacitive coupling is shown in Fig. 7.24(c). Here, the output common-mode level, V_{CM}, is equal to the equilibrium gate-source voltage of M_1 and M_2, i.e., when they carry equal currents. Since V_{CM} can be set to approximately $V_{DD}/2$, the MOS varactors sustain an average voltage of $-V_{DD}/2$ to $+V_{DD}/2$ as V_{cont} goes from 0 to V_{DD}.

A drawback of this oscillator is that the noise current of the top current source flows through M_1 and M_2, modulating the voltage across each capacitor and hence raising the phase noise. The topologies of Figs. 7.24(a) and (b) are free from this effect because L_1 and L_2 provide a low-impedance path to V_{DD} for the low-frequency noise of the tail current source.

Another differential VCO configuration is depicted in Fig. 7.26. Here, the cross-coupled

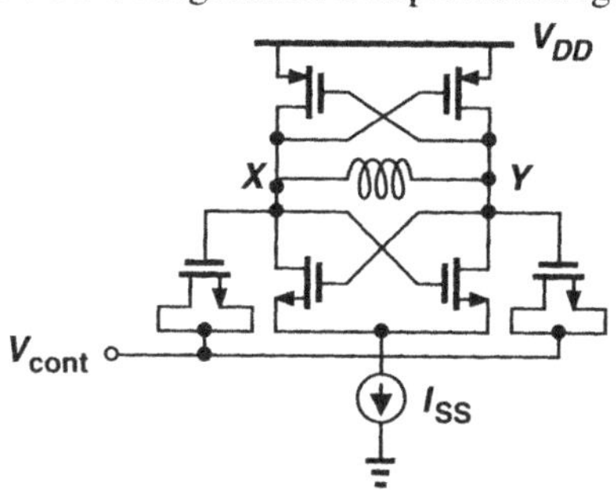

Figure 7.26 VCO using PMOS and NMOS devices to equalize rise and fall times.

PMOS devices allow more symmetry between the rise time and fall time of the waveforms at nodes X and Y, reducing the upconversion of flicker noise [12]. However, the low mobility of PMOS transistors translates to wide devices, especially at low supply voltages, thereby increasing the device capacitances and limiting the tuning range. Note that when $V_X = V_Y$, we have $V_{DD} = |V_{GSP}| + V_{GSN} + V_{ISS}$, where V_{ISS} is the voltage required across I_{SS}. Thus, the NMOS and PMOS transistors must be wide enough to exhibit a low gate-source voltage.

7.3.1 Differential Control

As the supply voltage of each new generation of submicron technologies scales down, the gain of LC VCOs must scale *up*. Illustrated in Fig. 7.27, this behavior arises because the

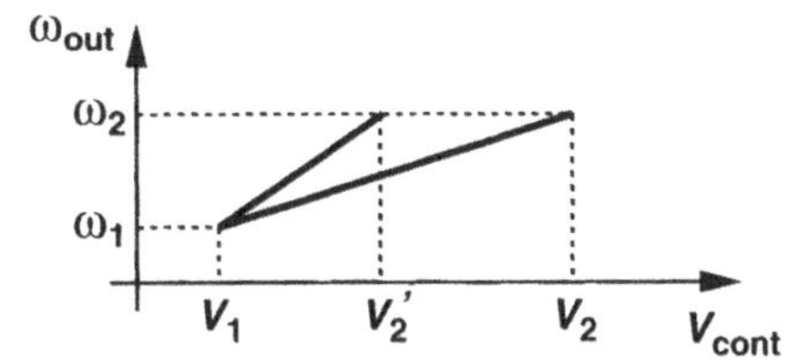

Figure 7.27 Increase in VCO gain as a result of supply scaling.

VCO must provide a tuning range commensurate with the standard and the expected process and temperature variations. In other words, while supply scaling lowers the maximum value of V_{cont} from V_2 to V_2', the range $[\omega_1 \;\; \omega_2]$ is fixed, yielding a greater K_{VCO}.

The inevitable increase in K_{VCO} results in a higher sensitivity to the noise that may be injected onto the control line by the environment. For example, in CDR and PLL circuits used in OC transceivers, the control line is typically connected to a large off-chip capacitor, potentially picking up noise through the external interconnects. It is therefore desirable to design VCOs with differential control. Figure 7.28 shows an example [9], where two sets of MOS varactors control the frequency. From the C-V characteristics of Fig. 7.22(d),

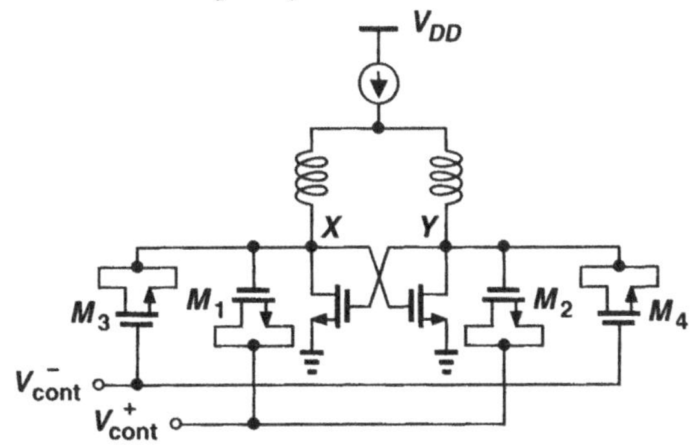

Figure 7.28 LC VCO with differential control.

we note that the capacitance of M_1 and M_2 decreases as V_{cont}^+ rises and that of M_3 and M_4 also decreases as V_{cont}^- falls. In other words, the oscillation frequency increases with $V_{cont}^+ - V_{cont}^-$.

To study the common-mode rejection of the circuit, we first assume $V_{cont}^{+} = V_{cont}^{-} = V_{CM,XY}$, where $V_{CM,XY}$ denotes the CM level at nodes X and Y. Under this condition, the average gate-source voltage of each varactor is zero [point A in Fig. 7.29(a)] and a small change, ΔV, in $V_{cont}^{+} = V_{cont}^{-}$ yields equal and opposite changes in the capacitances of M_1-M_2 and M_3-M_4. Thus, the oscillation frequency remains constant.

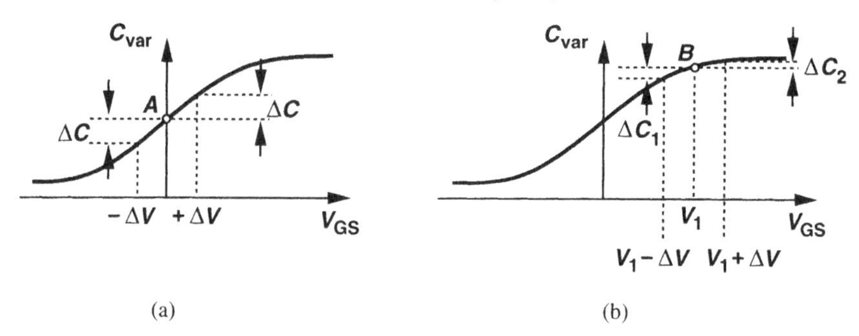

Figure 7.29 Effect of nonlinearity on CM rejection of a differentially-controlled oscillator.

Now suppose $V_{cont}^{+} = V_{CM,XY} + V_1$ and $V_{cont}^{-} = V_{CM,XY} - V_1$, placing the varactors at point B in Fig. 7.29(b). In this case, if the CM level of V_{cont}^{+} and V_{cont}^{-} changes by ΔV, then the varactor capacitances change by *unequal* amounts, ΔC_1 and ΔC_2, perturbing the oscillation frequency.

The above study indicates that the CM rejection of differentially-controlled oscillators degrades as $|V_{cont}^{+} - V_{cont}^{-}|$ increases and the C-V characteristic becomes more nonlinear. In practice, the circuit may reject the CM noise by only about 10 dB if $|V_{cont}^{+} - V_{cont}^{-}|$ is raised to maximize the tuning range.

Differential oscillator control typically necessitates the use of a differential charge pump, a circuit difficult to design if it must provide a nearly rail-to-rail output voltage range. This is because the feedback loop defining the output CM level of the charge pump cannot properly sense large differential voltages. Shown in Fig. 7.30 is an example. Here,

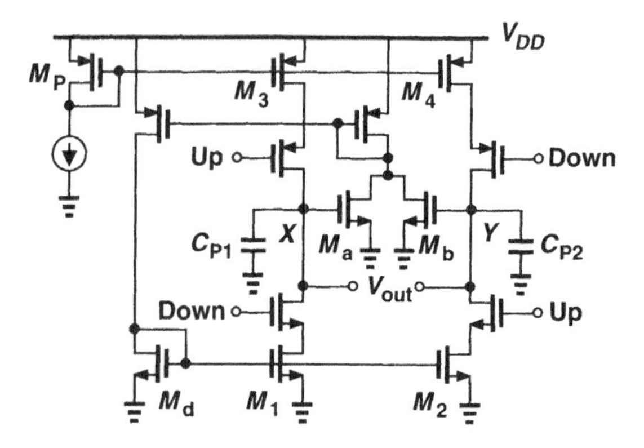

Figure 7.30 Differential charge pump with CM feedback.

two single-ended charge pumps are driven by Up and Down pulses, with their output CM level sensed by M_a and M_b and fed back to M_d. Determined by the charactristics of M_a and M_b, the CM level settles such that the average current drawn by M_1 and M_2 is equal to that provided by M_3 and M_4.

The principal issue in the CM feedback of Fig. 7.30 is that, if V_X or V_Y falls below the threshold voltage of M_a or M_b, then the corresponding transistor turns off, failing to sense the CM level. As a result, low supply voltages yield a narrow range for V_X and V_Y, limiting the tuning range of the VCO.

In order to alleviate the above difficulty, a PMOS pair with an NMOS current mirror can be added to the CM sense circuit to ensure that, as V_X or V_Y approach 0 or V_{DD}, they are sensed properly by a PMOS or an NMOS device, respectively.

7.3.2 Design Procedure

With the above overview of different VCO topologies, let us now outline a typical design procedure specific to optical communication systems. The design of VCOs seeks to maximize the tuning range and minimize the phase noise with the knowledge of four parameters: the load capacitance that the VCO must drive, the required output voltage swing, the center frequency, and the power budget. The first two parameters are dictated by the environment, often leading to the use of buffers [Fig. 7.31(a)]. In an optical receiver, for example, the VCO must drive several flipflops in the CDR's phase detector (Chapter 9), one flipflop in a

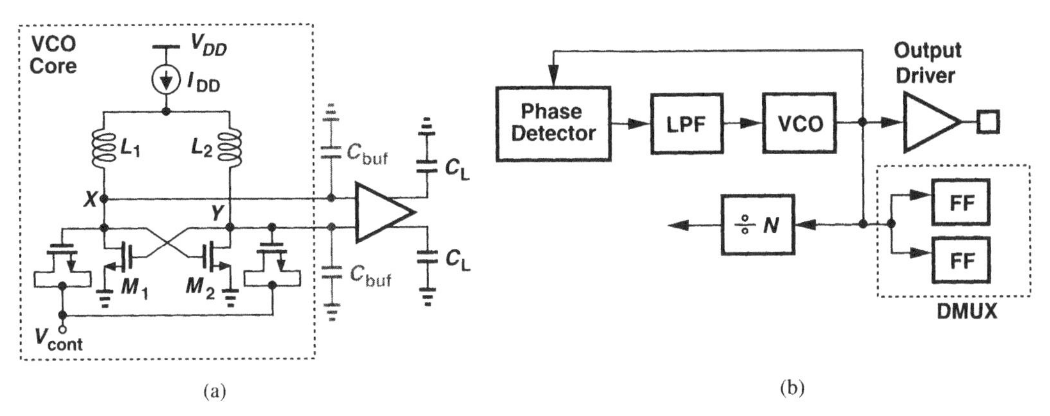

Figure 7.31 (a) Buffer capacitance seen by a VCO, (b) other circuits driven by a VCO.

frequency divider chain, two flipflops in the demultiplexer, and a 50-Ω output driver [Fig. 7.31(b)]. Moreover, the voltage swings provided by the VCO must be large enough to allow relatively complete switching of the differential pairs in the above circuits—especially at high temperatures and slow corners of the technology. Consequently, the buffer in Fig. 7.31(a) must utilize high current levels and wide transistors as well, thus presenting a large capacitance and imposing a minimum voltage swing on the oscillator core.

With the power budget and hence the value of I_{DD} in Fig. 7.31(a) known, the width of M_1 and M_2 is chosen so as to yield an average CM level of approximately $V_{DD}/2$ at nodes X and Y. Note that when $V_X = V_Y$, M_1 and M_2 have equal gate and drain voltages, sharing I_{DD} as if they were diode-connected devices. Under this condition, the transistor width is selected such that $V_X = V_Y \approx V_{DD}/2$. The small-signal transconductance of M_1 and M_2, $g_{m1,2}$, is therefore known.

The next step is to determine the value of the tank inductors. To maximize the tuning range, the inductance must be minimized. Recall from Chapter 6 that the loop gain must exceed unity to ensure oscillation startup. Thus, the minimum allowable inductance, L_{min}, exhibits an equivalent parallel resistance, $R_{p,min} = QL_{min}\omega_{osc}$, that places the circuit at the edge of oscillation:

$$(g_{m1,2}QL_{min}\omega_{osc})^2 = 1. \tag{7.17}$$

Assuming that for a given ω_{osc}, Q is relatively independent of the inductance, we obtain

$$L_{min} = \frac{1}{g_{m1,2}Q\omega_{osc}}. \tag{7.18}$$

With $L = L_{min}$, however, the oscillation amplitude is quite small to maintain the unity loop gain. If the amplitude grows, the transistor nonlinearities reduce the loop gain, violating Barkhausen's gain criterion. Another difficulty arises from the variation of g_m and Q with process and temperature, possibly prohibiting oscillation at some corners.

From the above observations, we conclude that L and R_p must sufficiently exceed L_{min} and $R_{p,min}$, respectively, to provide (a) the required voltage swings and (b) startup under worst-case conditions. In most LC VCOs used in optical communication transceivers, the former dictates the design, automatically guaranteeing the latter. To relate the output swing to R_p, we recognize from the equivalent circuit shown in Fig. 7.32 that each of M_1 and M_2

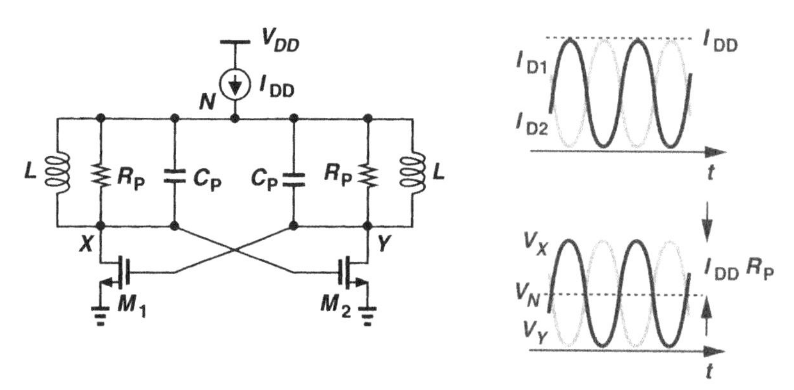

Figure 7.32 Voltage and current swings in a VCO.

carries an average current of $I_{DD}/2$. Thus, if the drain current waveforms are approximated by sinusoids varying between zero and I_{DD}, then V_X and V_Y swing from $V_N - I_{DD}R_p$ to $V_N + I_{DD}R_p$. With $V_N \approx V_{DD}/2$, and for largest swing, we have $I_{DD}R_p = V_{DD}/2$ and hence

$$R_{p,swing} = \frac{V_{DD}}{2I_{DD}}, \tag{7.19}$$

where $R_{p,swing}$ denotes the minimum tank resistance yielding the maximum swing. The corresponding inductance value is therefore equal to

$$L_{opt} = \frac{V_{DD}}{2I_{DD}} \cdot \frac{1}{Q\omega_{osc}}. \tag{7.20}$$

With the transistor dimensions and tank inductance value known, we must now compute the varactor capacitance. Since

$$\omega_{osc} = \frac{1}{\sqrt{L_{opt}C_{tot}}}, \tag{7.21}$$

we identify all contributions to C_{tot}: (1) the capacitances of M_1 and M_2, $C_{GS} + C_{DB} + 4C_{GD}$;[7] (2) the parasitic capacitance of each inductor, C_{ind}; and (3) the input capacitance of the buffer, C_{buf}. It follows that

$$C_{var} = \frac{1}{\omega_{osc}^2 L_{opt}} - (C_{GS} + C_{DB} + 4C_{GD}) - C_{ind} - C_{buf}. \tag{7.22}$$

This equation provides the required center value of the varactor capacitance, but the dynamic range of C_{var} yields a certain tuning range that may or may not be adequate.

The foregoing procedure allows a rapid estimate of the oscillator design parameters, leading to a unique value for C_{var} in Eq. (7.22). But what if the resulting tuning range is insufficient or C_{var} is *negative*?! Such a case occurs if the inductance and/or the fixed capacitances of the tank are excessively large. To widen the tuning range, either the inductance must be lowered, sacrificing the output voltage swing, or the buffer input capacitance must be decreased, degrading the drive capability.

In the above methodology, the other important VCO parameter, namely, phase noise, does not appear to influence the design. Indeed, the requirements imposed by the environment and the simplicity of the circuit topology greatly tighten the design space, leaving little flexibility in the choice of transistor and tank parameters. If the final design exhibits unacceptably high phase noise, raising the power dissipation may become inevitable.

The calculation of phase noise in oscillators is an important task. The reader is referred to the literature for a detailed treatment of the subject [10, 11, 12].

7.4 Quadrature Oscillators

A quadrature oscillator produces outputs having a phase difference of 90°. Used in "half-rate" clock recovery circuits as well as in frequency detectors for random data (Chapter 9), such oscillators can be realized with or without LC tanks. In this section, we study the LC implementations. It is helpful to bear in mind that all voltage and current symbols in this section denote small-signal quantities.

Let us first study the concept of "coupling" an external signal to an oscillator. Also called "injection" and illustrated in Fig. 7.33(a), coupling entails perturbing the natural

[7] Note that each gate-drain overlap capacitance experiences a large-signal Miller multiplication factor of 2.

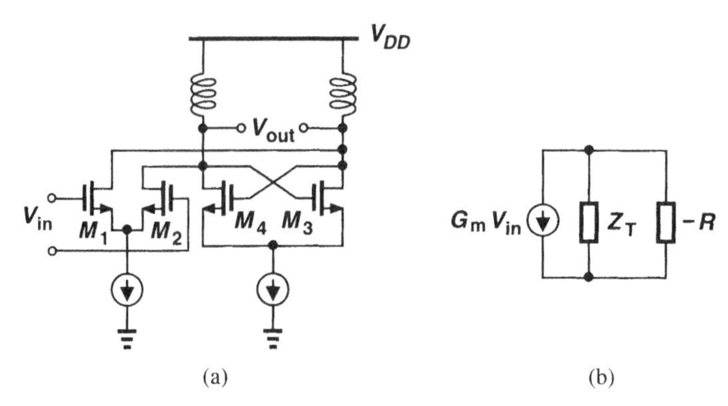

Figure 7.33 (a) Injection of a signal into an oscillator, (b) small-signal model.

oscillation by an external stimulus. Here, M_1 and M_2 convert V_{in} to a differential current and inject the result into the tanks. If V_{in} has the same frequency as the natural frequency of the oscillator, then the coupling may simply shift the phase of V_{out} because the vector summations $I_{D1} + I_{D3}$ and $I_{D2} + I_{D4}$ are affected by V_{in}. This property is exploited in quadrature oscillators. The topology of Fig. 7.33(a) is an example of "unilateral" injection because the coupling from V_{in} to V_{out} is much greater than that from V_{out} to V_{in}. Note that the addition of signals in the current domain leads to the simple model shown in Fig. 7.33(b), where G_m denotes the transconductance of the differential pair, Z_T represents the tank impedance, and $-R$ is the negative resistance provided by the cross-coupled pair. At the edge of oscillation, $-R$ exactly cancels the loss in Z_T. We call the ratio I_{D2}/I_{D4} (or I_{D1}/I_{D3}) the "coupling factor."[8]

The principle behind quadrature generation is to couple two identical oscillators such that they operate with a 90° phase shift. As depicted in Fig. 7.34(a), a fraction of each oscillator's output is injected into the other. Figure 7.34(b) shows the transistor-level implementation, depicting unilateral coupling.

To understand the operation, we exploit the coupling model of Fig. 7.33(b) to construct the equivalent circuit shown in Fig. 7.35. Writing the parallel combination of $-R$ and Z_T as $-RZ_T/(Z_T - R)$, we have

$$G_{m1} V_1 \frac{-RZ_T}{Z_T - R} = V_2 \tag{7.23}$$

$$G_{m2} V_2 \frac{-RZ_T}{Z_T - R} = V_1. \tag{7.24}$$

Assuming $V_1, V_2 \neq 0$, and dividing (7.23) by (7.24), we arrive at:

$$G_{m1} V_1^2 - G_{m2} V_2^2 = 0. \tag{7.25}$$

[8]Since these are small-signal quantities, they depend on both bias currents and device dimensions.

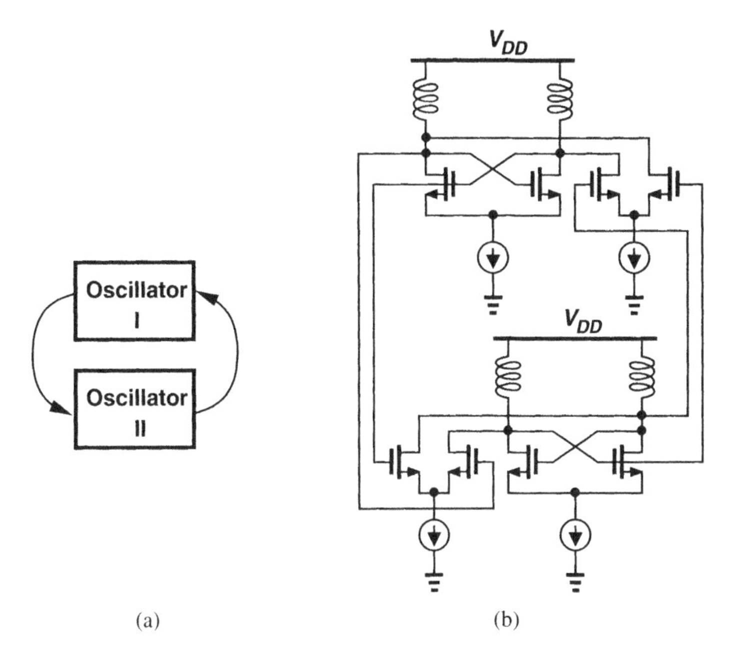

Figure 7.34 (a) Two oscillators coupled to each other, (b) implementation of (a).

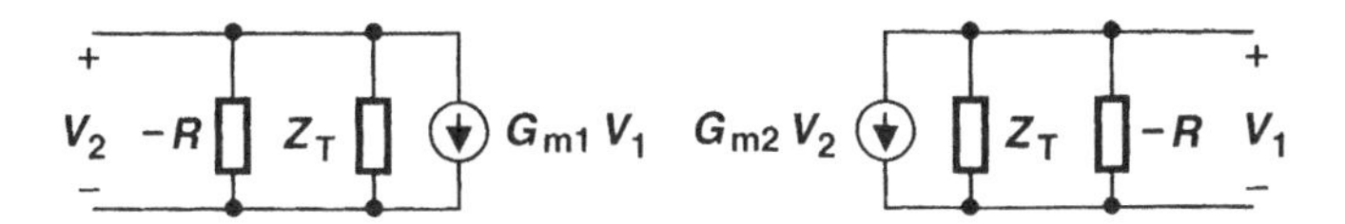

Figure 7.35 Small-signal model of coupled oscillators.

This result predicts two important cases: (1) If $G_{m1} = G_{m2}$, then $V_1 = \pm V_2$; i.e., the two oscillators operate with a phase difference of zero or 180°; (2) If $G_{m1} = -G_{m2}$, then $V_1^2 + V_2^2 = 0$ and hence $V_1 = \pm jV_2$; i.e., the phase difference is $-90°$ or $+90°$. We call the two cases "in-phase coupling" and "antiphase coupling," respectively. Figure 7.36 depicts the two cases at transistor level. Note that the above conditions do not include the Barkhausen's criteria; they simply arise from the mutual coupling of the oscillators.

7.4.1 In-Phase Coupling

While not directly of interest here, in-phase coupling provides a good starting point toward understanding the operation of coupled oscillators. To this end, we represent the voltages and currents by phasors. As depicted in Fig. 7.37, the differential voltages V_A, V_B, V_C, and V_D are denoted by equal and opposite phasors, and the drain currents of the transistors are

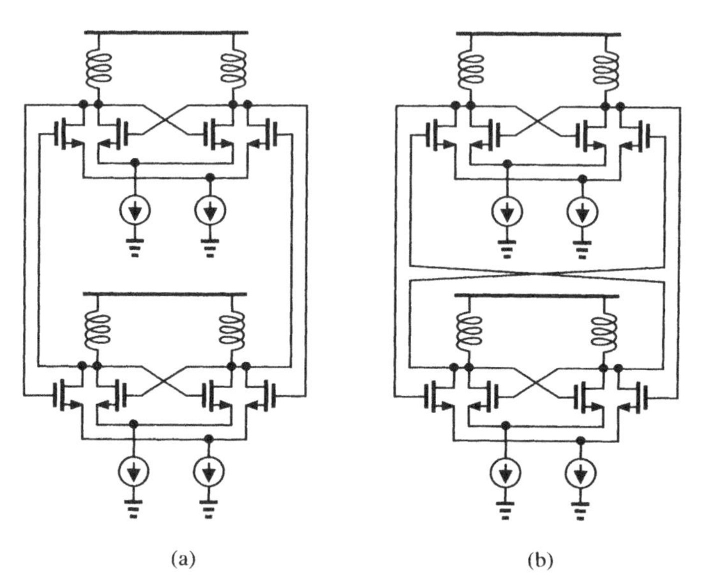

Figure 7.36 Illustration of (a) in-phase and (b) antiphase coupling.

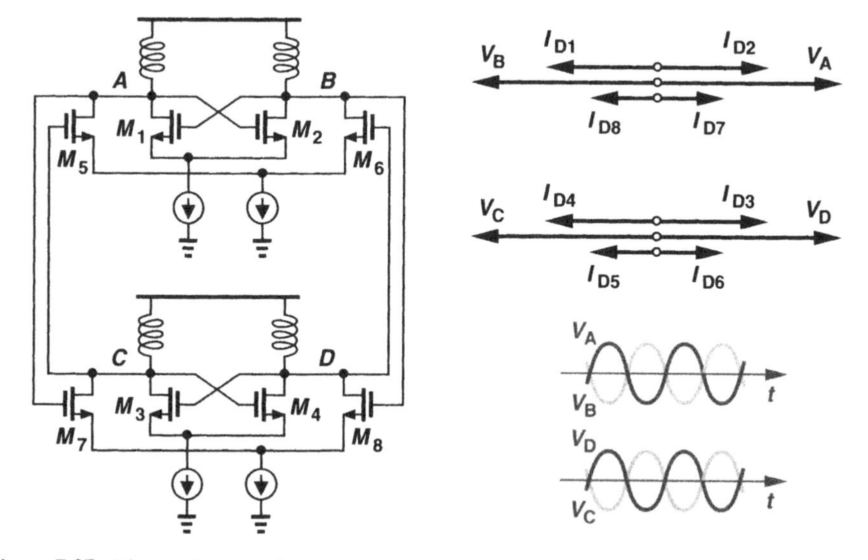

Figure 7.37 Phasor diagrams for in-phase coupling.

aligned with these voltages because $I_D = g_m V_{GS}$.[9] We note that I_{D5} and I_{D1} add at node

[9]We lump all transistor capacitances in the tanks and neglect phase shifts introduced in the channel itself. Also, we assume the common-source point of cross-coupled and differential pairs experiences negligible voltage excursions.

A and flow through the tank, thereby producing V_A. In other words, since $I_{D5} = g_m V_C$ and $I_{D1} = g_m V_B$, we have $V_A \propto -g_m(V_C + V_B)$. Since M_5 inverts the signal from its gate to its drain, the two oscillators approach a steady state where $V_A = -V_C$ and $V_B = -V_D$. That is, the injection from each oscillator augments the other.

In summary, in-phase coupling of identical oscillators forces their outputs to remain in phase. As depicted in Fig. 7.38, the idea can be extended to multiple oscillators as well

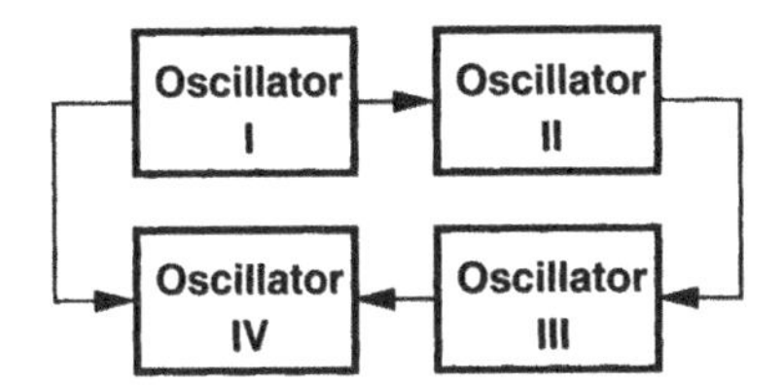

Figure 7.38 In-phase coupling of several oscillators.

[13].

7.4.2 Antiphase Coupling

Let us now study the case of antiphase coupling (Fig. 7.39). As with in-phase coupling, the drain current of each transistor can be approximated as a scalar transconductance multiplied by its gate-source voltage, yielding the phasor diagram shown in Fig. 7.39(a). Each tank exhibits an impedance equal to Z_T and carries the drain currents of two transistors, e.g.,

$$I_{TA} = I_{D1} + I_{D5}, \tag{7.26}$$

where I_{TA} denotes the current through the tank connected to node A. Illustrated in Fig. 7.39(b), the resultant bears an angle with respect to I_{D1} equal to

$$\theta = \tan^{-1} \frac{I_{D5}}{I_{D1}}. \tag{7.27}$$

How can the product of I_{TA} and Z_T create a voltage aligned with I_{D1}? This is possible if Z_T contributes a phase shift of $-\theta$. Equating the phase shift introduced by $Z_T = Ls||(CS)^{-1}||R_P$ to θ, we have:

$$\frac{\pi}{2} - \tan^{-1} \frac{L\omega_{osc}}{R_p(1 - LC\omega_{osc}^2)} = -\tan^{-1} \frac{I_{D5}}{I_{D1}}. \tag{7.28}$$

We call θ the "deviation angle." For example, if $I_{D5} = 0.2 I_{D1}$, then ω_{osc} must deviate from $\omega_0 = 1/\sqrt{LC}$ to

$$\omega_{osc} = \frac{1}{\sqrt{LC}} \sqrt{1 - \frac{1}{5Q}} \tag{7.29}$$

so as to provide the phase shift.

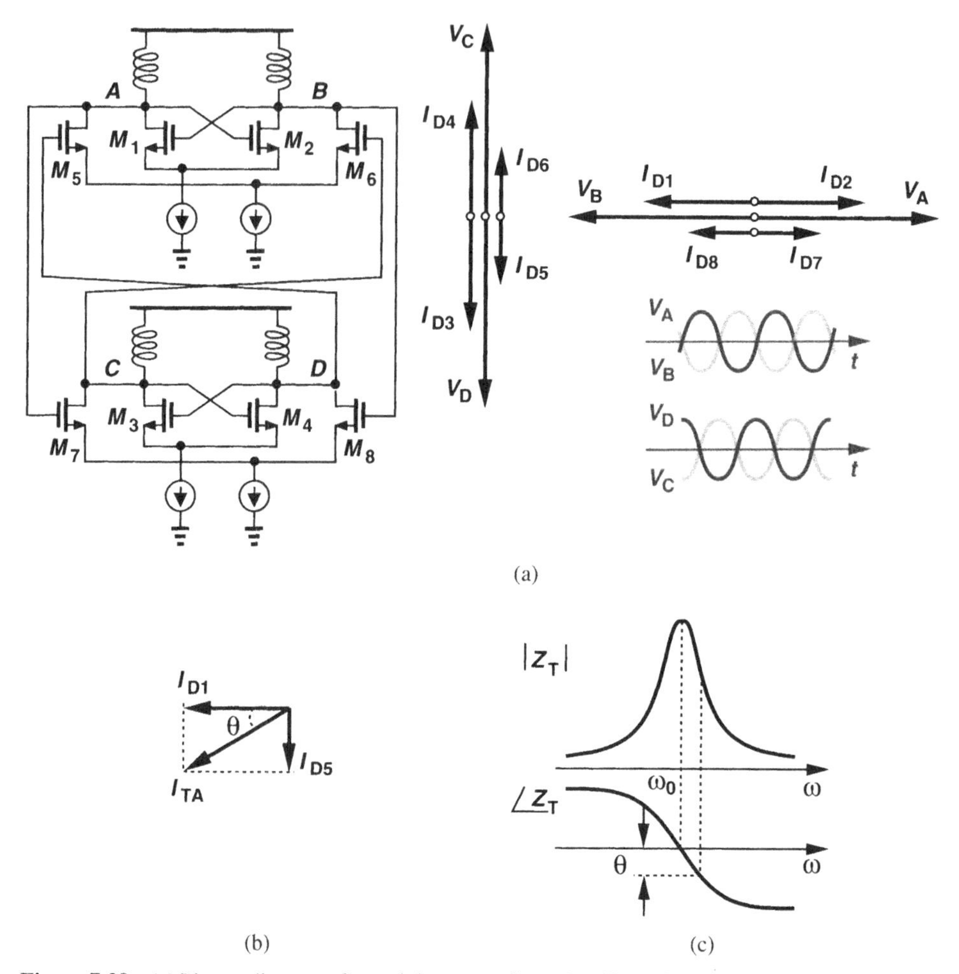

Figure 7.39 (a) Phasor diagrams for antiphase coupling, (b) effect of current addition, (c) deviation from resonance frequency.

The above analysis reveals another interesting attribute of antiphase coupling. Since ω_{osc} depends on θ and this angle depends on the coupling factor, I_{D5}/I_{D1}, the oscillation frequency can be *tuned* by varying this ratio.[10] Also evident from (7.23) and (7.24), this property leads to the tuning mechanism shown in Fig. 7.40 [14], where the control voltage varies the tail current of the coupling differential pairs. As the coupling factor is decreased, the current is steered to the cross-coupled pairs so as to maintain a relatively constant output swing.

The behavior illustrated in Figs. 7.39(b) and (c) indicates that, as the bias current of the differential pairs increases (and so does the coupling factor), the oscillation frequency

[10]Note that in this case, $I_{D5}/I_{D1} = g_{m5}/g_{m1}$.

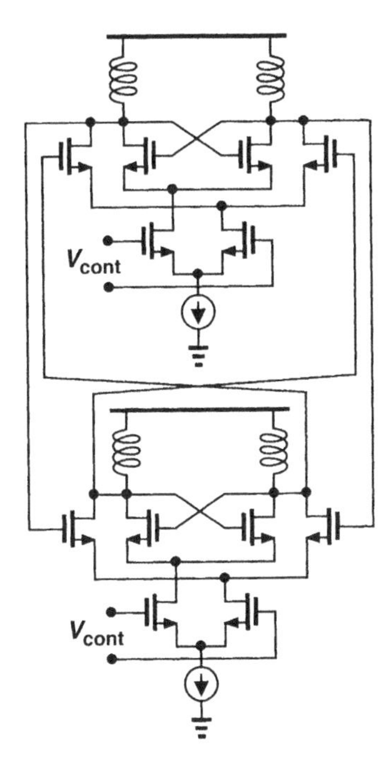

Figure 7.40 Oscillator tuning through variation of coupling factor.

must deviate from ω_0 by a greater amount so that each tank provides the required phase shift. Since the tanks operate at a frequency that is increasingly farther from the resonance frequency, the Q falls, raising the phase noise. From this point of view, it is desirable to *minimize* the coupling factor.

What determines the minimum allowable coupling factor? In the presence of *mismatches* between the two oscillators, insufficient coupling fails to force them to equal frequencies. As a result, the outputs do not exhibit a well-defined phase relationship. Moreover, the interaction between the oscillators may lead to spurious oscillations [15]. A coupling factor of approximately 25% typically provides a reasonable compromise between Q degradation and oscillation reliability.

The concept of multiple coupled oscillators shown in Fig. 7.38 can also incorporate antiphase coupling to create output phases that differ by less than 90°. For example, if four oscillators are placed in a loop with antiphase coupling, then the output phases are 45° apart. Other examples of quadrature oscillators are described in [16, 17].

7.5 Distributed Oscillators

A distributed amplifier can be placed in a feedback loop to form an oscillator (Fig. 7.41) [18, 19]. To this end, the total delay of the inverting amplifier must translate to a phase shift of 180° at the frequency of interest, f_{osc}. This is guaranteed if $l/v = 1/(2f_{osc})$, which, for

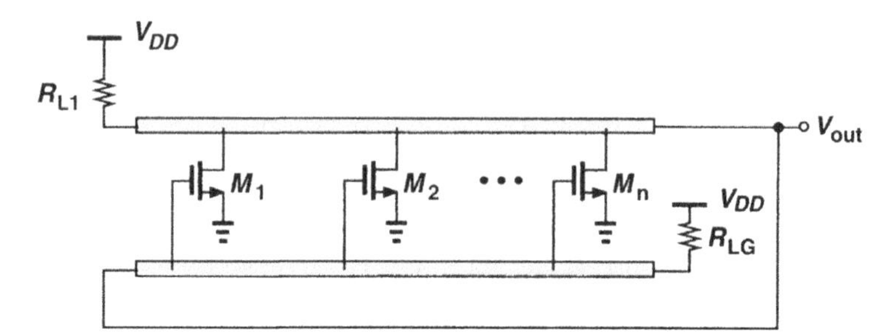

Figure 7.41 Distributed oscillator.

the idealized scenario described in Chapter 5, yields

$$A_v = \frac{\pi f_T}{2 f_{osc}}. \tag{7.30}$$

For a minimum loop gain of unity, (7.30) gives

$$f_{osc} = \frac{\pi}{2} f_T, \tag{7.31}$$

indicating potential for very high oscillation frequencies. For example, a 0.13-μm NMOS device with reasonable overdrive voltage exhibits an f_T of roughly 70 GHz, enabling distributed oscillators to run at speeds greater than 100 GHz.

Equally-spaced points along the output T line of Fig. 7.41 exhibit equally-spaced phases, seemingly allowing a single oscillator to provide multiple phases. However, since these waveforms suffer from unequal amplitudes, it is difficult to use such phases. As explained in Chapter 5, AM/PM conversion in the stages sensing the waveforms may give rise to substantial phase error.

Differential realization of distributed oscillators requires the use of four T lines, leading to a complex and cumbersome layout. Furthermore, the limiting factors outlined in Chapter 5 for DAs also impact the performance of distributed oscillators. With typical T lines available in CMOS technology, oscillation frequencies on the order of half of that predicted by (7.31) can be achieved.

References

1. H. M. Greenhouse, "Design of Planar Rectangular Microelectronic Inductors," *IEEE Trans. on Parts, Hybrids, and Packaging,* vol. 10, pp. 101-109, June 1974.
2. F. E. Terman, *Radio Engineers' Handbook,* New York: McGraw-Hill, 1943.
3. F. W. Grover, *Inductance Calculations,* New Jersey: Van Nostrand, 1946.
4. S. Jenei, B. K. J. C. Nauwelaers, and S. Decoutere, "Physics-Based Closed-Form Inductance Expressions for Compact Modeling of Integrated Spiral Inductors," *IEEE Journal of Solid-State Circuits,* vol. 37, pp. 77–80, Jan. 2002.
5. C. S. Yen, Z. Fazarinc, and R. Wheeler, "Time-Domain Skin-Effect Model for Transient Analysis of Lossy Transmission Lines," *Proc. of IEEE*, vol. 70, pp. 750–757, July 1982.

6. M. Danesh et al., "A Q-Factor Enhancement Technique for MMIC Inductors," *Proc. IEEE Radio Frequency Integrated Circuits Symp.*, pp. 217–220, April 1998.
7. A. Zolfaghari, A. Y. Chan, and B. Razavi, "Stacked Inductors and Transformers in CMOS Technology," *IEEE Journal of Solid-State Circuits,* vol. 36, pp. 620–628, April 2001.
8. O. E. Akcasu, "High-Capacity Structures in a Semiconductor Device," US Patent 5,208,725, May 1993.
9. L. Lin, L. Tee, and P. R. Gray, "A 1.4-GHz Differential Low-Noise CMOS Frequency Synthesizer Using a Wideband PLL Architecture," *ISSCC Dig. of Tech. Papers*, pp. 204–205, Feb. 2000.
10. J. Craninckx and M. Steyaert, "Low-Noise Voltage-Controlled Oscillators Using Enhanced LC Tanks," *IEEE Trans. Circuits and Systems II*, volume 42, pp. 794–804, Dec. 1995.
11. B. Razavi, "A Study of Phase Noise in CMOS Oscillators," *IEEE Journal of Solid-State Circuits,* vol. 31, pp. 331–343, March 1996.
12. A. Hajimiri and T. H. Lee, "A General Theory of Phase Noise in Electrical Oscillators," *IEEE Journal of Solid-State Circuits,* vol. 33, pp. 179–194, Feb. 1998.
13. J. J. Kim and B. Kim, "A Low Phase Noise CMOS LC Oscillator with a Ring Structure," *ISSCC Dig. Tech. Papers,* pp. 430–475, Feb. 2000.
14. T. P. Liu, "A 6.5-GHz Monolithic CMOS Voltage-Controlled Oscillator," *ISSCC Dig. Tech. Papers,* pp. 404–405, Feb. 1999.
15. C. Lam and B. Razavi, "A 2.6-GHz/5.2-GHz CMOS Voltage-Controlled Oscillator," *ISSCC Dig. Tech. Papers,* pp. 402–403, Feb. 1999.
16. J. Savoj and B. Razavi, "A 10-Gb/s CMOS Clock and Data Recovery Circuit with Frequency Detection," *ISSCC Dig. of Tech. Papers*, pp. 78–79, Feb. 2001.
17. J. van der Tang et al., "Analysis and Design of an Optimally Coupled 5-GHz Quadrature LC Oscillator," *IEEE J. Solid-State Circuits,* vol. 37, pp. 657–661, May 2002.
18. B. Kleveland et al., "Monolithic CMOS Distributed Amplifier and Oscillator," *ISSCC Dig. of Tech. Papers*, pp. 70–71, Feb. 1999.
19. H. Wu and A. Hajimiri, "Silicon-Based Distributed Voltage-Controlled Oscillators," *IEEE J. Solid-State Circuits,* vol. 36, pp. 493–502, March 2001.

CHAPTER 8

Phase-Locked Loops

The concept of phase locking was invented in the 1930s and swiftly found wide usage in electronics and communication. While the basic phase-locked loop has remained nearly the same since then, its implementation in different technologies and for different applications continues to challenge designers. A PLL serving the task of clock generation in a microprocessor appears quite similar to a frequency synthesizer used in a cellphone, but the actual circuits are designed quite differently.

This chapter deals with the analysis and design of PLLs with particular attention to implementations in VLSI technologies. A thorough study of PLLs would require an entire book by itself, but our objective here is to lay the foundation for more advanced work. Beginning with a simple PLL architecture, we study the phenomenon of phase locking and analyze the behavior of PLLs in the time and frequency domains. We then address the problem of lock acquisition and describe charge-pump PLLs (CPPLLs) and their nonidealities. Finally, we examine jitter in PLLs, study delay-locked loops (DLLs), and present a number of PLL applications.

8.1 Simple PLL

A PLL is a feedback system that compares the output phase with the input phase. The comparison is performed by a "phase comparator" or "phase detector" (PD). It is therefore beneficial to define the PD rigorously.

8.1.1 Phase Detector

A phase detector is a circuit whose average output, $\overline{V_{out}}$, is linearly proportional to the phase difference, $\Delta\phi$, between its two inputs (Fig. 8.1). In the ideal case, the relationship between $\overline{V_{out}}$ and $\Delta\phi$ is linear, crossing the origin for $\Delta\phi = 0$. Called the "gain" of the PD, the slope of the line, K_{PD}, is expressed in V/rad.

A familiar example of phase detector is the exclusive OR (XOR) gate. As shown in Fig. 8.2, as the phase difference between the inputs varies, so does the width of the output pulses, thereby providing a dc level proportional to $\Delta\phi$. While the XOR circuit produces

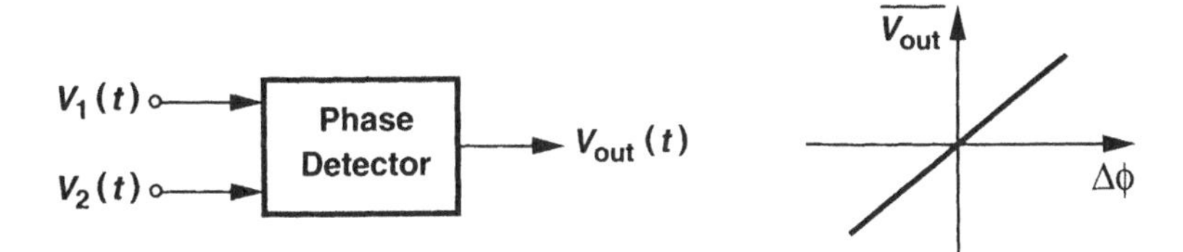

Figure 8.1 Definition of phase detector.

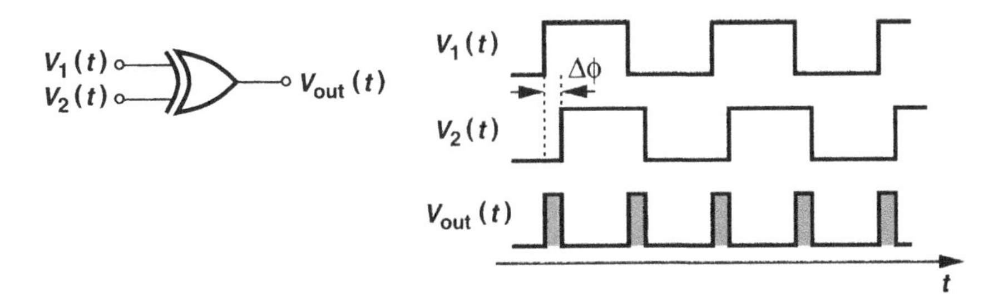

Figure 8.2 Exclusive OR gate as phase detector.

error pulses on both rising and falling edges, other types of PDs may respond only to positive or negative transitions.

As an example, suppose the output swing of the XOR in Fig. 8.2 is V_0 volts. We determine the gain of the circuit as a phase detector and plot its input-output characteristic. If the phase difference increases from zero to $\Delta\phi$ radians, the area under each pulse increases by $V_0 \cdot \Delta\phi$. Since each period contains *two* pulses, the average value rises by $2[V_0 \cdot \Delta\phi/(2\pi)]$, yielding a gain of V_0/π. Note that the gain is independent of the input frequency.

To construct the input-output characteristic, we examine the circuit's response to various input phase differences. As illustrated in Fig. 8.3, the average output voltage rises to $[V_0/\pi] \times \pi/2 = V_0/2$ for $\Delta\phi = \pi/2$ and V_0 for $\Delta\phi = \pi$. For $\Delta\phi > \pi$, the average begins to *drop*, falling to $V_0/2$ for $\Delta\phi = 3\pi/2$ and zero for $\Delta\phi = 2\pi$. The characteristic is therefore periodic, exhibiting both negative and positive gains.

The operation of phase detectors is similar to that of differential amplifiers in that both sense the *difference* between the two inputs, generating a proportional output.

8.1.2 Basic PLL Topology

To arrive at the concept of phase locking, let us consider the problem of aligning the output phase of a VCO with the phase of a reference clock. As illustrated in Fig. 8.4(a), the rising edges of V_{VCO} are "skewed" by Δt seconds with respect to V_{CK}, and we wish to eliminate this error. Assuming that the VCO has a single control input, V_{cont}, we note that to vary the phase, we *must* vary the frequency and allow the integration $\phi = \int(\omega_0 + K_{VCO}V_{cont})dt$ to take place. For example, suppose as shown in Fig. 8.4(b), the VCO frequency is stepped

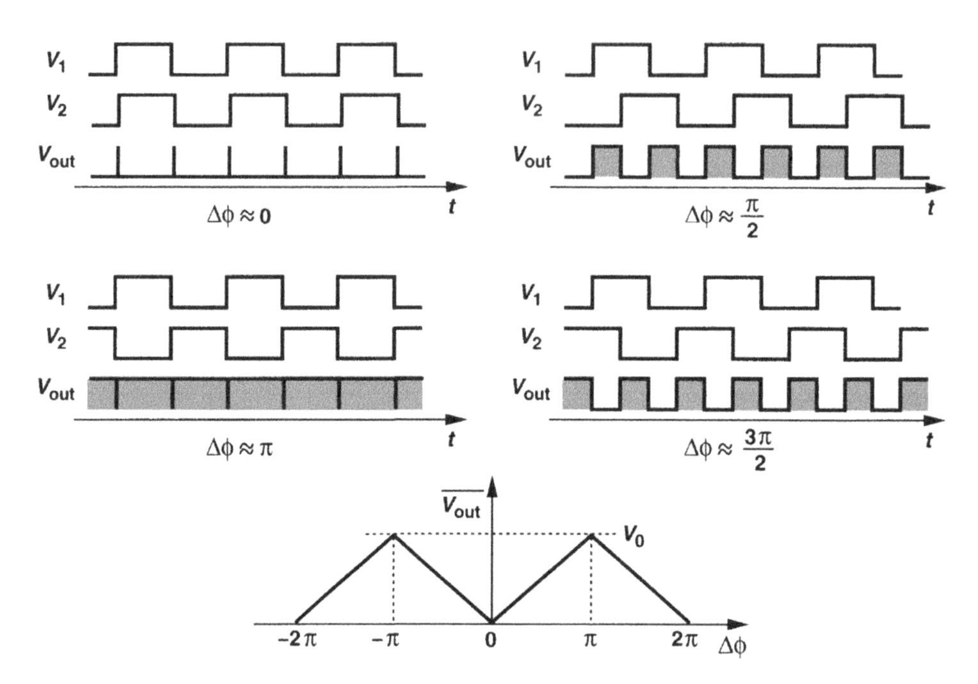

Figure 8.3 Response of XOR as a phase detector and the resulting characteristic.

to a higher value at $t = t_1$. The circuit then accumulates phase faster, gradually decreasing the phase error. At $t = t_2$, the phase error drops to zero and, if V_{cont} returns to its original value, V_{VCO} and V_{CK} remain aligned. Interestingly, the alignment can be accomplished by stepping the VCO frequency to a *lower* value for a certain time interval as well. Thus, phase alignment can be achieved only by a (temporary) frequency change.

The foregoing experiment suggests that the output phase of a VCO can be aligned with the phase of a reference if (1) the frequency of the VCO is changed momentarily, and (2) a means of comparing the two phases, i.e., a phase detector, is used to determine when the VCO and reference signals are aligned. The task of aligning the output phase of the VCO with the phase of the reference is called "phase locking."

From the above observations, we surmise that a PLL simply consists of a PD and a VCO in a feedback loop [Fig. 8.5(a)]. The PD compares the phases of V_{out} and V_{in}, generating an error that varies the VCO frequency until the phases are aligned, i.e., the loop is locked. This topology, however, must be modified because (1) as exemplified by the waveforms of Fig. 8.2, the PD output, V_{PD}, consists of a dc component (desirable) and high-frequency components (undesirable), and (2) as mentioned in Chapter 6, the control voltage of the oscillator must remain quiet in the steady state, i.e., the PD output must be filtered. We therefore interpose a low-pass filter (LPF) between the PD and the VCO [Fig. 8.5(b)], suppressing the high-frequency components of the PD output and presenting the dc level to the oscillator. This forms the basic PLL topology. For now, we assume the LPF has a gain of unity at low frequencies (e.g., as in a first-order RC section).

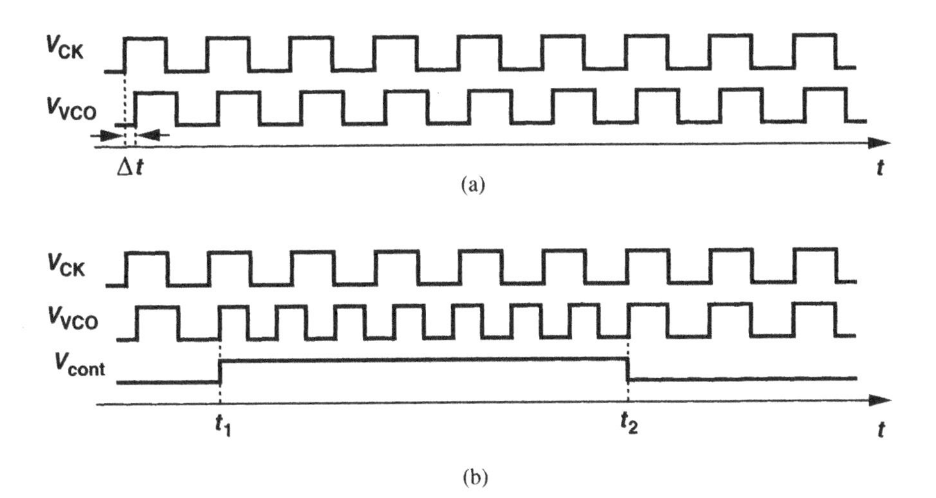

Figure 8.4 (a) Two waveforms with a skew, (b) change of VCO frequency to eliminate the skew.

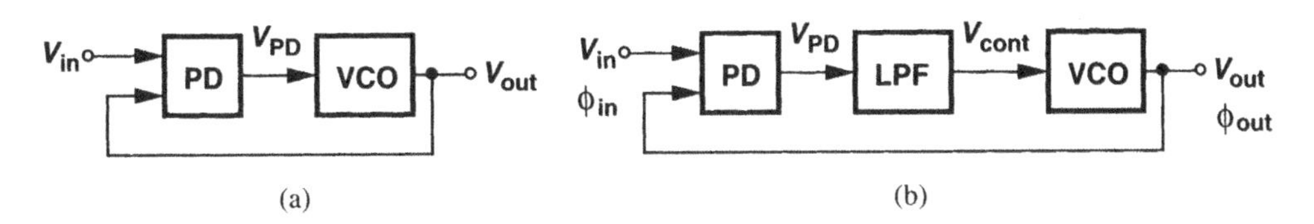

Figure 8.5 (a) Feedback loop comparing input and output phases, (b) simple PLL.

It is important to bear in mind that the feedback loop of Fig. 8.5(b) compares the *phases* of the input and output. Unlike the feedback topologies studied in the previous chapters, PLLs typically require no knowledge of voltages or currents in their feedback operation. If the loop gain is large enough, the difference between the input phase, ϕ_{in}, and the output phase, ϕ_{out}, falls to a small value in the steady state, providing phase alignment.

For subsequent analyses of PLLs, we must define the phase-lock condition carefully. If the loop of Fig. 8.5(b) is locked, we postulate that $\phi_{out} - \phi_{in}$ is constant and preferably small. We therefore define the loop to be locked if $\phi_{out} - \phi_{in}$ does not change with time. An important corollary of this definition is that

$$\frac{d\phi_{out}}{dt} - \frac{d\phi_{in}}{dt} = 0 \tag{8.1}$$

and hence

$$\omega_{out} = \omega_{in}. \tag{8.2}$$

This is a unique property of PLLs and will be revisited more closely later.

In summary, when locked, a PLL produces an output that has a small phase error with respect to the input but exactly the same frequency. The reader may then wonder why a PLL is used at all. A short piece of wire would seem to perform the task even better! We answer this question in Section 8.5.

We now construct a simple PLL in CMOS technology. Figure 8.6 illustrates an implementation employing an XOR gate as the phase detector. The VCO is configured as a

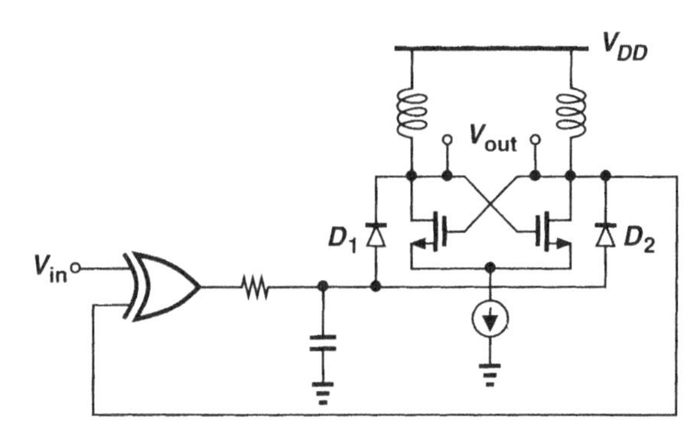

Figure 8.6 Simple PLL.

negative-G_m LC oscillator whose frequency is tuned by varactor diodes.

PLL Waveforms in Locked Condition To familiarize ourselves with the behavior of PLLs, we begin with the simplest case: the circuit is locked and we wish to examine the waveforms at each point around the loop. As illustrated in Fig. 8.7(a), V_{in} and V_{out} exhibit a small phase difference but equal frequencies. The PD therefore generates pulses

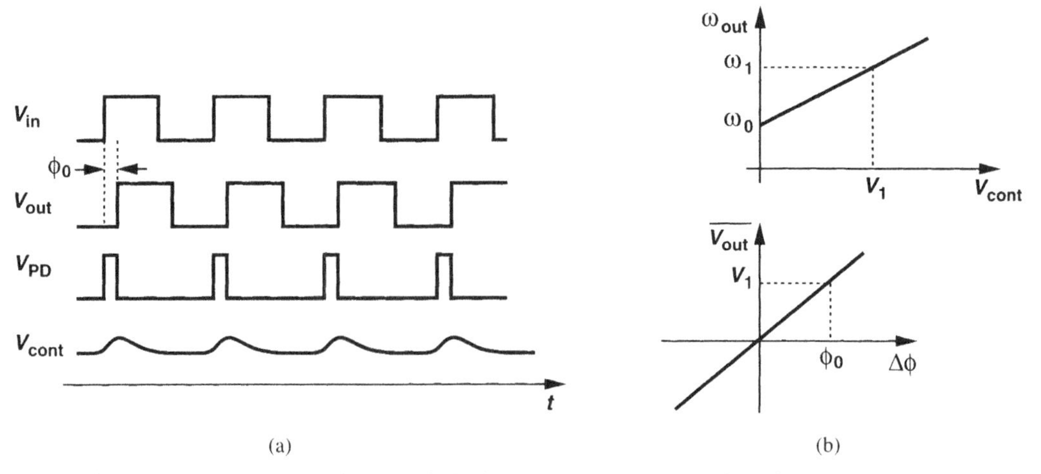

Figure 8.7 (a) Waveforms in a PLL in locked condition, (b) calculation of phase error.

as wide as the skew between the input and the output,[1] and the low-pass filter extracts the dc component of V_{PD}, applying the result to the VCO. We assume the LPF has a gain of unity at low frequencies. The small pulses in V_{LPF} are called "ripple."

In the waveforms of Fig. 8.7(a), two quantities are unknown: ϕ_0 and the dc level of V_{cont}. To determine these values, we construct the VCO and PD characteristics [Fig. 8.7(b)]. If the input and output frequencies are equal to ω_1, then the required oscillator control voltage is unique and equal to V_1. This voltage must be produced by the phase detector, demanding a phase error determined by the PD characteristic. More specifically, since $\omega_{out} = \omega_0 + K_{VCO}V_{cont}$ and $\overline{V_{PD}} = K_{PD}\Delta\phi$, we can write

$$V_1 = \frac{\omega_1 - \omega_0}{K_{VCO}}, \tag{8.3}$$

and

$$\phi_0 = \frac{V_1}{K_{PD}} \tag{8.4}$$

$$= \frac{\omega_1 - \omega_0}{K_{PD}K_{VCO}}. \tag{8.5}$$

Equation (8.5) reveals two important points: (1) as the input frequency of the PLL varies, so does the phase error; (2) to minimize the phase error, $K_{PD}K_{VCO}$ must be maximized.

As an example, suppose a PLL incorporates a VCO and a PD having the characteristics shown in Fig. 8.8. We determine what happens as the input frequency varies in the locked

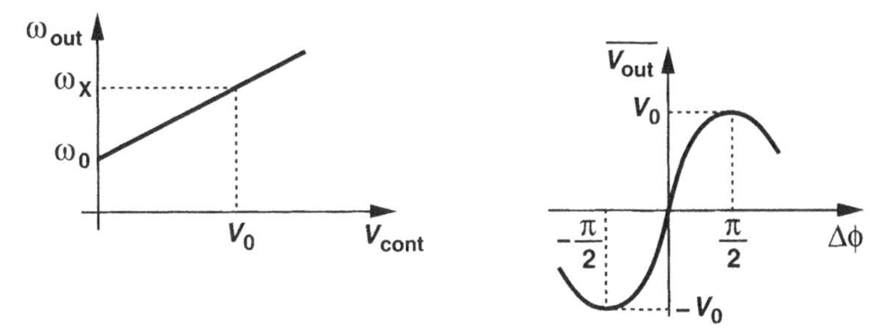

Figure 8.8 VCO and PD characteristics in a PLL.

condition. The PD characteristic is relatively linear near the origin but exhibits a small-signal gain of zero if the phase difference equals $\pm\pi/2$, at which point the average output is equal to $\pm V_0$. Now suppose the input frequency increases from ω_0, requiring a greater control voltage. If the frequency is high enough ($= \omega_x$) to mandate $V_{cont} = V_0$, then the PD must operate at the peak of its characteristic. However, the PD gain drops to zero here and the feedback loop fails. Thus, the circuit cannot lock if $\omega_{in} = \omega_X$.

With the basic understanding of PLLs developed thus far, we now return to Eq. (8.2). The exact equality of the input and output frequencies of a PLL in the locked condition

[1] In this example, the PD produces pulses only on the rising transitions.

is a critical attribute. The significance of this property can be seen from two observations. First, in many applications, even a very small (deterministic) frequency error may prove unacceptable. For example, if a data stream is to be processed synchronously by a clocked system, even a slight difference between the data rate and the clock frequency results in a "drift," creating errors (Fig. 8.9). Second, the equality would *not* exist if the PLL compared

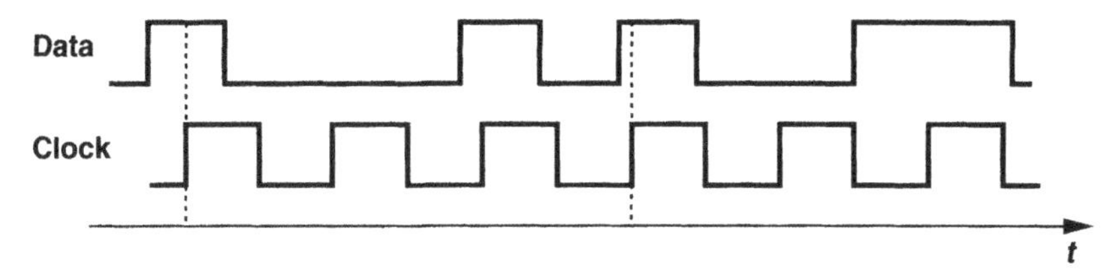

Figure 8.9 Drift of data with respect to clock in the presence of small frequency error.

the input and output frequencies rather than phases. As illustrated in Fig. 8.10(a), a loop employing a frequency detector (FD) would suffer from a finite difference between ω_{in}

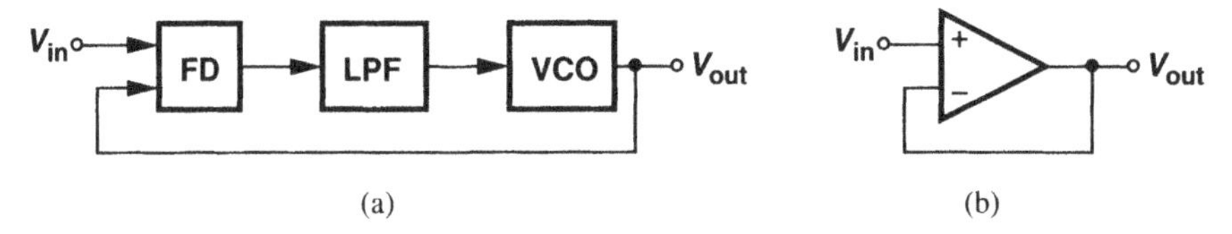

Figure 8.10 (a) Frequency-locked loop, (b) unity-gain feedback amplifier.

and ω_{out} due to various mismatches and other nonidealities. This can be understood by an analogy with the unity-gain feedback circuit of Fig. 8.10(b). Even if the op amp's open-loop gain is infinity, the input-referred offset voltage leads to a finite error between V_{in} and V_{out}.

Small Transients in Locked Condition Let us now analyze the response of a PLL in locked condition to small phase or frequency transients at the input.

Consider a PLL in the locked condition and assume the input and output waveforms can be expressed as

$$V_{in}(t) = V_A \cos\omega_1 t \tag{8.6}$$

$$V_{out}(t) = V_B \cos(\omega_1 t + \phi_0), \tag{8.7}$$

where higher harmonics are neglected and ϕ_0 is the static phase error. Suppose, as shown in Fig. 8.11, the input experiences a phase step of ϕ_1 at $t = t_1$, i.e., $\phi_{in} = \omega_1 t + \phi_1 u(t - t_1)$.[2] Since the output of the LPF does not change instantaneously, the VCO initially continues to oscillate at ω_1. The growing phase difference between the input and the output then creates wide pulses at the output of the PD, forcing V_{LPF} to rise gradually. As a result, the

[2]In this example, ϕ_{in} and ϕ_{out} denote the *total* phases of the input and output, respectively.

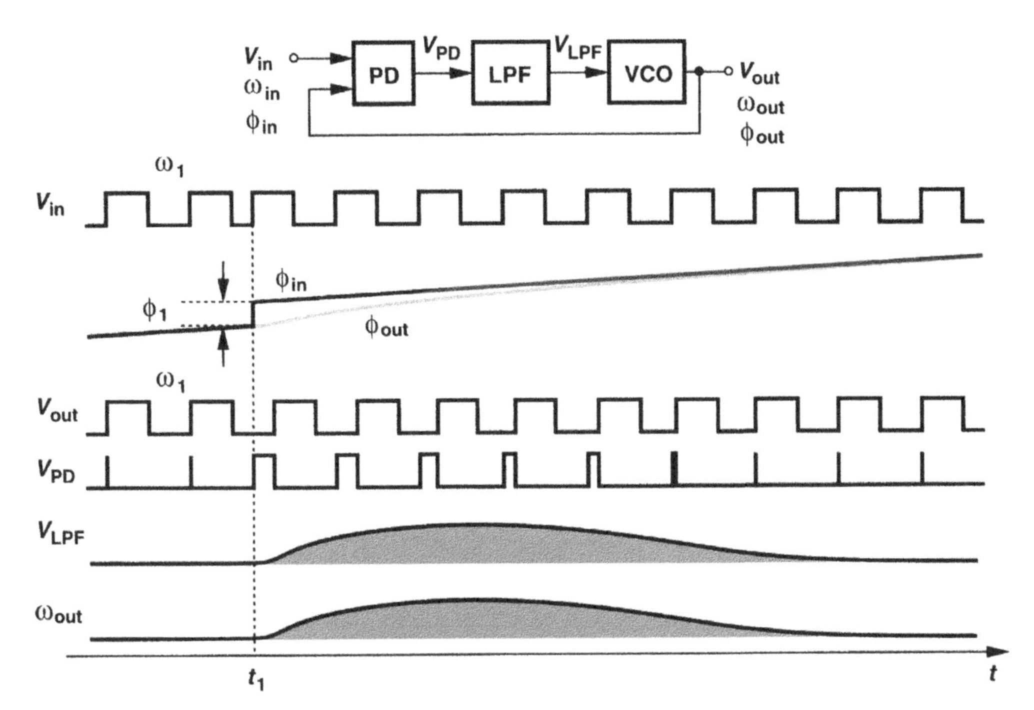

Figure 8.11 Response of a PLL to a phase step.

VCO frequency begins to change, attempting to minimize the phase error. Note that the loop is not locked during the transient because the phase error varies with time.

What happens after the VCO frequency begins to change? If the loop is to return to lock, ω_{out} must eventually go back to ω_1, requiring that V_{LPF} and hence $\phi_{out} - \phi_{in}$ also return to their original values. Since ϕ_{in} has changed by ϕ_1, the variation in the VCO frequency is such that the *area* under ω_{out} provides an additional phase of ϕ_1 in ϕ_{out}:

$$\int_{t1}^{\infty} \omega_{out} dt = \phi_1. \tag{8.8}$$

Thus, when the loop settles, the output becomes equal to

$$V_{out}(t) = V_B \cos[\omega_1 t + \phi_0 + \phi_1 u(t - t_1)]. \tag{8.9}$$

Consequently, as shown in Fig. 8.11, ϕ_{out} gradually "catches up" with ϕ_{in}.

It is important to make two observations. (1) After the loop returns to lock, *all* of the parameters (except for the total input and output phases) assume their original values. That is, $\phi_{in} - \phi_{out}$, V_{LPF}, and the VCO frequency remain unchanged—an expected result because these three parameters bear a one-to-one relationship and the input frequency has stayed the same. (2) The control voltage of the oscillator can serve as a suitable test point in the analysis of PLLs. While it is difficult to measure the time variations of phase and frequency in Fig. 8.11, $V_{cont}(= V_{LPF})$ can be readily monitored in simulations and measurements.

The reader may wonder whether an input phase step always gives rise to the response shown in Fig. 8.11. For example, is it possible for V_{LPF} to ring before settling to its final value? Such behavior is indeed possible and will be quantified in Section 8.1.3.

Let us now examine the response of PLLs to a small input frequency step $\Delta\omega$ at $t = t_1$ (Fig. 8.12). As with the case of a phase step, the VCO initially continues to oscillate at

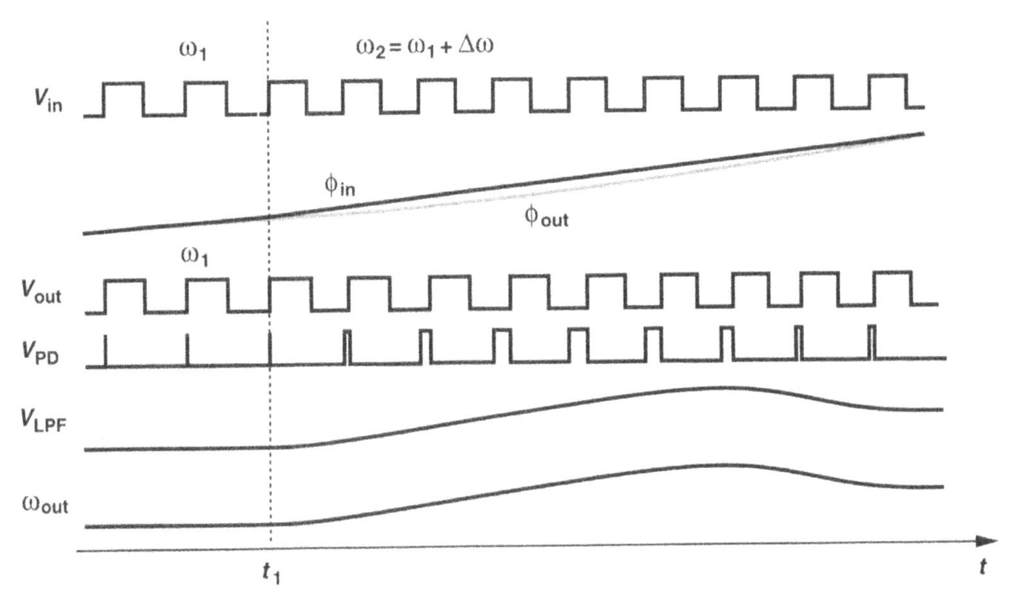

Figure 8.12 Response of a PLL to a small frequency step.

ω_1. Thus, the PD generates increasingly wider pulses, and V_{LPF} rises with time. As ω_{out} approaches $\omega_1 + \Delta\omega$, the width of the pulses generated by the PD decreases, eventually settling to a value that produces a dc component equal to $(\omega_1 + \Delta\omega - \omega_0)/K_{VCO}$. In contrast to the case of phase step, the response of a PLL to a frequency step entails a permanent change in both the control voltage and the phase error. If the input frequency is varied slowly, ω_{out} simply "tracks" ω_{in}.

The exact settling behavior of PLLs depends on the various loop parameters and will be studied in Section 8.1.3. But to arrive at an important observation, we consider the phase step response depicted in Fig. 8.13, where V_{cont} rings before settling to its final value. Consider the state of the loop at $t = t_2$. At this point, the output frequency is equal to its final value (because V_{cont} is equal to its final value), but the loop continues the transient because the phase error deviates from the required value. Similarly, at $t = t_3$, the phase error is equal to its final value but the output frequency is not. In other words, for the loop to settle, both the phase and the frequency must settle to proper values.

Now consider the PLL shown in Fig. 8.14, where an external voltage V_{ex} is added to the output of the low-pass filter.[3] Let us determine the phase error and V_{LPF} if the loop

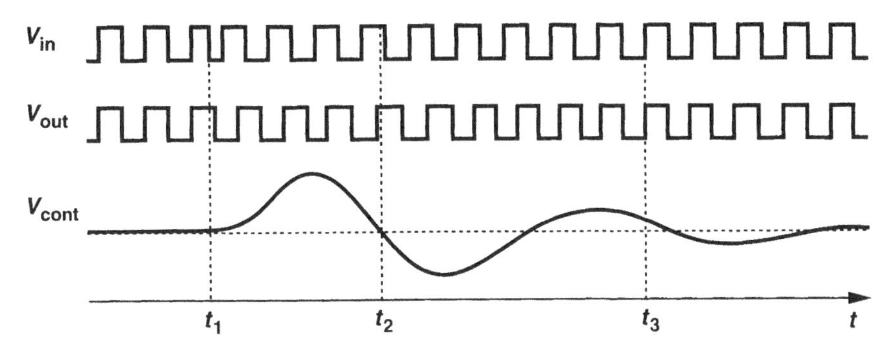

Figure 8.13 Example of phase step response.

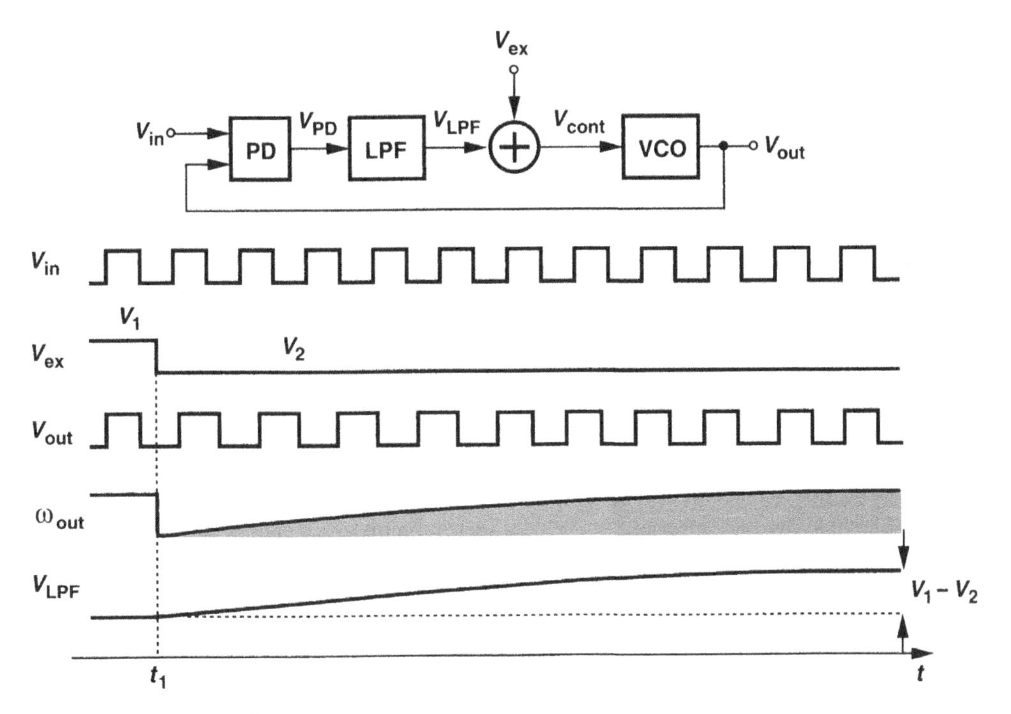

Figure 8.14 Response of a PLL to an external step on control line.

is locked and $V_{ex} = V_1$. Since the loop is locked, $\omega_{out} = \omega_{in}$ and $V_{cont} = (\omega_{in} - \omega_0)/K_{VCO}$. Thus, $V_{LPF} = (\omega_{in} - \omega_0)/K_{VCO} - V_1$ and $\Delta\phi = V_{LPF}/K_{PD} = (\omega_{in} - \omega_0)/(K_{PD}K_{VCO}) - V_1/K_{PD}$.

If V_{ex} steps from V_1 to V_2 at $t = t_1$, how does the loop respond? In this case, V_{cont} immediately goes from $(\omega_{in} - \omega_0)/K_{VCO}$ to $(\omega_{in} - \omega_0)/K_{VCO} + (V_2 - V_1)$, changing the VCO frequency to $\omega_{in} - K_{VCO}(V_1 - V_2)$. Since V_{LPF} cannot change instantaneously, the

[3]This topology is used for some types of frequency modulation in wireless communication.

PD begins to generate increasingly wider pulses, raising V_{LPF} and increasing ω_{out}. When the loop returns to lock, ω_{out} becomes equal to ω_{in} and $V_{LPF} = (\omega_{in} - \omega_0)/K_{VCO} - V_2$. The phase error also changes to $(\omega_{in} - \omega_0)/(K_{PD}K_{VCO}) - V_2/K_{PD}$. Note that the area under ω_{out} during the transient is equal to the change in the output phase and hence the change in the phase error:

$$\int_{t1}^{\infty} \omega_{out} dt = \frac{V_1 - V_2}{K_{PD}}. \tag{8.10}$$

From our study thus far, we conclude that phase-locked loops are "dynamic" systems, i.e., their response depends on the past values of the input and output. This is to be expected because the low-pass filter and the VCO introduce poles (and possibly zeros) in the loop transfer function. Moreover, we note that, so long as the input and the output remain perfectly periodic (i.e., $\phi_{in} = \omega_{in}t$ and $\phi_{out} = \omega_{in}t + \phi_0$), the loop operates in the steady state, exhibiting no transient. Thus, the PLL only responds to variations in the *excess* phase of the input or output. For example, in Fig. 8.11, $\phi_{in} = \omega_1 t + \phi_1 u(t - t_1)$ and in Fig. 8.12, $\phi_{in} = \omega_1 t + \Delta\omega \cdot tu(t - t_1)$.

8.1.3 Dynamics of Simple PLL

With the qualitative analysis of PLLs in Section 8.1.1, we can now study their transient behavior more rigorously. Assuming the loop is initially locked, we treat the PLL as a feedback system but recognize that the output quantity in this analysis must be the (excess) phase of the VCO because the "error amplifier" can only compare phases. Our objective is to determine the transfer function $\Phi_{out}(s)/\Phi_{in}(s)$ for both open-loop and closed-loop systems and subsequently study the time-domain response. Note that the dimensions change from phase to voltage through the PD and from voltage to phase through the VCO.

What does $\Phi_{out}(s)/\Phi_{in}(s)$ signify? An analogy with more familiar transfer functions proves useful here. A circuit having a transfer function $V_{out}(s)/V_{in}(s) = 1/(1 + s/\omega_0)$ is considered a low-pass filter because if V_{in} varies rapidly, V_{out} cannot fully track the input variations. Similarly, $\Phi_{out}(s)/\Phi_{in}(s)$ reveals how the output phase tracks the input phase if the latter changes slowly or rapidly.

To visualize the variation of the excess phase with time, consider the waveforms in Fig. 8.15. The period varies slowly in Fig. 8.15(a) and rapidly in Fig. 8.15(b). Thus, $y_2(t)$

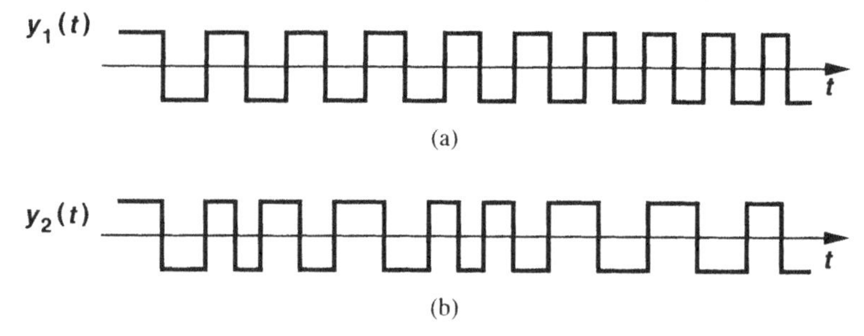

Figure 8.15 (a) Slow and (b) fast variation of the excess phase.

experiences faster phase variation than does $y_1(t)$.

Let us construct a linear model of the PLL, assuming a first-order low-pass filter for simplicity. The PD output contains a dc component equal to $K_{PD}(\phi_{out} - \phi_{in})$ as well as high-frequency components. Since the latter are suppressed by the LPF, we simply model the PD by a subtractor whose output is "amplified" by K_{PD}. Illustrated in Fig. 8.16, the overall PLL model consists of the phase subtractor, the LPF transfer function

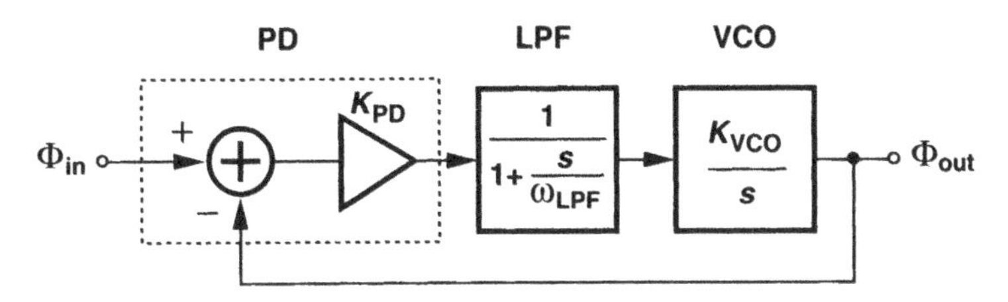

Figure 8.16 Linear model of type I PLL.

$1/(1+s/\omega_{LPF})$, where ω_{LPF} denotes the −3-dB bandwidth, and the VCO transfer function K_{VCO}/s. Here, Φ_{in} and Φ_{out} denote the excess phases of the input and output waveforms, respectively. For example, if the total input phase experiences a step change, $\phi_1 u(t)$, then $\Phi_{in}(s) = \phi_1/s$.

The open-loop transfer function is given by

$$H(s)|_{\text{open}} = \frac{\Phi_{out}}{\Phi_{in}}(s)|_{\text{open}} \tag{8.11}$$

$$= K_{PD} \cdot \frac{1}{1+\frac{s}{\omega_{LPF}}} \cdot \frac{K_{VCO}}{s}, \tag{8.12}$$

revealing one pole at $s = -\omega_{LPF}$ and another at $s = 0$. Note that the loop gain is equal to $H(s)|_{\text{open}}$ because of the unity feedback factor. Since the loop gain contains a pole at the origin, the system is called "type I."

Before computing the closed-loop transfer function, let us make an important observation. What is the loop gain if s is very small, i.e., if the input excess phase varies very slowly? Owing to the pole at the origin, the loop gain goes to infinity as s approaches zero, a point of contrast to the feedback circuits employing op amps. Thus, the phase-locked loop (under closed-loop, locked condition) ensures that the change in ϕ_{out} is *exactly* equal to the change in ϕ_{in} as s goes to zero. This result predicts two interesting properties of PLLs. First, if the input excess phase varies very slowly, the output excess phase "tracks" it. (After all, ϕ_{out} is "locked" to ϕ_{in}.) Second, if the transients in ϕ_{in} have decayed (another case corresponding to $s \to 0$), then the change in ϕ_{out} is precisely equal to the change in ϕ_{in}. This is indeed true in the example depicted in Fig. 8.11.

From (8.12), we can write the closed-loop transfer function as:

$$H(s)|_{\text{closed}} = \frac{K_{PD}K_{VCO}}{\frac{s^2}{\omega_{LPF}} + s + K_{PD}K_{VCO}}. \tag{8.13}$$

For the sake of brevity, we hereafter denote $H(s)|_{\text{closed}}$ simply by $H(s)$ or Φ_{out}/Φ_{in}. As expected, if $s \to 0$, $H(s) \to 1$ because of the infinite loop gain.

In order to analyze $H(s)$ further, we derive a relationship that allows a more intuitive understanding of the system. Recall that the instantaneous frequency of a waveform is equal to the time derivative of the phase: $\omega = d\phi/dt$. Since the frequency and the phase are related by a linear operator, the transfer function of (8.13) applies to variations in the input and output frequencies as well:

$$\frac{\omega_{out}}{\omega_{in}}(s) = \frac{K_{PD}K_{VCO}}{\dfrac{s^2}{\omega_{LPF}} + s + K_{PD}K_{VCO}}. \tag{8.14}$$

For example, this result predicts that if ω_{in} changes very slowly ($s \to 0$), then ω_{out} tracks ω_{in}, again an expected result because the loop is assumed locked. Equation (8.14) also indicates that if ω_{in} changes abruptly but the system is given enough time to settle ($s \to 0$), then the change in ω_{out} equals that in ω_{in} (as illustrated in the example of Fig. 8.12).

The above observation aids the analysis in two directions. First, some transient responses of the closed-loop system may be simpler to visualize in terms of changes in the frequency quantities rather than phase quantities. Second, since a change in ω_{out} must be accompanied by a change in V_{cont}, we have

$$H(s) = K_{VCO} \cdot \frac{V_{cont}}{\omega_{in}}(s). \tag{8.15}$$

That is, monitoring the response of V_{cont} to variations in ω_{in} indeed yields the response of the closed-loop system.

The second-order transfer function of (8.13) suggests that the step response of the type I system can be overdamped, critically damped, or underdamped. To derive the condition for each case, we rewrite the denominator in a familiar form used in control theory, $s^2 + 2\zeta\omega_n s + \omega_n^2$, where ζ is the "damping ratio" and ω_n is the "natural frequency." That is,

$$H(s) = \frac{\omega_n^2}{s^2 + 2\zeta\omega_n s + \omega_n^2}, \tag{8.16}$$

where

$$\omega_n = \sqrt{\omega_{LPF}K_{PD}K_{VCO}} \tag{8.17}$$

$$\zeta = \frac{1}{2}\sqrt{\frac{\omega_{LPF}}{K_{PD}K_{VCO}}}. \tag{8.18}$$

The two poles of the closed-loop system are given by

$$s_{1,2} = -\zeta\omega_n \pm \sqrt{(\zeta^2 - 1)\omega_n^2} \tag{8.19}$$

$$= (-\zeta \pm \sqrt{\zeta^2 - 1})\omega_n. \tag{8.20}$$

Thus, if $\zeta > 1$, both poles are real, the system is overdamped, and the transient response contains two exponentials with time constants $1/s_1$ and $1/s_2$. On the other hand, if $\zeta < 1$, the poles are complex and the response to an input frequency step $\omega_{in} = \Delta\omega u(t)$ is equal to

$$\omega_{out}(t) = \left\{1 - e^{-\zeta\omega_n t}[\cos(\omega_n\sqrt{1-\zeta^2}t) + \frac{\zeta}{\sqrt{1-\zeta^2}}\sin(\omega_n\sqrt{1-\zeta^2}t)]\right\}\Delta\omega u(t)$$

$$= [1 - \frac{1}{\sqrt{1-\zeta^2}}e^{-\zeta\omega_n t}\sin(\omega_n\sqrt{1-\zeta^2}t + \theta)]\Delta\omega u(t), \quad (8.21)$$

where ω_{out} denotes the change in the output frequency and $\theta = \sin^{-1}\sqrt{1-\zeta^2}$. Thus, as shown in Fig. 8.17, the step response contains a sinusoidal component with a frequency

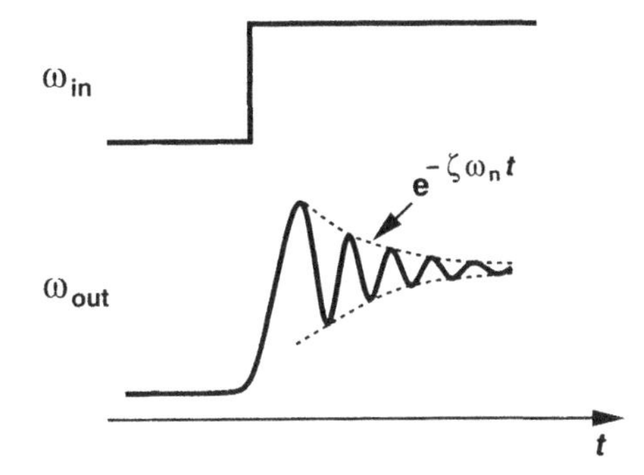

Figure 8.17 Underdamped response of PLL to a frequency step.

$\omega_n\sqrt{1-\zeta^2}$ that decays with a time constant $(\zeta\omega_n)^{-1}$. Note that the system exhibits the same response if a phase step is applied to the input and the output phase is observed.

The settling speed of PLLs is of great concern in most applications. Equation (8.21) indicates that the exponential decay determines how fast the output approaches its final value, implying that $\zeta\omega_n$ must be maximized. For the type I PLL under study here, (8.17) and (8.18) yield

$$\zeta\omega_n = \frac{1}{2}\omega_{LPF}. \quad (8.22)$$

This result reveals a critical trade-off between the settling speed and the ripple on the VCO control line: the lower ω_{LPF}, the greater the suppression of the high-frequency components produced by the PD but the longer the settling time constant.

As an example, suppose a cellular telephone incorporates a 900-MHz phase-locked loop to generate the carrier frequencies. If $\omega_{LPF} = 2\pi \times (20\text{ kHz})$ and the output frequency is to be changed from 901 MHz to 901.2 MHz, how long does the PLL output frequency take to settle within 100 Hz of its final value? Since the step size is 200 kHz, we have

$$[1 - e^{-\zeta\omega_n t_s}\sin(\omega_n\sqrt{1-\zeta^2}t_s + \theta)] \times 200\text{ kHz} = 200\text{ kHz} - 100\text{ Hz}. \quad (8.23)$$

Thus,

$$e^{-\zeta\omega_n t_s}\sin(\omega_n\sqrt{1-\zeta^2}t_s+\theta)=\frac{100\text{ Hz}}{200\text{ kHz}}. \tag{8.24}$$

In the worst case, the sinusoid is equal to unity and

$$e^{-\zeta\omega_n t_s}=0.0005. \tag{8.25}$$

That is,

$$t_s=\frac{7.6}{\zeta\omega_n} \tag{8.26}$$

$$=\frac{15.2}{\omega_{LPF}} \tag{8.27}$$

$$=0.12\text{ ms}. \tag{8.28}$$

In addition to the product $\zeta\omega_n$, the value of ζ itself is also important. Illustrated in Fig. 8.18 for several values of ζ and a constant ω_n, the step response exhibits severe ringing for

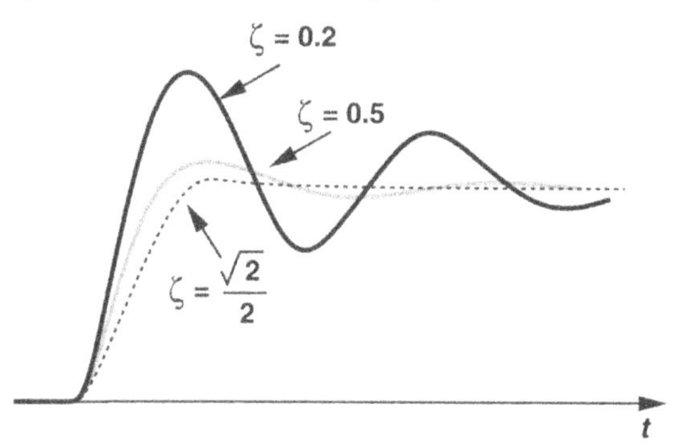

Figure 8.18 Underdamped response of a second-order system for various values of ζ.

$\zeta < 0.5$. In view of process and temperature variation of the loop parameters, ζ is usually chosen to be greater than $\sqrt{2}/2$ or even 1 to avoid excessive ringing.[4]

The choice of ζ entails other trade-offs as well. First, (8.18) implies that as ω_{LPF} is reduced to minimize the ripple on the control voltage, the stability degrades. Second, (8.5) and (8.18) indicate that both the phase error and ζ are inversely proportional to $K_{PD}K_{VCO}$; lowering the phase error inevitably makes the system less stable. In summary, the type I PLL suffers from trade-offs between the settling speed, the ripple on the control voltage (i.e., the quality of the output signal), the phase error, and the stability.

The stability behavior of PLLs can also be analyzed graphically, providing more insight. The Bode plots of the magnitude and phase of the loop gain readily yield the phase margin.

[4]The value of ζ may also yield peaking in the transfer function. Thus, some applications require a ζ of 5 to 10 to avoid peaking in the presence of higher order poles.

Let us utilize (8.12) to construct such plots. As shown in Fig. 8.19, the loop gain begins from infinity at $\omega = 0$ and falls at a rate of 20 dB/dec for $\omega < \omega_{LPF}$ and at a rate of 40

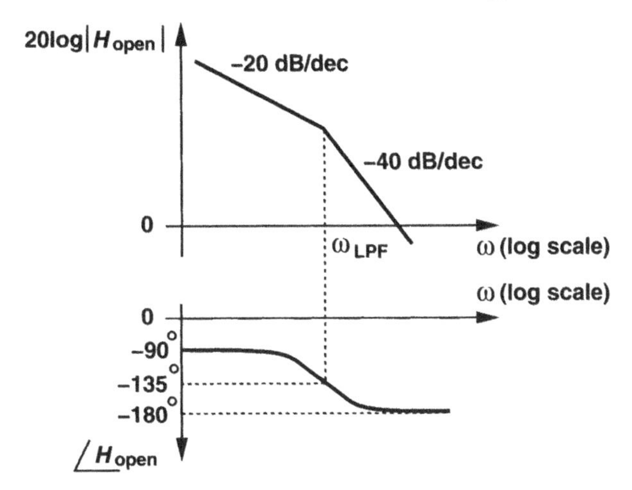

Figure 8.19 Bode plots of type I PLL.

dB/dec thereafter. The phase begins at $-90°$ and asymptotically reaches $-180°$.

What happens if a higher $K_{PD}K_{VCO}$ is chosen so as to minimize $\phi_{out} - \phi_{in}$? Since the entire gain plot in Fig. 8.19 is shifted up, the gain crossover moves to the right, thus degrading the phase margin. This is consistent with the dependence of ζ upon $K_{PD}K_{VCO}$.

As observed thus far, $K_{PD}K_{VCO}$ impacts many important parameters of PLLs. This quantity is sometimes called the loop gain (even though it is not dimensionless) due to the resemblance of $\Delta\phi = (\omega_{out} - \omega_0)/(K_{PD}K_{VCO})$ to the error equation in a feedback system.

The stability behavior of type I PLLs can also be analyzed by the locus of their poles in the complex plane as the parameter $K_{PD}K_{VCO}$ varies (Fig. 8.20). With $K_{PD}K_{VCO} = 0$, the loop is open, $\zeta = \infty$, and the two poles are given by $s_1 = -\omega_{LPF}$ and $s_2 = 0$. As $K_{PD}K_{VCO}$ increases (i.e., the feedback becomes stronger), ζ drops and the two poles, given by $s_{1,2} = (-\zeta \pm \sqrt{\zeta^2 - 1})\omega_n$, move toward each other on the real axis. For $\zeta = 1$

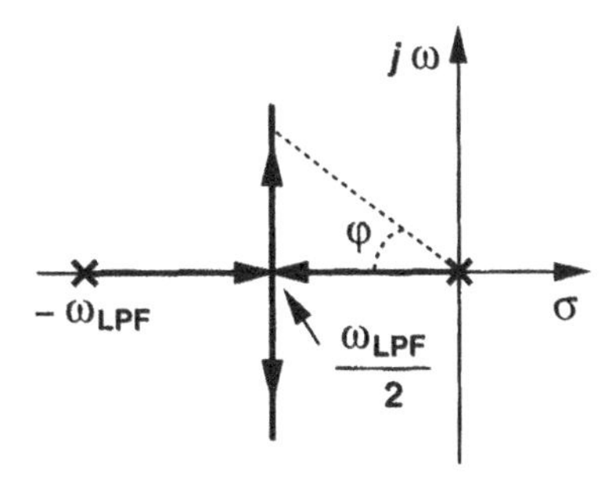

Figure 8.20 Root locus of type I PLL.

(i.e., $K_{PD}K_{VCO} = \omega_{LPF}/4$), $s_1 = s_2 = -\zeta\omega_n = -\omega_{LPF}/2$. As $K_{PD}K_{VCO}$ increases further, the two poles become complex, with a real part equal to $-\zeta\omega_n = -\omega_{LPF}/2$, moving in parallel with the $j\omega$ axis.

We recognize from Fig. 8.20 that, as s_1 and s_2 move away from the real axis, the system becomes less stable. In fact, the reader can prove that $\cos\psi = \zeta$, concluding that as ψ approaches $90°$, ζ drops to zero.

Another transfer function that reveals the settling behavior of PLLs is that of the error at the output of the phase subtractor in Fig. 8.16. Defined as $H_e(s) = (\phi_{in} - \phi_{out})/\phi_{in}$, this transfer function can be obtained by noting that $\phi_{out}/\phi_{in} = H(s)$ and, from (8.13),

$$H_e(s) = 1 - H(s) \tag{8.29}$$

$$= \frac{s^2 + 2\zeta\omega_n s}{s^2 + 2\zeta\omega_n s + \omega_n}. \tag{8.30}$$

As expected, $H_e(s) \to 0$ if $s \to 0$ because the output tracks the input when the input varies very slowly or the transient has settled.

As an example, suppose a type I PLL experiences a frequency step $\Delta\omega$ at $t = 0$. Let us calculate the change in the phase error. The Laplace transform of the frequency step equals $\Delta\omega/s$. Since $H_e(s)$ relates the phase error to the input phase, we write $\Phi_{in}(s) = (\Delta\omega/s)/s = \Delta\omega/s^2$. Thus, the Laplace transform of the phase error is

$$\Phi_e(s) = H_e(s) \cdot \frac{\Delta\omega}{s^2} \tag{8.31}$$

$$= \frac{s^2 + 2\zeta\omega_n s}{s^2 + 2\zeta\omega_n s + \omega_n^2} \cdot \frac{\Delta\omega}{s^2}. \tag{8.32}$$

From the final value theorem,

$$\phi_e(t = \infty) = \lim_{s \to 0} s\Phi_e(s) \tag{8.33}$$

$$= \frac{2\zeta}{\omega_n}\Delta\omega \tag{8.34}$$

$$= \frac{\Delta\omega}{K_{PD}K_{VCO}}, \tag{8.35}$$

which agrees with (8.5).

8.2 Charge-Pump PLLs

While type I PLLs have been realized widely in discrete form, their shortcomings often prohibit usage in high-performance integrated circuits. In addition to the trade-offs between ζ, ω_{LPF}, and the phase error, type I PLLs suffer from another critical drawback: limited acquisition range.

8.2.1 Problem of Lock Acquisition

Suppose when a PLL circuit is turned on, its oscillator operates at a frequency far from the input frequency, i.e., the loop is not locked. Under what conditions does the loop "acquire" lock? The transition of the loop from unlocked to locked condition is a very nonlinear phenomenon because the phase detector senses unequal frequencies. The problem of lock acquisition in type I PLLs has been studied extensively [1, 2], but we state without proof that the "acquisition range"[5] is on the order of ω_{LPF}; that is, the loop locks only if the difference between ω_{in} and ω_{out} is less than roughly ω_{LPF}.[6]

The problem of lock acquisition further tightens the trade-offs in type I PLLs. If ω_{LPF} is reduced to suppress the ripple on the control voltage, the acquisition range decreases. Note that even if the input frequency has a precisely-controlled value, a wide acquisition range is often necessary because the VCO center frequency may vary considerably with process and temperature. In most of today's applications, the acquisition range of the simple PLL studied thus far proves inadequate.

To remedy the acquisition problem, modern PLLs incorporate frequency detection in addition to phase detection. Called "aided acquisition" and illustrated in Fig. 8.21, the idea is to compare ω_{in} and ω_{out} by means of a frequency detector, generate a dc component

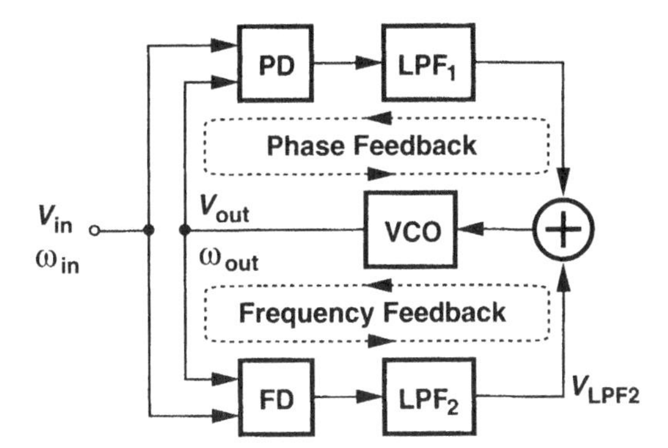

Figure 8.21 Addition of frequency detection to increase the acquisition range.

V_{LPF2} proportional to $\omega_{in} - \omega_{out}$, and apply the result to the VCO in a negative-feedback loop. At the beginning, the FD drives ω_{out} toward ω_{in} while the PD output remains "quiet." When $|\omega_{out} - \omega_{in}|$ is sufficiently small, the phase-locked loop takes over, acquiring lock. Such a scheme increases the acquisition range to the tuning range of the VCO.[7]

---

[5]Acquisition range, tracking range, lock range, capture range, and pull-in range are often used to describe the behavior of PLLs in the presence of input or VCO frequency variation. For our purposes, the acquisition range, the capture range, and the pull-in range are the same. The tracking range refers to the input frequency range across which a locked PLL can track the input. With the addition of frequency detection, the acquisition range becomes equal to the tracking range (for periodic signals).

[6]This is a very rough estimate. In practice, the acquisition range may be several times narrower or wider. It is also assumed that the tuning range of the VCO is large enough not to limit the acquisition range.

[7]This may not be true if the input is not periodic.

8.2.2 Phase/Frequency Detector and Charge Pump

For periodic signals, it is possible to merge the two loops of Fig. 8.21 by devising a circuit that can detect both phase and frequency differences. Called a "phase/frequency detector" (PFD) and illustrated conceptually in Fig. 8.22, the circuit employs sequential logic to create three states and respond to the rising (or falling) edges of the two inputs. If initially

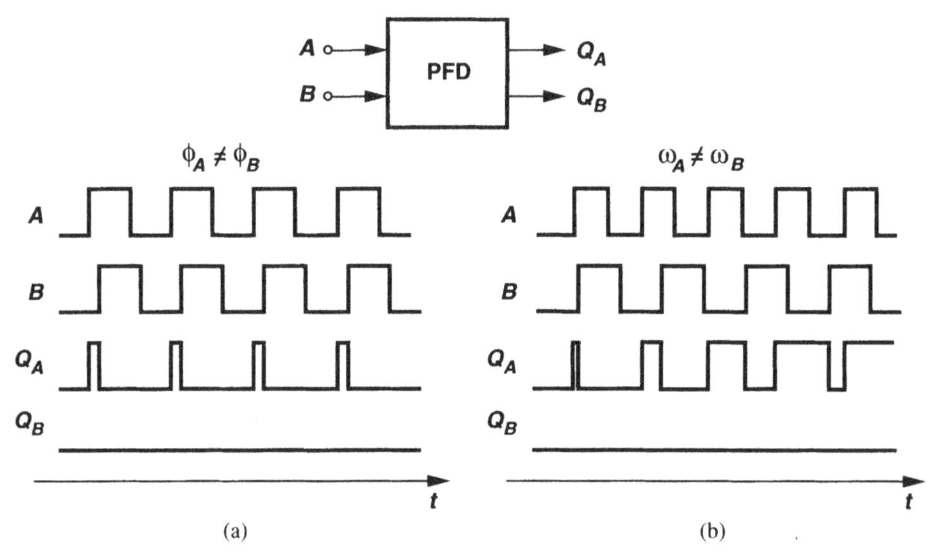

Figure 8.22 Conceptual operation of a PFD with (a) unequal phases, and (b) unequal frequencies.

$Q_A = Q_B = 0$, then a rising transition on A leads to $Q_A = 1, Q_B = 0$. The circuit remains in this state until B goes high, at which point Q_A returns to zero. The behavior is similar for the B input.

In Fig. 8.22(a), the two inputs have equal frequencies but A leads B. The output Q_A continues to produce pulses whose width is proportional to $\phi_A - \phi_B$ while Q_B remains at zero. In Fig. 8.22(b), A has a higher frequency than B, and Q_A generates pulses while Q_B does not. By symmetry, if A lags B or has a lower frequency than B, then Q_B produces pulses and Q_A remains quiet. Thus, the dc contents of Q_A and Q_B provide information about $\phi_A - \phi_B$ or $\omega_A - \omega_B$. The outputs Q_A and Q_B are called the UP and DOWN pulses, respectively.

Let us determine if a master-slave D flipflop can operate as a phase detector or a frequency detector. Assume the flipflop provides differential outputs. As shown in Fig. 8.23(a), we first apply inputs having equal frequencies and a finite phase difference, assuming the output changes on the rising edge of the clock input. If A leads B, then V_{out} remains at a logical ONE indefinitely because the flipflop continues to sample the high levels of A. Conversely, if A lags B, then V_{out} remains low. Plotted in Fig. 8.23(b), the input-output characteristic of the circuit displays a very high gain at $\Delta\phi = 0, \pm\pi, \cdots$ and a zero gain at other values of $\Delta\phi$. The D flipflop is sometimes called a "bang-bang" phase detector to emphasize that the average value of V_{out} jumps from $-V_1$ to $+V_1$ as $\Delta\phi$ varies from slightly below zero to slightly above zero. Now let us assume unequal frequencies

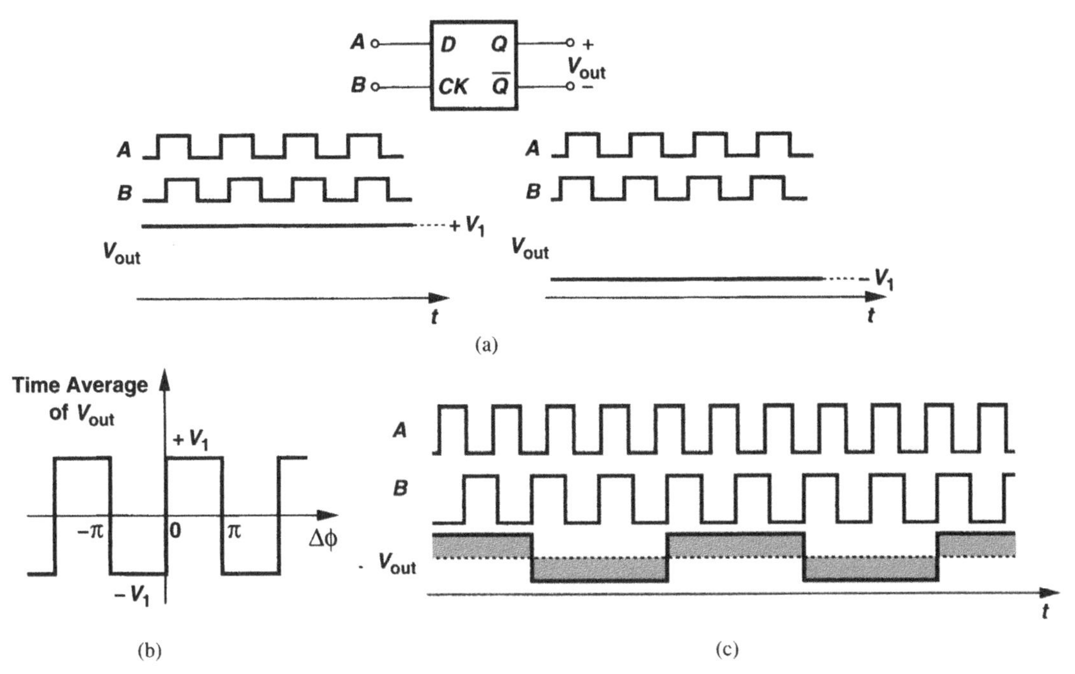

Figure 8.23 (a) D flipflop as a phase detector, (b) input-output characteristic, (c) response of D flipflop to unequal input frequencies.

for A and B. As illustrated in Fig. 8.23(c), the average output is zero, failing to provide information.

The circuit of Fig. 8.22 can be realized in various forms. Figure 8.24(a) shows a simple implementation consisting of two edge-triggered, resettable D flipflops with their D inputs tied to a logical ONE. The inputs of interest, A and B, serve as the clocks of the flipflops. If $Q_A = Q_B = 0$ and A goes high, Q_A rises. If this event is followed by a rising transition on B, Q_B goes high and the AND gate resets both flipflops. In other words, Q_A and Q_B are simultaneously high for a short time, but the difference between their average values still represents the input phase or frequency difference correctly. Each flipflop can be implemented as shown in Fig. 8.24(b), where two RS latches are cross-coupled. Latch 1 and Latch 2 respond to the rising edges of CK and Reset, respectively.

We now determine the width of the narrow reset pulses that appear in the Q_B waveform in Fig. 8.24(a). Figure 8.25(a) illustrates the overall PFD at the gate level. If the circuit begins with $A = 1, Q_A = 1$, and $Q_B = 0$, a rising edge on B forces $\overline{Q_B}$ to go low and, one gate delay later, Q_B to go high. As shown in Fig. 8.25(b), this transition propagates to Reset, $\overline{F}$, F, and Q_B. Thus, the width of the pulse on Q_B is approximately equal to 5 gate

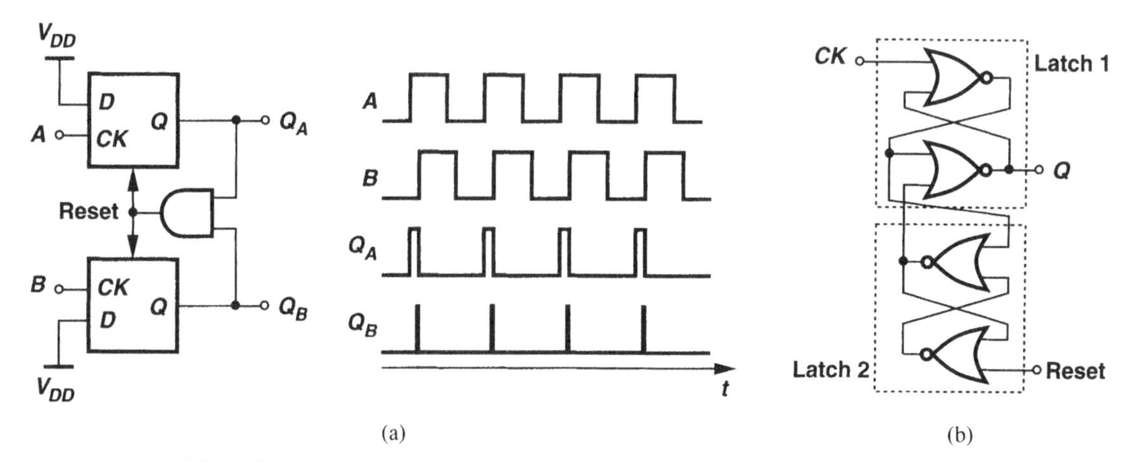

Figure 8.24 (a) Implementation of PFD, (b) implementation of D flipflop.

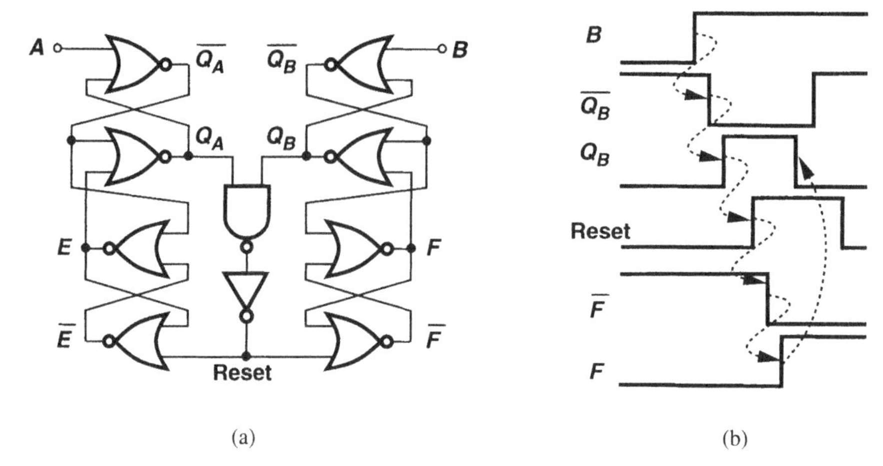

Figure 8.25 (a) PFD at gate level, (b) internal waveforms.

delays.[8]

It is instructive to plot the input-output characteristic of the above PFD. Defining the output as the difference between the average values of Q_A and Q_B when $\omega_A = \omega_B$ and neglecting the effect of the narrow reset pulses, we note that the output varies symmetrically as $|\Delta\phi|$ begins from zero (Fig. 8.26). For $\Delta\phi = \pm 360°$, V_{out} reaches its extrema and subsequently changes sign.

How is the PFD of Fig. 8.24(a) utilized in a phase-locked loop? Since the difference between the average values of Q_A and Q_B is of interest, the two outputs can be low-pass

[8]This is a rough approximation because the NAND gate, the inverter, and the NOR gates have different delays and fanouts.

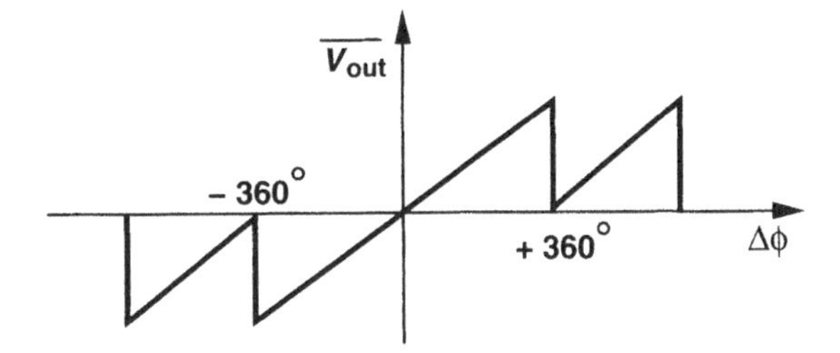

Figure 8.26 Input-output characteristic of the three-state PFD.

filtered and sensed differentially (Fig. 8.27). However, a more common approach is to interpose a "charge pump" (CP) between the PFD and the loop filter.

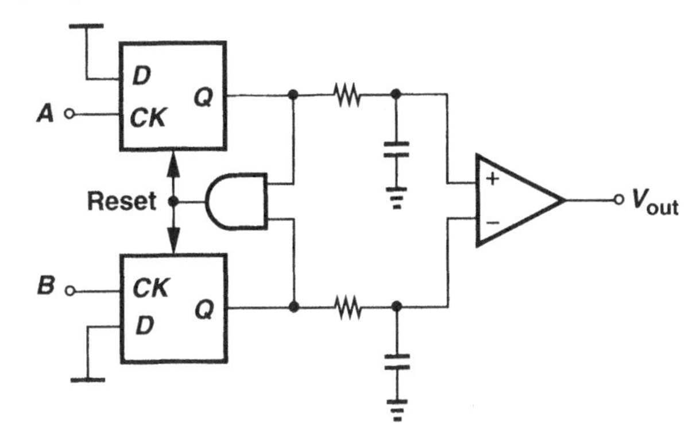

Figure 8.27 PFD followed by low-pass filters.

A charge pump consists of two switched current sources that pump charge into or out of the loop filter according to two logical inputs. Figure 8.28 illustrates a charge pump driven by a PFD and driving a capacitor. The circuit has three states. If $Q_A = Q_B = 0$, then S_1

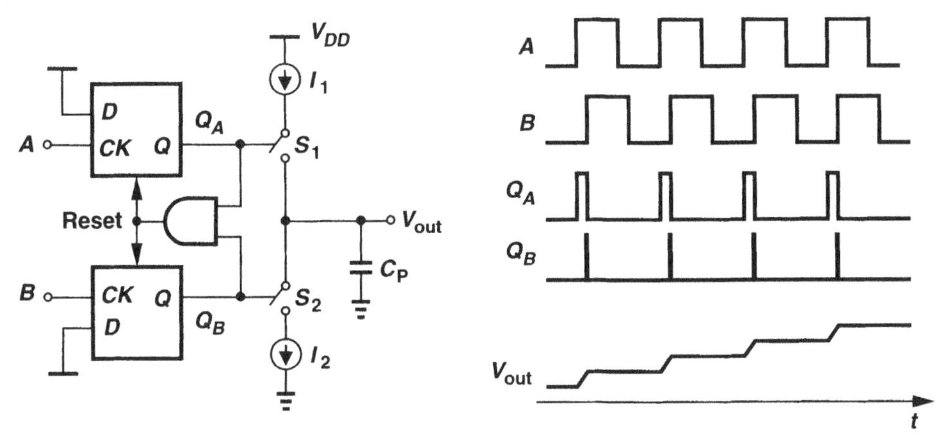

Figure 8.28 PFD with charge pump.

and S_2 are off and V_{out} remains constant. If Q_A is high and Q_B is low, then I_1 charges C_P. Conversely, if Q_A is low and Q_B is high, then I_2 discharges C_P. Thus, if, for example, A leads B, then Q_A continues to produce pulses and V_{out} rises steadily. Called UP and DOWN currents, respectively, I_1 and I_2 are nominally equal.

What is the effect of the narrow pulses that appear in the Q_B waveform in Fig. 8.28? Since Q_A and Q_B are simultaneously high for a finite period (approximately 5 gate delays), the current supplied by the charge pump to C_P is affected. In fact, if $I_1 = I_2$, the current through S_1 simply flows through S_2 during the narrow reset pulse, leaving no current to charge C_P. Thus, as shown in Fig. 8.29, V_{out} remains constant after Q_B goes high.

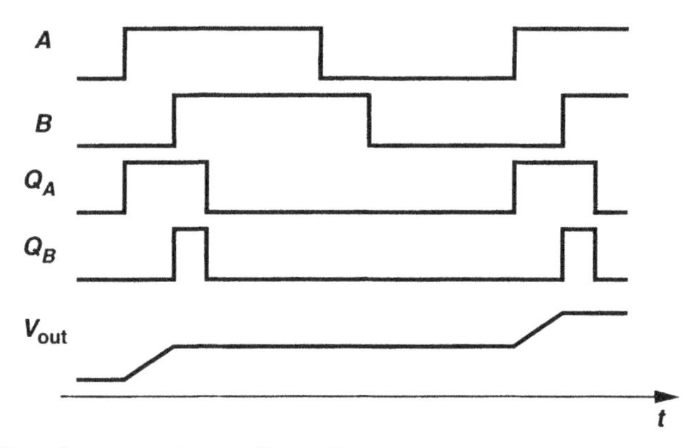

Figure 8.29 Effect of narrow pulses on Q_A or Q_B.

The circuit of Fig. 8.28 has an interesting property. If A, say, leads B by a finite amount, Q_A produces pulses indefinitely, allowing the charge pump to inject I_1 into C_P and forcing V_{out} to rise steadily. In other words, for a finite input error, the output eventually goes to $+\infty$ or $-\infty$, i.e., the "gain" of the circuit is infinity. The consequences of infinite gain are described below.

8.2.3 Basic Charge-Pump PLL

Let us now construct a PLL using the circuit of Fig. 8.28. Shown in Fig. 8.30 and called a "charge-pump PLL," such an implementation senses the transitions at the input and output, detects phase or frequency differences, and activates the charge pump accordingly. When the loop is turned on, ω_{out} may be far from ω_{in}, and the PFD and the charge pump vary the control voltage such that ω_{out} approaches ω_{in}. When the input and output frequencies are sufficiently close, the PFD operates as a phase detector, performing phase lock. The loop locks when the phase difference drops to zero and the charge pump remains relatively idle.

As observed above, the gain of the PFD/CP combination is infinite, i.e., a nonzero (deterministic) difference between ϕ_{in} and ϕ_{out} leads to indefinite charge buildup on C_P. What is the consequence of this attribute in a charge-pump PLL? When the loop of Fig. 8.30 is locked, V_{cont} is finite. Therefore, the input phase error must be exactly *zero*.[9] This

[9] As explained in Section 8.3.1, mismatches still yield a finite phase error.

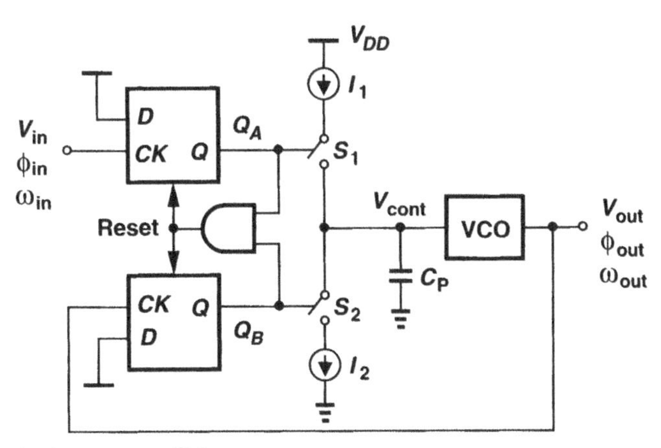

Figure 8.30 Simple charge-pump PLL.

is in contrast to the behavior of the type I PLL, in which the phase error is finite and a function of the output frequency.

To gain more insight into the operation of the PLL shown in Fig. 8.30, let us ignore the narrow reset pulses on Q_A and Q_B and assume that after $\phi_{out} - \phi_{in}$ drops to zero, the PFD simply produces $Q_A = Q_B = 0$. The charge pump thus remains idle and C_P sustains a constant control voltage. Does this mean that the PFD and the CP are no longer needed?! If V_{cont} remains constant for a long time, the VCO frequency and phase begin to drift. In particular, the noise sources in the VCO create random variations in the oscillation frequency that can result in a large accumulation of phase error. The PFD then detects the phase difference, producing a corrective pulse on Q_A or Q_B that adjusts the VCO frequency through the charge pump and the filter. This is why we stated earlier that the PLL responds only to the *excess* phase of waveforms. We also note that, since in Fig. 8.30 phase comparison is performed in every cycle, the VCO phase and frequency cannot drift substantially.

Dynamics of CPPLL To quantify the behavior of charge-pump PLLs, we must develop a linear model for the combination of the PFD, the charge pump, and the low-pass filter, thereby obtaining the transfer function. We therefore raise two questions: (1) Is the PFD/CP/LPF combination in Fig. 8.28 a linear system? (2) If so, how can its transfer function be computed?

To answer the first question, we test the system for linearity. For example, as illustrated in Fig. 8.31(a), we double the input phase difference and see if V_{out} exactly doubles. Interestingly, the flat sections of V_{out} double but not the ramp sections. After all, the current charging or discharging C_P is constant, yielding a constant slope for the ramp—an effect similar to slewing in op amps. Thus, the system is not linear in the strict sense. To overcome this quandary, we approximate the output waveform by a ramp [Fig. 8.31(b)], arriving at a linear relationship between V_{out} and $\Delta\phi$. In a sense, we approximate a discrete-time system by a continuous-time model.

To answer the second question, we recall that the transfer function is the Laplace transform of the impulse response, requiring that we apply a phase difference impulse and

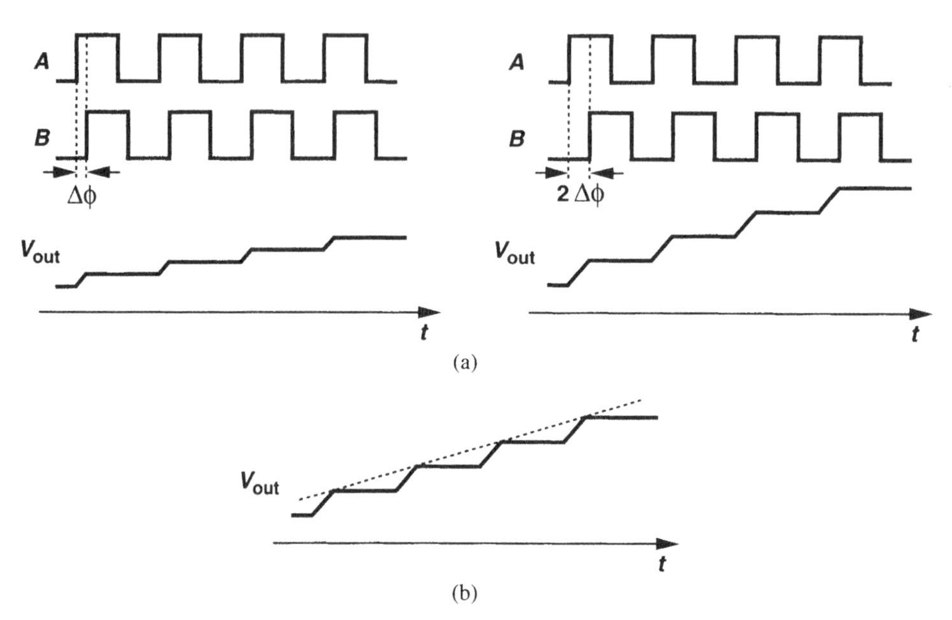

Figure 8.31 (a) Test of linearity of PFD/CP/LPF combination, (b) ramp approximation of the response.

compute V_{out} in the time domain. Since a phase difference impulse is difficult to visualize, we apply a phase difference step, obtain V_{out}, and differentiate the result with respect to time.

Let us assume the input period is T_{in} and the charge pump provides a current of $\pm I_P$ to the capacitor. As shown in Fig. 8.32, we begin with a zero phase difference and, at $t = 0$, step the phase of B by ϕ_0, i.e., $\Delta\phi = \phi_0 u(t)$. As a result, Q_A or Q_B contin-

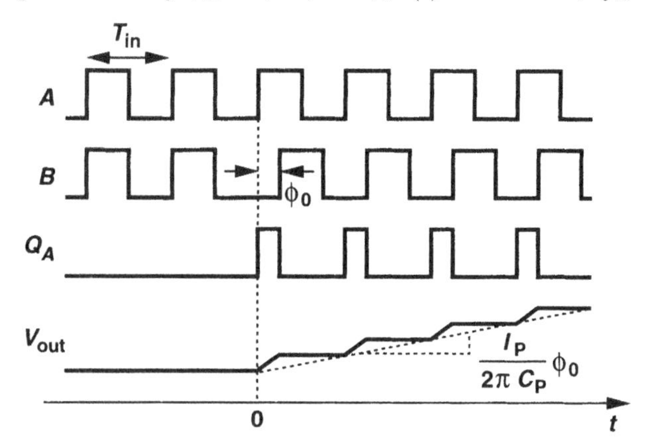

Figure 8.32 Step response of PFD/CP/LPF combination.

ues to produce pulses that are $\phi_0 T_{in}/(2\pi)$ seconds wide, raising the output voltage by

$(I_P/C_P)\phi_0 T_{in}/(2\pi)$ in every period.[10] Approximated by a ramp, V_{out} thus exhibits a slope of $(I_P/C_P)\phi_0/(2\pi)$ and can be expressed as

$$V_{out}(t) = \frac{I_P}{2\pi C_P} t \cdot \phi_0 u(t). \tag{8.36}$$

The impulse response is therefore given by

$$h(t) = \frac{I_P}{2\pi C_P} u(t), \tag{8.37}$$

yielding the transfer function

$$\frac{V_{out}}{\Delta\phi}(s) = \frac{I_P}{2\pi C_P} \cdot \frac{1}{s}. \tag{8.38}$$

Consequently, the PFD/CP/LPF combination contains a pole at the origin, a point of contrast to the PD/LPF circuit used in the type I PLL. In analogy with the expression K_{VCO}/s, we call $I_P/(2\pi C_P)$ the "gain" of the PFD and denote it by K_{PFD}.

Now suppose the output quantity of interest in the circuit of Fig. 8.28 is the current injected by the charge pump into the capacitor. Let us determine the transfer function from $\Delta\phi$ to this current, I_{out}. Since $V_{out}(s) = I_{out}/(C_P s)$, we have

$$\frac{I_{out}}{\Delta\phi}(s) = \frac{I_P}{2\pi}. \tag{8.39}$$

Let us now construct a linear model of charge-pump PLLs. Shown in Fig. 8.33, the model gives an open-loop transfer function

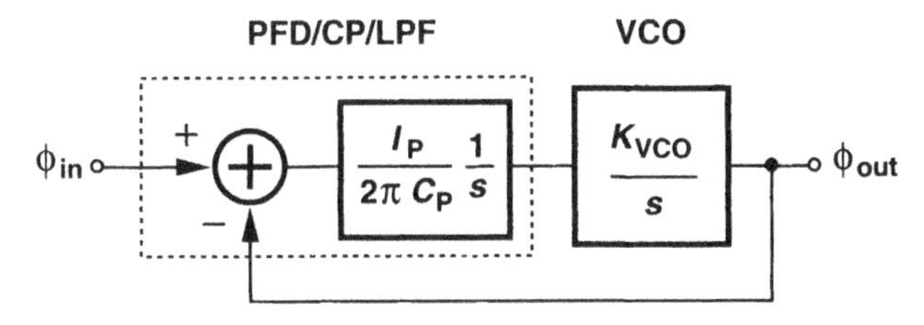

Figure 8.33 Linear model of simple charge-pump PLL.

$$\frac{\Phi_{out}}{\Phi_{in}}(s)|_{\text{open}} = \frac{I_P}{2\pi C_P}\frac{K_{VCO}}{s^2}. \tag{8.40}$$

[10] We neglect the effect of the narrow reset pulses that appear in the other output.

Since the loop gain has two poles at the origin, this topology is called a "type II" PLL. The closed-loop transfer function, denoted by $H(s)$ for the sake of brevity, is thus equal to

$$H(s) = \frac{\dfrac{I_P K_{VCO}}{2\pi C_P}}{s^2 + \dfrac{I_P K_{VCO}}{2\pi C_P}}. \tag{8.41}$$

This result is alarming because the closed-loop system contains two imaginary poles at $s_{1,2} = \pm j\sqrt{I_P K_{VCO}/(2\pi C_P)}$ and is therefore unstable. The instability arises because the loop gain has only two poles at the origin (i.e., two ideal integrators). As shown in Fig. 8.34(a), each integrator contributes a constant phase shift of 90°, allowing the system to

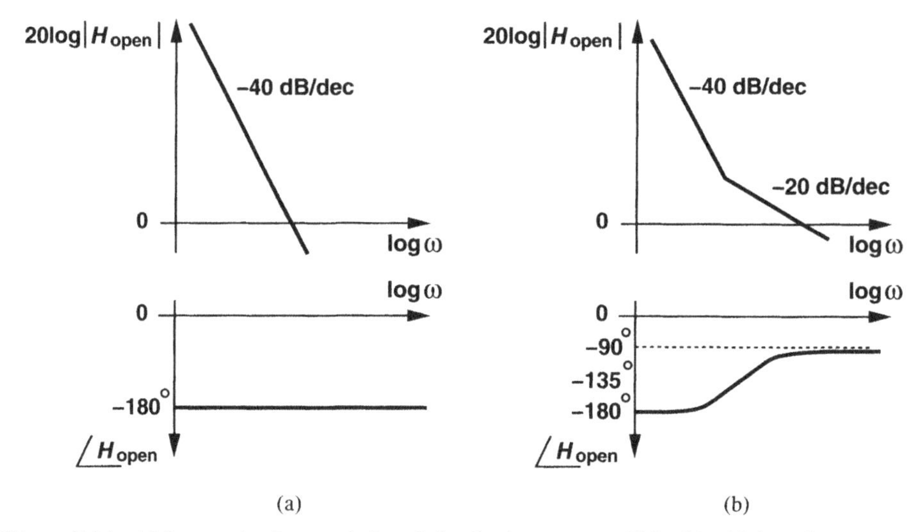

Figure 8.34 (a) Loop gain characteristics of simple charge-pump PLL, (b) addition of zero.

oscillate at the gain crossover frequency.

To stabilize the system, we must modify the phase characteristic such that the phase shift is less than 180° at the gain crossover. As shown in Fig. 8.34(b), this is accomplished by introducing a zero in the loop gain, i.e., by adding a resistor in series with the loop filter capacitor (Fig. 8.35). Using the result expressed by Eq. (8.39), the reader can prove that the PFD/CP/LPF now has a transfer function

$$\frac{V_{out}}{\Delta\phi}(s) = \frac{I_P}{2\pi}(R_P + \frac{1}{C_P s}). \tag{8.42}$$

It follows that the PLL open-loop transfer function is equal to

$$\frac{\Phi_{out}}{\Phi_{in}}(s)|_{\text{open}} = \frac{I_P}{2\pi}(R_P + \frac{1}{C_P s})\frac{K_{VCO}}{s}, \tag{8.43}$$

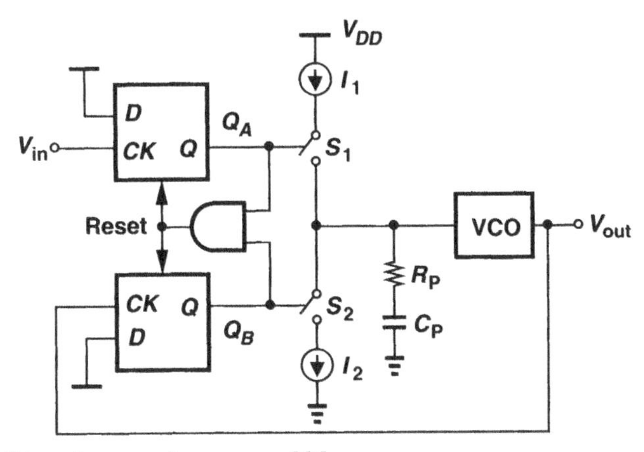

Figure 8.35 Addition of zero to charge-pump PLL.

and hence

$$H(s) = \frac{\frac{I_P K_{VCO}}{2\pi C_P}(R_P C_P s + 1)}{s^2 + \frac{I_P}{2\pi} K_{VCO} R_P s + \frac{I_P}{2\pi C_P} K_{VCO}}. \tag{8.44}$$

The closed-loop system contains a zero at $s_z = -1/(R_P C_P)$. Using the same notation as that for the type I PLL, we have

$$\omega_n = \sqrt{\frac{I_P K_{VCO}}{2\pi C_P}} \tag{8.45}$$

$$\zeta = \frac{R_P}{2}\sqrt{\frac{I_P C_P K_{VCO}}{2\pi}}. \tag{8.46}$$

As expected, if $R_P = 0$, then $\zeta = 0$. With complex poles, the decay time constant is given by $1/(\zeta\omega_n) = 4\pi/(R_P I_P K_{VCO})$.

Stability Issues The stability behavior of type II PLLs is quite different from that of type I PLLs. We begin the analysis with the Bode plots of the loop gain [Eq. (8.43)]. Shown in Fig. 8.36, these plots suggest that if $I_P K_{VCO}$ decreases, the gain crossover frequency moves toward the origin, *degrading* the phase margin. Predicted by (8.46), this trend is in sharp contrast to that expressed by (8.18) and illustrated in Fig. 8.19.

It is also possible to construct the root locus of the closed-loop system in the complex plane. For $I_P K_{VCO} = 0$ (e.g., $I_P = 0$), the loop is open and both poles lie at the origin. For $I_P K_{VCO} > 0$, we have $s_{1,2} = -\zeta\omega_n \pm \omega_n\sqrt{\zeta^2 - 1}$, and, since $\zeta \propto \sqrt{I_P K_{VCO}}$, the poles are complex if $I_P K_{VCO}$ is small. The reader can prove that as $I_P K_{VCO}$ increases, s_1 and s_2 move on a circle centered at $\sigma = -1/(R_P C_P)$ with a radius $1/(R_P C_P)$ (Fig. 8.37). The poles return to the real axis at $\zeta = 1$, assuming a value of $-2/(R_P C_P)$. For $\zeta > 1$, the poles remain real, one approaching $-1/(R_P C_P)$ and the other going to $-\infty$ as

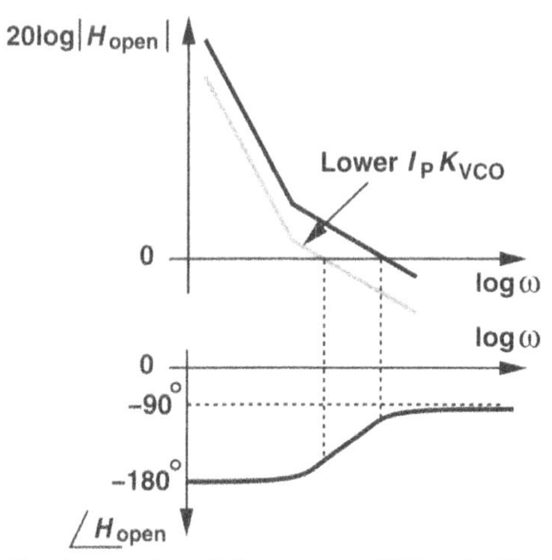

Figure 8.36 Stability degradation of charge-pump PLL as $I_P K_{VCO}$ decreases.

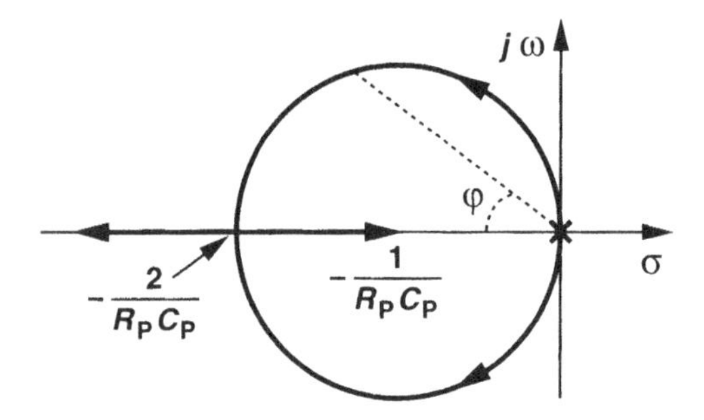

Figure 8.37 Root locus of type II PLL.

$I_P K_{VCO} \to +\infty$. Since for complex s_1 and s_2, $\zeta = \cos\psi$, we observe that as $I_P K_{VCO}$ exceeds zero, the system becomes more stable.

The compensated type II PLL of Fig. 8.35 suffers from a critical drawback. Since the charge pump drives the series combination of R_P and C_P, each time a current is injected into the loop filter, the control voltage experiences a large jump. Even in the locked condition, the mismatches between I_1 and I_2 and the charge injection and clock feedthrough of S_1 and S_2 introduce voltage jumps in V_{cont}. The resulting ripple severely disturbs the VCO, corrupting the output phase. To relax this issue, a second capacitor is usually added in parallel with R_P and C_P (Fig. 8.38), suppressing the initial step. The loop filter now is of *second* order, yielding a third-order PLL and creating stability difficulties [4]. Nonetheless, if C_2 is about one-fifth to one-tenth of C_P, the closed-loop time and frequency responses remain relatively unchanged.

Equation (8.46) implies that the loop becomes more stable as R_P increases. In reality, as R_P becomes very large, the stability degrades again. This effect is not predicted by the foregoing derivations because we have approximated the discrete-time system by a continuous-time loop. A more accurate analysis is given in [2], but simulations are often

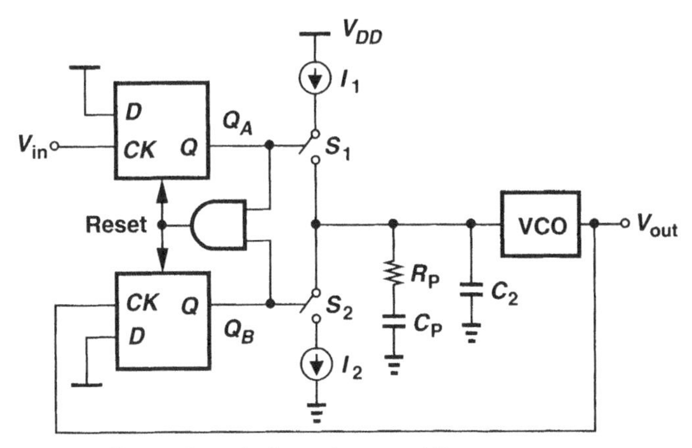

Figure 8.38 Addition of C_2 to reduce ripple on the control line.

necessary to determine the stability bounds of CPPLLs.

8.3 Nonideal Effects in PLLs

8.3.1 PFD/CP Nonidealities

Several imperfections in the PFD/CP circuit lead to high ripple on the control voltage even when the loop is locked. As mentioned earlier, the ripple modulates the VCO frequency, producing a waveform that is no longer periodic. In this section, we study these nonidealities.

The PFD implementation of Fig. 8.24(a) generates narrow, coincident pulses on both Q_A and Q_B even when the input phase difference is zero. As illustrated in Fig. 8.39, if A and B rise simultaneously, so do Q_A and Q_B, thereby activating the reset. That is, even

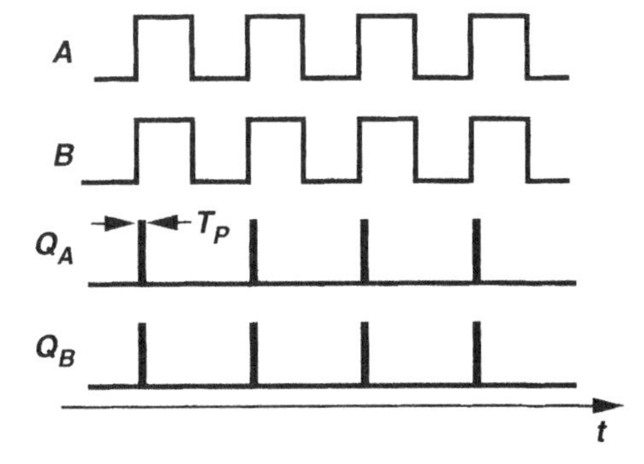

Figure 8.39 Coincident pulses generated by PFD with zero phase difference.

when the PLL is locked, Q_A and Q_B simultaneously turn on the charge pump for a finite period $T_P \approx 5T_D$, where T_D denotes the gate delay.

What are the consequences of the reset pulses on Q_A and Q_B? To understand why these pulses are *desirable*, we consider a hypothetical PFD that produces no pulses for a zero input phase difference [Fig. 8.40(a)]. How does such a PFD respond to a small phase error? As shown in Fig. 8.40(b), the circuit generates very narrow pulses on Q_A or Q_B.

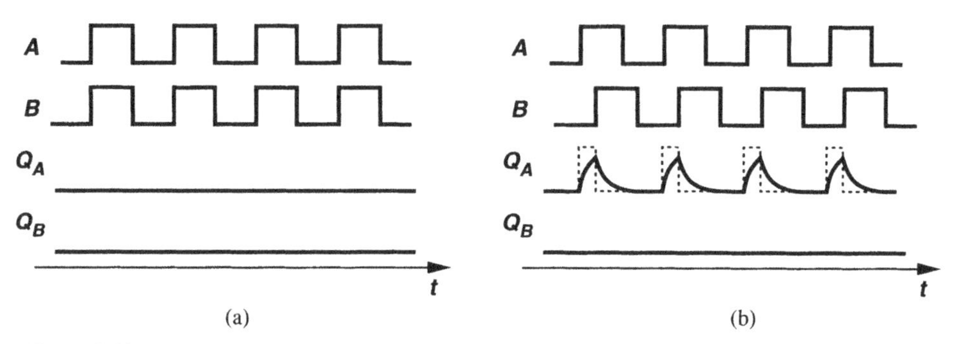

Figure 8.40 Output waveforms of a hypothetical PD with (a) zero input phase difference, and (b) a small input phase difference.

However, owing to the finite risetime and falltime resulting from the capacitance seen at these nodes, the pulse may not find enough time to reach a logical high level, failing to turn on the charge pump switches. In other words, if the input phase difference, $\Delta\phi$, falls below a certain value, ϕ_0, then the output voltage of the PFD/CP/LPF combination is no longer a function of $\Delta\phi$. Since, as depicted in Fig. 8.41, for $|\Delta\phi| < \phi_0$ the charge pump injects no current, Eq. (8.40) implies that the loop gain drops to zero and the output phase is not

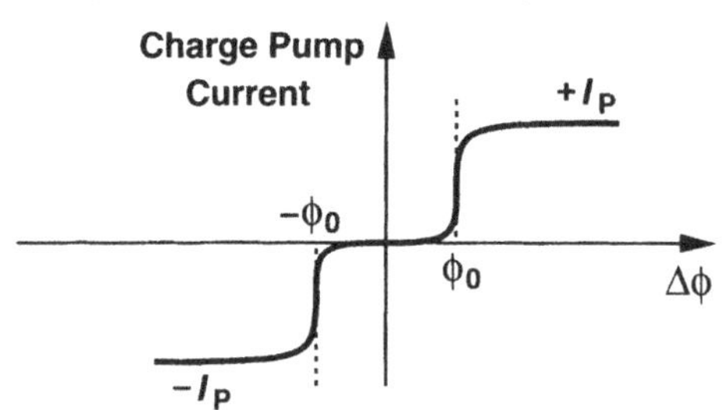

Figure 8.41 Dead zone in the charge pump current.

locked. We say the PFD/CP circuit suffers from a dead zone equal to $\pm\phi_0$ around $\Delta\phi = 0$.

The dead zone is highly undesirable because it allows the VCO to accumulate as much random phase error as ϕ_0 with respect to the input while receiving no corrective feedback. Thus, as illustrated in Fig. 8.42, the zero-crossing points of the VCO output experience substantial random variations, an effect called "jitter."

Interestingly, the coincident pulses on Q_A and Q_B can eliminate the dead zone. This is because, for $\Delta\phi = 0$, the pulses always turn on the charge pump if they are sufficiently wide. Consequently, as shown in Fig. 8.43, an infinitesimal increment in the phase difference results in a proportional increase in the net current produced by the charge pump. In other words, the dead zone vanishes if T_P is long enough to allow Q_A and Q_B to reach a valid logical level and turn on the switches in the charge pump.

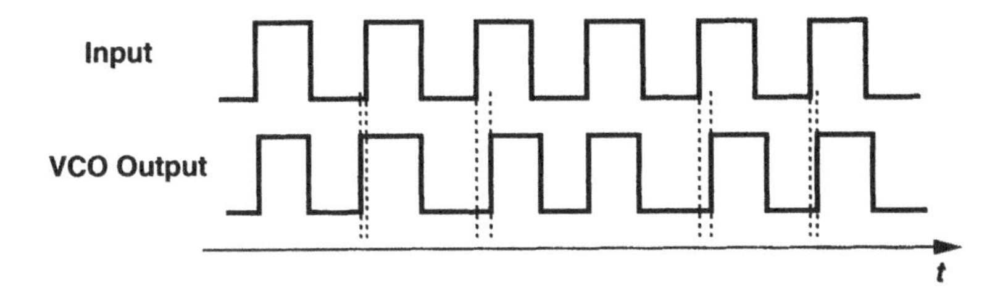

Figure 8.42 Jitter resulting from the dead zone.

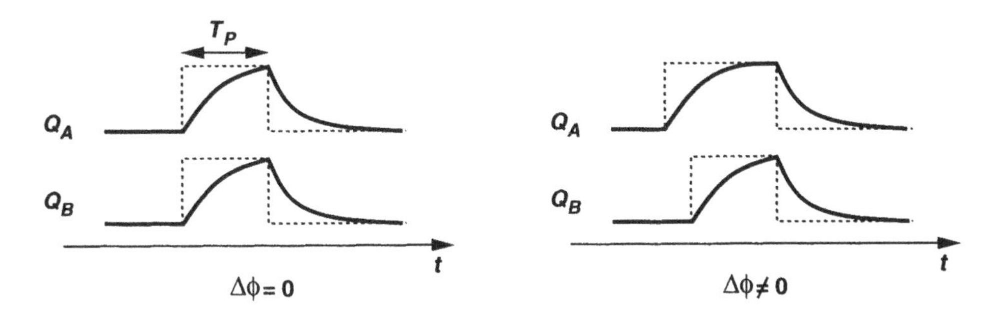

Figure 8.43 Response of actual PD to a small input phase difference.

While eliminating the dead zone, the reset pulses on Q_A and Q_B introduce other difficulties. Let us first implement the charge pump using MOS transistors [Fig. 8.44(a)]. Here, M_1 and M_2 operate as current sources and M_3 and M_4 as switches. The output Q_A is inverted so that when it goes high, M_4 turns on.

The first issue in the circuit of Fig. 8.44(a) stems from the delay difference between $\overline{Q_A}$ and Q_B in turning on their respective switches. As shown in Fig. 8.44(b), the net current injected by the charge pump into the loop filter jumps to $+I_P$ and $-I_P$, disturbing the oscillator control voltage periodically even if the loop is locked. To suppress this effect, a complementary pass gate can be interposed between Q_B and the gate of M_3, equalizing the delays [Fig. 8.44(c)].

The second issue in the CP of Fig. 8.44(c) relates to the mismatch between the drain currents of M_1 and M_2. As depicted in Fig. 8.45(a), even with perfect alignment of the UP and DOWN pulses, the net current produced by the charge pump is nonzero, changing V_{cont} by a constant increment at each phase comparison instant. How does the PLL respond to this error? For the loop to remain locked, the average value of the control voltage must remain constant. The PLL therefore creates a phase error between the input and the output such that the net current injected by the CP in every cycle is zero [Fig. 8.45(b)]. It is important to note that (1) the control voltage still experiences a periodic ripple, (2) owing to the low output impedance of short-channel MOSFETs, the current mismatch *varies* with the output voltage (i.e., with the VCO frequency), and (3) the clock feedthrough and charge injection mismatch between M_3 and M_4 further increases both the phase error and the ripple.

The third issue in the circuit of Fig. 8.44(c) originates from the finite capacitance seen

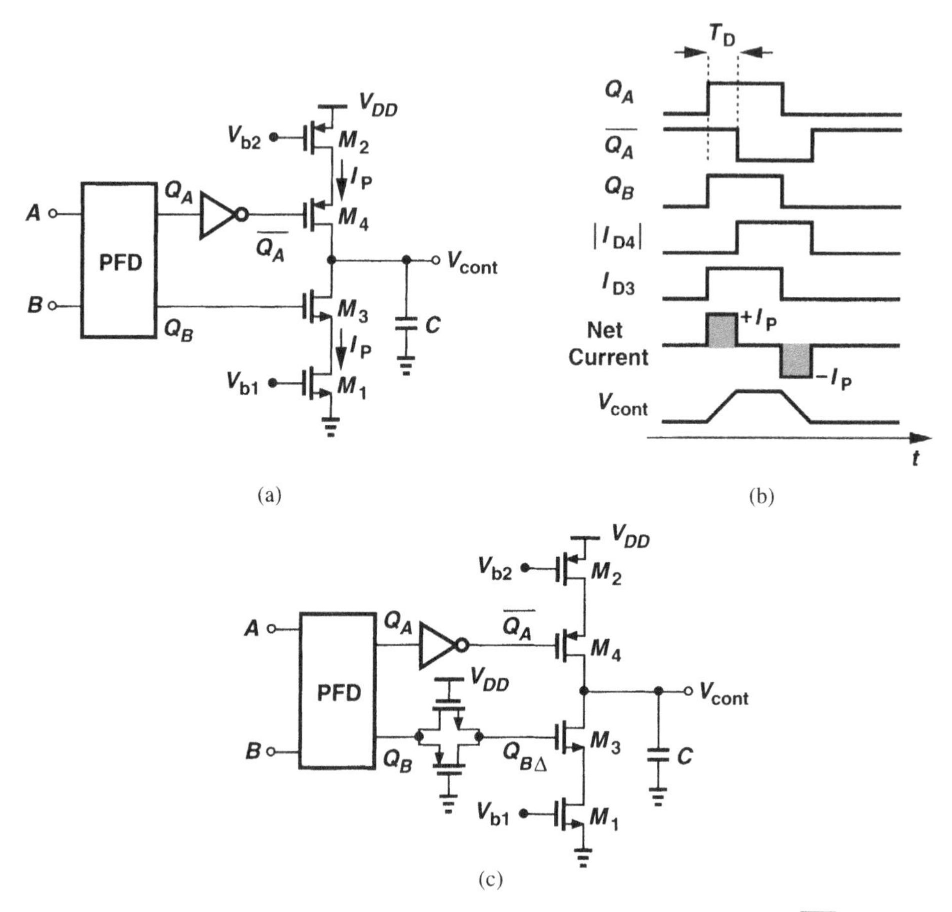

Figure 8.44 (a) Implementation of charge pump, (b) effect of skew between $\overline{Q_A}$ and Q_B, (c) suppression of skew by a pass gate.

at the drains of the current sources. Suppose, as illustrated in Fig. 8.46(a), S_1 and S_2 are off, allowing M_1 to discharge X to ground and M_2 to charge Y to V_{DD}. In the next phase comparison instance, both S_1 and S_2 turn on, V_X rises, V_Y falls, and $V_X \approx V_Y \approx V_{cont}$ if the voltage drop across S_1 and S_2 is neglected [Fig. 8.46(b)]. If the phase error is zero and $I_{D1} = |I_{D2}|$, does V_{cont} remain constant after the switches turn on? Even if $C_X = C_Y$, the change in V_X is not equal to that in V_Y. For example, if V_{cont} is relatively high, V_X changes by a large amount and V_Y by a small amount. The difference between the two changes must therefore be supplied by C_P, leading to a jump in V_{cont}.

The above charge sharing phenomenon can be suppressed by "bootstrapping." Illustrated in Fig. 8.47 [3], the idea is to "pin" V_X and V_Y to V_{cont} after phase comparison is finished. When S_1 and S_2 turn off, S_3 and S_4 turn on, allowing the unity-gain amplifier to hold nodes X and Y at a potential equal to V_{cont}. Note that the amplifier need not provide much current because $I_1 \approx I_2$. At the next phase comparison instant, S_1 and S_2 turn on, S_3 and S_4 turn off, and V_X and V_Y begin with a value equal to V_{cont}. Thus, no charge

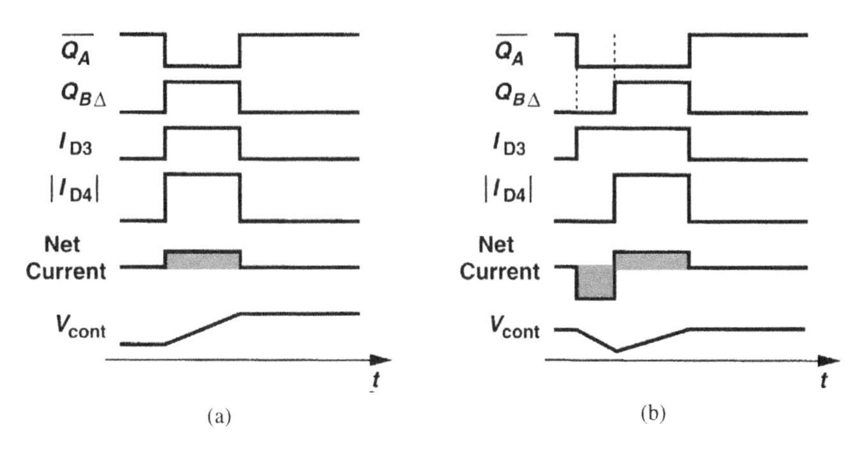

Figure 8.45 Effect of UP and DOWN (a) current mismatch, and (b) pulsewidth mismatch.

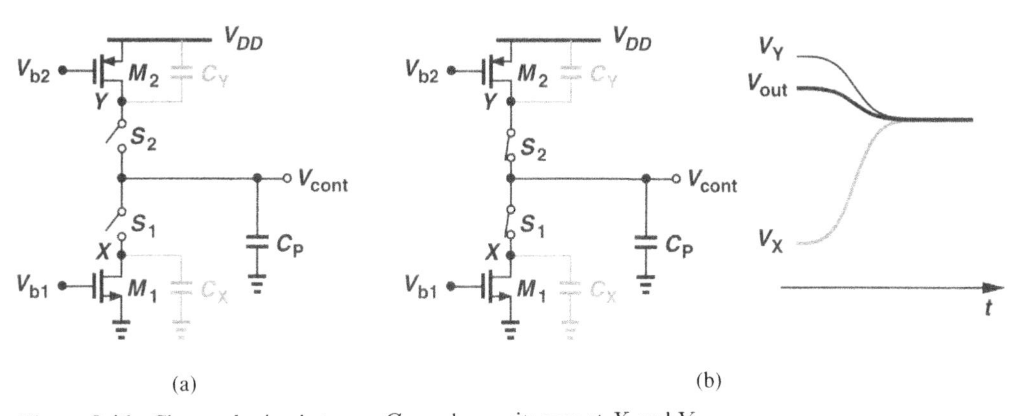

Figure 8.46 Charge sharing between C_P and capacitances at X and Y.

sharing occurs between C_P and the capacitances at X and Y.

8.3.2 Jitter in PLLs

The response of phase-locked loops to jitter is of extreme importance in most applications. We first describe the concepts of jitter and the rate of change of jitter.

As shown in Fig. 8.48, a strictly periodic waveform, $x_1(t)$, contains zero crossings that are evenly spaced in time. Now consider the nearly periodic signal $x_2(t)$, whose period experiences small changes, deviating the zero crossings from their ideal points. We say the latter waveform suffers from jitter.[11] Plotting the total phase, ϕ_{tot}, and the excess phase, ϕ_{ex}, of the two waveforms, we observe that jitter manifests itself as variation of the excess phase with time. In fact, ignoring the harmonics above the fundamental, we can write $x_1(t) = A\cos\omega t$ and $x_2(t) = A\cos[\omega t + \phi_n(t)]$, where $\phi_n(t)$ models the variation of the

[11] Jitter is quantified by several different mathematical definitions, e.g., as in [5].

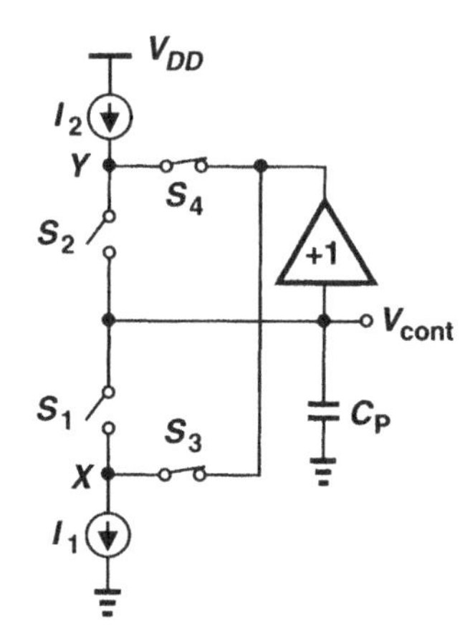

Figure 8.47 Bootstrapping X and Y to minimize charge sharing.

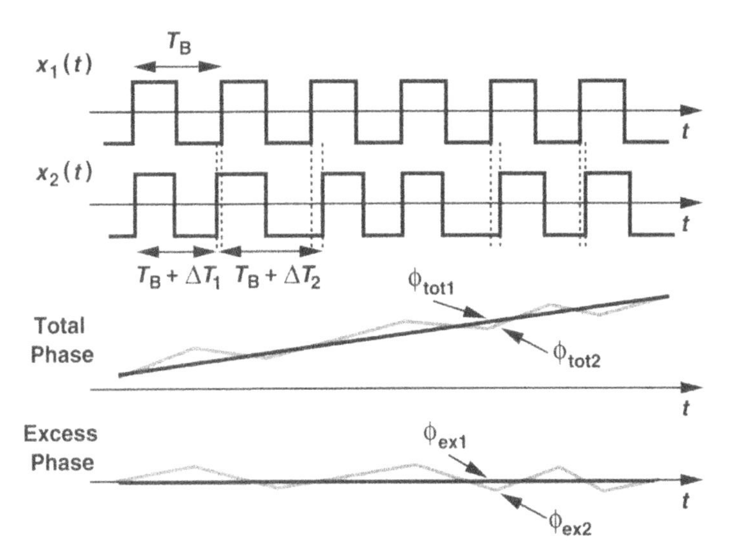

Figure 8.48 Ideal and jittery waveforms.

period.[12]

The rate at which the jitter varies is also important. Consider the two jittery waveforms depicted in Fig. 8.49. The first signal, $y_1(t)$, experiences "slow jitter" because its instantaneous frequency varies slowly from one period to the next. The second signal, $y_2(t)$, experiences "fast jitter." The rate of change is also evident from the excess phase plots of the two waveforms.

[12]The quantity $\phi_n(t)$ (or more commonly its spectrum) is called the "phase noise." In this book, we assume the jitter is uniquely represented by $\phi_n(t)$.

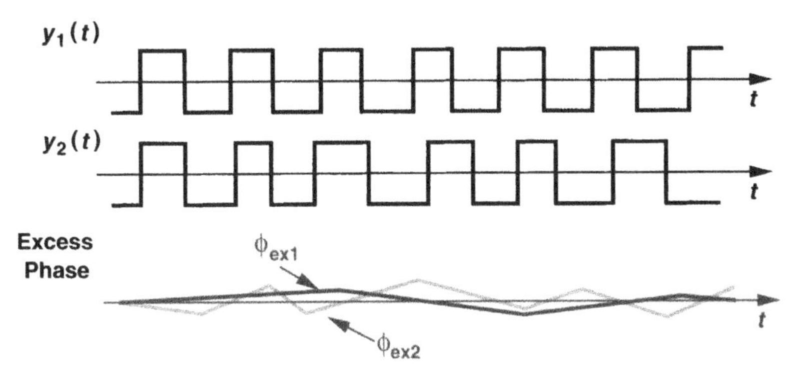

Figure 8.49 Illustration of slow and fast jitter.

Two jitter phenomena in phase-locked loops are of great interest: (a) the input exhibits jitter, and (b) the VCO produces jitter. Let us study each case, assuming the input and output waveforms are expressed as $x_{in}(t) = A\cos[\omega t + \phi_{in}(t)]$ and $x_{out}(t) = A\cos[\omega t + \phi_{out}(t)]$.

The transfer functions derived for type I and type II PLLs have a low-pass characteristic, suggesting that if $\phi_{in}(t)$ varies rapidly, then $\phi_{out}(t)$ does not fully track the variations. In other words, slow jitter at the input propagates to the output unattenuated, but fast jitter does not. We say the PLL low-pass filters $\phi_{in}(t)$.

Now suppose the input is strictly periodic but the VCO suffers from jitter. Viewing jitter as random phase variations, we construct the model depicted in Fig. 8.50, where the input excess phase is set to zero [i.e., $x_{in}(t) = A\cos\omega t$] and a random component Φ_{VCO} is

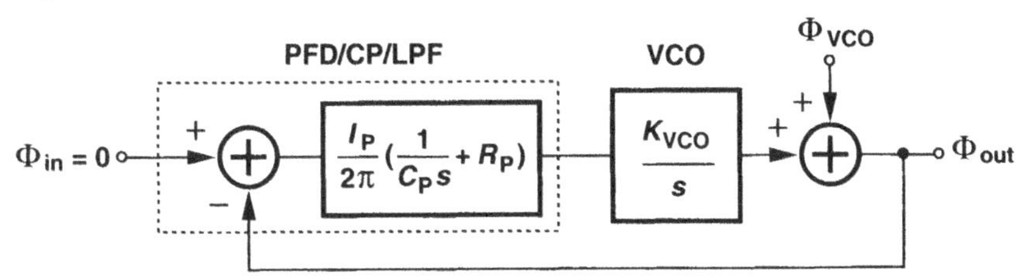

Figure 8.50 Effect of VCO jitter.

added to the output of the VCO to represent its jitter. The reader can show that the transfer function from Φ_{VCO} to Φ_{out} for a type II PLL is equal to

$$\frac{\Phi_{out}}{\Phi_{VCO}}(s) = \frac{s^2}{s^2 + 2\zeta\omega_n s + \omega_n^2}. \tag{8.47}$$

Interestingly, the characteristic has a high-pass nature, indicating that slow jitter components generated by the VCO are suppressed but fast jitter components are not. This can be understood with the aid of Fig. 8.50: if $\phi_{VCO}(t)$ changes slowly (e.g., the oscillation period drifts with temperature), then the comparison with $\phi_{in} = 0$ (i.e., a perfectly periodic signal) generates a slowly-varying error that propagates through the LPF and adjusts the

VCO frequency, thereby counteracting the change in ϕ_{VCO}. On the other hand, if ϕ_{VCO} varies rapidly (e.g., high-frequency noise modulates the oscillation period), then the error produced by the phase detector is heavily attenuated by the poles in the loop, failing to correct for the change.

Figure 8.51 conceptually summarizes the response of PLLs to input jitter and VCO jitter. Depending on the application and the environment, one or both sources may be significant, requiring an optimum choice of the loop bandwidth.

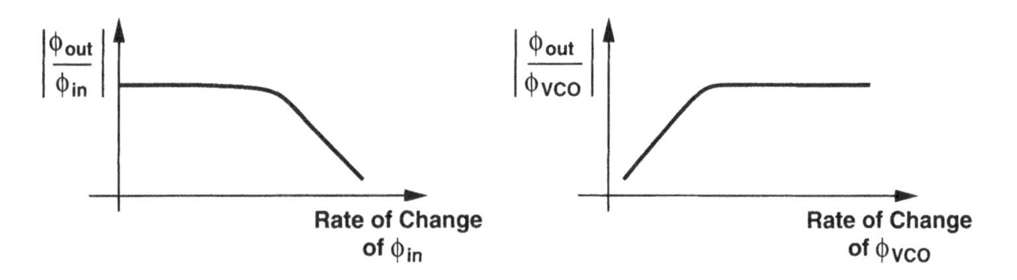

Figure 8.51 Transfer functions of jitter from input and VCO to the output.

8.4 Delay-Locked Loops

A variant of PLLs that has become popular in the past 10 years is the delay-locked loop. To arrive at the concept, let us begin with an example. Suppose an application requires four clock phases with a precise spacing of $\Delta T = 1$ ns between consecutive edges [Fig. 8.52(a)]. How should these phases be generated? We can use a two-stage differential ring

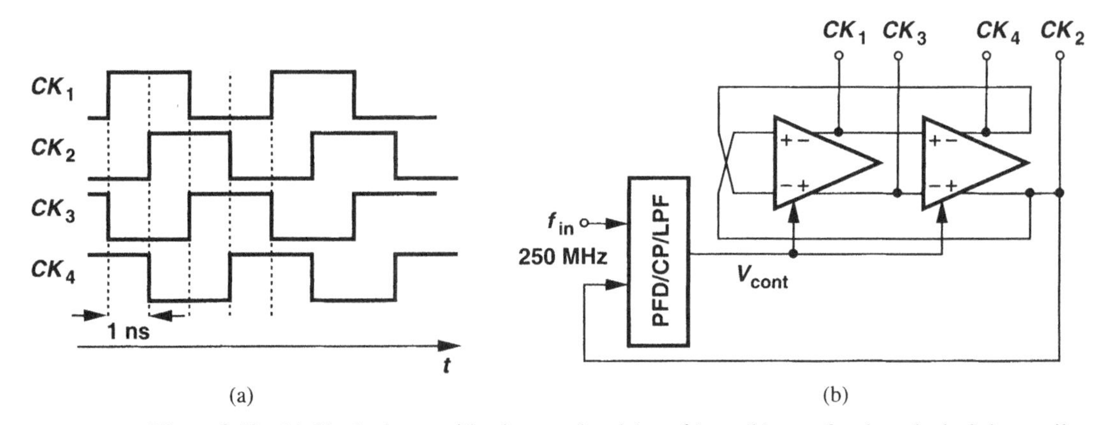

Figure 8.52 (a) Clock phases with edge-to-edge delay of 1 ns, (b) use of a phase-locked ring oscillator to generate the clock phases.

oscillator[13] to produce the four phases, but how do we guarantee that $\Delta T = 1$ ns despite process and temperature variations? This requires that the oscillator be locked to a 250-MHz reference so that the output period is exactly equal to 4 ns [Fig. 8.52(b)].

An alternative approach to generating the clock phases of Fig. 8.52(a) is to apply the input clock to four delay stages in a cascade. Illustrated in Fig. 8.53(a), this technique nonetheless does not produce a well-defined edge spacing because the delay of each stage

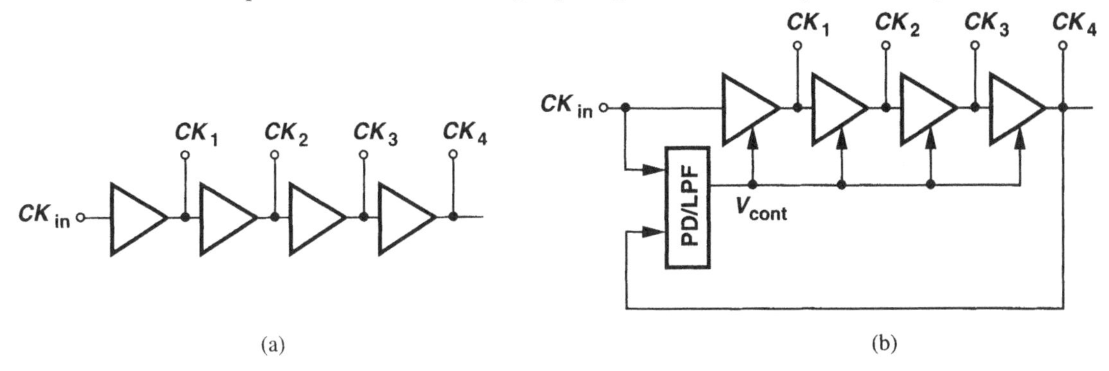

Figure 8.53 (a) Generation of clock edges by delay stages, (b) simple delay-locked loop.

varies with process and temperature. Now consider the circuit shown in Fig. 8.53(b), where the phase difference between CK_{in} and CK_4 is sensed by a phase detector, a proportional average voltage, V_{cont}, is generated, and the delay of the stages is adjusted with negative feedback. For a large loop gain, the phase difference between CK_{in} and CK_4 is small; that is, the four stages delay the clock by almost exactly one period, thereby establishing precise edge spacing.[14] This topology is called a "delay-locked loop" to emphasize that it incorporates a voltage-controlled delay line (VCDL) rather than a VCO. In practice, a charge pump is interposed between the PD and the LPF to achieve an infinite loop gain. Each delay stage may be based on one of the ring oscillator stages described in Chapter 6.

The reader may wonder about the advantages of DLLs over PLLs. First, delay lines are generally less susceptible to noise than oscillators are because corrupted zero crossings of a waveform disappear at the end of a delay line whereas they are recirculated in an oscillator, thereby experiencing more corruption. Second, in the VCDL of Fig. 8.53(b), a change in the control voltage immediately changes the delay; that is, the transfer function $\Phi_{out}(s)/V_{cont}(s)$ is simply equal to the gain of the VCDL, K_{VCDL}. Thus, the feedback system of Fig. 8.53(b) has the same order as the LPF, and its stability and settling issues are more relaxed than those of a PLL.

Let us determine the closed-loop transfer function of the DLL shown in Fig. 8.54. We write the transfer function of the PD/CP/LPF combination as

$$\frac{V_{cont}}{\Delta\Phi}(s) = \frac{I_P}{2\pi}[(R_P + \frac{1}{C_P s})||\frac{1}{C_2 s}] \tag{8.48}$$

[13]As explained in Chapter 6, a simple two-stage CMOS ring oscillator may not oscillate. This example is merely for illustration purposes.

[14]The total delay through the four stages may be equal to two or more periods. We return to this issue later.

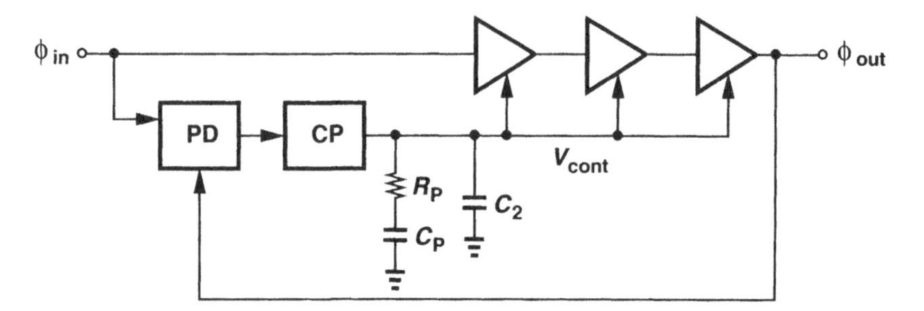

Figure 8.54 Simple DLL.

$$= \frac{I_P}{2\pi} \frac{R_P C_P s + 1}{(R_P C_P C_2 s + C_P + C_2)s}. \tag{8.49}$$

The closed-loop transfer function is thus equal to

$$\frac{\Phi_{out}}{\Phi_{in}}(s)|_{\text{closed}} = \frac{\frac{I_P K_{VCDL}}{2\pi}(R_P C_P s + 1)}{R_P C_P C_2 s^2 + [C_P + C_2 + I_P K_{VCDL} R_P C_P/(2\pi)]s + I_P K_{VCDL}/(2\pi)}. \tag{8.50}$$

This transfer function can be used to determine how ϕ_{out} settles if ϕ_{in} experiences a change. Note that in practice R_P may not be needed because the loop contains only one pole at the origin.

The principal drawback of DLLs is that they cannot generate a variable output frequency. This issue becomes clearer when we study the frequency synthesis capabilities of PLLs in Section 8.5.1. DLLs may also suffer from locked delay ambiguity. That is, if the total delay of the four stages in Fig. 8.53(b) can vary from below T_{in} to above $2T_{in}$, then the loop may lock with a CK_{in}-to-CK_4 delay equal to either T_{in} or $2T_{in}$. This ambiguity proves detrimental if the DLL must provide precisely-spaced clock edges because the edge-to-edge delay may settle to $2T_{in}/4$ rather than $T_{in}/4$. In such cases, additional circuitry is necessary to avoid the ambiguity. Also, mismatches between the delay stages and their load capacitances introduce error in the edge spacing, requiring large devices and careful layout.

8.5 Applications

After nearly 70 years since its invention, phase locking continues to find new applications in electronics, communication, and instrumentation. Examples include memories, microprocessors, hard disk drive electronics, RF and wireless transceivers, and optical fiber receivers.

The reader may recall from Section 8.1.2 that a PLL appears no more useful than a short piece of wire because both guarantee a small phase difference between the input and the output. In this section, we present a number of applications that demonstrate the versatility of phase locking. The concepts described below have been the topic of numerous books and papers, e.g., [6, 7].

8.5.1 Frequency Multiplication and Synthesis

Frequency Multiplication A PLL can be modified such that it multiplies its input frequency by a factor of M. To arrive at the implementation, we exploit an analogy with voltage multiplication. As depicted in Fig. 8.55(a), a feedback system amplifies the input

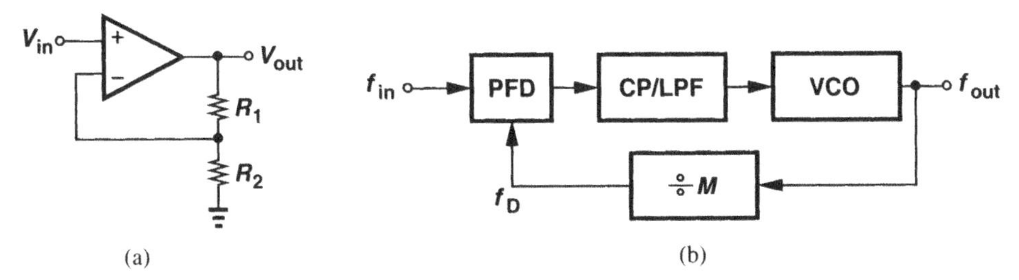

Figure 8.55 (a) Voltage amplification and (b) frequency multiplication.

voltage by a factor of M if the output voltage is divided by M [i.e., if $R_2/(R_1+R_2)=1/M$] and the result is compared with the input. Thus, as shown in Fig. 8.55(b), if the output *frequency* of a PLL is divided by M and applied to the phase detector, we have $f_{out}=Mf_{in}$. From another point of view, since $f_D=f_{out}/M$ and f_D and f_{in} must be equal in the locked condition, the PLL multiplies f_{in} by M. The $\div M$ circuit is realized as a counter that produces one output pulse for every M input pulses.

As with voltage division in Fig. 8.55(a), the feedback divider in the loop of Fig. 8.55(b) alters the system characteristics. Using (8.43), we rewrite (8.44) as

$$H(s)=\frac{\frac{I_P}{2\pi}(R_P+\frac{1}{C_Ps})\frac{K_{VCO}}{s}}{1+\frac{1}{M}\frac{I_P}{2\pi}(R_P+\frac{1}{C_Ps})\frac{K_{VCO}}{s}} \tag{8.51}$$

$$=\frac{\frac{I_PK_{VCO}}{2\pi C_P}(R_PC_Ps+1)}{s^2+\frac{I_P}{2\pi}\frac{K_{VCO}}{M}R_Ps+\frac{I_P}{2\pi C_P}\frac{K_{VCO}}{M}}. \tag{8.52}$$

Note that $H(s)\rightarrow M$ as $s\rightarrow 0$, i.e., phase or frequency changes at the input result in an M-fold change in the corresponding output quantity. Comparing the denominators of (8.44) and (8.52), we observe that frequency division in the loop manifests itself as division of K_{VCO} by M. In other words, as far as the poles of the closed-loop system are concerned, we can assume the oscillator and the divider form a VCO with an equivalent gain of K_{VCO}/M. This is, of course, to be expected because, for the VCO/divider cascade shown in Fig. 8.56, we have

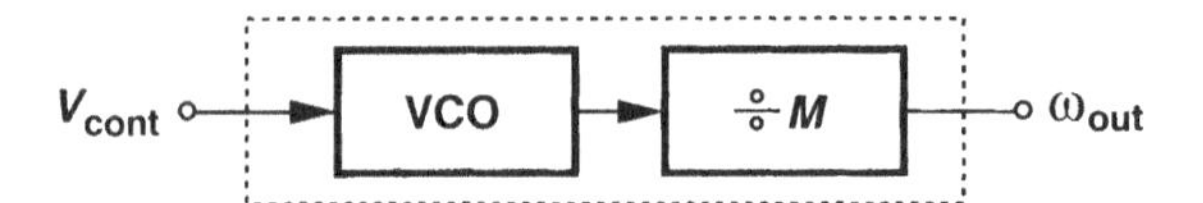

Figure 8.56 Equivalency of VCO/divider combination to a single VCO.

$$\omega_{out} = \frac{\omega_0 + K_{VCO}V_{cont}}{M} \tag{8.53}$$

$$= \frac{\omega_0}{M} + \frac{K_{VCO}}{M}V_{cont}. \tag{8.54}$$

Thus, the combination cannot be distinguished from a VCO having an intercept frequency of ω_0/M and a gain of K_{VCO}/M.

The foregoing discussion suggests that (8.45) and (8.46) can be respectively rewritten as

$$\omega_n = \sqrt{\frac{I_P}{2\pi C_P}\frac{K_{VCO}}{M}} \tag{8.55}$$

$$\zeta = \frac{R_P}{2}\sqrt{\frac{I_P C_P}{2\pi}\frac{K_{VCO}}{M}}. \tag{8.56}$$

Also, the decay time constant is modified to $(\zeta\omega_n)^{-1} = 4\pi M/(R_P I_P K_{VCO})$. It follows that inserting a divider in a type II loop degrades both the stability and the settling speed, requiring a proportional increase in the charge pump current.

The frequency-multiplying loop of Fig. 8.55(b) exhibits two interesting properties. First, unlike the voltage amplifier of Fig. 8.55(a), the PLL provides a multiplication factor *exactly* equal to M, an attribute resulting from the infinite loop gain and expressed by Eq. (8.52). Second, the output frequency can be varied by changing the divide ratio M, an extremely useful property in synthesizing frequencies. Note that DLLs cannot perform such synthesis.

Frequency Synthesis Some systems require a periodic waveform whose frequency (a) must be very accurate (e.g., exhibit an error less than 10 ppm), and (b) can be varied in very fine steps (e.g., in steps of 30 kHz from 900 MHz to 925 MHz). Commonly encountered in wireless transceivers, such requirements can be met through frequency multiplication by PLLs.

Figure 8.57 shows the architecture of a phase-locked frequency synthesizer. The channel control input is a digital word that varies the value of M. Since $f_{out} = M f_{REF}$, the relative accuracy of f_{out} is equal to that of f_{REF}. For this reason, f_{REF} is derived from a stable, low-noise crystal oscillator. Note that f_{out} varies in steps equal to f_{REF} if M changes by one each time.

CMOS frequency synthesizers achieving gigahertz output frequencies have been reported. Issues such as noise, sidebands, settling speed, frequency range, and power dissipation continue to challenge synthesizer designers.

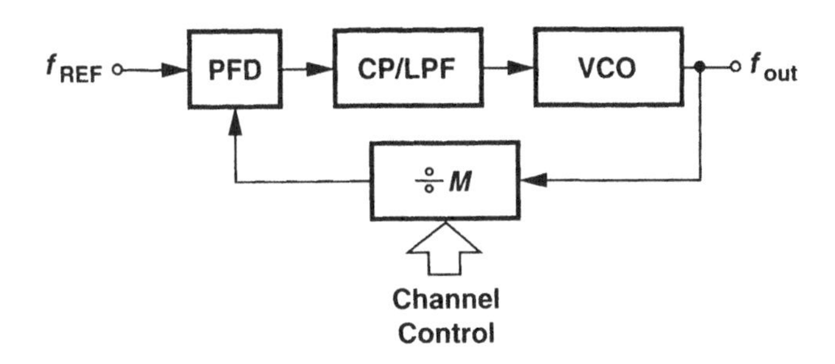

Figure 8.57 Frequency synthesizer.

8.5.2 Skew Reduction

The earliest usage of phase locking in digital systems was for skew reduction. Suppose a synchronous pair of data and clock lines enter a large digital chip as shown in Fig. 8.58. Since the clock typically drives a large number of transistors and long interconnects, it

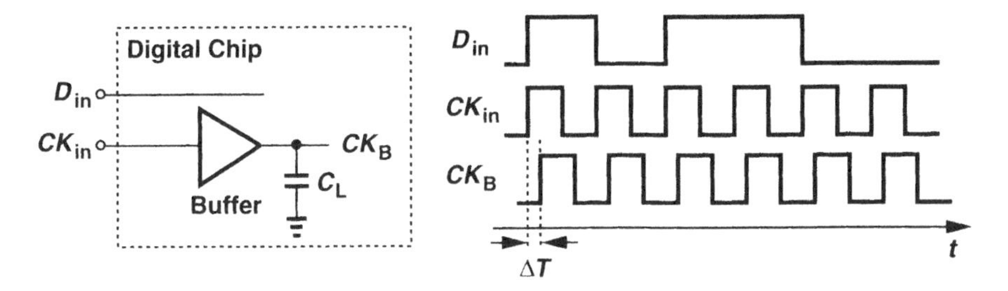

Figure 8.58 Skew between data and buffered clock.

is first applied to a large buffer. Thus, the clock distributed on the chip may suffer from substantial skew with respect to the data, an undesirable effect because it reduces the timing budget for on-chip operations.

Now consider the circuit shown in Fig. 8.59, where CK_{in} is applied to an on-chip PLL and the buffer is placed *inside* the loop. Since the PLL guarantees a nominally-zero phase

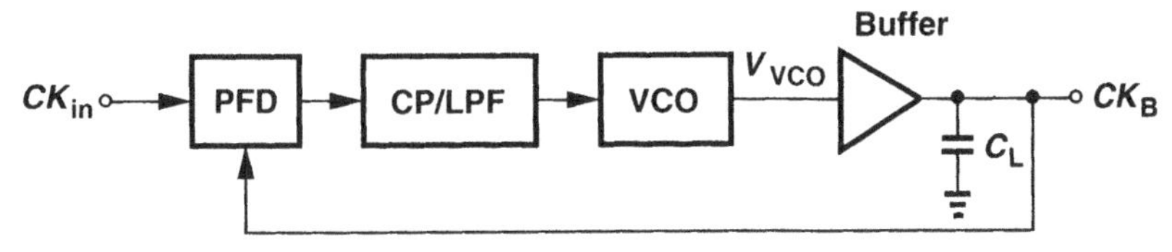

Figure 8.59 Use of a PLL to eliminate skew.

difference between CK_{in} and CK_B, the skew is eliminated. From another point of view, the constant phase shift introduced by the buffer is divided by the infinite loop gain of the feedback system. Note that the VCO output, V_{VCO}, may not be aligned with CK_{in}, a nonetheless unimportant issue because V_{VCO} is not used.

As an example, we construct the voltage-domain counterpart of the loop shown in Fig. 8.59. The buffer creates a constant phase shift in the signal generated by the VCO. The voltage-domain counterpart therefore assumes the topology shown in Fig. 8.60. We have

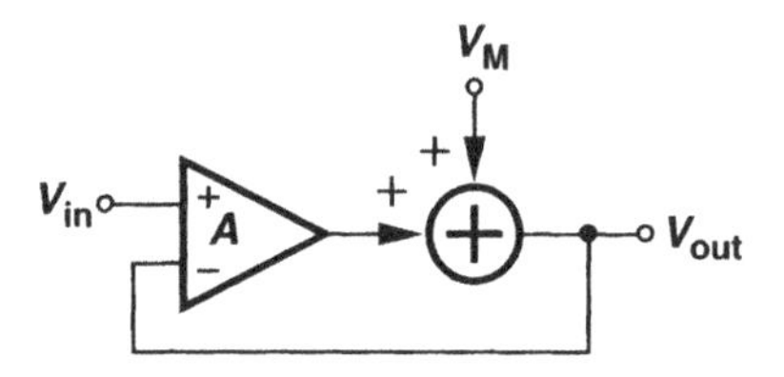

Figure 8.60 Linear model of PLL.

$$(V_{in} - V_{out})A + V_M = V_{out} \tag{8.57}$$

and hence

$$V_{out} = \frac{AV_{in} + V_M}{1 + A}. \tag{8.58}$$

As $A \to \infty$, $V_{out} \to V_{in}$.

We should note that the skew can be suppressed by a delay-locked loop as well. In fact, if frequency multiplication is not required, DLLs are preferred because they are less susceptible to noise.

8.5.3 Jitter Reduction

Recall from Section 8.3.2 that PLLs suppress fast jitter components at the input. For example, if a 1-GHz jittery signal is applied to a PLL having a bandwidth of 10 MHz, then input jitter components that vary faster than 10 MHz are attenuated. In a sense, the phase-locked loop operates as a narrowband filter centered around 1 GHz with a total bandwidth of 20 MHz. This is another important and useful property of PLLs.

Many applications must deal with jittery waveforms. Random binary signals experience jitter because of (a) crosstalk on the chip and in the package, (b) package parasitics, (c) additive electronic noise of devices, etc. Such waveforms are typically "retimed" by a low-noise clock so as to reduce the jitter. Illustrated in Fig. 8.61(a), the idea is to resample the midpoint of each bit by a D flipflop that is driven by the clock. However, in many applications, the clock may not be available independently. For example, an optical fiber carries only the random data stream, providing no separate clock waveform at the receive end. The circuit of Fig. 8.61(a) is therefore modified as shown in Fig. 8.61(b), where a "clock recovery circuit" (CRC) produces the clock from the data. Employing phase locking with a relatively narrow loop bandwidth, the CRC minimizes the effect of the input jitter on the recovered clock. This concept is studied in Chapter 9.

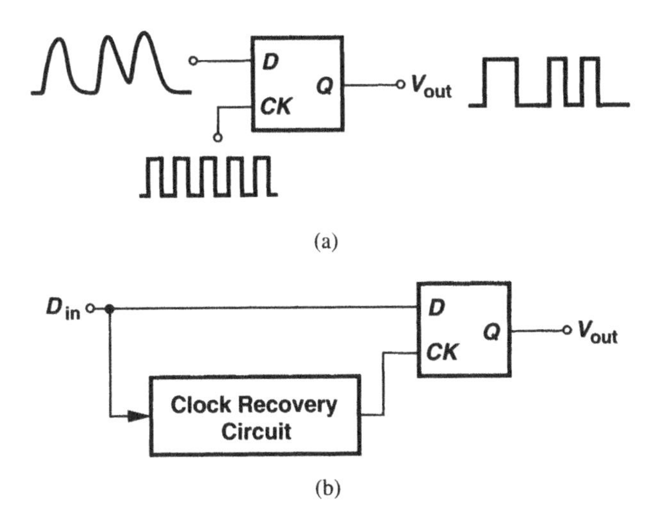

Figure 8.61 (a) Retiming data with D flipflop driven by a low-noise clock, (b) use of a phase-locked clock recovery circuit to generate the clock.

References

1. R. E. Best, *Phase-Locked Loops*, Second Ed., New York: McGraw-Hill, 1993.
2. F. M. Gardner, *Phaselock Techniques*, Second Ed., New York: Wiley & Sons, 1979.
3. M. G. Johnson and E. L. Hudson, "A Variable Delay Line PLL for CPU-Coprocessor Synchronization," *IEEE Journal of Solid-State Circuits*, vol. 23, pp. 1218–1223, Oct. 1988.
4. F. M. Gardner, "Charge-Pump Phase-Locked Loops," *IEEE Trans. Comm.*, vol. COM-28, pp.1849–1858, Nov. 1980.
5. F. Herzel and B. Razavi, "A Study of Oscillator Jitter Due to Supply and Substrate Noise," *IEEE Transactions on Circuits and Systems, Part II*, vol. 46, pp. 56–62, Jan. 1999.
6. W. F. Egan, *Frequency Synthesis by Phase Lock*, New York: Wiley & Sons, 1981.
7. J. A. Crawford, *Frequency Synthesizer Design Handbook*, Artech House, 1994.

CHAPTER 9

Clock and Data Recovery

The data stream received and amplified by an optical receiver is both asynchronous and noisy. For subsequent processing, timing information, e.g., a clock, must be extracted from the data so as to allow synchronous operations. Furthermore, the data must be "retimed" such that the jitter accumulated during transmission is removed. The task of clock extraction and data retiming is called "clock and data recovery"(CDR). CDR circuits must satisfy stringent specifications defined by optical standards, presenting difficult challenges to system and circuit designers.

Following our study of oscillators and PLLs in Chapters 6, 7, and 8, we deal with CDR circuits in this chapter. We begin with a generic example, addressing its shortcomings. We then introduce various phase and frequency detectors suited to random data. Finally, we present a number of CDR architectures and describe CDR jitter characteristics.

9.1 General Considerations

In order to perform synchronous operations such as retiming and demultiplexing on random data, optical receivers must generate a clock. As illustrated in Fig. 9.1, a clock recov-

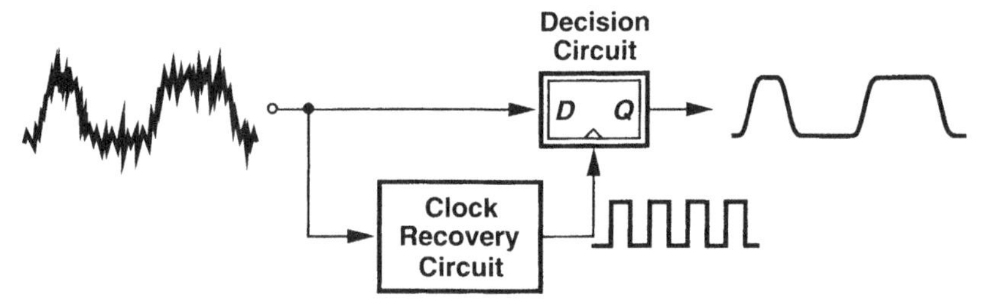

Figure 9.1 Role of a CDR circuit in retiming data.

ery circuit senses the data and produces a periodic clock. A D flipflop driven by the clock then retimes the data; i.e., it samples the noisy data, yielding an output with less jitter. As such, the flipflop is sometimes called a "decision circuit." Note that the zero crossings of the received data are corrupted by noise and jitter whereas those at the flipflop output area as "clean" as the recovered clock itself. This removal of timing errors forms the essence of data retiming.

The clock generated in the circuit of Fig. 9.1 must satisfy three important conditions. (1) It must have a frequency equal to the data rate; e.g., a data rate of 10 Gb/s (each bit 100 ps wide) translates to a clock frequency of 10 GHz (with a period of 100 ps). (2) It must bear a certain phase relationship with respect to data, allowing optimum sampling of the bits by the clock. As exemplified by the waveforms shown in Fig. 9.2, if the rising edges of the clock coincide with the midpoint of each bit, then the sampling occurs farthest from the

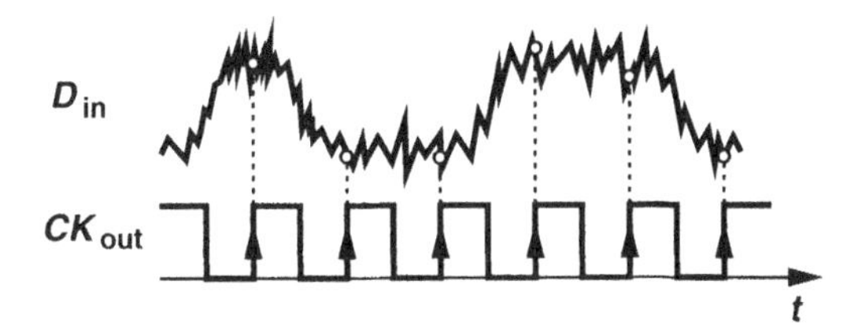

Figure 9.2 Optimum sampling of noisy data.

preceding and following data transitions, providing maximum margin for jitter and other timing uncertainties.[1] (3) It must exhibit a small jitter as it is the principal contributor to the retimed data jitter. We follow the above conditions to develop the foundation for CDR design, beginning with the first.

Edge Detection How can a circuit produce a clock that has a frequency, f_{CK}, equal to the bit rate, R_b? Recall from Chapter 2 that random binary data contains no spectral line at the bit rate, providing no direct information for clock extraction. To create such a spectral line, the data may undergo "edge detection." Illustrated in Fig. 9.3(a) is an attempt at edge detection, where the data waveform is differentiated in the time domain. Unfortunately, however, the positive and negative pulses still yield no spectral line at R_b because differentiation is a linear operation and hence incapable of generating new frequency components. More specifically, since a differentiator has a transfer function $H(s) = s$, we multiply the input power spectral density, $S_{in}(f)$, by $|H(j\omega)|^2 = \omega^2$ to obtain the output spectrum, $S_{out}(f)$:

$$S_{out}(f) = S_{in}(f) \cdot (2\pi f)^2 \tag{9.1}$$

$$= \left[\frac{\sin(\pi f T_b)}{\pi f T_b}\right]^2 (2\pi f)^2 \tag{9.2}$$

$$= \left[\frac{2\sin(\pi f T_b)}{T_b}\right]^2. \tag{9.3}$$

[1]We say the clock has a "phase margin" of nearly 180° because its rising edge can tolerate an offset of almost ±180° before the data is sampled incorrectly. Of course, jitter and other timing errors lower this margin.

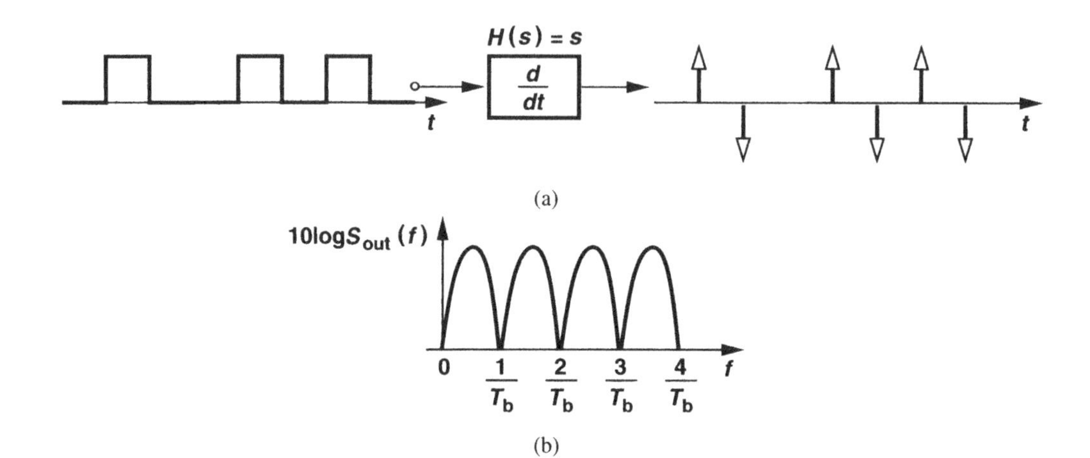

Figure 9.3 (a) Differentiation of data, (b) resulting spectrum.

Interestingly, the spectrum of the differentiated waveform displays a sinusoidal behavior, still falling to zero at $f = n/T_b$ [Fig. 9.3(b)].

This result can also be obtained by the reasoning described in Section 2.1; to determine whether a signal contains a spectral line at a certain frequency, f_0, we multiply the signal by $\cos(2\pi f_0 t + \theta)$ and compute the time average (i.e., we correlate the signal with the cosine). If the average assumes a nonzero value for some θ, then the signal contains such a component. As shown in Fig. 9.4, a sinusoid with period T_b and arbitrary phase is multiplied by two equal and opposite impulses that are T_b seconds apart, yielding a zero

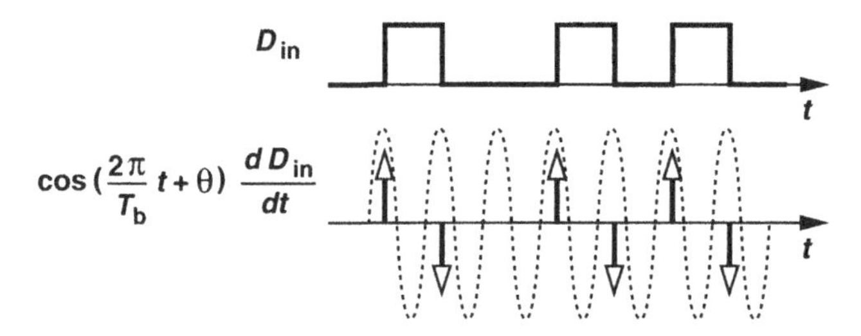

Figure 9.4 Correlation of differentiated data with a sinusoid.

average in each bit.

The foregoing view suggests that, if the negative impulses corresponding to the falling edges of data are converted to positive impulses, then the product of the data waveform and $\cos(2\pi t/T_b + \theta)$ may exhibit a nonzero average. Indeed, as depicted in Fig. 9.5(a), if all of the data transitions are represented by impulses having the same polarity, the correlation is nonzero.

The waveforms of Fig. 9.5(a) reveal that differentiation and *rectification* of a random binary sequence (also called edge detection) generates a frequency component at the bit rate. The result in Fig. 9.5(a) is obtained by "full-wave" rectification, i.e., converting negative

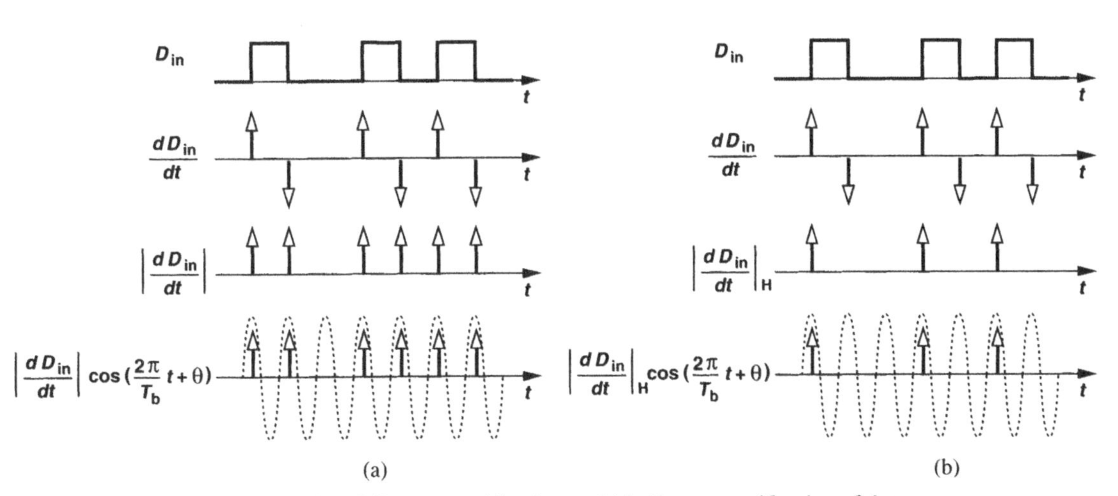

Figure 9.5 Differentiation and (a) full-wave rectification or (b) half-wave rectification of data.

impulses to positive or vice versa. Alternatively, we can perform "half-wave" rectification, whereby only positive (or negative) impulses are retained. Illustrated in Fig. 9.5(b), where $|dD_{in}/dt|_H$ denotes the half-wave rectified waveform, this technique ignores half of the data transitions, but it is used in some cases if data suffers from pulsewidth distortion. The combination of differentiation and rectification is called edge detection. It is also worth noting that edge detection need not generate *impulses*; any stream of narrow unipolar pulses corresponding to the data transitions still yields a nonzero correlation.[2]

The reader may wonder how differentiation and rectification are realized in integrated circuits. Shown in Fig. 9.6(a) is a simple, yet efficient differentiator employing capacitive

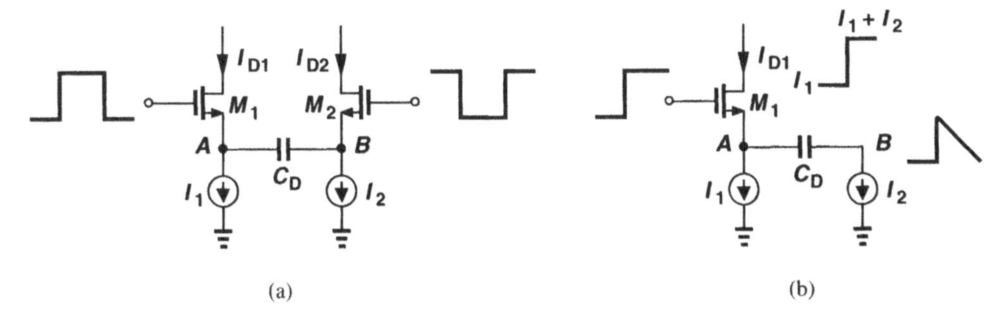

Figure 9.6 (a) Differentiator circuit, (b) pulse shapes for one data edge.

degeneration. Each rising transition at the gate of M_1 momentarily turns M_2 off, allowing both I_1 and I_2 to flow through M_1. The source voltage of M_2, V_B, then falls at a rate of I_2/C_D until M_2 turns on [Fig. 9.6(b)]. Similarly, rising transitions at the gate of M_2

[2]Note that edge detected data is *not* the same as RZ data.

momentarily turn M_1 off. Thus, each data transition at the input creates short pulses of equal and opposite amplitudes in I_{D1} and I_{D2}.

The differentiated signal produced by the circuit of Fig. 9.6(a) can be rectified if it is multiplied by the original random binary sequence. Illustrated in Fig. 9.7(a), the overall circuit resembles a Gilbert cell but with capacitive degeneration added. The level shift stages

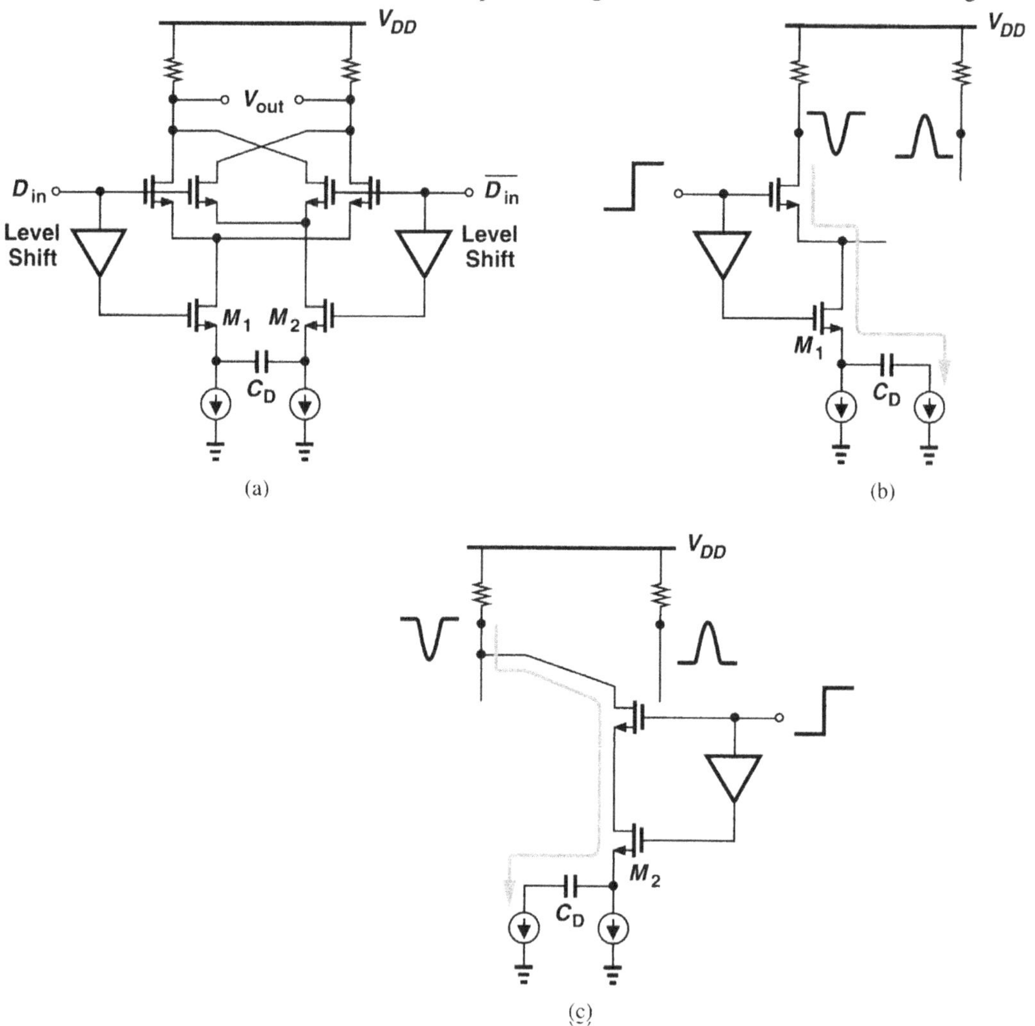

Figure 9.7 (a) Edge detector circuit, (b) simplified circuit for rising data edge, (c) simplified circuit for falling data edge.

(e.g., source followers) ensure that M_1 and M_2 remain in saturation. For each positive or negative data edge at the input, the circuit reduces to that in Fig. 9.7(b) or (c), respectively, generating only unipolar impulses at the output.

The circuit of Fig. 9.7(a) exemplifies analog implementation of edge detectors. This is suited to extremely high speeds, i.e., speeds at which flipflops fail but such a simple realization still provides a frequency component at the bit rate. At lower speeds, digital edge detectors prove more robust and versatile. Figure 9.8(a) depicts an example, where

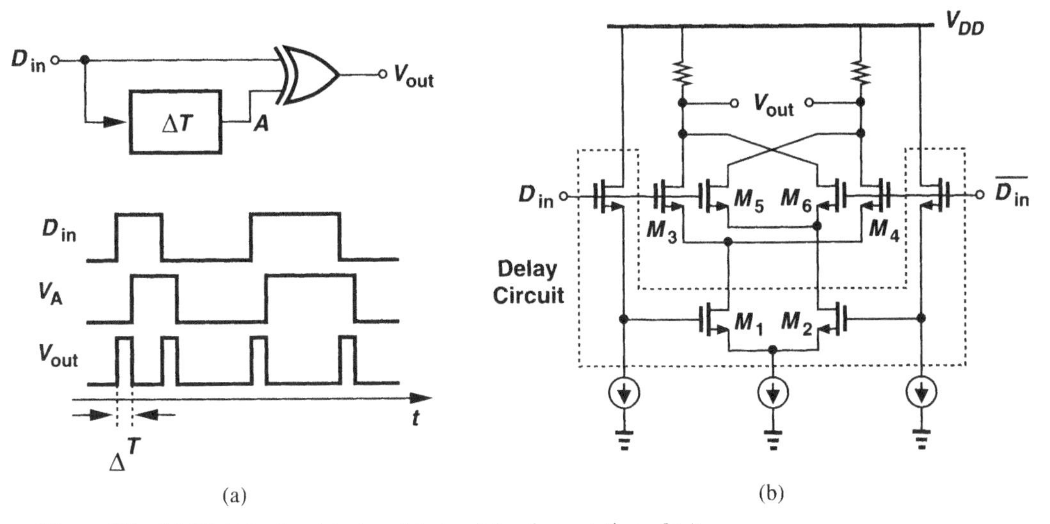

Figure 9.8 (a) Digital edge detector, (b) circuit implementation of (a).

the data sequence and its delayed version are XORed, thereby producing a positive pulse of width ΔT on each data edge. The circuit lends itself to a compact implementation: as shown in Fig. 9.8(b), a differential current-mode XOR (i.e., a Gilbert cell) employing level-shift source followers for the bottom input introduces significant delay in one signal path before XORing the drain currents of M_1 and M_2 with the gate voltages of M_3-M_6.

The choice of the delay, ΔT, in Fig. 9.8(a) becomes a critical issue at high speeds. First, suppose ΔT is not sufficiently large. Then, as illustrated in Fig. 9.9(a), the finite time constants in the circuit prohibit the XOR output from reaching a full level. For example, if

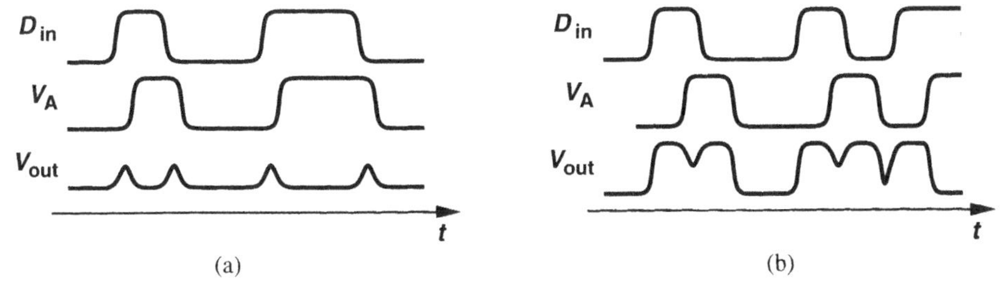

Figure 9.9 Insufficient pulse height due to (a) excessively short and (b) excessively long delay ΔT.

$\Delta T = 10$ ps and the time constants in the XOR are on the order of 50 ps, then the output pulses suffer from a small amplitude, providing little energy at a frequency equal to the bit rate. Next, suppose ΔT is excessively large, i.e., it is approximately equal to the bit period. Consequently, as shown in Fig. 9.9(b), the time overlap between each data pulse and its delayed version is small, prohibiting edge detection. Thus, if the data rate is near the speed limit of the technology, no value of ΔT may produce adequate pulse height at the output.

We study other methods of frequency detection in Section 9.3. It is nevertheless important to note that clock extraction from random data requires edge detection because oscillators can be locked to only *impulsive* frequency content (i.e., a spectral "line").

Phase Detection Following edge detection, we must generate a clock that samples the data bits at their peak values. This is accomplished by measuring the phase difference between the data and the clock and driving it toward the desired value.

Since the spectrum of random data exhibits a null at the bit rate, the phase detectors used for periodic signals may completely fail if they sense random data. For example, as shown in Fig. 9.10, an XOR gate computing the phase difference between CK and D_{in} produces

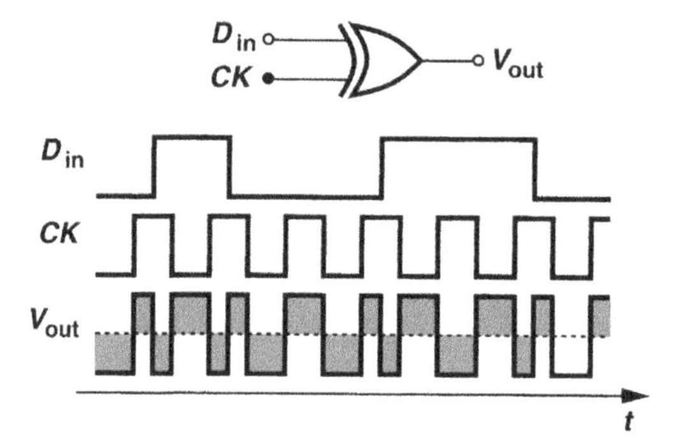

Figure 9.10 Failure of XOR gate as a phase detector with random data.

a zero average at the output if the ONEs and ZEROs in D_{in} occur with equal probabilities. (The XOR output is assumed to be differential). This result can also be reached from an analog viewpoint: the XOR gate is represented by an analog multiplier (mixer) and the clock is replaced by its first harmonic,

$$x_{out}(t) = D_{in} \cdot \cos \omega_0 t, \tag{9.4}$$

where $\omega_0 = 2\pi/T_b$. Recall from Chapter 2 that this product displays a zero average. The reader can prove that the phase/frequency detector described in Chapter 8 also fails with random data.

Let us now consider a D flipflop (DFF) for the task of phase detection. First, suppose the flipflop senses the random data at its D input and the recovered clock at its clock input. As shown in Fig. 9.11(a), each rising transition of the clock samples the level of the data waveform, thereby producing a delayed replica of D_{in}. Unfortunately, if the random sequence contains equal numbers of ONEs and ZEROs, then the average output is still zero. Next, suppose the data and clock ports are swapped so that each rising edge of data samples the clock waveform [Fig. 9.11(b)]. In this case, if the data lags the clock, the flipflop continues to sample the high levels of CK, generating a positive output. Conversely, if the data leads the clock, the flipflop produces a negative output. Thus, the average output signifies the

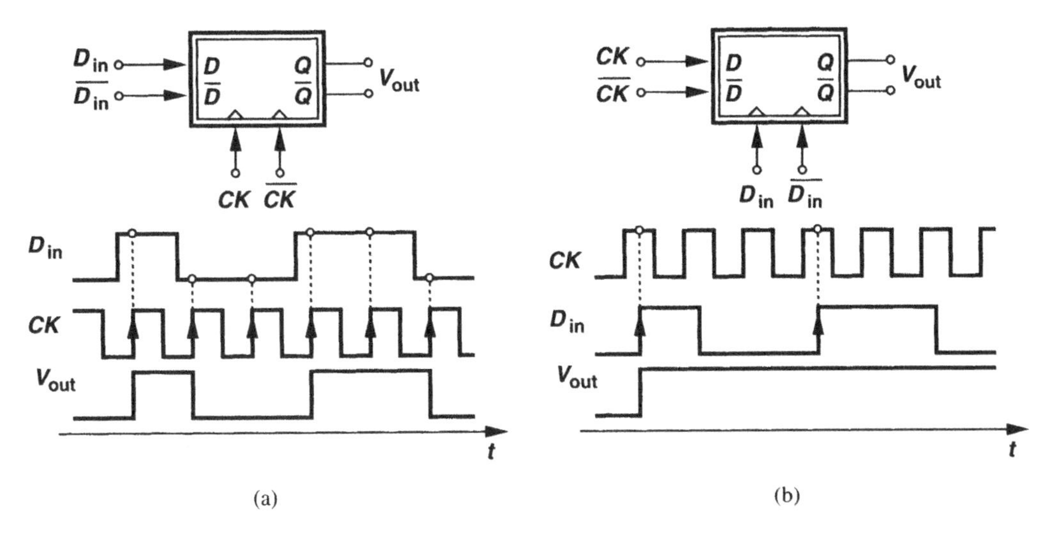

Figure 9.11 Use of a D flipflop as a phase detector: (a) data sampled by clock, (b) clock sampled by data.

polarity of the phase difference between D_{in} and CK. We call the arrangement of Fig. 9.11(b) a DFF phase detector.

It is instructive to study the input-output characteristic of a DFF PD. As illustrated in Fig. 9.11(b), the flipflop output assumes only one of two values depending on whether the data lags or leads the clock.[3] As a result, the characteristic appears as shown in Fig. 9.12, exhibiting an extremely nonlinear behavior. In contrast to the phase detectors studied in

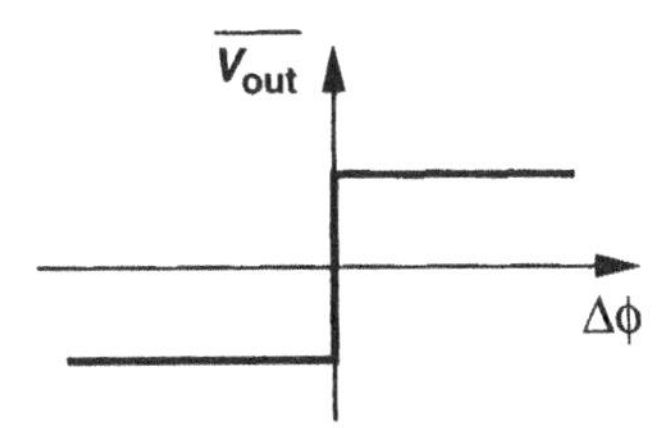

Figure 9.12 Bang-bang PD characteristic.

Chapter 8, a D flipflop provides a very high gain in the vicinity of $\Delta\phi = 0$ and a zero gain elsewhere. This is an example of a "bang-bang" characteristic, where the output jumps from one extreme to another with a slight change in the input.

Another important property of the DFF PD relates to edge detection. Consider the waveforms shown in Fig. 9.13(a), where each rising transition of data samples the clock. The operation is equivalent to multiplying the clock waveform by impulses that occur at each positive edge of data, a behavior similar to that of an edge detector using a half-wave recti-

[3]For now, we ignore the case where the data samples the clock very close to its zero crossings (metastability).

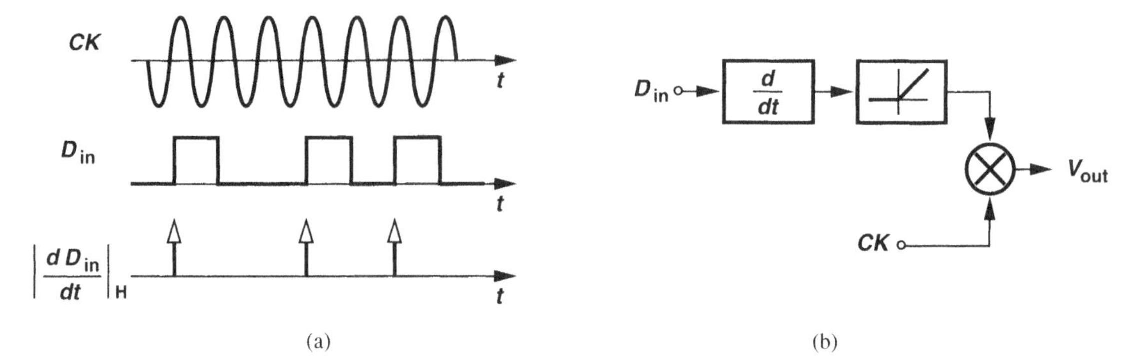

Figure 9.13 (a) Equivalence of DFF operation to impulse sampling, (b) analog implementation of the concept.

fier [Fig. 9.5(b)]. Modeled as shown in Fig. 9.13(b), a DFF PD therefore differentiates and half-wave rectifies the data and multiplies the result by the clock. If the data and the clock exhibit unequal rates, then the output contains a "beat" equal to the difference between the two. The beat component is in fact necessary for phase locking. Note that a DFF does not provide such a property if it is configured as shown in Fig. 9.11(a).

The reader may wonder if a D flipflop can perform differentiation and *full*-wave rectification. This is indeed possible if the circuit samples the clock on both rising and falling edges of data. Depicted in Fig. 9.14 is a double-edge-triggered (DET) flipflop consisting of two latches that operate on opposite phases of D_{in} and a MUX that selects each latched output.

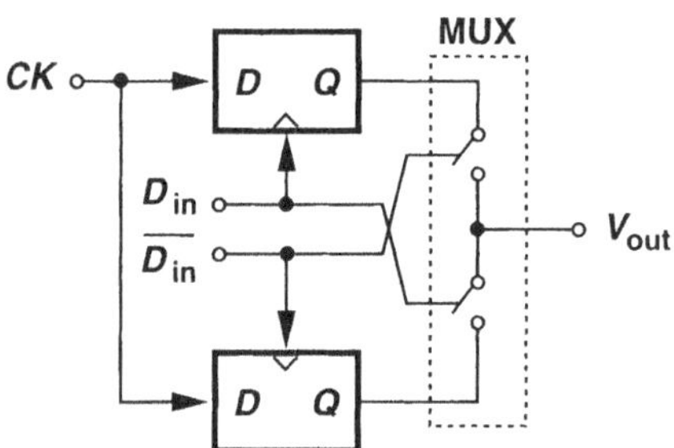

Figure 9.14 Double-edge-triggered flipflop.

Simple CDR Circuit With the foundation formed thus far, we construct a simple CDR circuit, analyzing its behavior and shortcomings. As illustrated in Fig. 9.15, to generate the clock waveform, we employ a VCO, and to define its frequency and phase, we phase-lock the VCO to the input data. Also, to retime the data, we add another DFF that is clocked by the VCO output. Note that the recovered clock, CK_{out}, drives the D input of the phase detector and the clock input of the retimer.

The circuit of Fig. 9.15 operates as follows. Upon turn-on, the DFF multiplies the edge-detected data by the VCO output, generating a beat that drives the VCO frequency toward

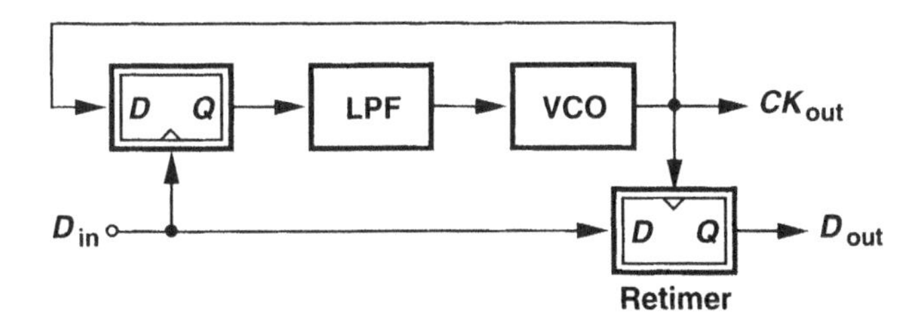

Figure 9.15 Simple CDR circuit using DFF PD.

the input bit rate. If the initial difference between the VCO frequency and the data rate is sufficiently small, then the loop locks, establishing a well-defined phase relationship between D_{in} and CK_{out}. In fact, with the PD bang-bang characteristic, the data edges must sample zero-crossing points of the clock. Even for a slight phase error, the PD generates a large output, driving the loop toward lock.

We now consider a more realistic bang-bang characteristic, where the gain in the vicinity of $\Delta\phi = 0$ is finite [Fig. 9.16(a)]. In a DFF PD, the finite slope arises from metastability:

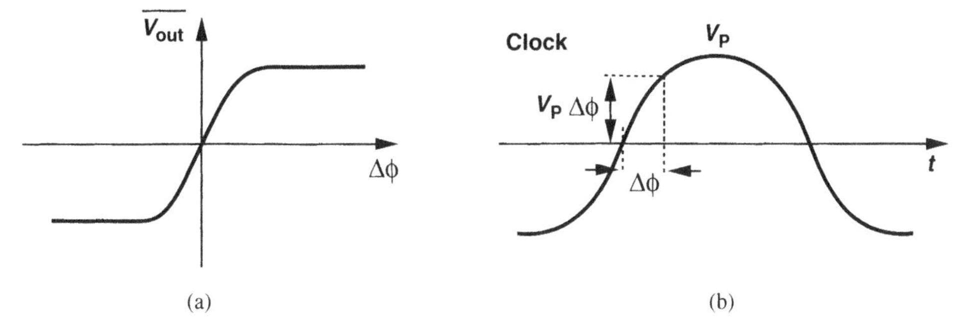

Figure 9.16 (a) Actual bang-bang characteristic, (b) realistic clock waveform.

if the clock waveform is sampled near its zero crossings, the output may not reach the full logical level in one bit period. In other words, as the phase difference between the clock and the data approaches zero, the PD output developed in one bit period also decreases. More specifically, we assume both the master and the slave latches in the flipflop are metastable and write the output after one bit period as

$$V_{out} = V_0 \exp \frac{T_b}{\tau_R}, \tag{9.5}$$

where V_0 denotes the initial voltage and τ_R the regeneration time constant. Approximating the clock by a sinusoid, $V_{CK} = V_p \sin \omega t$, we note that a small phase difference of $\Delta\phi$ translates to a voltage $V_{CK}(\Delta\phi) \approx V_p \Delta\phi$ [Fig. 9.16(b)]. Substituting this value for V_0 in Eq. (9.5) yields:

$$V_{out} = V_p \Delta\phi \exp \frac{T_b}{\tau_R} \tag{9.6}$$

and hence

$$\frac{V_{out}}{\Delta\phi} = V_p \exp\frac{T_b}{\tau_R}. \tag{9.7}$$

That is, the gain of the phase detector after one bit period is finite and constant in the metastable region, yielding a *linear* characteristic. The PD characteristic remains linear for phase differences that are small enough to produce an "unsaturated" output. Denoting the saturated output by V_{sat}, we can determine the maximum phase difference for linear operation as:

$$\Delta\phi_{max} = \frac{V_{sat}}{V_p} \exp\frac{-T_b}{\tau_R}. \tag{9.8}$$

In a typical design, $V_{sat} \approx 0.5V_p$ and $T_b \approx 4\tau_R$, thereby giving $\Delta\phi_{max} \approx 0.0092$ rad $\approx 0.525°$.

Considering the finite-gain region in Fig. 9.16(a), we can now exploit the concepts described for linear PDs in Chapter 8 to predict the behavior of the bang-bang loop in locked condition. Repeating the graphical method of Chapter 8 in Fig. 9.17, we note that the loop must lock with such a phase difference that produces the required control voltage at the

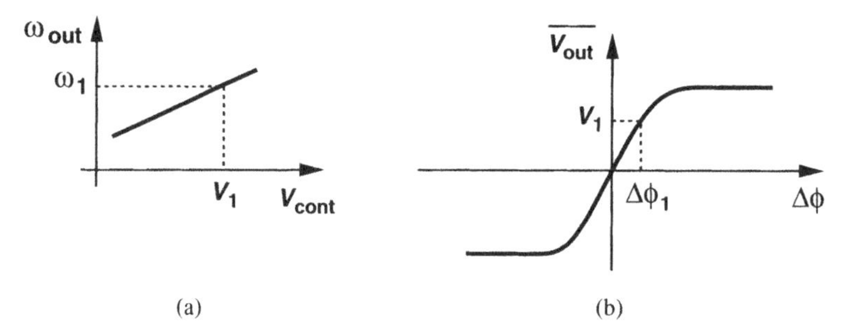

Figure 9.17 Computation of phase error in terms of CDR output frequency.

output of the low-pass filter. Thus, the clock zero crossings lock to the data edges, fluctuating in the metastability region most of the time. If the rising edge of the clock is locked to data, then its falling edge appears in the center of the data eye and can be used to strobe the decision circuit.

The foregoing study has tacitly assumed a *periodic* data stream so that the phase detector becomes metastable on each rising (or falling) data edge when the loop is locked. With random data, on the other hand, the time for regeneration reaches nT_b, where n denotes the number of consecutive ONEs or ZEROs. As a result, with a string of several consecutive ONEs or ZEROs, the output is likely to reach a logical level, leading to a critical drawback. In the absence of data transitions, the PD continues to produce a high or low level, thus forcing the oscillator voltage to vary. As a result, for long runs, the VCO frequency changes substantially, creating a large jitter.

The simple CDR circuit of Fig. 9.15 suffers from a number of drawbacks. First, as mentioned above, the PD may produce full digital outputs for run lengths greater than one, thus creating substantial ripple on the oscillator control voltage and hence jitter at the output. Second, since the PD samples the clock by the data whereas the decision circuit samples the data by the clock, data retiming exhibits significant phase offset at high speeds. In typical latches, the delay from the D input to the output (called the "D-to-Q delay") is unequal to that from the CK input to the output (called the "CK-to-Q delay"). Figure 9.18 illustrates an example where the CK-to-Q delay is longer than the D-to-Q delay by ΔT. Note that the skew is explicitly referred to the CK input of each flipflop. Here, the PD locks such that the data *leads* the clock by ΔT, sampling the clock closer to the zero crossing after the data experiences the intrinsic delay of the PD. The VCO output suffers

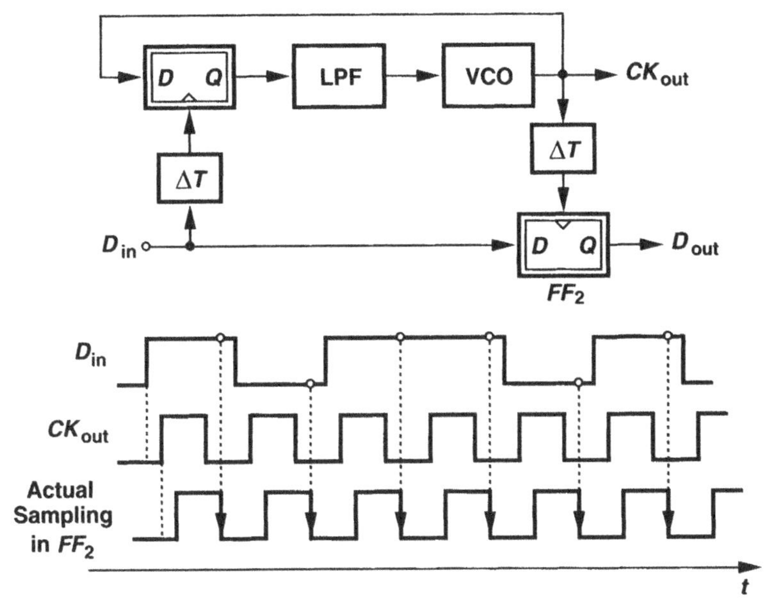

Figure 9.18 Problem of skew resulting from DFF delays.

from even more delay as it propagates through the decision circuit, sampling the data far from the middle of the eye. In other words, if the difference between the CK-to-Q and D-to-Q delays is equal to ΔT, then the retiming suffers from a skew of $2\Delta T$.

The third drawback of the simple CDR architecture of Fig. 9.15 relates to the feedthrough of data to the VCO output through both flipflops. Figure 9.19 illustrates this phenomenon for the PD and the decision circuit. In Fig. 9.19(a), data transitions turn M_1-M_2 on and off, drawing transient currents from the VCO. Similarly, in Fig. 9.19(b), data edges couple through the gate-source capacitance of M_1 and the gate-drain overlap capacitance of M_3 to the oscillator. The random arrival of the data transitions therefore introduces jitter at the VCO output. For this reason, the VCO must be followed by a buffer stage exhibiting a high reverse isolation.

In the remainder of this chapter, we present various phase and frequency detectors and CDR architectures that alleviate the issues described above.

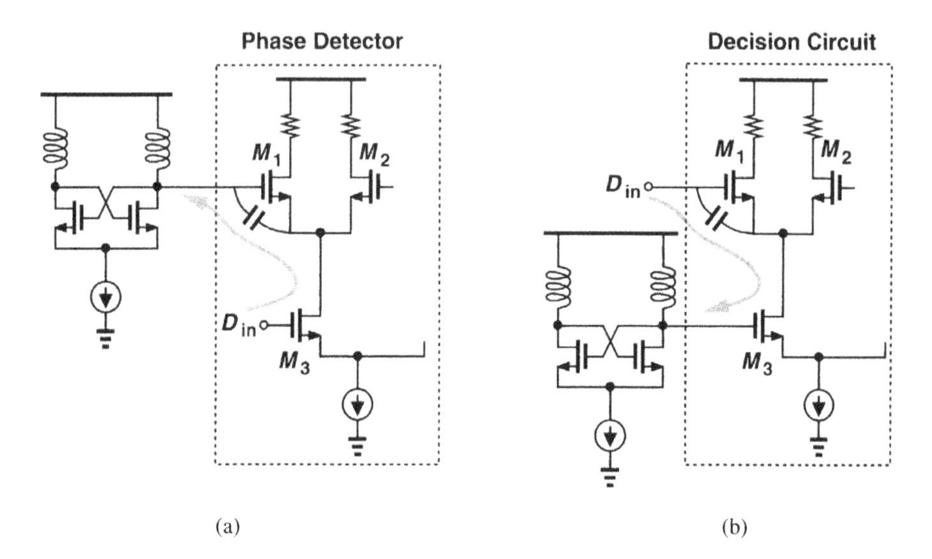

Figure 9.19 Feedthrough of random data to VCO.

9.2 Phase Detectors for Random Data

As mentioned in Section 9.1, PDs for random data must provide two essential functions: (1) data transition detection and (2) phase difference detection. Furthermore, the skew effect illustrated in Fig. 9.18 makes it desirable to retime the data *inside* the phase detector, thereby eliminating the explicit decision circuit and the skew associated with it. This observation immediately leads to another: if the data is to be retimed by the VCO output, then the flipflop(s) in the phase detector must be strobed by the latter rather than the former. In other words, unlike the DFF PD of Fig. 9.11(b), the phase detector must sample the incoming data by the VCO signal.

9.2.1 Hogge Phase Detector

How can a PD detect data transitions if it samples the data by the VCO output? Recall from Section 9.1 that a single DFF fails to operate as a phase detector if it is used in such a mode. However, recognizing that a DFF produces a *delayed* replica of the input data, we can replace the delay element in the edge detector of Fig. 9.8(a) with a DFF [Fig. 9.20(a)], arriving at a synchronous counterpart. Since sample B changes only on the CK edges, $Y = D_{in} \oplus B$ contains pulses whose width represents the phase difference between D_{in} and CK [Fig. 9.20(b)]. It is important to note that (a) the circuit produces a pulse for each data transition, thereby providing edge detection, and (b) the width of the output pulses varies *linearly* with the input phase difference, suggesting that the circuit can operate as a linear PD. We call this type of output "proportional pulses."

It may appear that the topology of Fig. 9.20(a) satisfies the requirements of a phase detector and can therefore be used as such. Unfortunately, however, the average value of the output is a function of the data transition density, failing to uniquely represent the phase

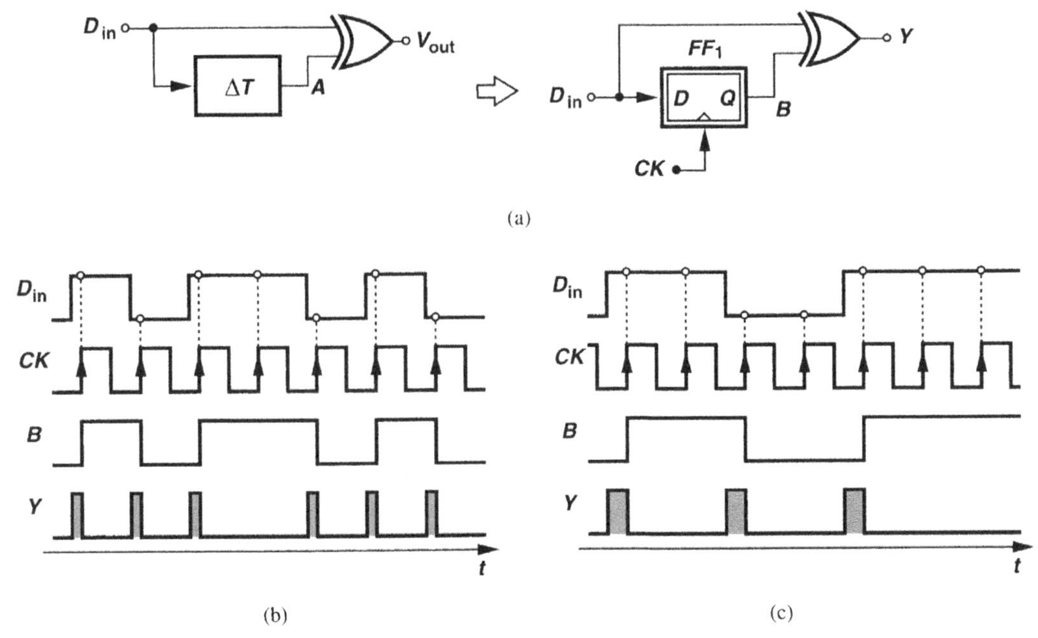

Figure 9.20 (a) Simple PD using synchronous edge detection, (b) PD output for a data pattern, (c) PD output for a different data pattern.

difference for various data patterns. For example, as illustrated in Figs. 9.20(b) and (c), the average output remains unchanged if the transition density falls by a factor of two and the phase difference rises by the same factor. In other words, two different phase errors may result in the same dc output, leading to false lock.

To overcome the above ambiguity, the proportional pulses must be accompanied by "reference pulses." The latter are pulses that appear on data edges but exhibit a *constant* width, thus eliminating the pattern dependency.

How can reference pulses be generated? We note that, if the retimed data at point B in Fig. 9.20(a) is delayed by half a clock cycle, $T_{CK}/2$, and XORed with itself [Fig. 9.21(a)], then pulses of width $T_{CK}/2$ are produced for each data transition. As depicted in Fig. 9.21(b), the *difference* between the areas under X and Y can be viewed as the PD output, eliminating the ambiguity due to transition density. Note that under locked condition, X and Y produce equal pulsewidths. This circuit is called the "Hogge phase detector" [1].

Let us summarize our thought process thus far. To avoid skews in the decision circuit, we choose to sample the data by the clock even in the phase detector. This in turn requires explicit edge detection, carried out by a DFF and an XOR gate. Finally, we produce a reference pulse to eliminate ambiguity for different data transition densities.

Recall that the principal motivation behind the above development is to retime the data inside the PD. Does the Hogge PD accomplish this? Indeed, both flipflops in Fig. 9.21 operate as decision circuits as well, thereby providing retimed data.

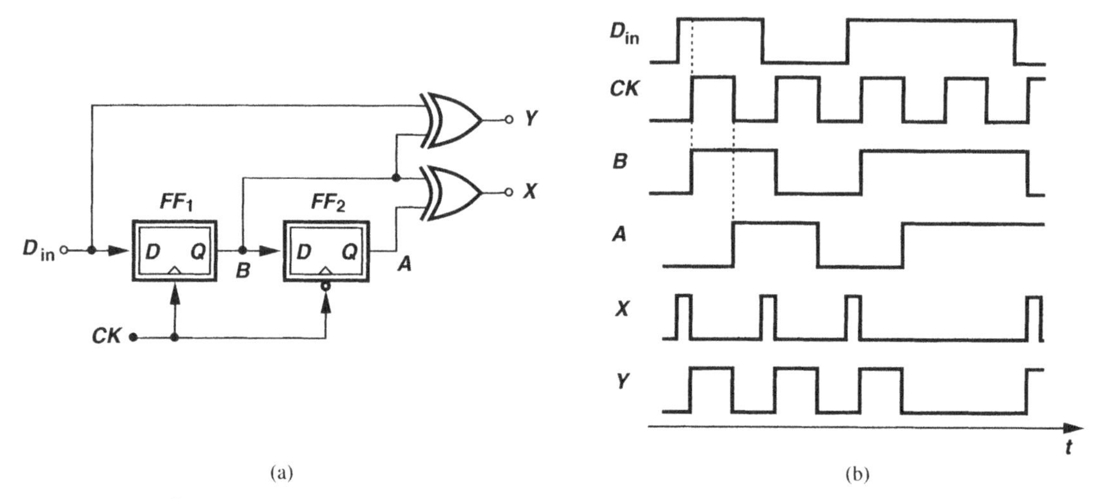

Figure 9.21 (a) Hogge phase detector and (b) its waveforms.

Nonidealities We now examine the behavior of the Hogge PD in the presence of finite delays in the flipflops. As illustrated in Fig. 9.22, owing to the CK-to-Q delay, ΔT, of

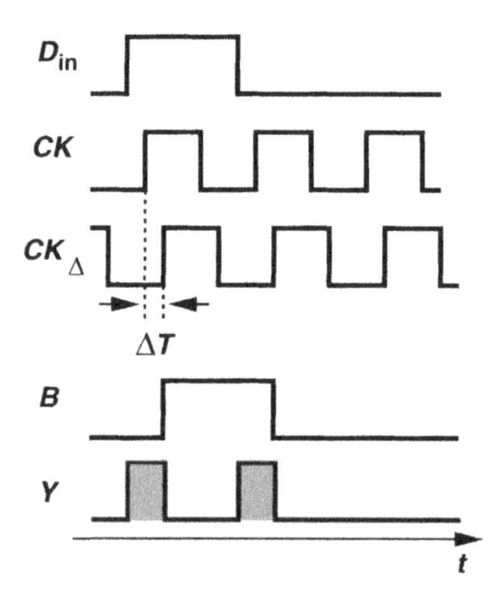

Figure 9.22 Performance of Hogge PD in the presence of FF delays.

FF_1, B changes ΔT seconds after the clock rises, yielding a pulse at the Y output that is ΔT seconds wider than the actual phase difference between D_{in} and CK. On the other hand, the CK-to-Q delay of FF_2 simply *shifts* the pulses at A by ΔT, still producing a pulsewidth equal to one clock period. As a result, X continues to produce pulses of width $T_{CK}/2$ for each data transition. This means, with a zero input phase difference, the proportional pulses are wider than the reference pulses by ΔT seconds. Thus, under locked condition, D_{in} and CK must sustain a skew of ΔT so as to equalize the widths of the X and Y pulses.

The above skew effect becomes a serious issue at high speeds. Since ΔT can be a significant fraction of the clock period, a systematic phase offset of several tens of degrees may arise after the loop is locked, degrading the clock phase margin (Section 9.1) and hence the jitter tolerance (Section 9.5). To resolve this difficulty, we can either narrow the proportional pulses by ΔT or widen the reference pulses by the same amount. We consider each approach.

The skew phenomenon in Fig. 9.22 appears simply because the clock experiences a delay that data does not. Thus, if an equal delay is inserted in the data path, the skew is removed and the pulsewidth restored. Depicted in Fig. 9.23(a) [1], the idea is to ensure that the edges of D_{in} and B arrive at the inputs of the XOR gate simultaneously. Of course,

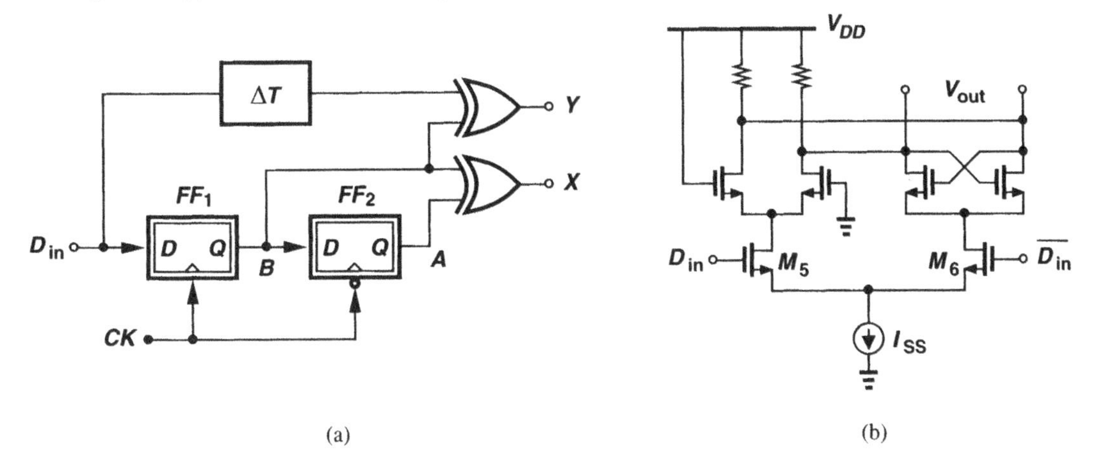

Figure 9.23 (a) FF delay compensation in Hogge PD, (b) realization of delay cell.

to avoid process and temperature variabilities, the delay stage must be an exact replica of the CK-to-Q path of the flipflop. Figure 9.23(b) depicts a current-steering example, where the D input of the latch is set to a logical ONE and the circuit simply provides a delay corresponding to the CK-to-Q path.

The second method involves widening the reference pulses by ΔT [2]. As shown in Fig. 9.24, this is accomplished by inserting a delay stage between A and the input of the second XOR gate. We note that, since A lags B by $T_{CK}/2$, A_Δ lags B by $T_{CK}/2 + \Delta T$. Thus, $B \oplus A_\Delta$ exhibits a pulsewidth of $T_{CK}/2 + \Delta T$.

Another drawback of the Hogge phase detector stems from the half-cycle skew between the two XOR outputs [3, 4]. As illustrated in Fig. 9.25(a) for locked conditions, the PD produces the reference pulse *after* the proportional pulse, thereby creating a skew of $T_{CK}/2$ between the two. What is the effect of such a skew? Suppose the XOR gates drive a charge pump and a loop filter [Fig. 9.25(b)]. When the proportional pulse appears, the top current source turns on for half a clock period, forcing V_{out} to rise by $(I_P/C_P)(T_{CK}/2)$. Next, the reference pulse turns on the bottom current source, bringing V_{out} to its original value. Thus, for each data transition, a triangular pulse is generated that exhibits a nonzero

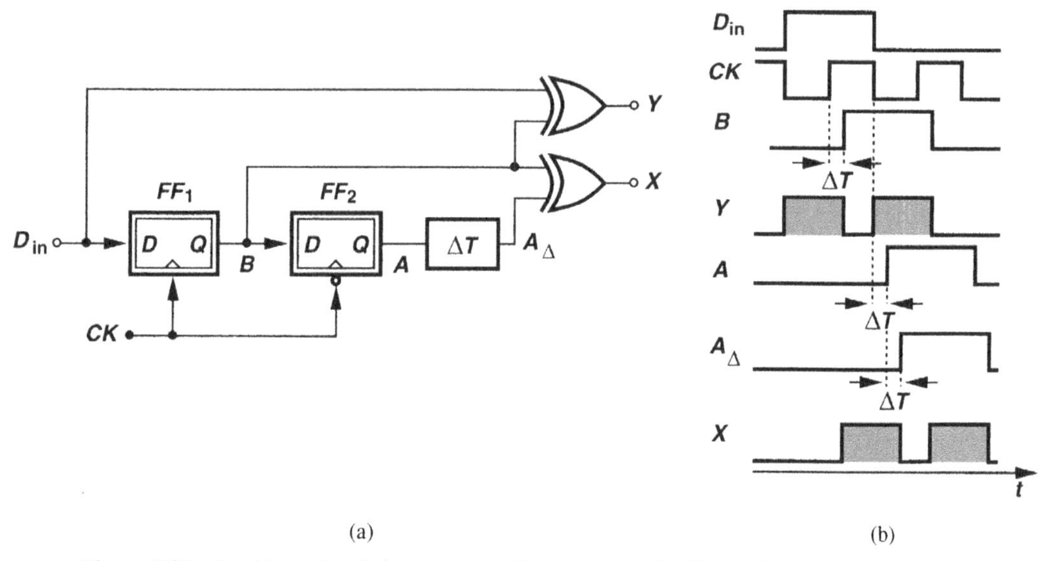

Figure 9.24 (a) Alternative delay compensation technique for Hogge PD, (b) waveforms of the circuit.

net area. As a result, the VCO phase (which is proportional to the integral of the control voltage) is severely disturbed.

The Hogge PD can be modified so as to ameliorate the above issue [4]. Shown in Fig. 9.26 is an example where one more latch and two more XOR gates are added. Note that, in contrast to the Hogge PD, here X_2 senses the inputs and outputs of a latch (rather than a flipflop). The additional XOR gates create two reference pulses offset by $T_{CK}/2$ and T_{CK} with respect to the original reference pulse, thereby producing a *negative* triangular pulse whose area is *constant*. Consequently, the two triangular pulses yield a net zero area under locked condition, lowering the effect on the VCO phase.

The Hogge topology is a linear phase detector, generating a vanishing average as the phase difference approaches zero. Thus, a charge pump driven by a Hogge PD experiences little "activity" when the CDR loop is locked. This behavior is in contrast to that of the bang-bang PD of Fig. 9.11.

Figure 9.27 shows a CDR loop incorporating a Hogge PD. The XOR outputs drive a charge pump and a loop filter. As with type II PLLs, a charge pump is necessary here to provide an infinite loop gain and hence a zero static phase error.

The need for a charge pump in linear CDR loops poses serious speed limitations. Under the locked conditon, the XOR outputs contain pulses only half a bit period wide, requiring a very broad bandwidth at these nodes to ensure complete switching of the charge pump and hence avoid a dead zone (Chapter 8).

9.2.2 Alexander Phase Detector

The Alexander configuration is another example of PDs providing inherent data retiming. Following our reasoning for the Hogge PD, we note that this property requires that the

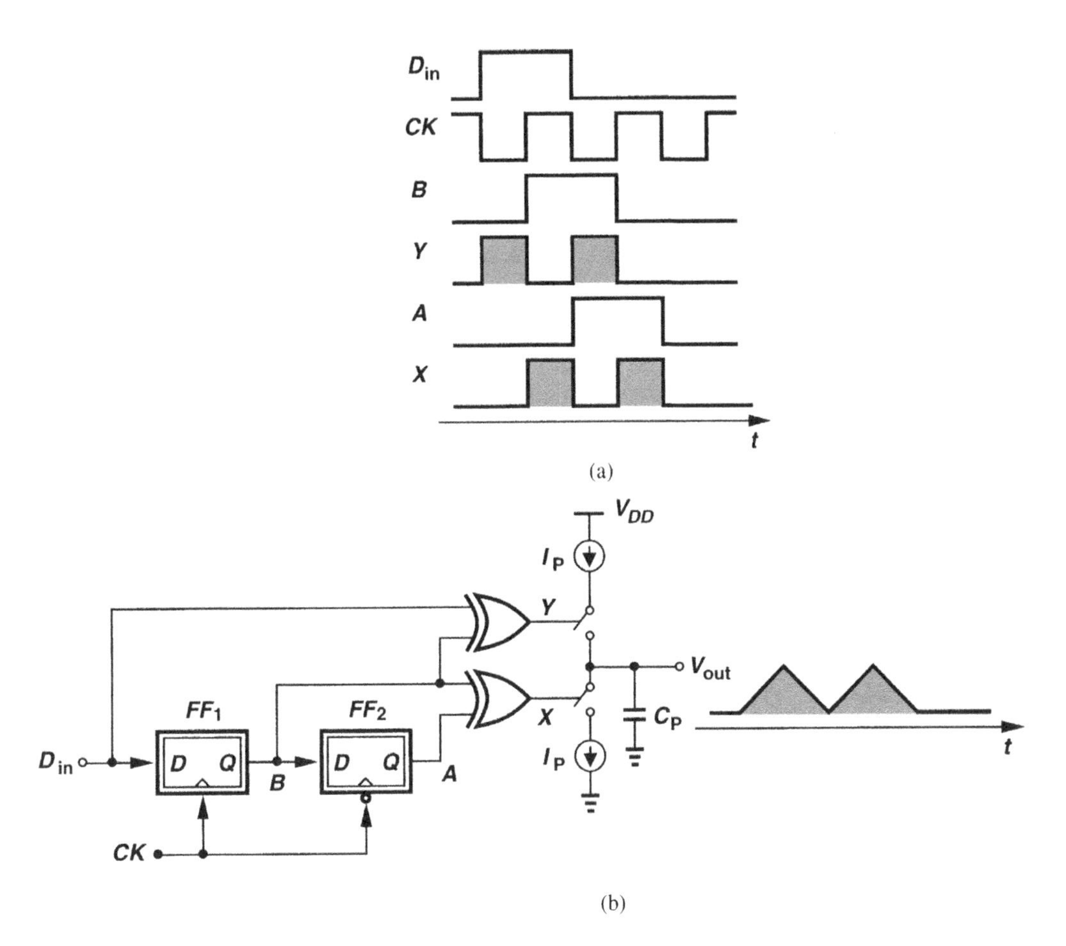

Figure 9.25 (a) Hogge PD waveforms, (b) resulting triwave produced by a charge pump.

data be sampled by the clock, but a single DFF does not suffice. Nonetheless, if the clock strobes the data waveform at *multiple* points in the vicinity of expected transitions, then the resulting samples can provide the necessary information.

Figure 9.28(a) illustrates the Alexander PD principle, also known as the "early-late" detection method. Using three data samples, S_1-S_3, taken by three consecutive clock edges, the PD can determine whether a data transition is present and whether the clock leads or lags the data. In the absence of data transitions, all three samples are equal and no action is taken. If the falling edge of the clock leads (is "early"), then the last sample, S_3, is unequal to the first two. Conversely, if the clock lags (is "late"), then the last two samples, S_2 and S_3, are equal but unequal to the first sample, S_1. Thus, $S_1 \oplus S_2$ and $S_2 \oplus S_3$ provide the early-late information: (a) if $S_1 \oplus S_2$ is high and $S_2 \oplus S_3$ is low, then the clock is late; (b) if $S_1 \oplus S_2$ is low and $S_2 \oplus S_3$ is high, then the clock is early; (c) if $S_1 \oplus S_2 = S_2 \oplus S_3$, then no data transition is present.

The foregoing observations lead to the circuit topology shown in Fig. 9.28(b) [5].

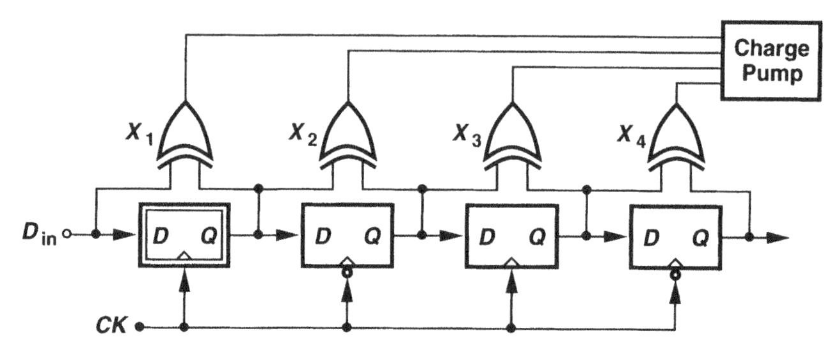

Figure 9.26 Modified Hogge PD to remedy the triwave issue.

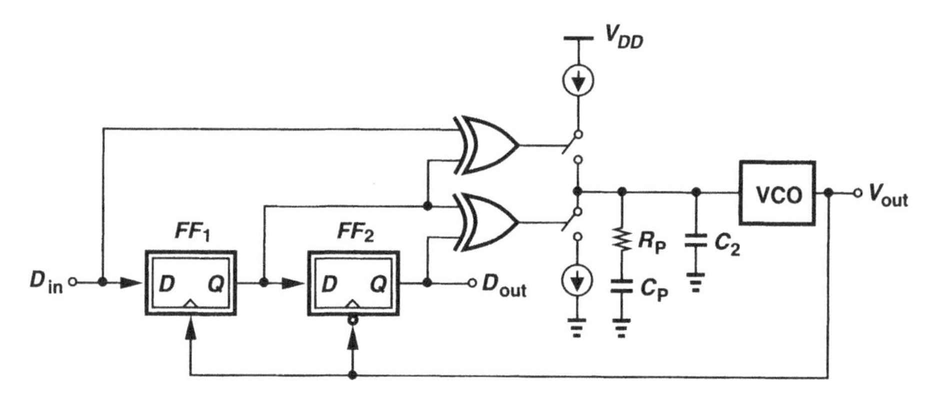

Figure 9.27 CDR circuit using Hogge PD.

Flipflop FF_1 samples S_1 and S_3 on the rising edge of CK and flipflop FF_2 delays the result by one clock cycle. Flipflop FF_3 samples S_2 on the falling edge of CK and flipflop FF_4 delays this sample by half a clock cycle.

Let us examine the waveforms at various points in the Alexander PD to gain more insight into its operation. As depicted in Fig. 9.28(c), the first rising edge of CK samples a high data level. The second rising edge of CK then accomplishes two tasks: it produces a delayed version of the first sample at the output of FF_2, and it samples the low level on the input data. The values of S_1 and S_2 are therefore valid for comparison at $t = T_1$, remaining constant for one clock period.

On the first falling edge of CK in Fig. 9.28(c), FF_3 samples a high level on the input data and on the next rising edge, FF_4 reproduces this level. The key point here is that the choice of clock phases for the four flipflops ensures that S_1, S_2, and S_3 reach valid levels for comparison at $t = T_1$, and remain at these levels for one clock period. As a result, the XOR gates generate valid outputs simultaneously.

How does a CDR loop using an Alexander PD lock? Let us first study the PD's behavior as the phase difference between D_{in} and CK, $\Delta\phi$, varies from a negative value to a positive value. As shown in Fig. 9.29(a), if the clock is late ($\Delta\phi < 0$), X assumes a high level for each data transition and Y a low level. Thus, the average PD output, $(X - Y)_{avg}$,

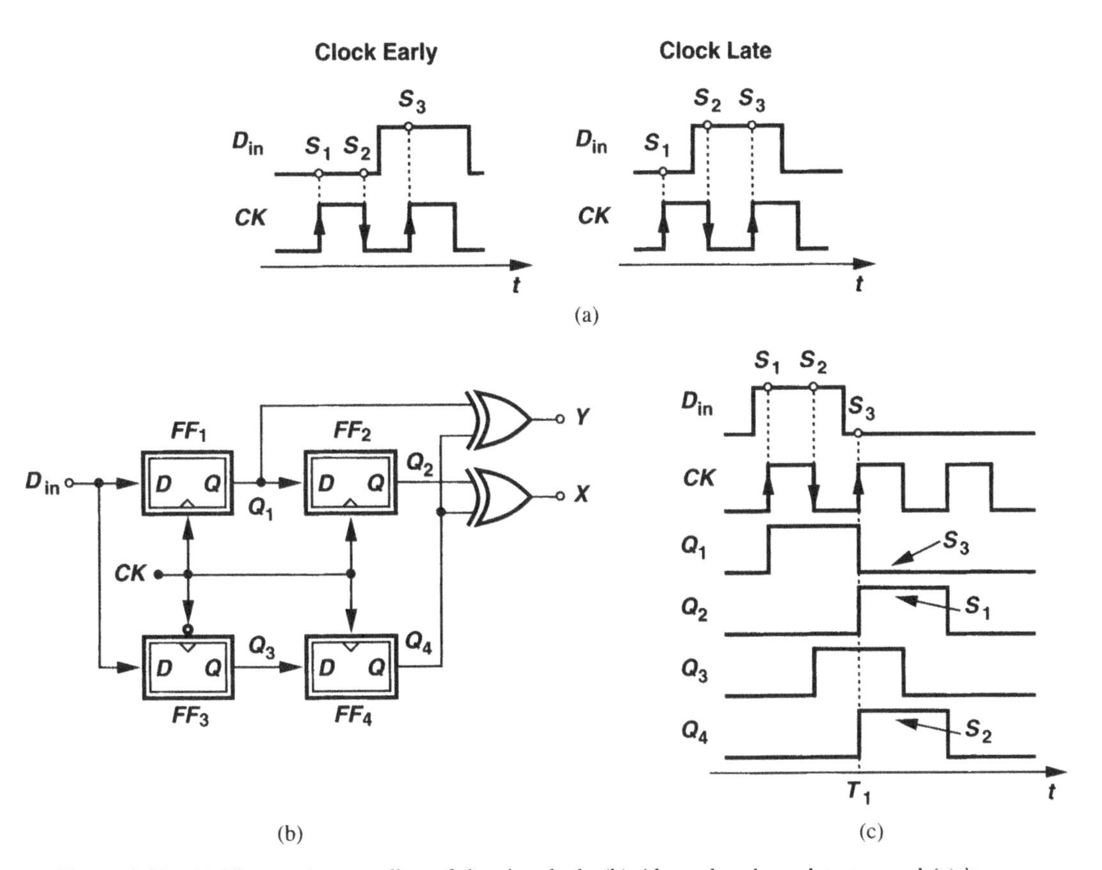

Figure 9.28 (a) Three-point sampling of data by clock, (b) Alexander phase detector, and (c) its waveforms.

remains at a high positive value. Conversely, if $\Delta\phi > 0$ [Fig. 9.29(b)], then $(X - Y)_{avg}$ assumes a high negative value. As $\Delta\phi$ approaches zero, the second sample, S_2, falls in the vicinity of the data zero crossing, thereby driving FF_3 and FF_4 into metastability—a behavior similar to the DFF PD of Section 9.1.

The above study implies that the Alexander PD is a bang-bang system, exhibiting a very high gain in the vicinity of $\Delta\phi = 0$. Consequently, the CDR loop locks such that S_2 coincides with the data zero crossings. However, the use of the XOR gates to generate the final output requires a closer examination of the circuit in metastability. Suppose the differential outputs of FF_3 and FF_4 are near zero while the other flipflops produce valid logical outputs. How do the XOR gates respond to such inputs? Considering the typical XOR implementation shown in Fig. 9.30, we note that if one differential input is large and the other small, then the differential output is near zero. From a small-signal point of view, the XOR operates as an analog multiplier, generating a zero output if one of the inputs falls to zero.

Sensing Q_4, both XOR gates in Fig. 9.28(b) produce small differential outputs under

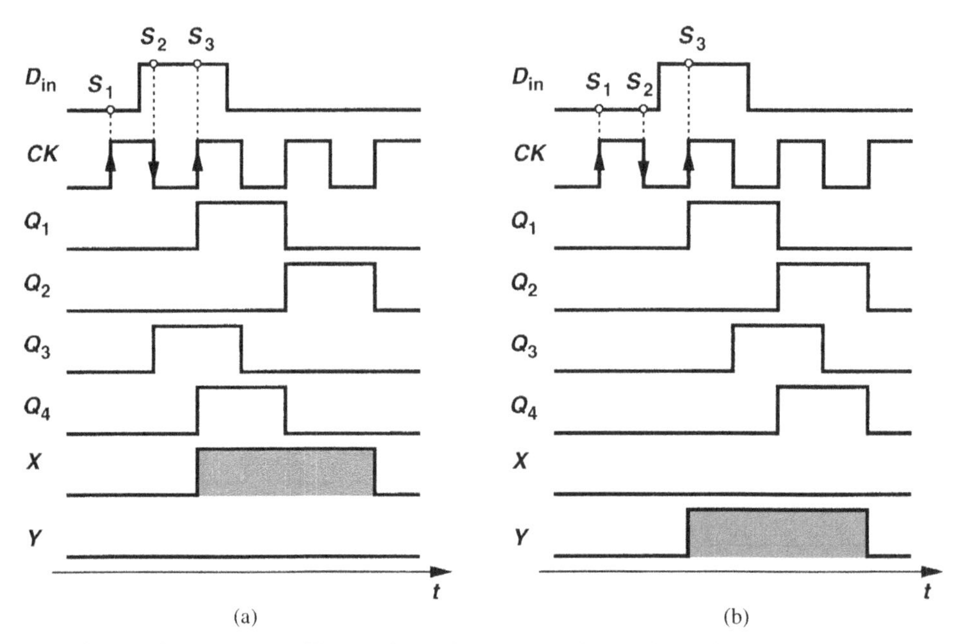

Figure 9.29 Alexander PD waveforms for (a) late and (b) early clock.

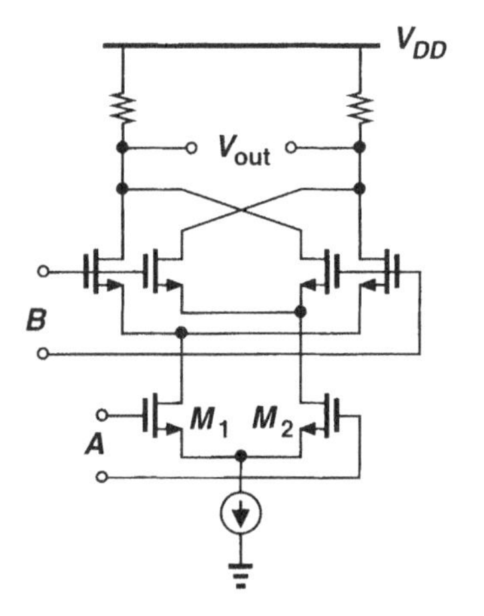

Figure 9.30 XOR gate realized by a Gilbert cell.

metastability, yielding a small average output for the overall PD. Thus, the CDR loop can, in principle, lock such that the XOR gates experience metastable inputs most of the time. But what happens if the XOR circuits suffer from random offset voltages? For example, what if asymmetries in the differential pair M_1-M_2 of Fig. 9.30 result in an input-referred offset of +20 mV for one XOR gate? To compensate for such an offset, sample S_2 must slightly deviate from the data zero crossings.

Another important effect in the Alexander PD relates to systematic delay mismatches in the XOR gates, i.e., the skew between the A and B paths in Fig. 9.30. It is left for the reader to study various combinations of the XOR delay asymmetries in the PD and determine if they result in a phase offset.

While exhibiting a bang-bang characteristic, the Alexander PD offers two critical advantages over a simple DFF PD. First, it retimes the data automatically, producing a valid data waveform at the output of FF_1 and FF_2 in Fig. 9.28(b). Second, in the absence of data transitions, it generates a zero dc output, leaving the oscillator control undisturbed. As a result, for long data runs, the VCO frequency drifts only due to device electronic noise rather than due to a high or low level on the control line.

Figure 9.31 shows a CDR circuit employing an Alexander PD. The XOR gates drive

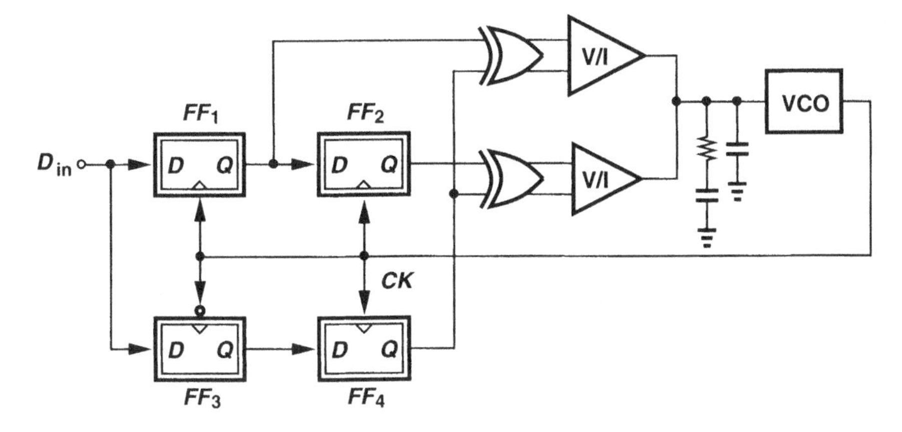

Figure 9.31 CDR circuit using Alexander PD.

voltage-to-current (V/I) converters, the two outputs are summed in the current domain, and the result is applied to the loop filter. In contrast to the architecture of Fig. 9.27, this CDR circuit need not incorporate a charge pump because the high gain of the Alexander PD yields a small phase offset under locked condition. This is an important advantage because the XOR ouputs in Fig. 9.31 need not switch at high speeds; the V/I converters sense only the average value.

CDR circuits using similar phase detectors are described in [6, 7, 8]. Note that, with Hogge or Alexander PDs, the loop gain is a function of the data transition density because the oscillator control line remains idle in the absence of data edges.

9.2.3 Half-Rate Phase Detectors

At very high speeds, it may be difficult to design oscillators that provide an adequate tuning range with reasonable jitter. For this reason, CDR circuits may sense the input random data at full rate but employ a VCO running at *half* the input rate. This technique also relaxes the speed requirements of the phase detector and, in some CDR configurations, the frequency dividers. Called "half-rate" architectures, such CDR topologies require a phase detector that provides a valid output while sensing a full-rate random data stream and a half-rate clock.

It is important to note that none of the three phase detectors studied thus far in this chapter can operate with a half-rate clock. For example, a DFF PD sensing a half-rate clock responds as shown in Fig. 9.32, creating both negative and positive outputs for a

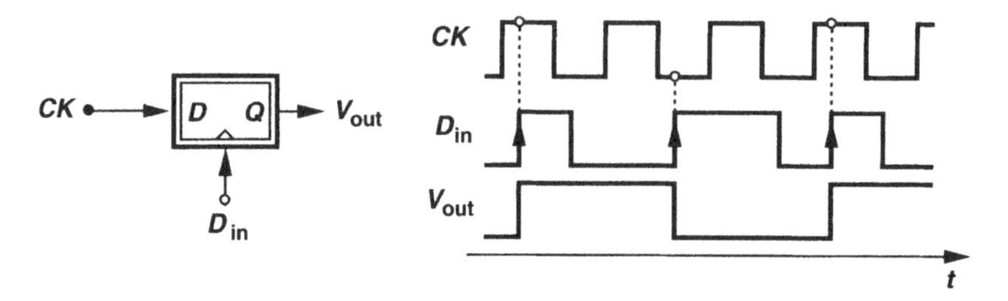

Figure 9.32 Failure of a DFF as a half-rate PD.

constant phase difference.

To arrive at a linear half-rate PD, let us first examine the Hogge configuration and understand its failure mechanism. As illustrated in Fig. 9.33, since half of the clock transitions are absent, the random sequence can be chosen such that the half-rate clock continues to

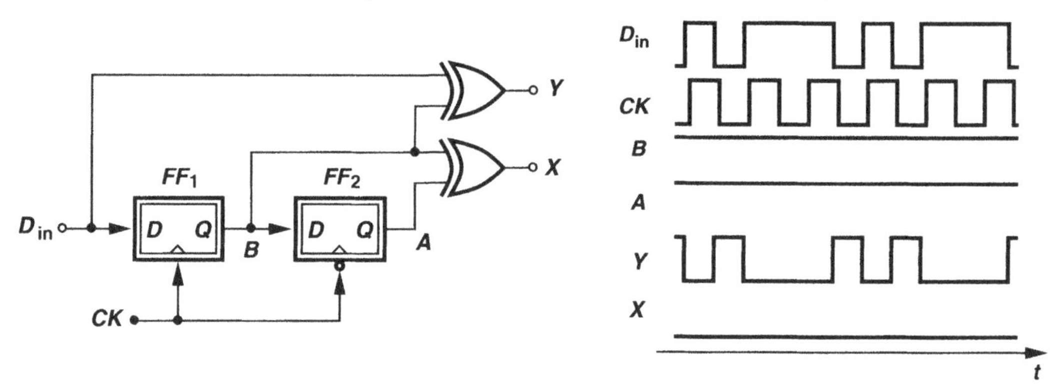

Figure 9.33 Failure of Hogge circuit as a half-rate PD.

sample a high level on the data waveform, failing to recognize the data transitions. As a result, the proportional and reference pulses are invalid.

The above observation suggests that data transitions may be detected if *both edges* of the half-rate clock are utilized to sample the input data. Consider the topology shown in Fig. 9.34, where two D latches, L_1-L_2, operate on the opposite edges of the clock. Note that each latch is transparent for half of the clock cycle, passing data transitions to its output. Assuming D_{in} leads CK by ΔT and L_1 is transparent when CK is high, we observe that A goes high *before* CK falls and remains high until CK rises. In other words, L_1 produces a pulsewidth equal to $T_{CK}/2 + \Delta T$. On the other hand, if L_2 is transparent when CK is low, then B goes high when CK falls and remains high only until D_{in} falls. That is, L_2 generates a pulsewidth of $T_{CK}/2 - \Delta T$. Thus, $A \oplus B$ exhibits a pulse of width ΔT for each data transition.

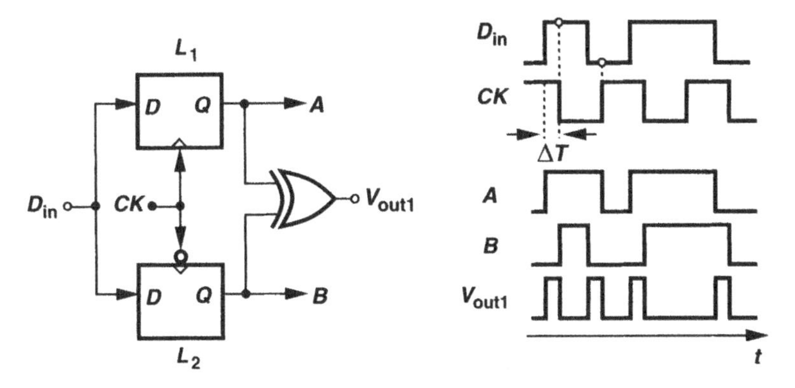

Figure 9.34 Simple linear half-rate PD.

The above study implies that the simple topology of Fig. 9.34 can indeed operate as a linear phase detector because it (a) detects data edges and (b) produces proportional pulses. However, as with the Hogge topology, this circuit must also provide a reference output so as to uniquely represent the phase error for different data transitions. To this end, let us follow the latches with two more (creating a master-slave flipflop in each path) and subsequently perform XOR on the outputs (Fig. 9.35). In the presence of data transitions, the outputs of

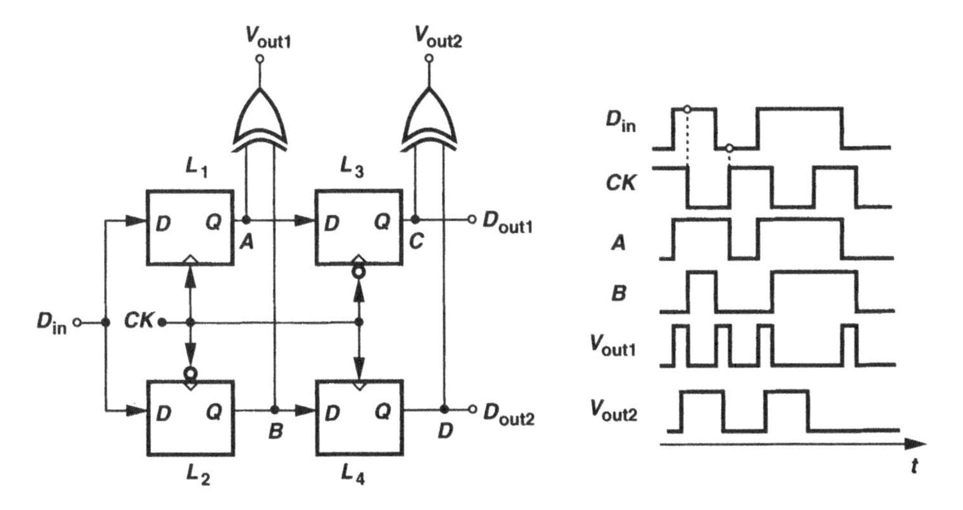

Figure 9.35 Complete linear half-rate PD.

L_3 and L_4 change on the falling and rising edges of the clock, respectively. As a result, $C \oplus D$ contains a pulse of width $T_{CK}/2$ for each input data edge, serving as the reference output [9].

How does a CDR loop employing the PD of Fig. 9.35 lock to random data? If the clock edge is to strobe the data in the middle of the eye, then the proportional pulses are $T_{CK}/4$ seconds wide whereas the reference pulses are $T_{CK}/2$ seconds wide. The disparity between the average values of these outputs is removed by scaling down the *effect* of the output of the second XOR by a factor of two, i.e., by halving the corresponding current source in the charge pump.

The half-rate PD of Fig. 9.35 also retimes and demultiplexes the data, producing two streams at the outputs of L_3 and L_4. The linear characteristic of the circuit allows simple formulation of the loop dynamics.

Let us now consider the early-late method for half-rate operation. Since the Alexander PD already requires sampling on both clock edges for full-rate detection, it must employ additional phases of the clock if it is to operate in the half-rate mode. As illustrated in Fig. 9.36, the rising and falling edges of a half-rate clock sample only two points, detecting data

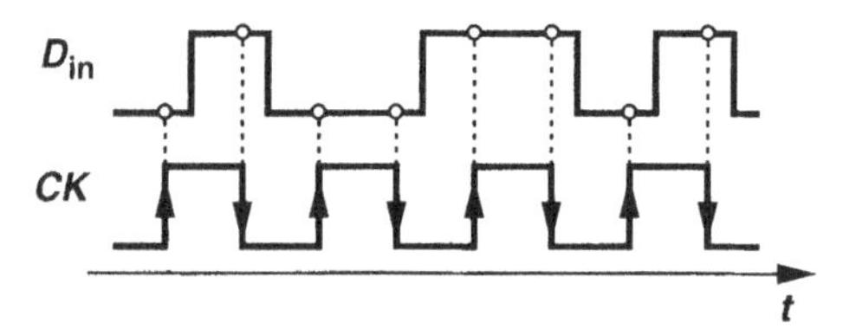

Figure 9.36 Failure of Alexander circuit as a half-rate PD.

transitions but failing to provide the phase difference information. Shown in Fig. 9.37(a), the solution involves sampling the data by both the in-phase and quadrature phases of the

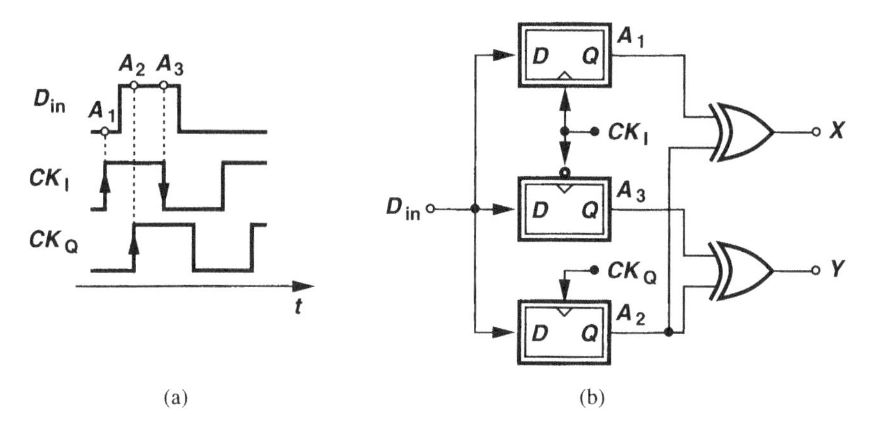

Figure 9.37 (a) Use of quadrature clocks for half-rate phase detection, (b) half-rate binary phase detector.

clock, CK_I and CK_Q, respectively. Now, A_1, A_2, and A_3 play the same role as the consecutive samples in a full-rate counterpart. As depicted in Fig. 9.37(b), the implementation incorporates three flipflops sampling the data by CK_I and CK_Q and two XOR gates producing $A_1 \oplus A_2$ and $A_2 \oplus A_3$. Under the locked condition, the rising edge of CK_Q occurs in the vicinity of the data zero crossings.

The flipflops in the circuit of Fig. 9.37(b) generate skewed outputs, requiring additional retiming latches before the results are applied to the XOR gates. The effect of the skews must be examined for locked condition, i.e., when CK_Q samples the data zero crossings, driving FF_3 into metastability. As depicted in Fig. 9.38, if A_2 is metastable, so are $A_1 \oplus A_2$

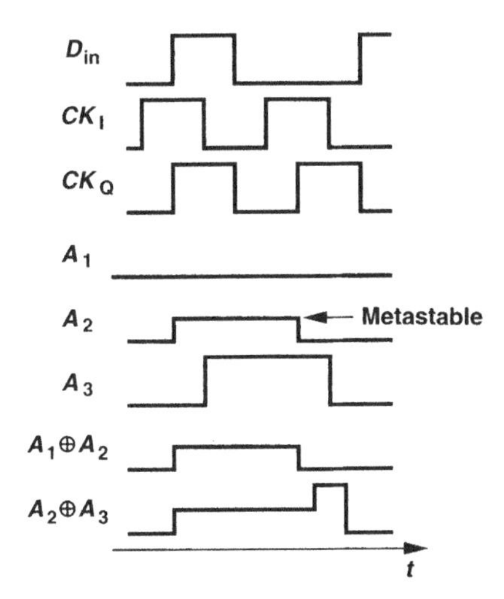

Figure 9.38 Effect of skews under locked condition.

and $A_2 \oplus A_3$, but since A_3 holds its value for $T_{CK}/4$ seconds after A_2 changes, $A_2 \oplus A_3$ produces an extraneous pulse of width $T_{CK}/4$, disturbing the VCO. It is therefore desirable to delay A_1 by $T_{CK}/2$ and A_2 by $T_{CK}/4$.

9.3 Frequency Detectors for Random Data

Most CDR circuits require a means of frequency detection in addition to phase detection. As mentioned in Chapter 8, the idea is to drive the VCO frequency toward the desired value by a frequency-locked loop and, when the frequency error reaches a sufficiently small value, allow the PLL to take over and perform phase locking. This "aided acquisition" is necessary because the capture range of typical PLLs is quite small, especially if they operate with random data. With the narrow loop bandwidths specified by optical standards, the capture range may not exceed a few percent of the data rate. Thus, if the VCO center frequency varies by, say, $\pm 10\%$ with process and temperature, a simple PLL fails to lock.

A frequency detector (FD) must generate an output whose average represents the polarity and (possibly) the magnitude of the input frequency difference. Let us first assume both inputs are periodic and consider a multiplier (mixer) as a possible FD. Denoting the inputs by $x_1(t) = A_1 \cos \omega_1 t$ and $x_2(t) = A_2 \cos \omega_2 t$, we express the product as

$$x_1(t)x_2(t) = \frac{A_1 A_2}{2}[\cos(\omega_1 + \omega_2)t + \cos(\omega_1 - \omega_2)t]. \tag{9.9}$$

The component at $\omega_1 + \omega_2$ can be removed by high-pass filtering, but the term $(A_1 A_2/2)\cos(\omega_1 - \omega_2)t$ still exhibits a zero average. It is possible to differentiate the result with respect to time (Fig. 9.39),

$$\frac{d}{dt}[\frac{A_1A_2}{2}\cos(\omega_1-\omega_2)t] = -\frac{A_1A_2}{2}(\omega_1-\omega_2)\sin(\omega_1-\omega_2)t, \tag{9.10}$$

thereby obtaining an *amplitude* proportional to the frequency difference. Nevertheless, the

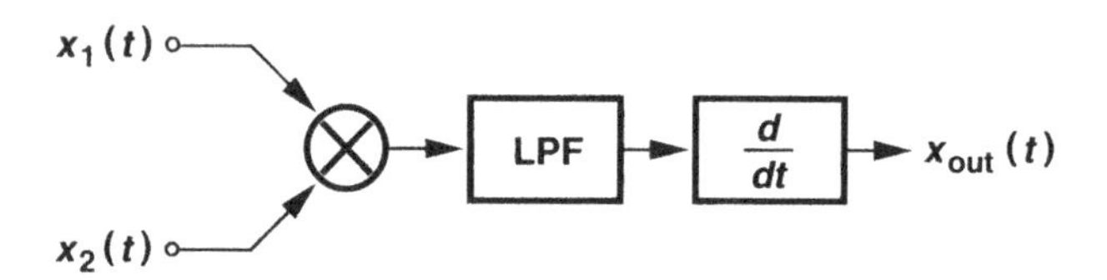

Figure 9.39 Mixing and differentiation for frequency detection.

average is still zero and the polarity depends on the initial phases of $x_1(t)$ and $x_2(t)$. In other words, a single multiplier, even followed by a differentiator, fails to detect the frequency difference.

How can we derive a unique average from $(\omega_1-\omega_2)\sin(\omega_1-\omega_2)t$? If this term is multiplied by $\sin(\omega_1-\omega_2)t$, the product yields a dc component proportional to $\omega_1-\omega_2$, retaining the polarity as well. However, the topology of Fig. 9.39 provides only a $\cos(\omega_1-\omega_2)t$ term (at the output of the LPF), whose product with $(\omega_1-\omega_2)\sin(\omega_1-\omega_2)t$ gives a zero average. We must therefore modify the circuit so as to obtain a $\sin(\omega_1-\omega_2)t$ component.[4] Illustrated in Fig. 9.40(a), a possible solution is to multiply one of the inputs, $x_1(t)$ by both the in-phase and quadrature phases of the other, $x_{2,I}(t) = A_2\cos\omega_2 t$ and $x_{2,Q}(t) = A_2\sin\omega_2 t$. We now have

$$x_A(t) = \frac{A_1A_2}{2}\cos(\omega_1-\omega_2)t \tag{9.11}$$

$$x_B(t) = \frac{A_1A_2}{2}\sin(\omega_1-\omega_2)t, \tag{9.12}$$

and hence

$$x_C(t) \propto \left(\frac{A_1A_2}{2}\right)^2(\omega_1-\omega_2). \tag{9.13}$$

The key point is that $x_C(t)$ changes sign with $\omega_1-\omega_2$. The reader can show that the initial phases of $x_1(t)$ and $x_2(t)$ do not affect $x_C(t)$ so long as $x_{2,I}(t)$ and $x_{2,Q}(t)$ remain in quadrature.

The topology of Fig. 9.40(a) is called a "quadricorrelator" [10]. Requiring that $x_1(t)$ contain a spectral line, the circuit must be preceded by an edge detector for operation with random NRZ data [Fig. 9.40(b)].

[4]The reader may wonder if $\cos(\omega_1-\omega_2)t$ can be shifted by 90° to generate $\sin(\omega_1-\omega_2)t$. However, near lock, $\omega_1 \approx \omega_2$, and shifting such a low frequency signal is difficult.

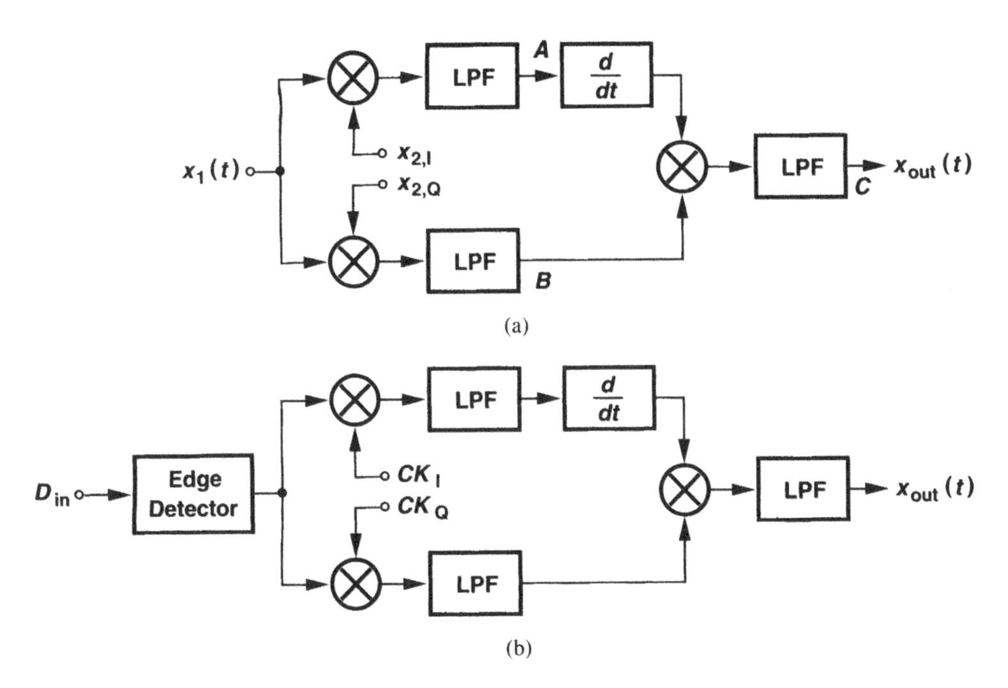

Figure 9.40 (a) Quadricorrelator, (b) FD including edge detection.

In the above analysis, we assumed that the LPF at the output of the quadricorrelator suppresses the other cross product, namely, $\cos 2(\omega_1 - \omega_2)t$. In an FLL, however, as the VCO frequency is driven toward the input frequency, $\omega_1 - \omega_2$ may fall to arbitrarily small values, eventually creating a large "ripple" at the output. The ripple may lead to difficulty in the final settling of the loop or the transition to phase locking. Figure 9.41 shows a "balanced" version [11], which suppresses the component at $2(\omega_1 - \omega_2)$ in the output

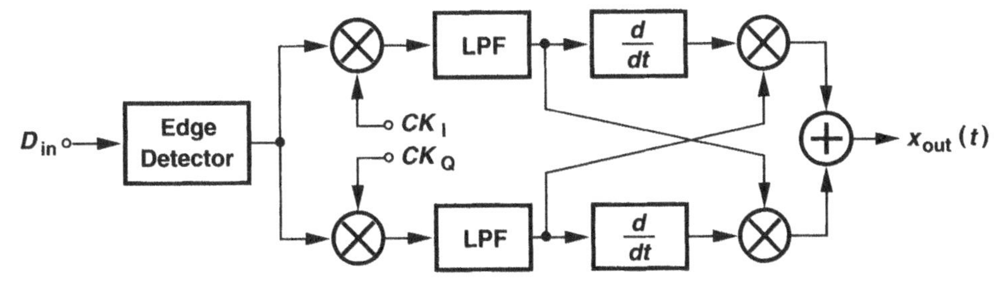

Figure 9.41 Modified quadricorrelator.

adder, thus improving the lock behavior.

It is possible to construct a digital version of the quadricorrelator, obviating the need for an analog edge detector and achieving robust operation. Recall from Section 9.1 that a DFF that samples the clock by the data can be viewed as an edge detector followed by a multiplier. In fact, a DFF PD operating on both edges of the data produces a beat output if the data rate and the clock frequency are unequal.[5] Moreover, similar to a multiplier, such

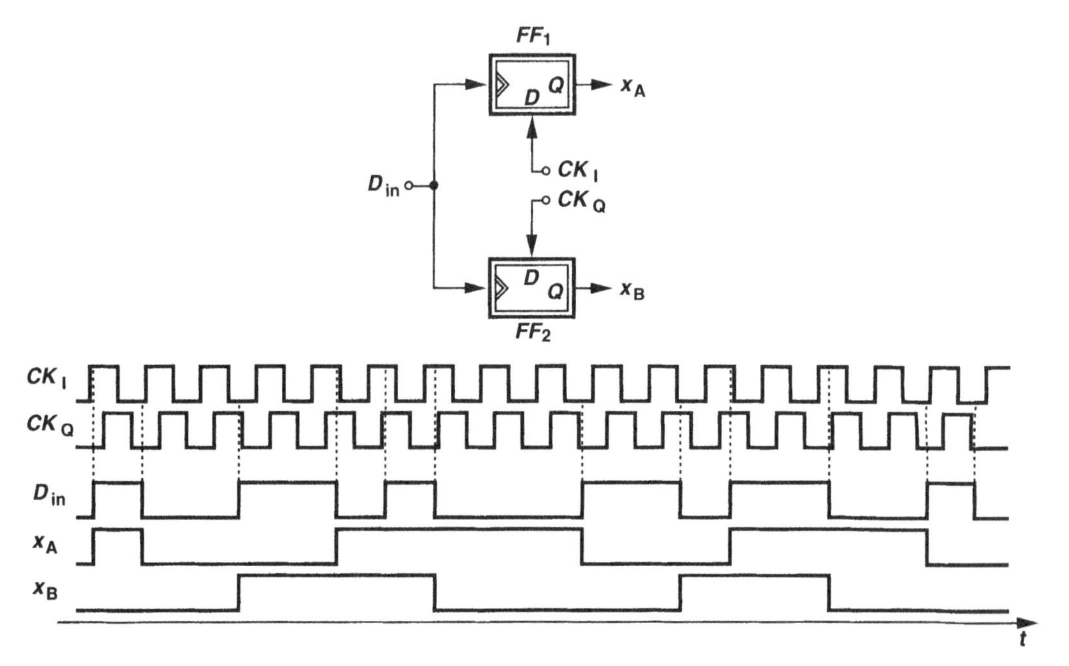

Figure 9.42 Sampling of quadrature clocks by data to determine frequency difference.

a PD translates a phase shift in the clock to an equal shift in the beat waveform. We can therefore employ two DFF PDs that sample the quadrature phases of the clock by the data edges (Fig. 9.42), thereby generating two beat waveforms that bear a 90° phase difference. Note the resemblance between these outputs and those at ports A and B in Fig. 9.40(a).

We must now make an important observation in the circuit of Fig. 9.42. As the difference between the data rate and the clock frequency goes from negative to positive, the phase difference between x_A and x_B jumps from $+90°$ to $-90°$. To understand this phenomenon, let us consider the output waveforms for both a fast clock and a slow clock (Fig. 9.43). If CK is fast and the rising data transition at $t = t_1$ samples a ONE on CK_I and a ZERO on CK_Q, then the next rising edge of data appears somewhat *later* than $t = T_{CK} + t_1$, sampling a point on CK_I closer to its next falling edge and a point on CK_Q closer to its next rising edge. Thus, after some cycles, at $t = t_2$, D_{in} begins to sample a high level on both CK_I and CK_Q; i.e., $x_B(t)$ goes high while $x_A(t)$ is high.

Now suppose the clock is slow. As illustrated in Fig. 9.43(b), the next rising edge of data samples closer to the falling edge of CK_Q. Thus, after data edges drift enough, $x_B(t)$ goes high while $x_A(t)$ is low.

The foregoing observation proves essential to completing the circuit of Fig. 9.42. If the rising edges of $x_B(t)$ are used to sample the levels on $x_A(t)$, then for $f_{VCO} > R_b$,

[5]The double arrow inside the flipflop symbol denotes double-edge triggering.

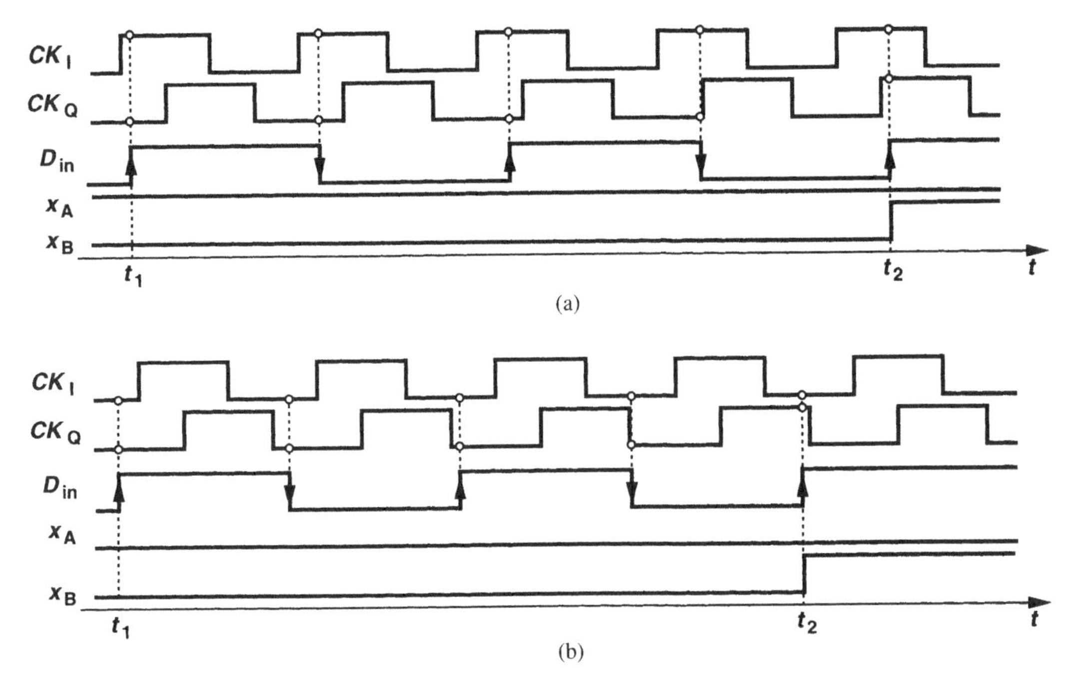

Figure 9.43 FD waveforms for fast and slow clocks.

the result is high and for $f_{VCO} < R_b$, it is low. Illustrated in Fig. 9.44 [12], the overall

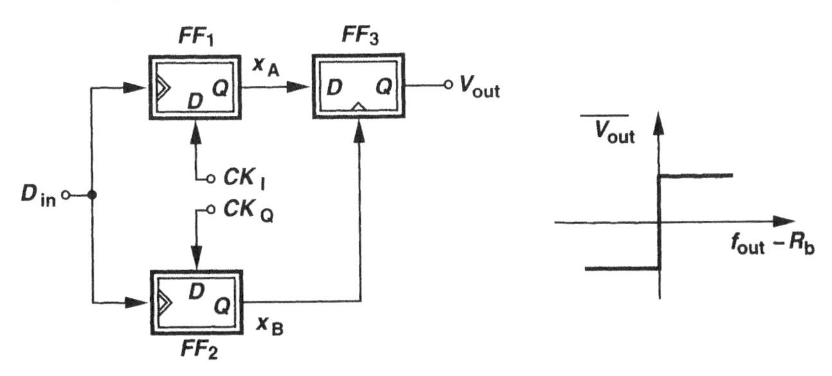

Figure 9.44 Complete frequency detector.

frequency detector exhibits a bang-bang characteristic.

Other examples of frequency detectors along with their properties are described in [13] and [4]. A half-rate FD is presented in [14].

The use of a frequency detector does not increase the capture range indefinitely. In fact, a brief examination of the FD of Fig. 9.44 reveals its capture limitations. Suppose $f_{CK} - R_b$ is so large that the positive data edges in Fig. 9.43(a) drift considerably from one bit to the next. Then, the outputs of FF_1 and FF_2 appear as shown in Fig. 9.45(a), indicating that FF_3 is metastable for most of the samples. In essence, the bang-bang characteristic of Fig.

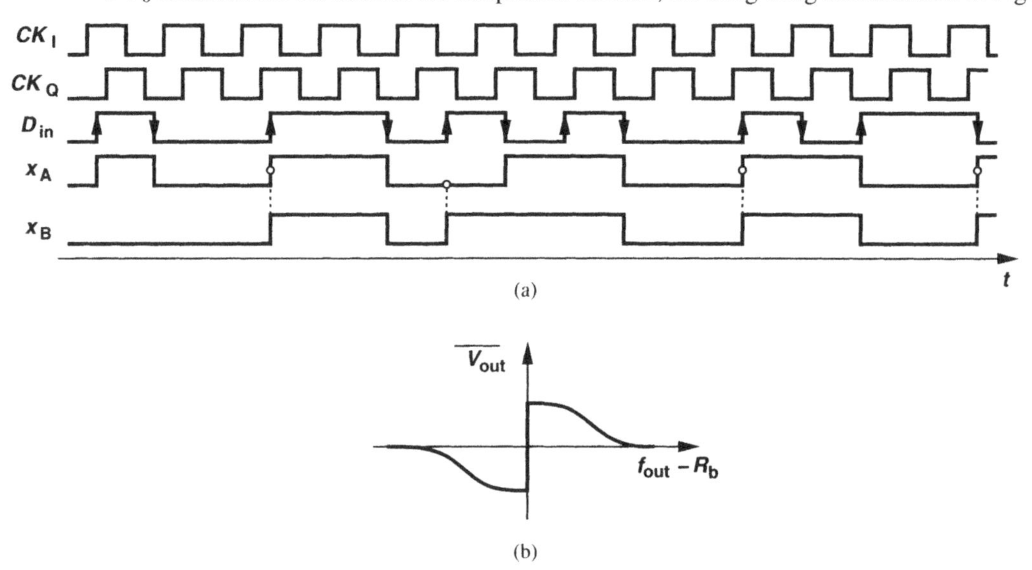

Figure 9.45 (a) Failure of the FD of Fig. 9.44 in the presence of a large frequency error, (b) actual FD characteristic.

9.44 is modified as depicted in Fig. 9.45(b) to reflect the lack of a reliable average as the frequency error increases. In practice, the capture range of an FLL incorporating the FD of Fig. 9.44 falls below approximately 15%.

9.4 CDR Architectures

With our study of phase and frequency detectors in Sections 9.2 and 9.3, we can now develop complete CDR architectures. Each architecture must include (a) frequency and phase acquisition to ensure lock despite process and temperature variations of the VCO frequency, and (b) data retiming inside the PD to avoid systematic skews.

A critical difficulty in modern CDR circuits stems from the use of low supply voltages. As mentioned in Chapter 6, the gain of VCOs must *increase* as the supply is scaled down because the tuning range must remain a constant percentage of the center frequency. As a result, for a given ripple on its control line, the VCO suffers from greater jitter. A method of alleviating this issue is to decompose the VCO control into "fine" and "coarse" inputs, allowing the latter to remain quiet after the system is phase-locked. This concept is described in the context of some CDR architectures.

9.4.1 Full-Rate Referenceless Architecture

Frequency detectors capable of handling random data obviate the need for external reference frequencies. Figure 9.46 depicts a referenceless architecture, where Loop I employs

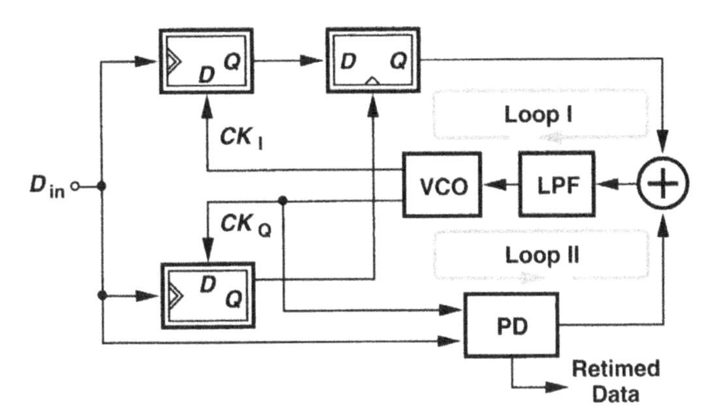

Figure 9.46 CDR architecture incorporating FD and PD.

the FD of Fig. 9.44 and Loop II incorporates one of the phase detectors studied in Section 9.2. Upon start-up or loss of phase lock, the FD produces a dc level that drives the VCO frequency toward the input data rate. When the frequency error falls within the capture range of Loop I, the PD takes over, phase-locking the clock to the data.

The above architecture entails three issues. First, as the CDR circuit transfers the control of the VCO from the FD to the PD, the two loops may interact so heavily that the overall system fails to phase-lock. In this case, the two loops continue to "fight" indefinitely. Second, with the actual random data produced in a network, short-term spectral lines close but unequal to the nominal data rate may appear occasionally, possibly confusing the frequency detector. Third, if, in phase-lock, the edges of CK_I or CK_Q happen to be aligned with those of D_{in}, the FD may still produce extraneous pulses. For these reasons, the bandwidth of the frequency-locked loop is typically chosen to be much smaller than that of the PLL. These issues must be studied carefully for each specific design and application.

Note that this architecture employs only one control line for the oscillator and must therefore ensure a very small ripple. It is possible to modify the system as shown in Fig. 9.47, where Loop II drives only the coarse control. This modification also permits independent choice of the PLL and FLL bandwidths.

9.4.2 Dual-VCO Architecture

Recall that it is desirable to decompose the VCO control into fine and coarse inputs. A CDR architecture using such a scheme is shown in Fig. 9.48 [15]. Here, Loop I phase-locks VCO_1 to the input data through the fine control. Since the gain of VCO_1 with respect to the fine input is relatively low, ripple on this line translates to a small jitter at the output. Of course, the fine control may not provide enough tuning range to encompass process and temperature variations. Loop II is therefore added to lock VCO_2, a replica of VCO_1, to Nf_{REF}, with the resulting control voltage applied to the coarse input of VCO_1 as well. If Nf_{REF} is exactly equal to the input data rate, the two VCOs match perfectly, and the total gain of VCO_1 is equal to the gain of VCO_2, then the fine control of VCO_1 stabilizes at a voltage equal to that on the coarse control. The low-pass filter consisting of R_1 and C_1

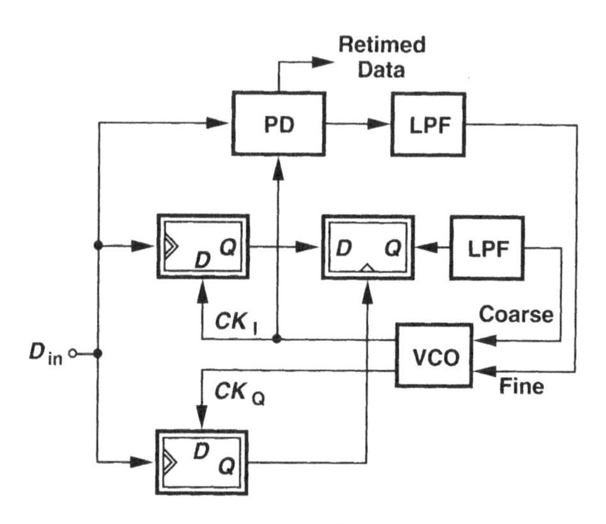

Figure 9.47 CDR architecture with coarse and fine VCO control.

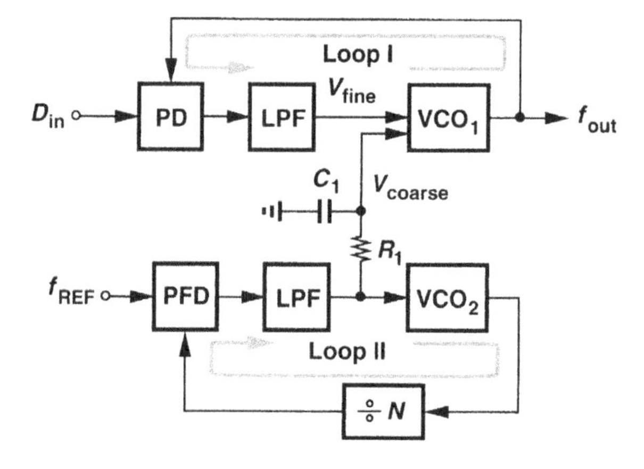

Figure 9.48 CDR architecture using two VCOs.

suppresses the ripple generated in Loop II, presenting a low-noise control to VCO_2.

While reducing the effect of ripple, the architecture of Fig. 9.48 faces two issues. First, inevitable random mismatches between the two VCOs lead to a substantial center frequency mismatch even though the oscillators share the same coarse input. For this reason, Loop I must still achieve a sufficiently wide capture range to guarantee lock despite the initial frequency error. With a typical frequency mismatch of a few percent, random-data PDs may not provide adequate capture range.

Second, even if the oscillators match perfectly, the incoming data rate is not exactly equal to Nf_{REF} because the reference frequency in the far-end transmitter is derived from a crystal oscillator that may suffer from an error of 5 to 10 ppm with respect to the f_{REF} generator in the receiver. Consequently, VCO_1 and VCO_2 operate at slightly different frequencies, possibly pulling each other through the substrate or supply lines. The use of differential swings for both VCOs alleviates this issue, but random asymmetries in each circuit still give rise to a finite amount of crosstalk. It is interesting to note that the large bandwidth of Loop II partially corrects for the pulling experienced by VCO_2. Loop I, on the other hand, has a bandwidth commensurate with optical standards and hence cannot counteract pulling effects on VCO_1.

Another general concern in the architecture of Fig. 9.48 relates to the layout of the two VCOs. If both oscillators incorporate LC tanks, the large area occupied by on-chip inductors creates difficulties in routing the signal and power lines.

9.4.3 Dual-Loop Architecture with External Reference

Figure 9.49 shows a relatively simple CDR architecture that acquires frequency and phase

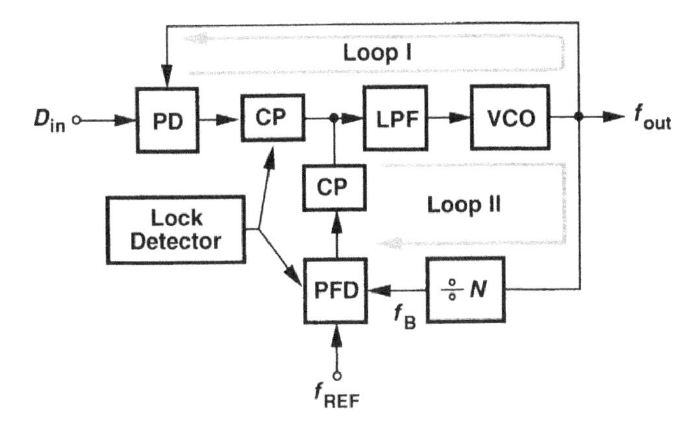

Figure 9.49 CDR architecture using external reference.

in two steps. Using a single VCO, the circuit first enables Loop II, thereby beginning to lock the oscillator to Nf_{REF}. The lock detector monitors the difference between f_B and f_{REF}, disabling Loop II and enabling Loop I when the frequency error drops to a sufficiently small value (e.g., 0.1%). Thus, Loop I begins with a frequency error well within its capture range, locking the VCO to the data. The lock detector continues to operate so that Loop II can be activated again if Loop I loses lock due to unexpected noise.

Let us compare the CDR architectures of Figs. 9.48 and 9.49. Both topologies require an external reference but the latter need not deal with frequency mismatches or oscillator pulling. However, the former allows the use of fine and coarse controls for the oscillator whereas the latter does not.

Another issue in the architecture of Fig. 9.49 relates to the transition from Loop II to Loop I. If the switches that perform this transition disturb the control voltage significantly, then the VCO frequency may jump by a large amount, falling out of the capture range of Loop I. Thus, the charge injection and clock feedthrough of the switches must be examined carefully.

The transition from Loop II to Loop I also requires attention to the design of the low-pass filter(s) preceding the VCO. Consider the filter shown in Fig. 9.50(a), where $C_2 <$

$0.2C_P$ so that it has negligible effect on the stability. The damping factor of Loop II is

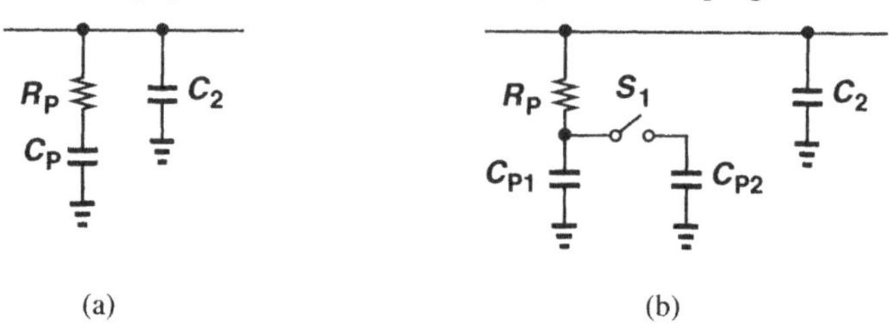

Figure 9.50 (a) Simple and (b) switchable loop filters.

given by $\zeta = (R_P/2)\sqrt{I_P C_P K_{VCO}/(2\pi N)}$, typically necessitating large values for R_P, I_P, or C_P to compensate for the effect of N. However, such values of R_P and C_P may not yield the required transfer function for Loop I, making it desirable to change R_P or C_P when this loop takes over.

Interestingly, C_P must remain the same for both loops, and only R_P can assume different values. Consider the scenario illustrated in Fig. 9.50(b), where C_{P1} stores the required oscillator control voltage when Loop II is active. Capacitor C_{P2} is added in parallel with C_{P1} upon transition to Loop I because this loop requires a greater capacitor. However, charge sharing between C_{P1} and C_{P2} heavily corrupts the stored voltage, possibly driving the VCO frequency out of the capture range of Loop I.

9.4.4 Quarter-Rate Phase Detectors

The concept of half-rate phase detection can be extended to afford quarter-rate operation. We simply require that the VCO generate enough edges (phases) to ensure that every data transition is sampled by the phase detector. A fringe benefit of quarter-rate operation is that the circuit inherently provides 1-to-4 demultiplexing as well. We consider a binary PD here.

Consider the NRZ data and the quarter-rate clock shown in Fig. 9.51(a) and assume we employ flipflops that sample the data on only the rising edge of the clock. We note that CK samples D_{in} at t_1 and $t_1 + t_{CK}$ but misses a number of transitions in between. Recall from the Alexander PD analysis that we must also sample D_{in} at the points denoted by solid circles and compare their values. Thus, we need another seven phases of the clock. Illustrated in Fig. 9.51(b), the eight clock phases sample all of the critical points on the data waveform, requiring eight flipflops.

How do we combine the eight samples to extract the phase information? In a manner similar to the Alexander PD, we compare each two consecutive samples by means of an XOR gate. To interpret the resulting outputs, we consider both late and early clock phases. As shown in Fig. 9.51(c), if the clock is late, the sample taken by CK_{45} (B) is unequal to that taken by CK_0 (A) but equal to that taken by CK_{90} (C). Thus, $A \oplus B = 1$ and $B \oplus C = 0$. This pattern continues for each data transition. Conversely, if the clock is early, [Fig. 9.51(d)], then the pattern $A \oplus B = 0$, $B \oplus C = 1$ repeats around data edges. In other words, the *sign* of the difference between $A \oplus B$ and $B \oplus C$ carries the late or early phase information. We can therefore sense these differences by voltage-to-current (V/I) converters and sum the output currents.

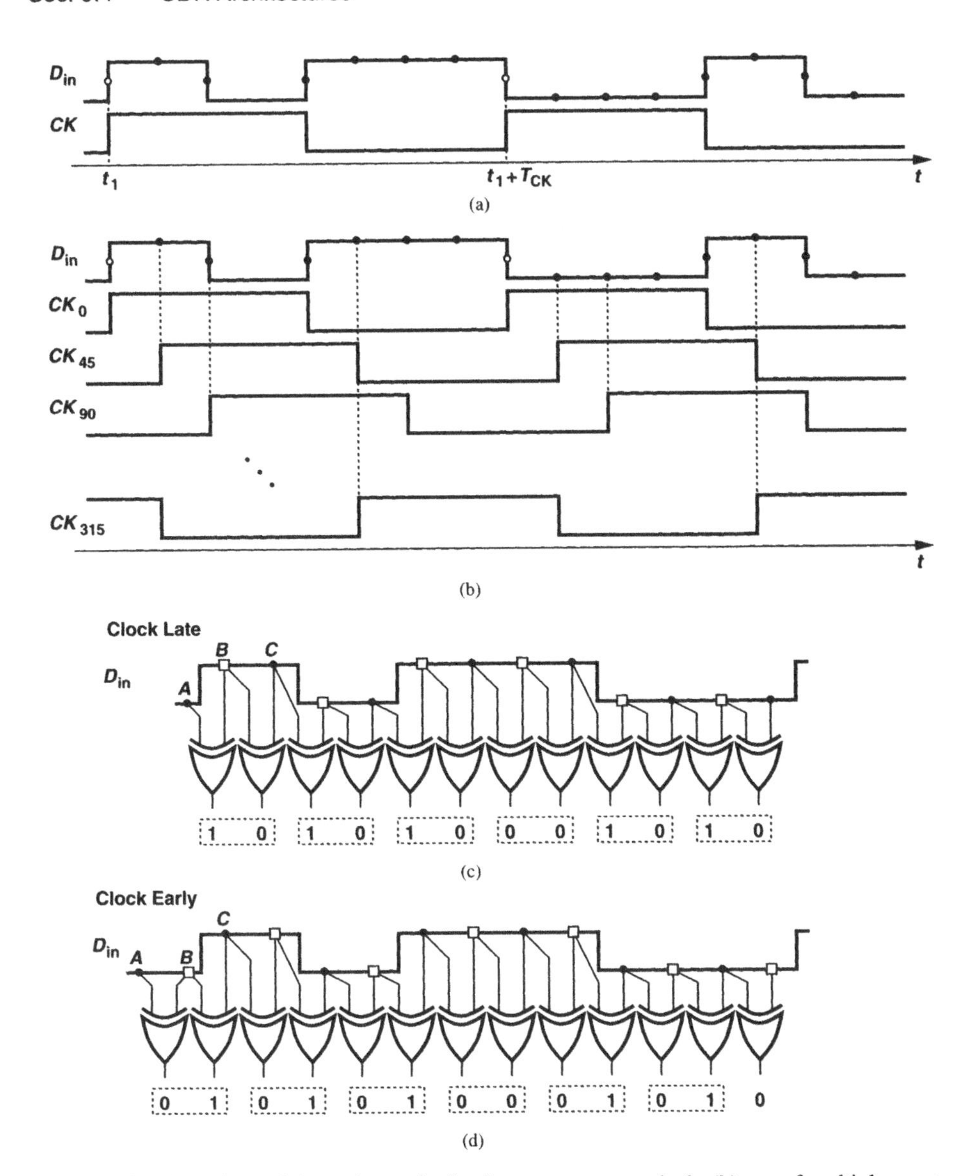

Figure 9.51 Sampling of data using a single-phase quarter-rate clock, (b) use of multiple quarter-rate clock phases, (c) outputs generated by XOR gates if clock is late, (d) outputs generated by XOR gates if clock is early.

Figure 9.52 shows the overall phase detector topology [19]. The circuit consists of eight flipflops, eight XOR gates, and four V/I converters. The samples taken by CK_{45}, CK_{135}, CK_{225}, and CK_{315} serve as the demultiplexed outputs. The total output current exhibits a bang-bang behavior with respect to the phase difference between the data and the clock. Note that in the absence of data transitions, the flipflops generate equal outputs, and each

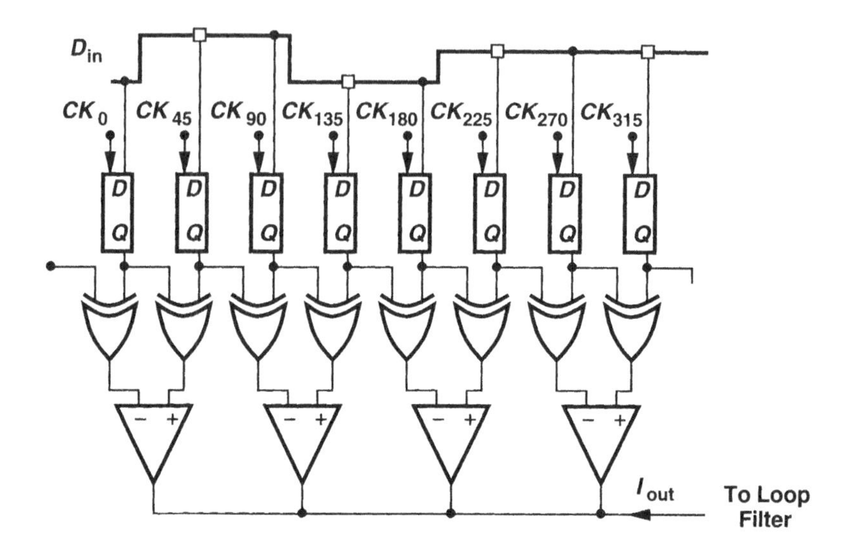

Figure 9.52 Example of quarter-rate phase detector.

V/I converter produces a zero current, in essence presenting a tristate (high) impedance to the loop filter [19].

It is interesting to compare the above quarter-rate architecture with a full-rate system in terms of power dissipation, hardware, and clock load capacitance [19]. The PD of Fig. 9.52 employs 16 latches to perform phase detection, data retiming, and 1-to-4 demultiplexing, with each clock phase driving two latches. Now consider a full-rate architecture. The reader can readily show that such an architecture requires seven latches to form an Alexander PD, four latches for two divide-by-two circuits, and (at least) nine latches for the demultiplexer [19]. That is, the full-rate clock drives the input capacitance of nine latches. We thus conclude that the quarter-rate and full-rate architectures consume comparable power levels but the latter presents a substantially larger capacitance to the full-rate clock.

9.5 Jitter in CDR Circuits

CDR circuits targeting optical communication standards must satisfy stringent and difficult jitter specifications. In this section, we describe the CDR jitter characteristics of interest in optical systems and present methods of estimating such characteristics. Optical standards often express jitter in terms of the bit period, also called the "unit interval" (UI). For example, a jitter of 0.01 UI (10 mUI) refers to 1% of the bit period.

Recall from Chapter 2 that jitter represents (1) the deviation of the zero crossings of a waveform from their ideal points on the time axis, or, from another perspective, (2) the deviation of each period from its ideal value. We also noted that the *rate of change* of jitter is reflected in the input-output transfer function of PLLs. CDR loops are characterized by "jitter transfer," "jitter generation," and "jitter tolerance."

9.5.1 Jitter Transfer

The jitter transfer function of a CDR circuit represents the output jitter as the input jitter is varied at different rates. This characteristic is indeed the same as the closed-loop transfer function defined for PLLs in Chapter 8.

What is the desirable jitter transfer function for CDR loops? We note that if the input jitter varies slowly, i.e., if the zero crossings wander from their ideal points at a low rate, then the output must follow the input to ensure phase-locking. On the other hand, if the input jitter varies rapidly, the CDR circuit must filter the jitter, i.e., the output must track the input to a lesser extent. Thus, the jitter transfer exhibits a low-pass characteristic, as is the case with phase-locked loops.

The jitter transfer required by optical standards must meet two difficult specifications. First, the CDR bandwidth is very small, i.e., approximately 120 kHz in OC-192. In other words, a 10-Gb/s CDR loop must attenuate jitter components above 120 kHz. The small loop bandwidth provides little suppression of the VCO phase noise, necessitating low-noise oscillator design. Second, "jitter peaking," i.e., the amount of peaking in the jitter transfer, must be less than 0.1 dB ($\approx 1.2\%$) (Fig. 9.53). This constraint demands careful attention

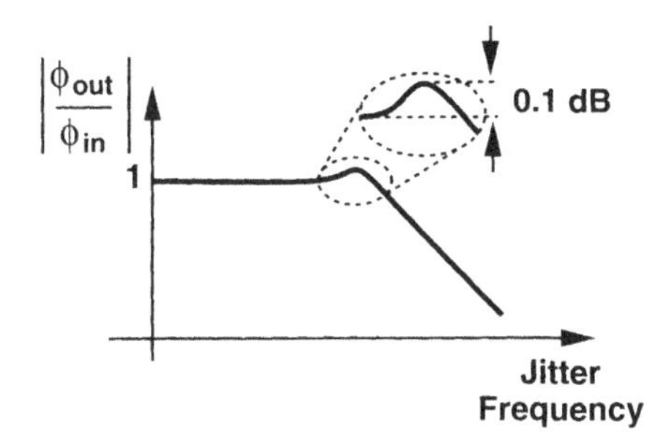

Figure 9.53 Jitter transfer function.

to the poles and zeros in the PLL.

The reader may wonder why SONET imposes the above requirements on CDR design. Long-haul networks employ many data regenerators in the signal path, suppressing the effect of fiber nonidealities periodically. It is therefore important to minimize the jitter accumulated through the chain of the repeaters, i.e., each repeater must provide a small jitter bandwidth. Furthermore, since the equivalent jitter transfer function of a cascade of regenerators is given by the *product* of the individual transfer functions, the amount of jitter peaking per regenerator must be so small that tens of them still yield acceptable peaking [16].

Loop Bandwidth Reduction How can the bandwidth of a CDR loop be reduced? Let us first determine the -3-dB bandwidth of a simple charge-pump PLL. Rewriting the transfer function from Chapter 8 as:

$$H(s) = \frac{2\zeta\omega_n s + \omega_n^2}{s^2 + 2\zeta\omega_n s + \omega_n^2}, \tag{9.14}$$

we must obtain the value of $s_0 = j\omega_{-3dB}$ such that

$$\left|\frac{2\zeta\omega_n s_0 + \omega_n^2}{s_0^2 + 2\zeta\omega_n s_0 + \omega_n^2}\right| = \frac{1}{\sqrt{2}}. \tag{9.15}$$

In other words,

$$\frac{4\zeta^2\omega_n^2\omega_{-3dB}^2 + \omega_n^2}{(-\omega_{-3dB}^2 + \omega_n^2)^2 + (2\zeta\omega_n\omega_{-3dB})^2} = \frac{1}{2}. \tag{9.16}$$

It follows that

$$\omega_{-3dB}^4 - 2\omega_n^2(2\zeta^2 + 1)\omega_{-3dB}^2 - \omega_n^4 = 0 \tag{9.17}$$

and hence

$$\omega_{-3dB}^2 = \left[(2\zeta^2 + 1) + \sqrt{(2\zeta^2 + 1)^2 + 1}\right]\omega_n^2. \tag{9.18}$$

To reduce ω_{-3dB}, either ζ or ω_n must be decreased. However, loop stability requires that ζ exceed $\sqrt{2}/2$, leaving ω_n as the only candidate. Since $\omega_n = \sqrt{I_P K_{VCO}/(2\pi C_P)}$ and $\zeta = (R_P/2)\sqrt{I_P C_P K_{VCO}/(2\pi)}$, we note that, if ω_n is reduced by lowering I_P or K_{VCO}, then ζ suffers. Thus, C_P is the principal parameter that can decrease the loop bandwidth without compromising the stability. In fact, as C_P increases, so does ζ.

The above observation leads to two important results. First, CDR loops must often incorporate large off-chip capacitors to achieve a small bandwidth. Second, such loops typically exhibit a large ζ (as high as 10 or 20), and their transfer functions can be simplified to develop more insight. Specifically, since ω_n is small, the term ω_n^2 can be neglected in Eq. (9.14) for some range of $s = j\omega$, yielding:

$$H(s) \approx \frac{2\zeta\omega_n}{s + 2\zeta\omega_n}. \tag{9.19}$$

That is, the -3-dB bandwidth of the loop is equal to $2\zeta\omega_n$ rad/s ($= \zeta\omega_n/\pi$ Hz). Interestingly, we obtain the same result by assuming $2\zeta^2 \gg 1$ in Eq. (9.18). We also note that

$$\omega_{-3dB} \approx 2\zeta\omega_n \tag{9.20}$$

$$= \frac{R_P I_P K_{VCO}}{2\pi}, \tag{9.21}$$

a value independent of the loop filter capacitor. This point should not be confusing: *if* C_P is large enough to create a small ω_n and a large ζ, *then* the -3-dB bandwidth becomes independent of C_P (Fig. 9.54). The reader may then wonder how ω_{-3dB} can be lowered. In practice, R_P is reduced and C_P is increased such that ζ remains constant.

We must now answer an important question: for what range of s is the approximation given by (9.19) valid? Considering the magnitude of $H(s)$ in (9.14), we note that

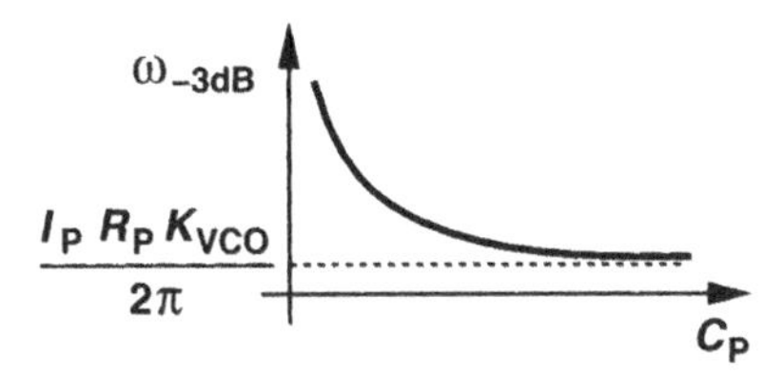

Figure 9.54 Variation of −3-dB bandwidth with loop filter capacitor.

a sufficient condition is $(2\zeta\omega_n\omega)^2 \gg \omega_n^4$, because it guarantees both the numerator and denominator of (9.14) are reduced to those of (9.19). That is,

$$\omega^2 \gg \left(\frac{\omega_n}{2\zeta}\right)^2. \tag{9.22}$$

Thus, if ω is greater than 3 to 4 times $\omega_n/(2\zeta)$, the magnitude of the transfer function can be approximated by that of a one-pole system. This falls well below the −3-dB frequency because $\omega_{-3dB} = 2\zeta\omega_n \gg \omega_n/(2\zeta)$. Consequently, the one-pole approximation holds for most of the range.

Another effect that we have thus far ignored relates to the input data transition density, D_T. Recall that the phase detectors studied in Section 9.2 generate a zero output in the absence of data transitions, thereby exhibiting a gain proportional to D_T. To take this effect into account, we simply multiply the charge pump current by D_T:

$$\omega_n = \sqrt{\frac{I_P D_T K_{VCO}}{2\pi C_P}} \tag{9.23}$$

$$\zeta = \frac{R_P}{2}\sqrt{\frac{I_P D_T C_P K_{VCO}}{2\pi}} \tag{9.24}$$

and hence

$$\omega_{-3dB} = \frac{R_P I_P D_T K_{VCO}}{2\pi}. \tag{9.25}$$

For example, if ONEs and ZEROs occur with equal probabilities and the phase detector operates with only the rising or falling edges of data, then $D_T = 0.5$.

Jitter Peaking As mentioned earlier, jitter peaking must not exceed 0.1 dB in SONET. Let us now examine the closed-loop transfer function given by Eq. (9.14) for this effect. The zero is given by

$$\omega_z = \frac{-\omega_n}{2\zeta} \tag{9.26}$$

$$= \frac{-1}{R_P C_P}, \tag{9.27}$$

a quantity independent of the data transition density. The two poles are equal to

$$\omega_{p1,2} = (-\zeta \pm \sqrt{\zeta^2 - 1})\omega_n \tag{9.28}$$

$$= (-1 \pm \sqrt{1 - \frac{1}{\zeta^2}})\zeta\omega_n. \tag{9.29}$$

Since $\zeta \gg 1$, the square root can be approximated as $\sqrt{1-\epsilon} \approx 1 - \epsilon/2 - \epsilon^2/8$:

$$\omega_{p1,2} \approx \left[-1 \pm (1 - \frac{1}{2\zeta^2} - \frac{1}{8\zeta^4})\right]\zeta\omega_n. \tag{9.30}$$

It follows that

$$\omega_{p1} = -\frac{\omega_n}{2\zeta} - \frac{\omega_n}{8\zeta^3} \tag{9.31}$$

$$\omega_{p2} = -2\zeta\omega_n + \frac{\omega_n}{2\zeta} + \frac{\omega_n}{8\zeta^3}. \tag{9.32}$$

Illustrated graphically in Fig. 9.55(a), the above expressions yield several interesting re-

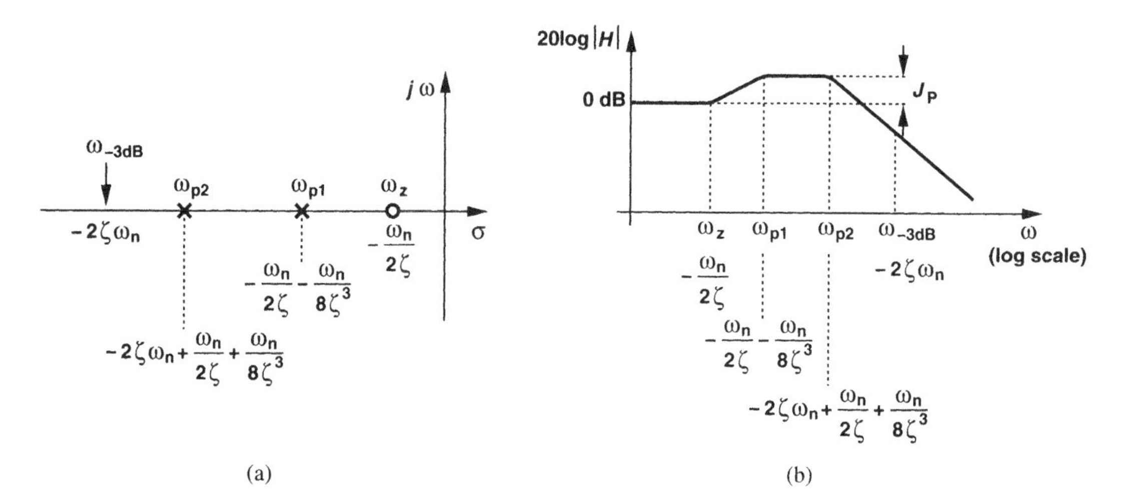

Figure 9.55 (a) Position of poles and zeros of a type II PLL, (b) corresponding jitter transfer function.

sults. First, the zero *always* appears before the poles, inevitably leading to peaking. Second, the zero and the first pole differ in magnitude by $\omega_n/8\zeta^3$, a small value because $\zeta \gg 1$. Third, the ω_z-ω_{p1} pair falls well below ω_{p2} because $\omega_n/(2\zeta) \ll 2\zeta\omega_n$. Fourth, ω_{p2} is slightly lower than ω_{-3dB}, by an amount equal to the magnitude of ω_{p1}.

Figure 9.55(b) shows a Bode plot of the magnitude of the closed-loop transfer function. The magnitude begins to rise at 20 dB/dec at $\omega = \omega_z$, assumes a constant value for $\omega >$

ω_{p1}, and begins to fall at 20 dB/dec at $\omega > \omega_{p2}$, dropping by -3 dB at $\omega = 2\zeta\omega_n$. With logarithmic scales, the value of jitter peaking, J_P, can be written as[6]

$$\frac{20 \log J_P}{20 \log \omega_{p1} - 20 \log \omega_z} = 1. \tag{9.33}$$

That is,

$$J_P = \frac{\omega_{p1}}{\omega_z} \tag{9.34}$$

$$\approx 1 + \frac{1}{4\zeta^2}. \tag{9.35}$$

To express J_P in decibels, we recognize that

$$10 \log J_P = 20(\ln J_P)(\log_{10} e) \tag{9.36}$$

$$= 8.686 \ln(1 + \frac{1}{4\zeta^2}), \tag{9.37}$$

which, since $4\zeta^2 \gg 1$, reduces to

$$20 \log J_P \approx \frac{8.686}{4\zeta^2} \tag{9.38}$$

$$= \frac{2.172}{\zeta^2}. \tag{9.39}$$

For example, to ensure $20 \log J_P < 0.1$ dB, ζ must exceed 4.66. We can also express jitter peaking in terms of the CDR parameters:

$$20 \log J_P = \frac{2.171}{\frac{R_P^2}{4} I_P C_P K_{VCO} D_T} \tag{9.40}$$

$$= \frac{8.686}{R_P^2 I_P C_P K_{VCO} D_T}. \tag{9.41}$$

If R_P is lowered to obtain a smaller jitter bandwidth, then C_P must be raised substantially to maintain J_P constant. This is indeed the same trend required to achieve a constant ζ.

9.5.2 Jitter Generation

Jitter generation refers to the jitter produced by a CDR circuit itself when the input random data contains no jitter. From our studies in Chapter 8 and in this chapter, we summarize the sources of jitter as follows: (1) VCO phase noise due to electronic noise of its constituent

[6]Since $\omega_{p2} \gg \omega_{p1}$, we assume that $20 \log |H|$ is flat between the poles. If ω_{p1} and ω_{p2} are comparable, J_P is smaller.

devices; (2) ripple on the control voltage; (3) coupling of data transitions to the VCO through the phase detector and retiming circuits; (4) supply and substrate noise.

Let us first consider the contribution of the VCO to CDR jitter. Our objective is to derive the jitter of a phase-locked CDR in terms of the jitter of the *free-running* VCO. Recall from Chapter 2 that the phase noise due to white noise sources and the cycle-to-cycle jitter of a free-running VCO are related by the following equation:

$$\Delta T_{cc}^2 \approx \frac{4\pi}{\omega_0^3} S_\phi(\Delta\omega)\Delta\omega^2, \tag{9.42}$$

where ω_0 denotes the oscillation frequency and $S_\phi(\Delta\omega)$ represents the relative phase noise power at an offset frequency of $\Delta\omega$ [17]. We must therefore take into account the effect of noise shaping by the PLL. It can be assumed that for a loop bandwidth of $2\pi f_u$, the jitter rises with the square root of time (as if the oscillator were free-running) until $t_1 = (2\pi f_u)^{-1}$ and "saturates" thereafter (Fig. 9.56) [18]. As proved in [17], the total jitter accumulated over time t_1 by a free-running oscillator is equal to

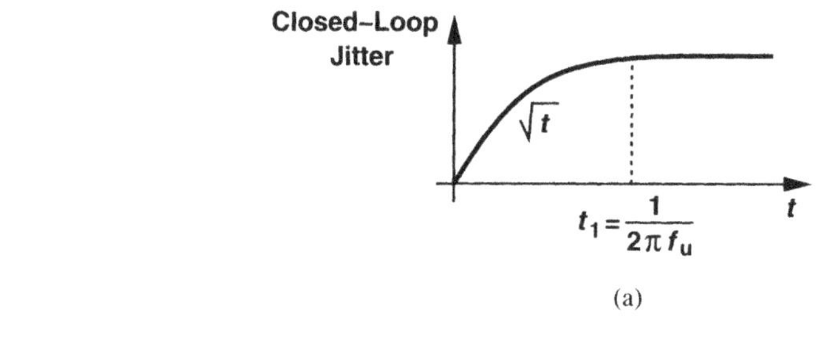

(a)

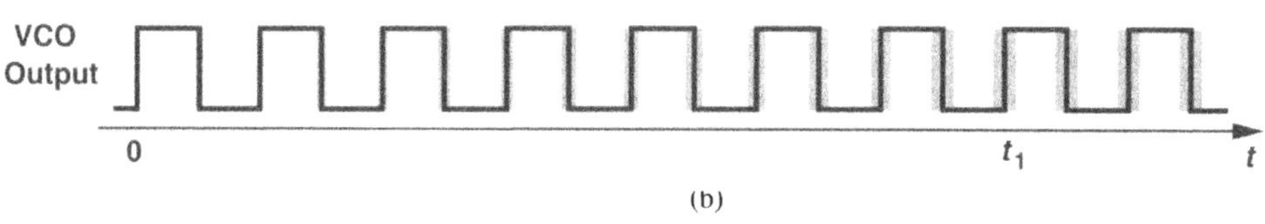

(b)

Figure 9.56 Accumulation of cycle-to-cycle jitter in a phase-locked oscillator: (a) actual behavior and (b) conceptual behavior.

$$\Delta T_1 = \sqrt{\frac{f_0}{2}}\Delta T_{cc}\sqrt{t_1}. \tag{9.43}$$

Substituting (9.42) in (9.43) yields the closed-loop jitter as

$$\Delta T_{PLL} = \frac{1}{\sqrt{2\pi f_u}}\sqrt{S_\phi(\Delta\omega)}\frac{\Delta\omega}{\omega_0}. \tag{9.44}$$

For example, if ΔT_{PLL} must be less than 0.25 ps,rms, at 40 GHz and $f_u = 20$ MHz, then $S_\phi(\Delta\omega)$ must not exceed -79 dBc/Hz at 1-MHz offset.

Another source of jitter in CDR circuits is the ripple on the control voltage. As explained in Chapter 8, various mismatches lead to a net injection into the loop filter on every phase comparison instant even if the loop is locked. With random data, the ripple is also generally random. Owing to the high gain of low-voltage VCOs, the frequency modulation resulting from the ripple may be significant, thus yielding large jitter.

For periodic modulation of a VCO, it is possible to calculate the output jitter. In particular, as proved in [17], with a modulating sinusoid $V_m \cos \omega_m t$, the cycle-to-cycle jitter is given by

$$\Delta T_{cc} = \frac{V_m K_{VCO}}{f_0^2} \sqrt{1 - \cos \frac{\omega_m}{f_0}}. \tag{9.45}$$

For $f_m \ll f_0$, this expression reduces to

$$\Delta T_{cc} \approx \frac{V_m K_{VCO} \omega_m}{\sqrt{2} f_0^3}. \tag{9.46}$$

9.5.3 Jitter Tolerance

Our analysis of jitter properties has thus far focused on their effect on the recovered clock. However, the quality of the retimed data is also important. How should a CDR circuit perform retiming if the input data contains substantial jitter? From our studies of CDR loops, we make the following observations: (a) for slowly-varying jitter at the input, the recovered clock tracks the phase variations, always sampling the data in the middle of the eye [Fig. 9.57(a)] and guaranteeing a low bit error rate; (b) for rapidly-varying jitter, the

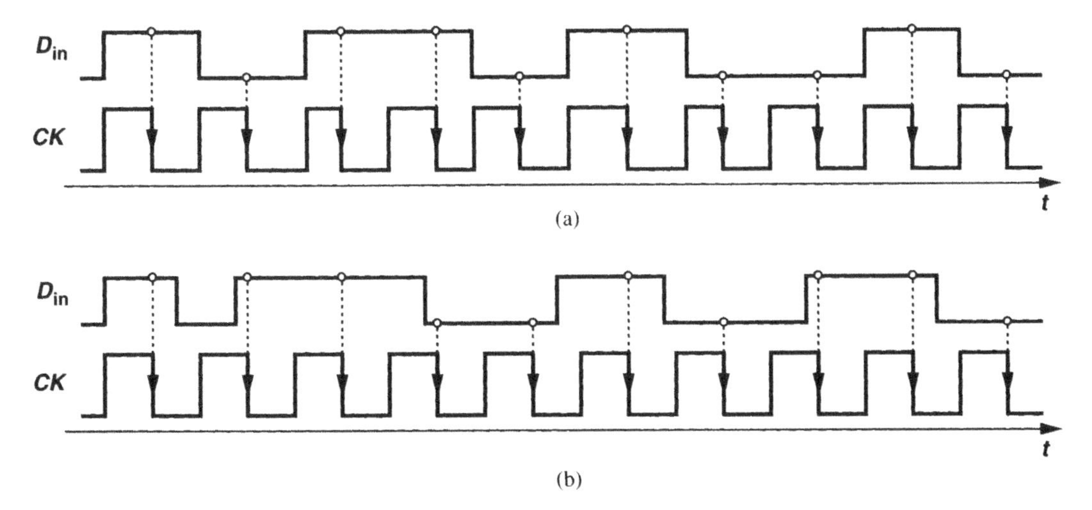

Figure 9.57 Effect of (a) slow and (b) fast jitter on data retiming.

clock cannot fully track the input phase variations, failing to sample the data optimally [Fig. 9.57(b)] and creating a greater BER. These are natural properties of CDR circuits but they must still conform to certain requirements posed by optical standards.

Jitter tolerance specifies how much input jitter a CDR loop must tolerate without increasing the bit error rate. Illustrated in Fig. 9.58, the specification is typically described by a mask as a function of the jitter frequency. For example, the CDR must withstand a

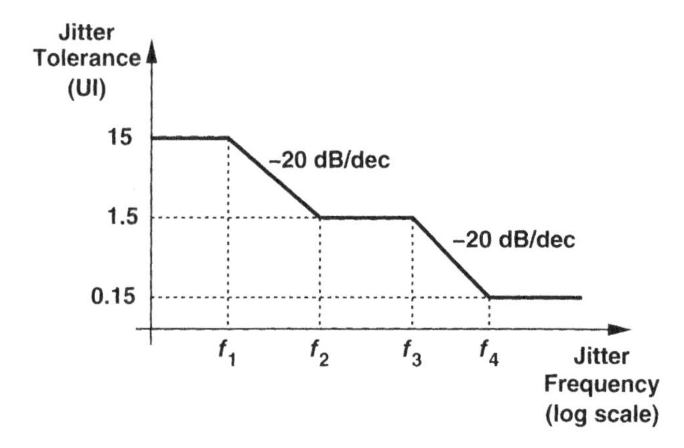

Figure 9.58 Example of jitter tolerance mask.

peak-to-peak jitter of 15 UI if the jitter varies at a rate below 100 Hz. Note that the tolerance test is performed with a random sequence whose phase is modulated at different rates for different parts of the mask.

It is instructive to quantify the jitter tolerance of a typical CDR loop and compare the result with the mask shown in Fig. 9.58. At a given jitter frequency, we must increase the magnitude of the input excess phase, ϕ_{in}, until the error rate begins to rise. This occurs when the phase error, $\phi_{in} - \phi_{out}$, approaches one-half unit interval, bringing the sampling edge of the clock close to the zero-crossing points of data. Thus, an approximate condition to avoid increasing the BER is

$$\phi_{in} - \phi_{out} < \frac{1}{2}\ \text{UI}. \tag{9.47}$$

Equivalently,

$$\phi_{in}[1 - H(s)] < \frac{1}{2}\ \text{UI}, \tag{9.48}$$

where $H(s) = \phi_{out}/\phi_{in}$, and hence

$$\phi_{in} < \frac{0.5\ \text{UI}}{1 - H(s)}. \tag{9.49}$$

We therefore express the jitter tolerance as:

$$G_{JT}(s) = \frac{0.5}{1 - H(s)}, \tag{9.50}$$

where $G_{JT}(s)$ denotes the largest phase modulation at the input that increases the BER negligibly. For a second-order CDR loop, we have from Eq. (9.14):

$$G_{JT}(s) = \frac{1}{2}\frac{s^2 + 2\zeta\omega_n s + \omega_n^2}{s^2}. \tag{9.51}$$

Equation (9.51) contains two poles at the origin and two zeros coincident with the poles of $H(s)$ (Section 9.5.1). Consequently, as depicted in Fig. 9.59, $|G_{JT}|$ falls at a rate of 40 dB/dec for $\omega < \omega_{p1}$ and at 20 dB/dec for $\omega_{p1} < \omega < \omega_{p2}$, approaching 1/2 UI for

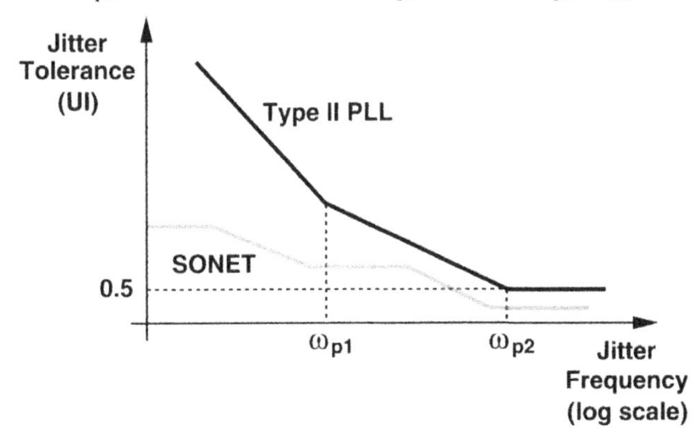

Figure 9.59 Jitter tolerance of a type II PLL.

$\omega > \omega_{p2}$. The SONET mask is also superimposed in this plot.

Let us make some interesting observations from the above results. First, the magnitude of G_{JT} at $s = j|\omega_{p1}|$ can be calculated as

$$|G_{JT}(s = j|\omega_{p1}|)|^2 = \frac{1}{4}\frac{(\omega_n^2 - \omega_{p1}^2)^2 + 4\zeta^2\omega_n^2\omega_{p1}^2}{\omega_{p1}^4}. \tag{9.52}$$

From Eq. (9.31), we have $\omega_n \approx 2\zeta|\omega_{p1}|$ because $\zeta \gg 1$ and hence

$$|G_{JT}(s = j|\omega_{p1}|)|^2 \approx 8\zeta^4. \tag{9.53}$$

That is,

$$|G_{JT}(s = j|\omega_{p1}|)| \approx 2\sqrt{2}\zeta^2 \text{ UI}. \tag{9.54}$$

Also, Eq. (9.32) yields $|\omega_{p2}| \approx 2\zeta\omega_n$ for $\zeta \gg 1$. Figure 9.60(a) plots $|G_{JT}|$ for two different values of ζ, revealing that if ζ increases and ω_n remains constant, jitter tolerance is more easily met. Moreover, Fig. 9.60(b) suggests that as ω_n increases while ζ remains constant, jitter tolerance improves.

The above observations point to an important trade-off in CDR design: if ω_n is decreased so as to lower the jitter transfer bandwidth, then jitter tolerance degrades substantially at high jitter frequencies. A difficult situation arises in OC-192, where the jitter

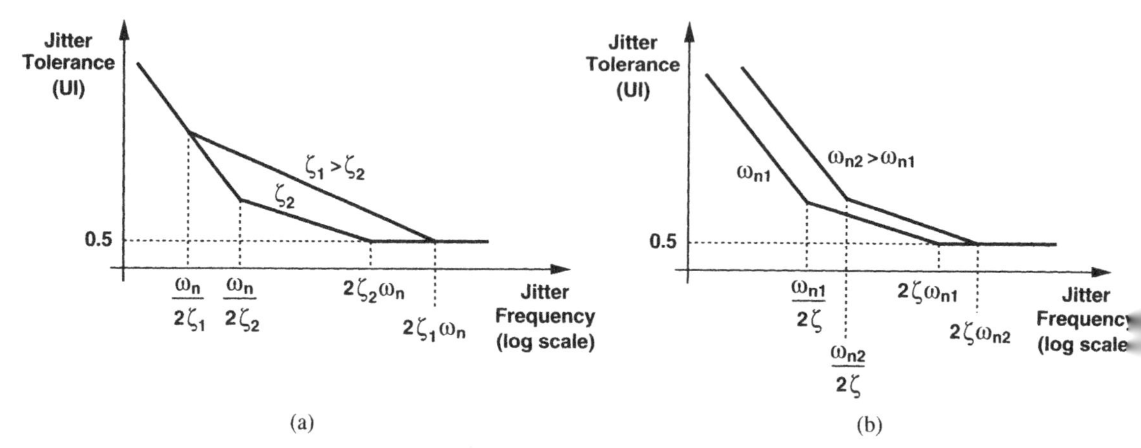

Figure 9.60 Jitter tolerance for different values of (a) ζ and (b) ω_n.

bandwidth is only 120 kHz but the fourth jitter tolerance corner frequency (f_4 in Fig. 9.58) is 4 MHz. If $2\zeta\omega_n = 2\pi \times 120$ kHz, then it may be impossible to satisfy the jitter tolerance mask by a single phase-locked CDR circuit.

References

1. C. R. Hogge, "A Self-Correcting Clock Recovery Circuit," *IEEE J. Lightwave Tech.*, vol. 3, pp. 1312–1314, Dec. 1985.
2. S. B. Anand and B. Razavi, "A 2.75–Gb/s CMOS Clock and Data Recovery Circuit with Broad Capture Range," *ISSCC Dig. of Tech. Papers*, pp. 214–215, Feb. 2001.
3. L. DeVito et al., "A 52 MHz and 155 MHz Clock Recovery PLL," *ISSCC Dig. of Tech. Papers*, pp. 142–143, Feb. 1991.
4. L. DeVito, "A Versatile Clock Recovery Architecture and Monolithic Implementation," in *Monolithic Phase-Locked Loops and Clock Recovery Circuits,* B. Razavi, Ed., New York: IEEE Press, 1996.
5. J. D. H. Alexander, "Clock Recovery from Random Binary Data," *Electronics Letters*, vol. 11, pp. 541–542, Oct. 1975.
6. R. C. Walker, et al., "A 1.5 Gb/s Link Interface Chipset for Computer Data Transmission," *IEEE J. of Selected Areas in Communications,* vol. 9, pp. 698–703, June 1991.
7. Y. M. Greshishchev and P. Schvan, "SiGe Clock and Data Recovery IC with Linear-Type PLL for 10-Gb/s SONET Application," *IEEE Journal of Solid-State Circuits,* vol. 35, pp. 1353–1359, Sept. 2000.
8. Y. M. Greshishchev, et al., "A Fully Integrated SiGe Receiver IC for 10-Gb/s Data Rate," *IEEE Journal of Solid-State Circuits,* vol. 35, pp. 1949–1957, Dec. 2000.
9. J. Savoj and B. Razavi, "A 10-Gb/s CMOS Clock and Data Recovery Circuit with a Half-Rate Linear Phase Detector," *IEEE Journal of Solid-State Circuits,* vol. 36, pp. 761–768, May 2001.
10. D. Richman, "Color-Carrier Reference Phase Synchronization Accuracy in NTSC Color Television," *Proc. IRE*, vol. 42, pp. 106–133, Jan. 1954.
11. F. M. Gardner, "Properties of Frequency Difference Detectors," *IEEE Trans. Comm.*, vol. 33, pp.

131–138, Feb. 1985.
12. A. Pottbacker, U. Langmann, and H. U. Schreiber, "A Si Bipolar Phase and Frequency Detector for Clock Extraction up to 8 Gb/s," *IEEE Journal of Solid-State Circuits,* vol. 27, pp. 1747–1751, Dec. 1992.
13. D. G. Messerschmitt, "Frequency Detectors for PLL Acquisition in Timing and Carrier Recovery," *IEEE Trans. Comm.*, vol. 27, pp. 1288–1295, Sept. 1979.
14. J. Savoj and B. Razavi, "A 10-Gb/s CMOS Clock and Data Recovery Circuit with Frequency Detection," *ISSCC Dig. of Tech. Papers*, pp. 78–79, Feb. 2001.
15. J. C. Scheytt, G. Hanke, and U. Langmann, "A 0.155, 0.622, and 2.488 Gb/s Automatic Bit Rate Selecting Clock and Data Recovery IC for Bit Rate Transparent SDH Systems," *ISSCC Dig. of Tech. Papers*, pp. 348–349, Feb. 1999.
16. P. Trischitta and E. Varma, *Jitter in Digital Transmission Systems,* Norwood, MA: Artech House, 1989.
17. F. Herzel and B. Razavi, "A Study of Oscillator Jitter Due to Supply and Substrate Noise," *IEEE Trans. Circuits and Systems, Part II,* vol. 46, pp. 56–62, Jan. 1999.
18. J. A. McNeill, "Jitter in Ring Oscillators," *IEEE Journal of Solid-State Circuits,* vol. 32, pp. 870–879, June 1997.
19. J. Lee and B. Razavi, "A 40-Gb/s clock and data recovery circuit in 0.18-.m CMOS technology," *IEEE Journal of Solid-State Circuits,* vol. 38, pp. 2181-2190, Dec. 2003.

CHAPTER 10

Multiplexers and Laser Drivers

The data path in an optical transmitter consists of a multiplexer (MUX), a retiming flipflop, and a laser driver. High-speed MUXes and drivers present many design challanges as they determine the quality of the serial data delivered to the fiber.

This chapter deals with the design of multiplexers and laser and modulator drivers[1], emphasizing low-voltage and high-speed implementations. We begin with 2-to-1 MUX topologies and study MUX architectures.[2] Next, we present frequency dividers as a building block used in both MUXes and PLLs. Finally, we describe the design of laser drivers and their power control circuits.

10.1 Multiplexers

A multiplexer combines many low-speed parallel channels into a high-speed stream of serial data. For example, most transmitters incorporate a 16-to-1 MUX, allowing the 16 inputs to be much slower than the output and hence simplifying the design of the package and the PC board.

10.1.1 2-to-1 MUX

We begin our study by first considering a 2-to-1 MUX. Such a circuit must accommodate two random binary inputs, routing each to the output according to the logical level of a "select" command [Fig. 10.1(a)]. Shown in Fig. 10.1(b) is an implementation example where differential pairs M_1-M_2 and M_3-M_4 sense the inputs D_1 and D_2, respectively, convert the signals to current, and apply the result to the load resistors. The *Select* command determines whether I_{SS} flows through M_1-M_2 or M_3-M_4, enabling one data path and disabling the other. Illustrated in Fig. 10.1(c), V_{out} therefore assumes a new value after each transition of *Select*.

[1] In this book, we use the term "laser driver" to refer to both types.

[2] The design of demultiplexers draws upon similar ideas and is typically less difficult.

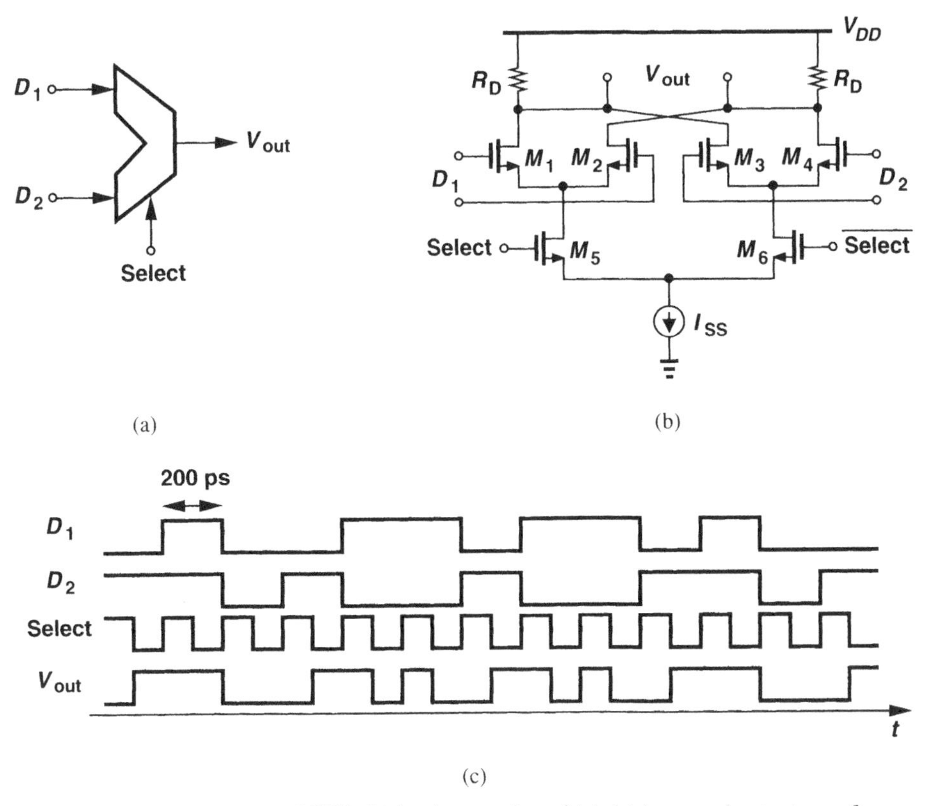

Figure 10.1 (a) Two-to-one MUX, (b) implementation of (a), (c) input and output waveforms.

Since the terminology used for MUXes may appear somewhat confusing, we give an example here. If the MUX of Fig. 10.1 is designed for 10 Gb/s, then each of the input streams has a rate of 5 Gb/s and the *output*, D_{out}, exhibits the specified rate of 10 Gb/s. Furthermore, the *Select* command is in fact a 5-GHz clock signal that routes D_1 to V_{out} for one half of a cycle (100 ps) and D_2 to V_{out} for the other half. We hereafter denote *Select* by clock (CK).

The MUX configuration of Fig. 10.1(b) is among the fastest circuits in a given technology as it employs current steering and contains no feedback loop requiring settling. However, the circuit demands certain timing relationships among D_1, D_2, and the clock. To understand this issue, let us redraw the waveforms of Fig. 10.1(c) more realistically, including finite rise and fall times but still assuming perfect edge alignment between D_1, D_2, and CK. As depicted in Fig. 10.2, at $t = t_1$, both D_1 and D_2 change levels but V_{out} must not because it must remain equal to D_1 immediately before t_1 and equal to D_2 immediately after. Unfortunately, since both D_1 and D_2 cross zero at $t = t_1$, the tail current in Fig. 10.1(b) is divided equally between the load resistors *regardless* of the value of the clock. As a result, V_{out} suffers from a large glitch.

The above analysis suggests that one of the input streams must be *offset* with respect to

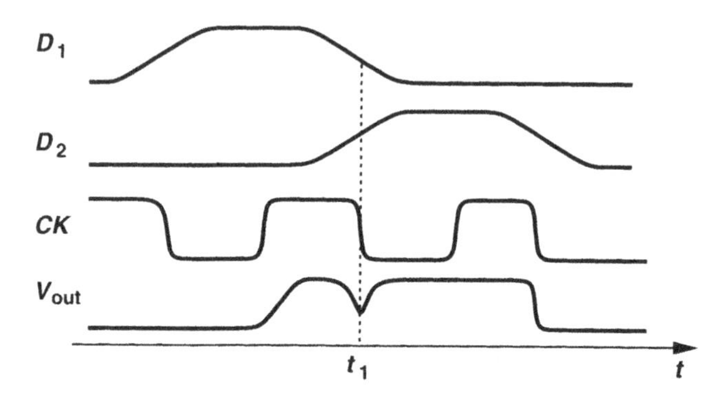

Figure 10.2 Effect of finite rise and fall times in the 2-to-1 MUX.

the other in the time domain so as to avoid simultaneous transitions in both inputs of the MUX. The optimum value of the offset is half a clock period, obtained by introducing a latch in one of the data paths (Fig. 10.3). Note that, when CK is high, the latch is "opaque," holding $D_{1\Delta}$ constant. When CK is low, the latch is "transparent," but the pair M_1-M_2 is disabled. If D_1 and D_2 are aligned with CK, then the edges of $D_{1\Delta}$ are shifted by $T_{CK}/2$ and, at each clock transition, only one of the signals sensed by M_1-M_2 and M_3-M_4 may change.

While avoiding simultaneous transitions at the two MUX inputs, the topology of Fig. 10.3 nonetheless requires that D_1 and D_2 be aligned with CK. If, for example, D_2 arrives at an arbitrary time, the MUX output still experiences glitches or pulsewidth distortion (Fig. 10.4). Thus, if the stages preceding this circuit do not guarantee such an alignment, D_1 and D_2 must be retimed by means of flipflops. Figure 10.5 depicts the resulting configuration in single-ended form. Here, when CK is high, latches L_1, L_3, and L_4 are opaque and the MUX selects $D_{1\Delta}$. During this time, L_2 senses the output of L_1 and L_5 senses the output of L_4. When CK goes low, $D_{1\Delta}$ begins to track the output of L_2 whereas L_5 becomes opaque, providing a stable level for the MUX. Depending on the environment, it is possible to eliminate L_1 and L_4, saving power consumption [1].

Even with ideal clock and input data waveforms, multiplexers may introduce substantial jitter in the output data. Such jitter arises from three sources: (1) Duty cycle distortion of the clock: as exemplified by Fig. 10.6, if $D_1 = 1$ and $D_2 = 0$, but CK exhibits a non-50% duty cycle, then D_{out} contains wider ONEs than ZEROs. Thus, the deviation of the clock duty cycle from 50% directly translates to jitter[3] at the output. (2) MUX dc offsets: each of the input differential pairs in the MUX of Fig. 10.1(c) may suffer from dc offsets, primarily because small device dimensions are chosen to allow high-speed operation. As a result, each data stream experiences pulsewidth distortion while traveling to the output (Section 10.3.1). (3) MUX propagation delay mismatches: even with a 50% clock duty cycle, the random mismatches between M_5 and M_6 and between M_1-M_2 and M_3-M_4 in

[3]The term jitter typically refers to a random effect, but it is also used here to indicate horizontal eye closure. We may even say "deterministic jitter" to emphasize the nature of the corruption.

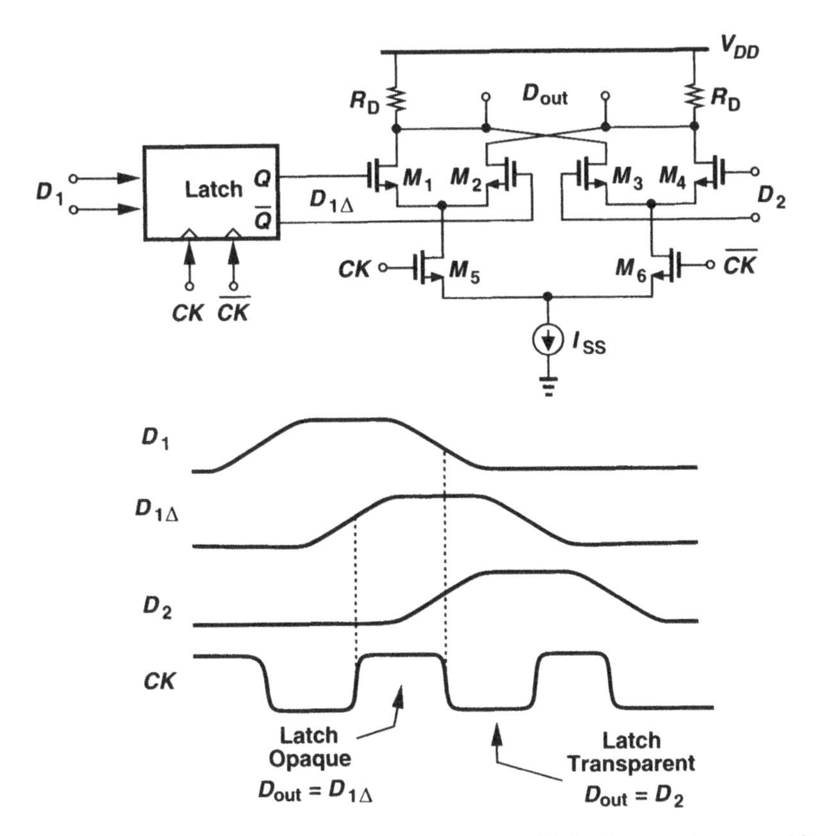

Figure 10.3 Preceding one input of MUX by a latch to avoid simultaneous input transitions.

Fig. 10.1(b) distort the arrival times of D_1 and D_2 at the output.

The jitter produced by the above mechanisms is often unacceptably large, requiring that the output data be retimed. Figure 10.7 shows a MUX employing both input and output retimers, with the main clock operating at the output data rate and $f_{CK}/2$ driving the MUX core and the preceding latches. Since FF_{out} and the $\div 2$ circuit must run at the full clock rate, their design poses difficult challenges. It is common to denote the half-rate clock by $CK/2$ rather than $f_{CK}/2$.

A critical issue in the architecture of Fig. 10.7 arises from the delay of the $\div 2$ circuit. Figure 10.8 illustrates the waveforms for zero and finite delays through the divider. In Fig. 10.8(a), the falling edges of the full-rate clock coincide with the midpoint of each bit of V_1, optimally retiming the data. In Fig. 10.8(b), on the other hand, $CK/2$ and hence the inputs and output of the MUX are delayed by ΔT with respect to CK. As a result, the sampling point of FF_{out} is shifted to the left of the data eye by ΔT, thereby reducing the time that the master latch in FF_{out} has for sensing its input. This effect severely degrades the setup and hold margins of FF_{out}, limiting its speed.

Another difficulty in the topology of Fig. 10.7 relates to the load seen by the $\div 2$ circuit.

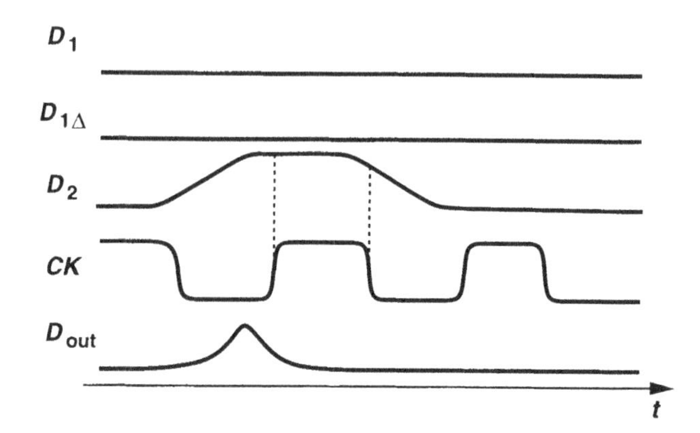

Figure 10.4 Glitch resulting from arbitrary arrival time of MUX input.

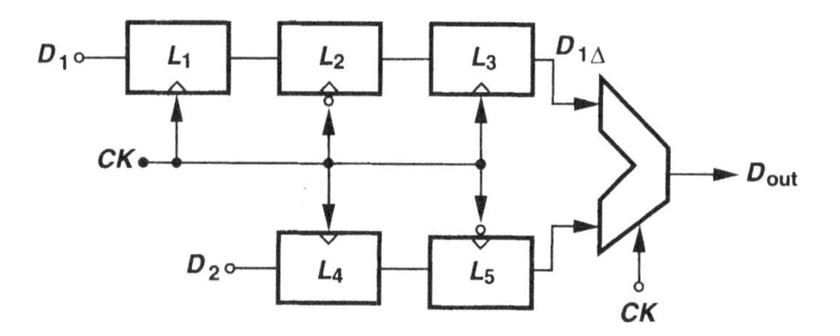

Figure 10.5 Five-latch MUX.

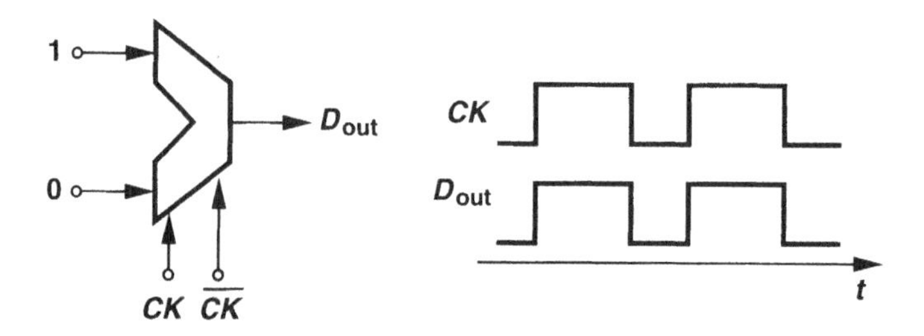

Figure 10.6 Pulsewidth distortion resulting from clock duty cycle distortion.

Here, the divider must drive seven differential pairs (the master latch within the divider, the MUX, and L_1-L_5), experiencing a large load capacitance.[4] This capacitance limits the maximum speed of the divider considerably. The circuit is therefore usually followed by a buffer, but the total delay resulting from both the divider and the buffer may be quite large, intensifying the effect illustrated in Fig. 10.8(b).

[4]In a 4-to-1 MUX, the divider must drive another divider as well.

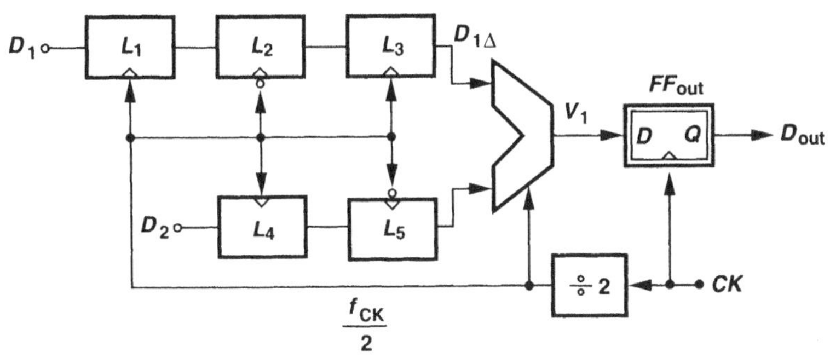

Figure 10.7 MUX followed by retimer.

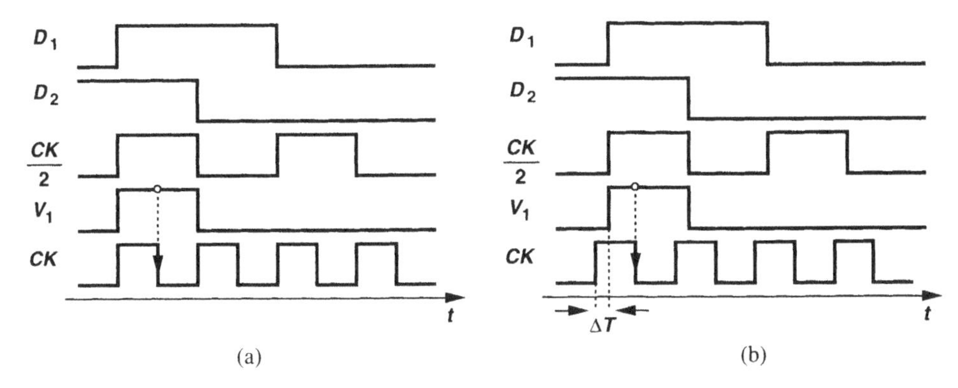

Figure 10.8 Operation of MUX/retimer chain with (a) zero circuit delays, (b) finite divider delay.

The architecture of Fig. 10.7 entails two other sources of delay that must be considered carefully. The reader can show that the delay from the clock input of the 2-to-1 MUX to its output exacerbates the above skew whereas the delay from the clock input of FF_{out} to its output alleviates the problem. In fact, inserting a proper delay in series with this flipflop's input can, in principle, cancel the divider delay.

10.1.2 MUX Architectures

With the 2-to-1 multiplexer developed above, we can now study architectures for a greater number of input channels. Most MUXes are based on one of two topologies: the "tree" structure and the shift-register configuration.

Tree Architecture The tree structure is a natural extension of the 2-to-1 MUX. Illustrated in Fig. 10.9, the idea is to group the input channels in pairs and multiplex each pair, reducing the number by a factor of two after each rank. In this example, Rank 1 is driven by $CK/4$, Rank 2 by $CK/2$, and FF_{out} by CK.

The divider delay issue described for the circuit of Fig. 10.7 applies to the entire tree structure. While lowering the frequency by a factor of two, each divider must drive twice as many devices as the preceding divider in the chain. Consequently, if all ranks incorporate identical 2-to-1 MUXes, then the nth divider produces an output frequency of $f_{CK}/2^n$ but

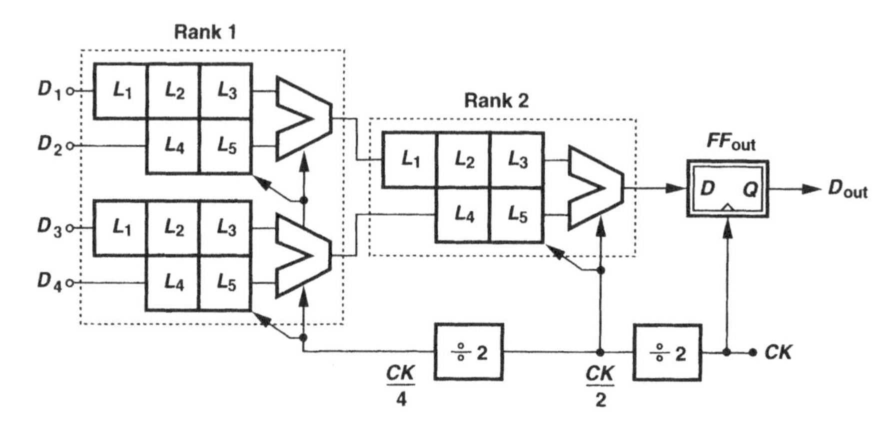

Figure 10.9 MUX using a tree structure.

experiences a total capacitance of $2^n C_M$, where C_M denotes the capacitance introduced by one complete 2-to-1 MUX. In other words, all of the divider stages must satisfy roughly equal speed and delay requirements.

Another difficulty in the architecture of Fig. 10.9 is the exponential growth of the hardware and power dissipation with the number of the input channels. For example, a 16-to-1 tree requires 15 2-to-1 MUXes and 4 dividers. If all of the building blocks employ current steering, such a tree includes roughly 100 current sources.

To remedy the issues related to power dissipation and divider loading, the 2-to-1 MUXes used in lower ranks can be scaled down in speed, power, and device dimensions. In Fig. 10.9, for example, the building blocks in Rank 1 can be scaled down by a factor of two with respect to those in Rank 2.

The tree architecture is the most commonly-used MUX topology. The modular nature and the ability to run at high speeds have made this configuration the dominant choice in transmitter design.

Shift-Register Architecture The shift-register architecture is based on viewing the multiplexing operation as parallel-to-serial conversion. As shown in Fig. 10.10, a shift register uses CK/N to load D_1-D_m in parallel once every NT_{CK} seconds. The main clock, CK, then allows the stored data to be shifted from FF_1 to FF_m, thereby generating the serial sequence at D_{out}.

Owing to its linear growth with the number of channels, the shift-register architecture is more compact and less power-hungry than the tree structure. For example, if $m = 16$, then the MUX requires 16 flipflops and four ÷2 circuits, i.e., a total of 20 flipflops and hence 40 current sources. However, the high-speed clock, CK, must now drive all of the data flipflops in addition to the first divider, i.e., 17 flipflops in this example. More importantly, the divider delay proves much more critical here than in a tree topology. The delay of the overall divider chain must be small with respect to NT_{CK} in the tree structure whereas it must be much less than T_{CK} in the register architecture. In the above example ($m = 16$), the chain of four ÷2 circuits may introduce a delay even exceeding T_{CK}, thus limiting the

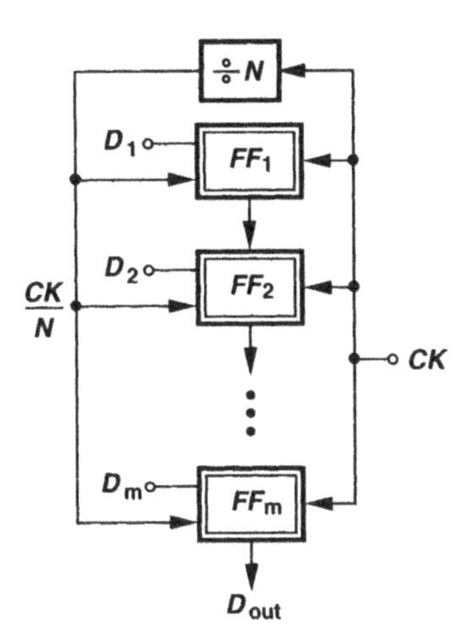

Figure 10.10 MUX based on parallel write and serial read.

utility of the architecture for high-speed operation.

A variant of the shift-register architecture incorporates a multiphase multiplexer. Illustrated in Fig. 10.11(a), the idea is to sense multiple channels by means of differential pairs, activating each pair by nonoverlapping pulses of width T_{CK}. As a result, V_{out} is equal to D_1 for $t_1 < t < t_2$, equal to D_2 for $t_2 < t < t_3$, etc. The nonoverlapping pulses can be generated by a ring counter that shifts a single logical ONE each clock cycle [Fig. 10.11(b)]. The overall N-to-1 MUX must therefore be configured as shown in Fig. 10.11(c). Here, the data is first loaded into the register by CK/N and subsequently multiplexed by the nonoverlapping pulses. [The switches represent the differential pairs in Fig. 10.11(a).]

While simplifying the design of the register, the multiphase architecture of Fig. 10.11(c) still requires precise alignment between CK and CK/N. Furthermore, as with the circuit of Fig. 10.1(b), it displays glitches at the output. Depicted in Fig. 10.12, this effect arises when the last pulse in V_4 and the pulse immediately after it in V_1 multiplex D_m and D_1, respectively. Since both data channels may contain edges during clock transition, the MUX output momentarily falls.

The architecture of Fig. 10.11(c) requires short, nonoverlapping pulses in the voltage domain, necessitating high slew rates for high-speed operation. It is therefore preferable to generate and apply the pulses in the current domain. We surmise that if various phases of a clock are ANDed, the output currents of the AND gates can yield the required pulses [2]. For example, let us consider the four clock phases shown in Fig. 10.13(a), recognizing that $CK_1 \cdot \overline{CK_2}$, $CK_2 \cdot \overline{CK_3}$, $CK_3 \cdot \overline{CK_4}$, and $CK_4 \cdot CK_1$ display nonoverlapping pulses [Fig. 10.13(b)]. With rail-to-rail clock signals, the circuit can be realized as shown in Fig. 10.13(c). Note that the input channels must be retimed by proper clock phases to avoid glitches at the output.

In practice, a combination of the above architectures can be used. For example, the low-speed ranks may employ a shift-register topology and the high-speed ranks a tree architecture.

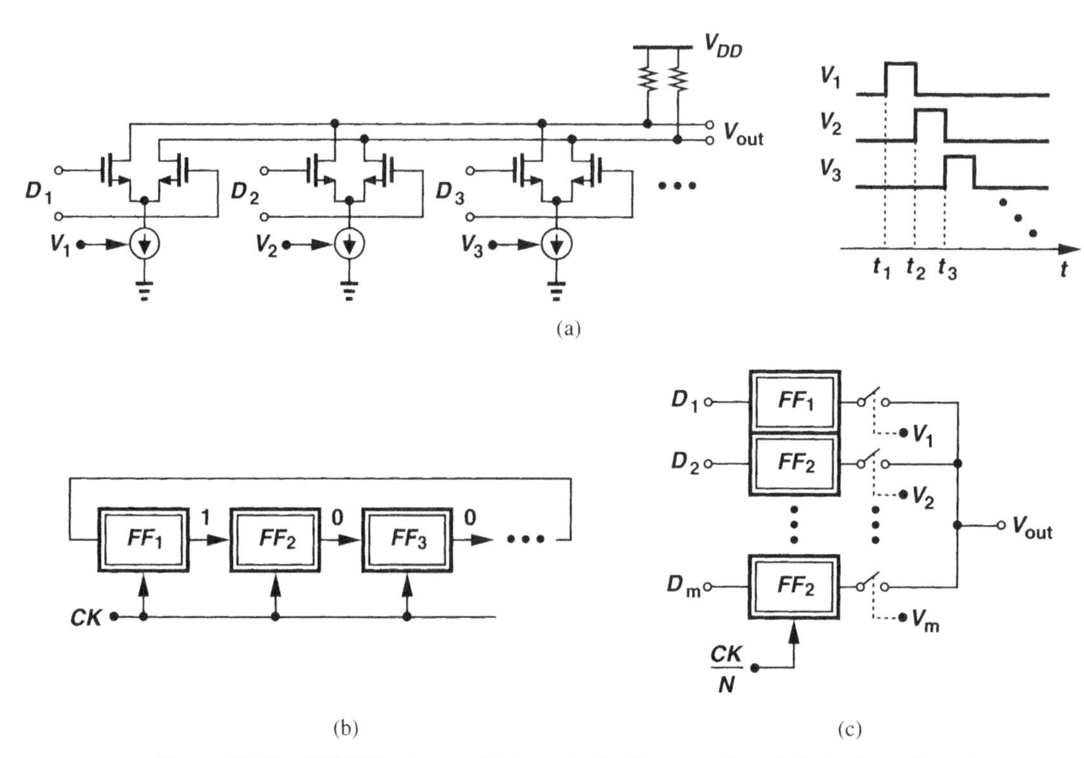

Figure 10.11 (a) MUX using multiphase clock, (b) generation of clock phases by a ring counter, (c) alternative architecture.

10.2 Frequency Dividers

Both the multiplexer and the phase-locked loop in an optical transmitter employ high-speed frequency dividers. Since in most optical standards, the output data rate is a power-of-two multiple of a reference frequency, the PLL typically incorporates a $\div 2$ circuit after the VCO. Fortunately, $\div 2$ circuits are generally much faster than dividers having other moduli.

Divide-by-two topologies fall under two categories: (1) flipflop-based circuits (also known as "static dividers") and (2) Miller configuration (also known as "dynamic dividers").

10.2.1 Flipflop Dividers

As with many other functions, dividers benefit from robust, flexible operation if they are implemented digitally. A digital $\div 2$ circuit incorporates a single master-slave flipflop in a negative feedback loop, forcing the state of each latch to toggle between ONE and ZERO in consecutive clock cycles (Fig. 10.14). Note that the slave latch sees the input capacitance of both the master latch and the output buffer.

The maximum speed of the divider is determined by the fanout of the circuit as well as the implementation of the latches. To achieve a high speed, the input capacitance of the buffer is typically chosen not to exceed that of each latch, leading to a fanout of two for the

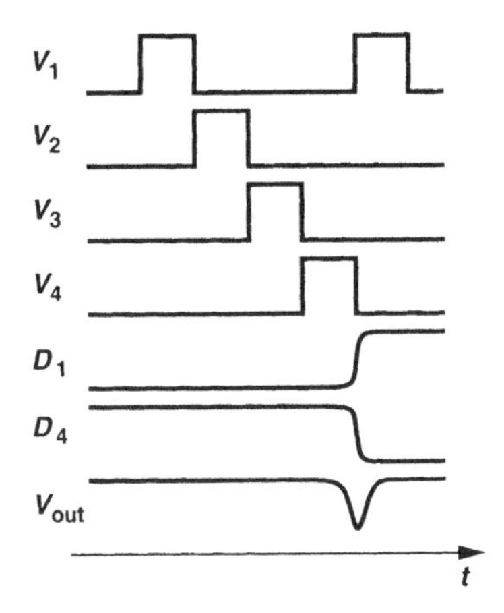

Figure 10.12 Glitch resulting from finite data transition times.

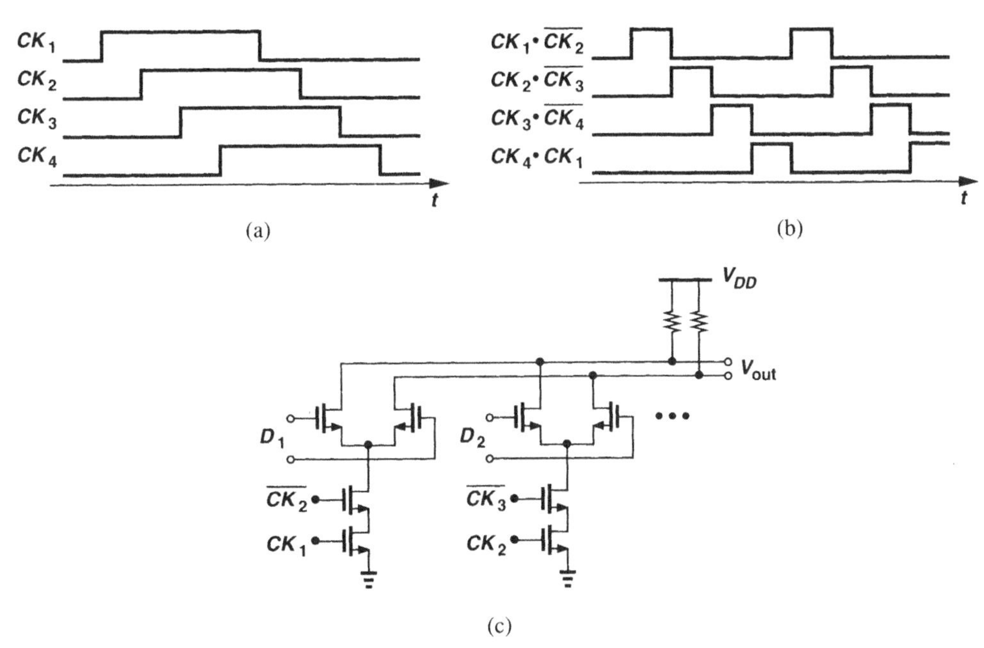

Figure 10.13 (a) Clock edges used in MUX, (b) generation of clock phases by logical operation, (c) circuit realization.

slave latch.

In the circuit of Fig. 10.14, the buffer doubles the load capacitance seen by the slave, thereby lowering the speed considerably. How can the drive capability of the flipflop be improved? Let us double all of the transistor widths and bias currents and halve the resistor

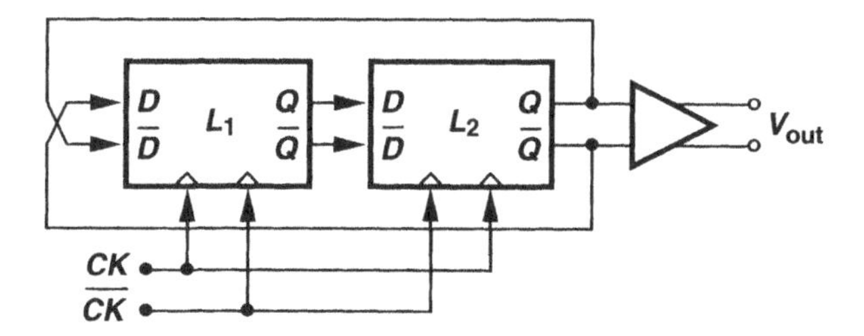

Figure 10.14 Flipflop-based divide-by-two circuit.

values (Fig. 10.15).[5] If the buffer remains unchanged, its input capacitance is now *one-*

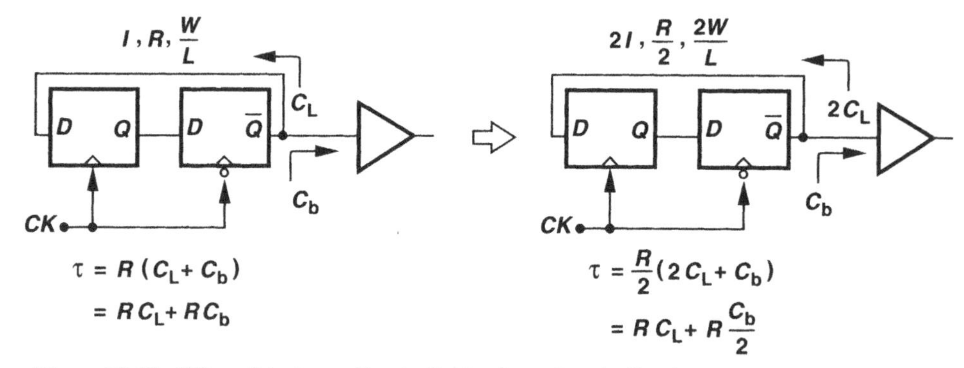

Figure 10.15 Effect of device scaling in divider for a given buffer size.

half that of each latch, yielding a fanout of 1.5 for the slave. In other words, doubling the flipflop power dissipation halves the contribution of the buffer to the time constant.

The above scaling scenario entails an important drawback: it doubles the load capacitance seen by the clock. Consequently, as illustrated in Fig. 10.16, the buffer interposed between the VCO and the divider must provide a low output impedance, inevitably ex-

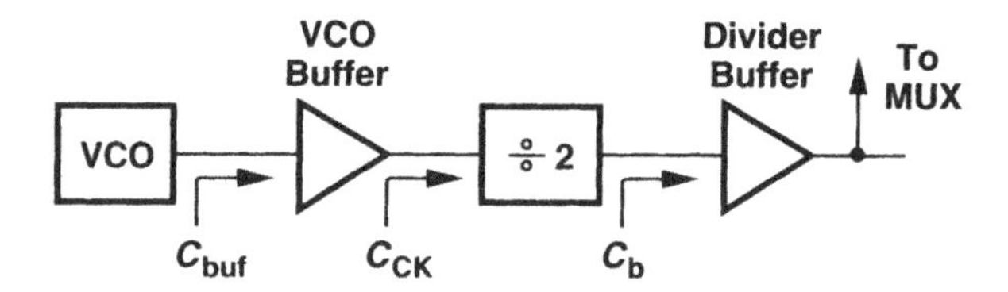

Figure 10.16 Chain of stages in a transceiver.

hibiting a large *input* capacitance. The effect of scaling in the divider therefore propagates

[5] Note that such scaling does not change the voltage swings or the voltage headroom.

back to the VCO. For this reason, the overall chain must be treated as one entity, perhaps requiring iterations in the design of the MUX as well.

Let us now study the implementation of the latches. A latch senses the input signal for half a clock cycle and stores the result for the other half. As such, the circuit must sense and track the signal rapidly so that the acquired level at the end of the sense mode represents the logical level correctly. Figure 10.17(a) shows an example, where M_1 and M_2 sense

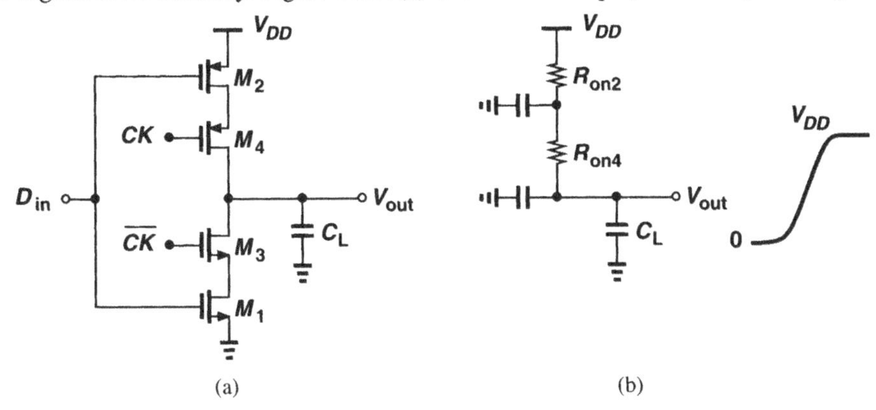

Figure 10.17 (a) Dynamic rail-to-rail latch, (b) low-to-high transition path.

the input and M_3 and M_4 control the signal path to the output. When CK goes from low to high, the circuit enters the store mode, retaining the level across the capacitance at the output node.

While simple, the latch topology of Fig. 10.17(a) suffers from several drawbacks at high speeds. First, the output node exhibits a long time constant, especially when going from low to high. Figure 10.17(b) depicts the current path using a simplified model for each transistor, revealing that the gate-channel and source and/or drain junction capacitances of M_2, M_3, and M_4 contribute to the time constant. Since in submicron MOS technologies, the mobility of PFETs is about one-fourth of that of NFETs, the output rise time severely limits the speed of the latch.

The second issue is the large voltage swings required at the input and output. The rail-to-rail clock swings necessary for fast switching of M_3 and M_4 make the design of the VCO and its buffer difficult. Moreover, since the output voltage must approach the rails so as to provide a reliable logical level, the output rise and fall times tend to be quite long.

The third shortcoming of the latch shown in Fig. 10.17(a) is that it remains "idle" in the store mode. If the circuit could somehow continue amplification of the signal in this mode, then the speed might improve. This point is clarified below.

The foregoing issues can be alleviated through the use of current-steering circuits, also known as "current-mode logic" (CML). Illustrated in Fig. 10.18(a), a CML latch consists of an input pair, M_1-M_2, for sensing the input and a regenerative pair, M_3-M_4, for storing the state. The sense and store modes are established by the clock pair, M_5-M_6. When CK is high, the tail current is steered to M_1-M_2, allowing V_{out} to track V_{in} while M_3 and M_4 are off. In the transition to the store mode, CK goes low, the input pair is disabled, and the cross-coupled pair is enabled, storing the logical levels at X and Y.

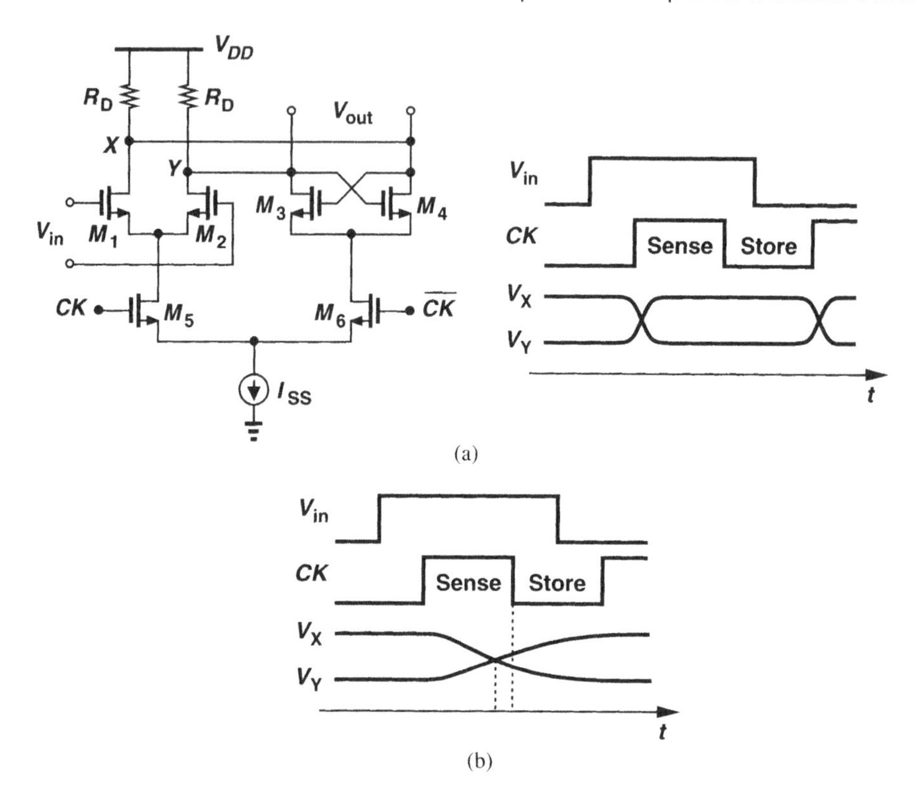

Figure 10.18 (a) Curent-steering latch, (b) effect of excessively long time constants.

In contrast to the rail-to-rail circuit of Fig. 10.17(a), the CML latch of Fig. 10.18(a) operates with relatively small swings at V_{in}, V_{out}, and CK—typically less than 1 V_{pp} (0.5 V_{pp} single-ended) in CMOS realizations. In bipolar implementations, the differential swings can be as small as 300 mV_{pp}. Furthermore, the regenerative pair provides gain in the store mode, allowing a shorter cycle for the sense mode. As illustrated in Fig. 10.18(b) for more realistic waveforms, even if V_X and V_Y do not reach full logical levels during the sense mode, the regenerative pair continues to amplify the difference in the store mode.

In the latch of Fig. 10.18(a), the loop gain provided by the cross-coupled pair must exceed unity so as to ensure the state is stored indefinitely. We therefore have $g_{m3,4}R_D > 1$, imposing a minimum value for R_D, I_{SS}, and the widths of M_3 and M_4. With a typical swing of $R_D I_{SS} = 400$ mV in CMOS implementations, M_3 and M_4 must be sufficiently wide to (a) provide a reasonable loop gain and (b) consume a moderate voltage headroom (i.e., V_{GS}). In low-voltage systems, the latter issue dominates the choice of $W_{3,4}$ and hence the capacitance contributed by M_3 and M_4 to nodes X and Y.

If a flipflop is configured as a $\div 2$ circuit, its constituent regenerative pairs need not exhibit a loop gain greater than unity. This is because the store mode is so short that even "weak" regeneration can hold the state. This observation is exploited in the "super-dynamic" flipflop topology [3]. Illustrated in Fig. 10.19, this method allows the currents flowing through the input pair and the cross-coupled pair to be different. Here, one tail cur-

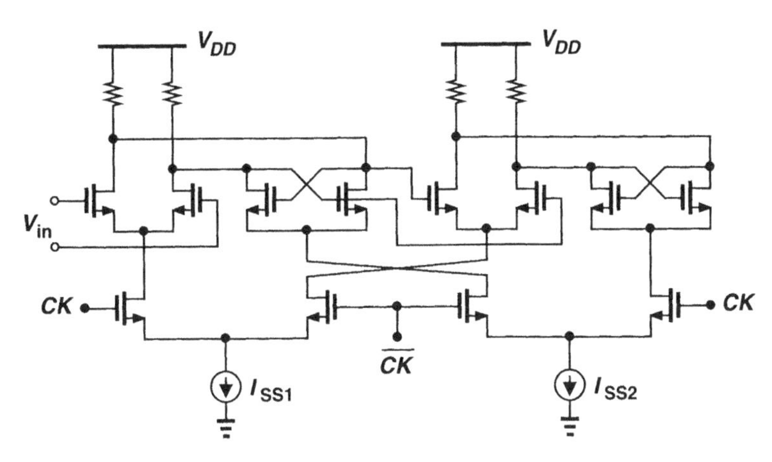

Figure 10.19 Superdynamic divider.

rent, I_{SS1}, is steered between the master and slave input pairs and the other, I_{SS2}, between their cross-coupled pairs. With $I_{SS2} < I_{SS1}$, the circuit can use smaller transistors for regeneration, thereby lowering the capacitance at each latch's output nodes. In other words, the positive feedback is only strong enough to retain the logical state for a short period of time—half a clock cycle. Note that without the cross-coupled pair, the load resistors rapidly charge the drain nodes to V_{DD}.

In CMOS technologies lacking well-controlled resistors, the loads of CML circuits must employ only MOSFETs. PMOS devices operating in the deep triode region can serve such a purpose [Fig. 10.20(a)], but their on-resistance varies significantly with process and tem-

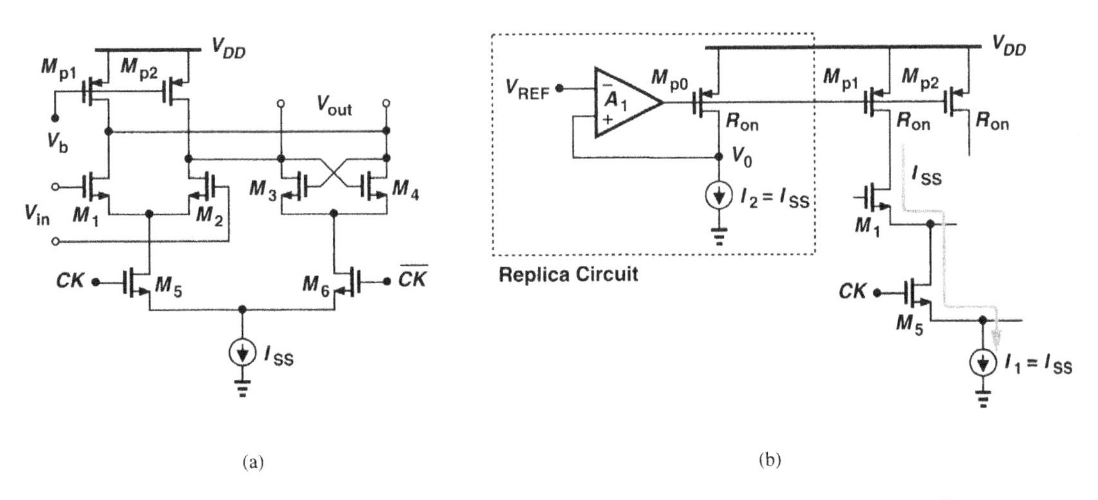

Figure 10.20 (a) Current-steering latch with PMOS loads operating in the triode region, (b) replica circuit to define voltage swings.

perature, thereby leading to large variation of the output swing. For this reason, a "swing control" circuit is often employed to define the output swing according to a reference. Illustrated in Fig. 10.20(b), this technique incorporates a replica circuit, M_{p0} and I_2, whose output voltage, V_0, is forced to be equal to V_{REF} by means of the error amplifier, A_1. For example, at high temperatures, R_{on} tends to increase, V_0 tends to fall, and A_1 lowers the gate voltage of M_{p0}, restoring V_0 to a value close to V_{REF}. With sufficient matching between the current sources and the PMOS devices, the output swing of the latch remains nearly equal to $V_{DD} - V_{REF}$ if M_1-M_6 experience complete switching.

The circuit of Fig. 10.20(b) is generally much slower than the topology shown in Fig. 10.18(a). This is because the load PMOS devices must be sufficiently wide to remain in deep triode region for the slow corner of the process at high temperatures. We note that under these extreme conditions, amplifier A_1 attempts to lower the gate voltage of M_{p0}-M_{p2} to near zero. If the PMOS transistors still exhibit an excessively large on-resistance with $|V_{GS}| \approx V_{DD}$, then I_1 and I_2 may force them out of the triode region, creating large, slow, and poorly-defined voltage swings in the latch.

The load devices in latches can alternatively be realized by diode-connected MOSFETs. To achieve a reasonable voltage gain during sense and regeneration modes, PMOS transistors are preferred [Fig. 10.21(a)] as their lower mobility yields a higher small-signal re-

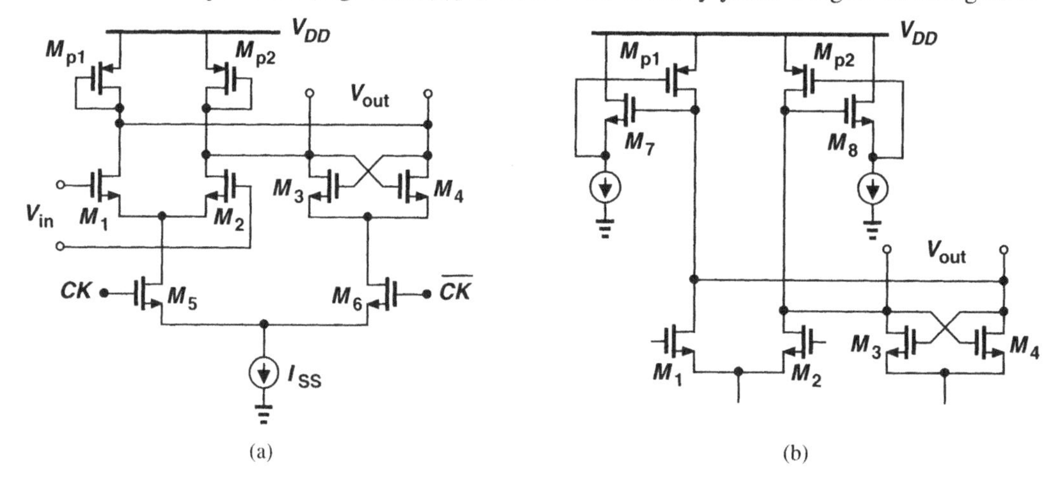

Figure 10.21 (a) Current-steering latch with diode-connected loads, (b) use of source followers to alleviate headroom limitation.

sistance for a given device capacitance. The principal difficulty, however, stems from the large headroom consumed by the loads, i.e., one PMOS threshold voltage plus the overdrive voltage. As described in Chapter 6, this issue can be resolved by inserting an NMOS source follower between the drain and the gate of each PMOS load [Fig. 10.21(b)].

Another critical drawback of diode-connected loads manifests itself with random data. Consider the circuit in Fig. 10.22(a), where I_M falls to zero at $t = 0$. The drain voltage of M_1 then begins to rise, exhibiting a decreasing transconductance as the drain current drops. As a result, the small-signal time constant continues to *increase*, slowing down the rate of charge considerably. This observation indicates that diode-connected MOSFETs turn off very slowly. Now suppose a differential pair using such loads senses high-speed random

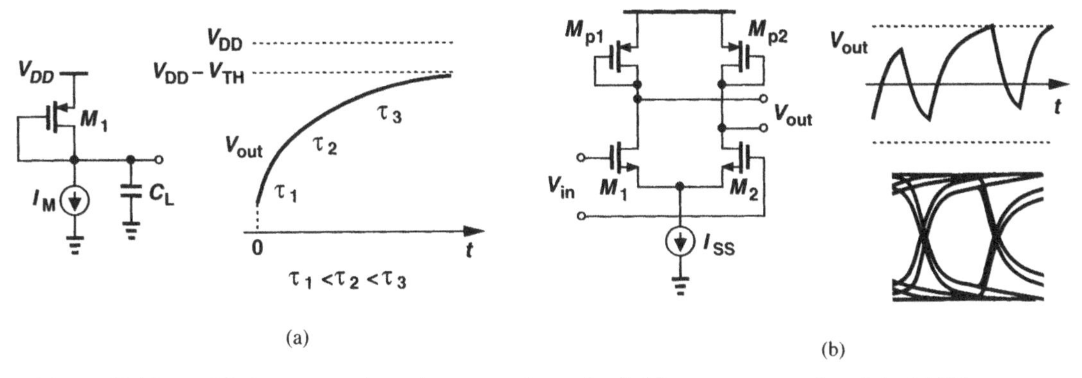

Figure 10.22 (a) Diode-connected transistor operating as load, (b) response to random data, (c) ISI resulting from incomplete switching of diode loads.

binary data [Fig. 10.22(b)]. For short runs, the loads do not approach turn-off whereas for long runs, they do. As a consequence, the voltage level from which each bit must start depends on the previous bits, leading to variable high and low levels. More importantly, a bit following a long run must bring the diode-connected load out of its "relaxed" mode, experiencing a longer time to cross zero and hence substantial jitter. The eye diagram in Fig. 10.22(c) illustrates both effects. Similar phenomena occur in latches, suggesting that diode-connected loads are ill-suited to circuits operating with random data.

An interesting phenomenon observed in dividers relates to the required clock swing. As illustrated in Fig. 10.23(a), the minimum clock swing for proper division at relatively

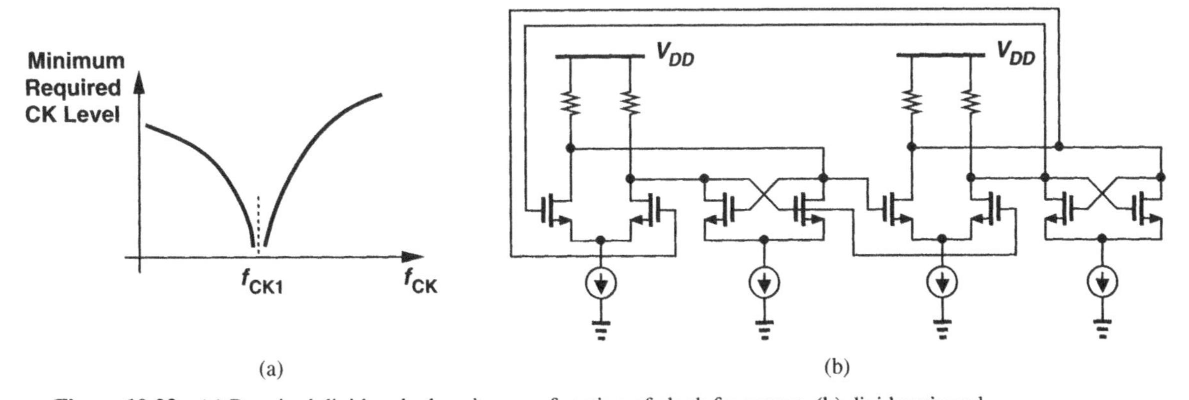

Figure 10.23 (a) Required divider clock swing as a function of clock frequency, (b) divider viewed as a ring oscillator.

low frequencies may be relatively small because, even if the current is not fully switched between the input and cross-coupled pairs, the circuit still has enough time to restore the levels. As f_{CK} increases, the required swing begins to fall sharply in the vicinity of a particular frequency, f_{CK1}, indicating that if CK and $\overline{CK}$ have *equal* dc voltages and little swing, the circuit continues to produce $f_{CK}/2$. Why does this happen? Under these conditions, the divider in fact *oscillates* at a frequency $f_{CK}/2$. Shown in Fig. 10.23(b) for equal CK and $\overline{CK}$ levels, the circuit resembles a two-stage ring oscillator, with the regenerative pairs creating enough hysterisis and hence phase shift to allow the loop to oscillate.[6]

As f_{CK} increases further, the divider requires greater clock slew rates so as to steer the tail current rapidly. For a sinusoidal clock waveform, this translates to larger swings.

10.2.2 Miller Divider

Suppose $x_1(t) = A_1 \cos\omega_1 t$ is multiplied by $x_2(t) = A_2 \cos(\omega_1/2)t$, generating two new components: $y_1(t) = (A_1 A_2/2)\cos(3\omega_1/2)t$ and $y_2(t) = (A_1 A_2/2)\cos(\omega_1/2)t$. Interestingly, $y_2(t)$ is simply a scaled version of $x_2(t)$ and can be used in its stead. This is the principle behind the Miller divider. Depicted in Fig. 10.24, the circuit incorporates

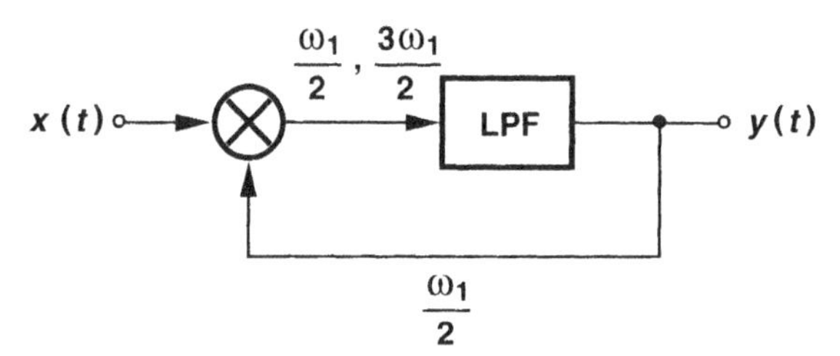

Figure 10.24 Miller divider.

a multiplier (mixer) and a low-pass filter, returning the output to one port of the mixer. If operating properly, the divider produces $\omega_1/2$ at the LPF output and hence $3\omega_1/2$ and $\omega_1/2$ at the mixer output. The LPF corner frequency, ω_{LPF}, is chosen to be

$$\frac{\omega_1}{2} < \omega_{LPF} < \frac{3\omega_1}{2} \tag{10.1}$$

so that it suppresses only the sum component. As a result, only the frequency $\omega_1/2$ survives around the loop. This is of course possible only if the loop gain at $\omega_1/2$ exceeds unity, demanding sufficient gain in the mixer. Figure 10.25 shows a transistor-level implementation using a Gilbert cell as a multiplier.

The conditions expressed by (10.1) imply that the divider operates across a limited frequency range. For example, if ω_1 is much less than $2\omega_{LPF}$, then the $3\omega_1/2$ component is not attenuated sufficiently, possibly distorting the output as well as creating other frequencies. As shown in Fig. 10.26, the existence of both $\omega_1/2$ and $3\omega_1/2$ at the output leads to a waveform with deep notches in the middle of each half cycle, thereby degrading the margin between the high and low levels. In the limit, if ω_1 is low enough, then the input and output of the LPF in Fig. 10.24 are equal, and we have

$$A\cos\omega_1 t \cdot y(t) \approx \alpha y(t), \tag{10.2}$$

[6]In fact, reducing the clock swing to zero in simulations and laboratory measurements serves as a quick sanity check: if the circuit fails to oscillate, it may also fail to divide.

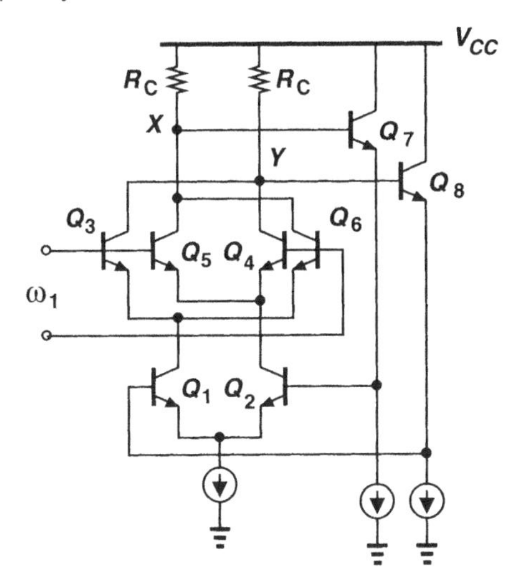

Figure 10.25 Implementation of Miller divider.

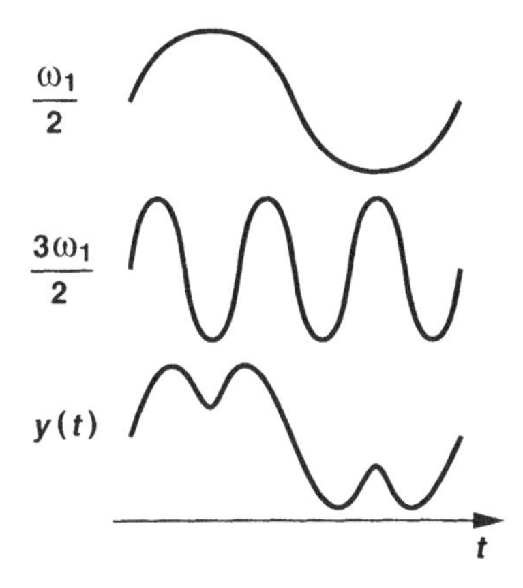

Figure 10.26 Effect of third harmonic in Miller divider.

where α denotes the mixer conversion gain. Thus, $y(t) = 0$; i.e., the divider fails.

What happens if the input frequency is well *above* $2\omega_{LPF}$? The $\omega_1/2$ component then experiences significant attenuation, the loop gain falls below unity, and the circuit fails to divide.

The Miller divider achieves high speeds because it exploits the natural low-pass filtering (poles) of the circuit in its operation. This can be seen in the implementation example of Fig. 10.25. The parasitic capacitances at nodes X and Y (resulting from Q_3-Q_8) along with the load resistors constitute the low-pass filter, allowing the input frequency to be twice as high as $1/(R_C C_X)$.

10.3 Laser and Modulator Drivers

A laser driver can be viewed as a simple current switch that turns the laser on and off according to the logical value of the data (Fig. 10.27). Recall from Chapter 3 that the output

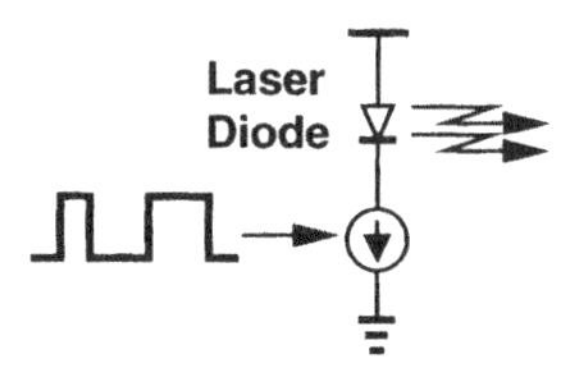

Figure 10.27 Conceptual diagram of a laser driver.

power of laser diodes is more accurately defined as a function of the input current rather than voltage. For this reason, and because of the speed advantages of current switching, lasers are driven by currents.

The driver of Fig. 10.27 must meet three important criteria: (a) adequate switching speed to ensure the light is modulated with minimal intersymbol interference; (b) high output current so as to generate the required optical power corresponding to logical ONEs; (c) tolerance of the voltage swing across the laser. For example, a driver operating at 10 Gb/s must exhibit a small-signal bandwidth on the same order while delivering as much as 100 mA and sustaining a swing of several volts.

The above three criteria point to a great difficulty in laser driver design: as transistor dimensions are scaled down to improve the speed, the inevitably lower breakdown voltage makes it increasingly more difficult to tolerate the voltage swings across lasers. Indeed, a great deal of research is expended on new lasers and modulators that operate with smaller swings.

The large output currents required of drivers mandate the use of very wide transistors, thus yielding high input and output capacitances for the output stage. Since such drivers must provide output currents five to ten times those delivered by 50-Ω buffers, the design of laser drivers is generally much more difficult. Nevertheless, most of the design principles described in Chapter 5 apply to laser drivers as well.

10.3.1 Performance Parameters

We now examine the performance parameters of laser drivers more carefully.

Speed The speed can be quantified by the output current rise and fall times, the sum of which must be well below one bit period so as to avoid significant ISI. Interestingly, if lasers are driven by excessively short rise times, then chirping (Chapter 3) becomes more pronounced.

Optical standards specify the speed requirements by a mask. Illustrated in Fig. 10.28(a), such a mask defines the maximum tolerable eye closure in both vertical or horizontal directions. Note that the test procedure includes the nonidealities of the laser itself, thereby requiring that the optical signal be converted to the electrical domain as shown in Fig. 10.28(b). The filter following the amplifier has specific transfer characteristics (e.g., a

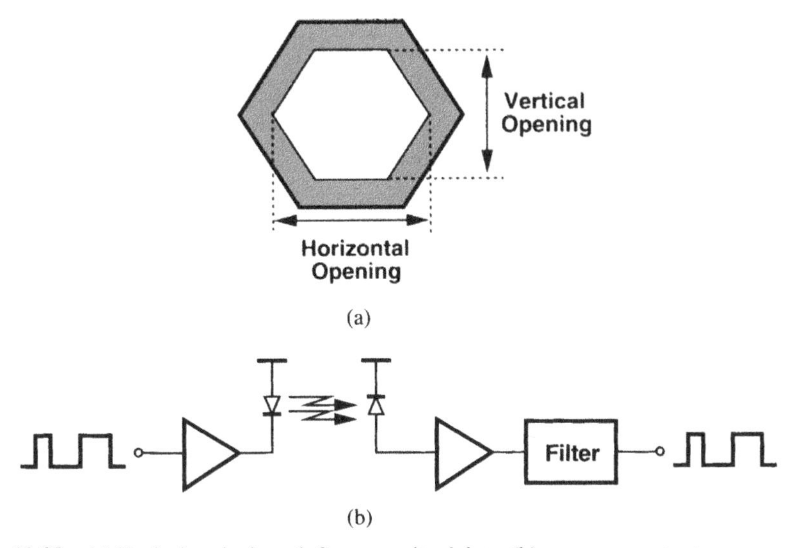

Figure 10.28 (a) Typical optical mask for transmitted data, (b) measurement setup.

fourth-order Bessel-Thompson circuit with a −3-dB bandwidth equal to 0.75 times the bit rate).

Output Current Recall from Chapter 3 that lasers must be biased in the vicinity of a threshold so that an increment in the current turns them on rapidly. Thus, the laser driver must provide both a "bias" current and a "modulation" current.

Temperature variations and aging of lasers generally demand an adaptive bias current, a task carried out by a power control circuit (Section 10.4.1). Since the *slope* of the laser characteristic is relatively independent of temperature and age, the modulation current itself is typically constant.[7]

Voltage Compliance Laser diodes experience a relatively large voltage swing as they turn on and off (Fig. 10.29). The voltage drop in the on state results from both the expo-

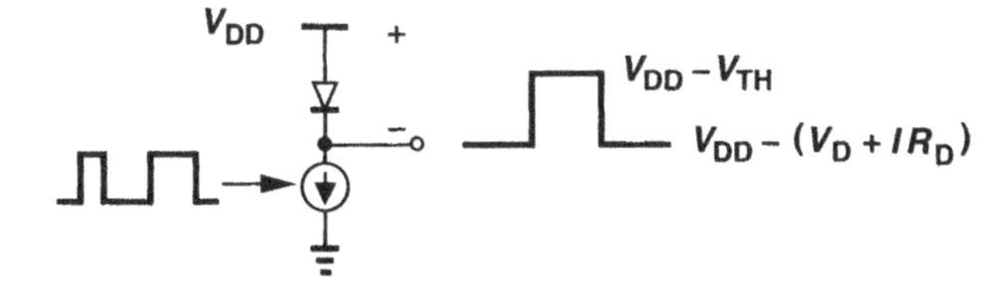

Figure 10.29 Voltage swings experienced by a laser diode.

nential I-V characteristic and the ohmic resistance, R_D, of the semiconductor materials comprising the diode. The drop in the off state corresponds to the bias current required to place the diode at the threshold.

[7]The slope of VCSEL characteristics does vary with temperature, requiring adaptive control of the modulation current.

As explained in Chapter 3, Mach-Zehnder modulators exhibit a nonmonotonic P-V behavior [Fig. 10.30(a)]. Thus, both the bias voltage and the voltage swing must be defined

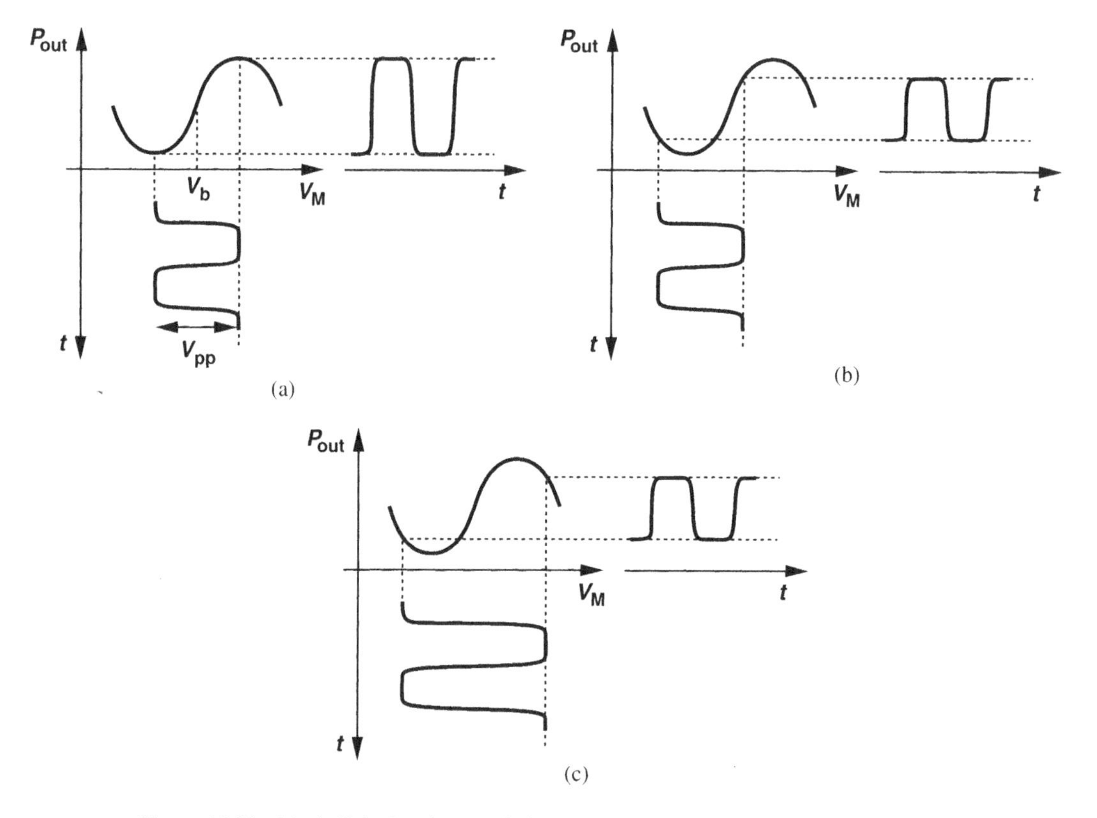

Figure 10.30 Mach-Zehnder characteristics, (a) ideal, (b) with input offset, (c) with input amplitude error.

accurately. As depicted in Figs. 10.30(b) and (c), errors in V_b or V_{pp} translate to a lower extinction ratio. Also, MZ modulators typically require greater voltage swings than laser diodes do, making the design of the driver more difficult.

Pulsewidth Distortion The logical ONEs and ZEROs produced by lasers or modulators may suffer from unequal periods. Illustrated in Fig. 10.31 and called "pulsewidth

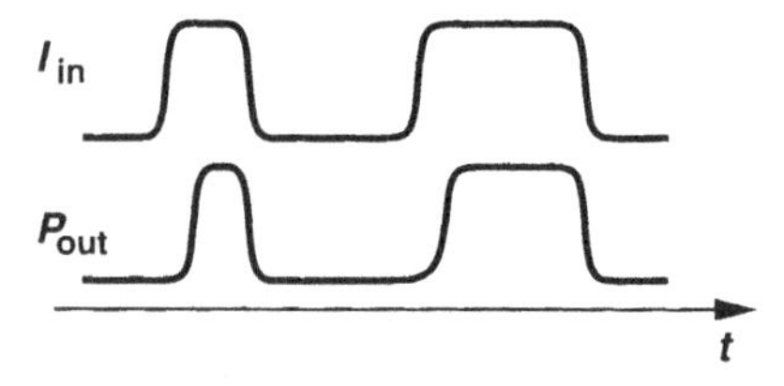

Figure 10.31 Pulsewidth distortion in a laser diode.

distortion," this effect closes the data eye horizontally, thereby leading to greater jitter and higher error rate. It may also create difficulties in clock and data recovery circuits.

Pulsewidth distortion arises from two nonidealities: offset voltages in the driver circuit and the turn-on delay in the laser.[8] To understand the effect of offsets, consider the data waveforms in Fig. 10.32, assuming that the driver exhibits such a high small-signal gain

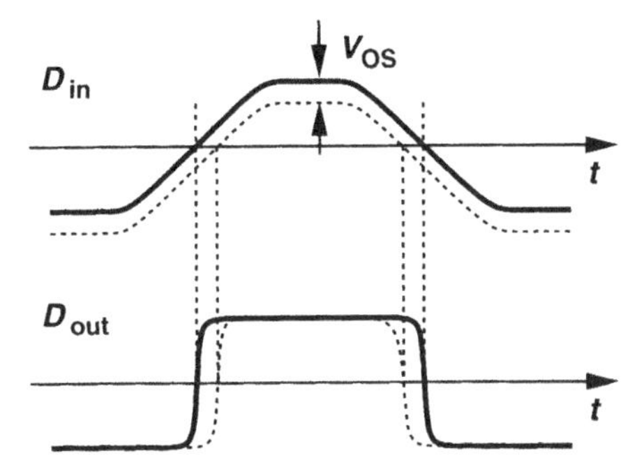

Figure 10.32 Pulsewidth distortion resulting from offsets.

that D_{out} trips abruptly on the zero crossings of D_{in}. An input-referred offset of $+V_{OS}$ in effect shifts the input signal down (or up), thereby altering the actual input zero crossings experienced by the driver and hence creating narrower ONEs than ZEROs. In fact, for an input slew rate of S_R V/s, the width of logical ONEs (or ZEROs) decreases by V_{OS}/S_R. It is therefore desirable to maximize the slew rate and minimize the asymmetries in the driver circuit. As mentioned in Chapter 3, lasers suffer from a significant delay when turning on but not so much when turning off, thus shortening the pulses in the optical domain.

Typical lasers and modulators exhibit a pulsewidth distortion of 0.03 to 0.05 UI. To reduce this effect, drivers often incorporate an external adjustment, deliberately introducing an offset in the signal path to compensate for this type of distortion.

Input and Output Impedances As with the output stages described in Chapter 5, laser drivers must employ a cascade of tapered stages so as to deliver large output currents while displaying a small input capacitance. Stand-alone drivers typically include on-chip 50-Ω termination at the input, but their input capacitance must be small enough to minimize the ISI resulting from reflections.

At speeds exceeding a few gigahertz, laser drivers incorporate on-chip termination at the output as well. This is because the impedance of lasers and modulators contains both a substantial imaginary component and a real part that may notably deviate from 50 Ω. On-chip termination is therefore necessary to suppress secondary reflections (Chapter 5).

Some lasers exhibit a low impedance, on the order of 5 Ω. Since it is difficult to construct low-impedance traces on printed-circuit boards, a resistor is placed in series with such lasers, raising the impedance to more manageable values. However, if the resistance is increased to 50 Ω, a peak current of 100 mA translates to a very large voltage swing at the driver output. Thus, the impedance of some lasers is in fact raised to 25 Ω, requiring a

[8] A third source of pulsewidth distortion is the unequal loading at the output, discussed in Section 10.4.

25-Ω transmission line (Fig. 10.33). Nevertheless, the impedance of lasers operating with lower currents is increased to 50 Ω.

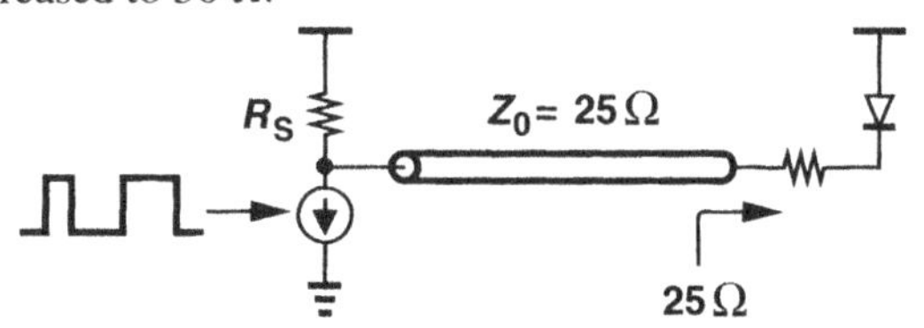

Figure 10.33 Transmission-line environment for a 25-Ω laser.

With 25-Ω lasers, impedance matching at the far end becomes more problematic: variations in the laser impedance are absorbed by only a 20-Ω series resistor. On the other hand, 50-Ω lasers provide more suppression.

10.4 Design Principles

As with the output buffers described in Chapter 5, the design of laser drivers begins from the last stage. Illustrated in Fig. 10.34, this stage typically employs a differential pair that

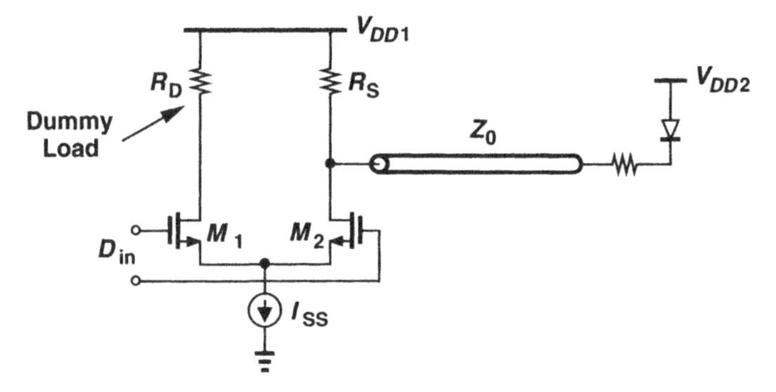

Figure 10.34 Last stage of a laser driver.

steers its tail current to the laser or to a dummy load. As decribed in Chapter 5, differential drivers such as this circuit provide many important advantages over single-ended circuits. First, despite large current swings, they maintain a relatively constant supply current, thus achieving low switching noise across package inductances.[9] Second, even with large output voltage swings, they introduce a small crosstalk if the signals remain symmetric. Third, in the presence of package parasitics, they are less prone to oscillation because their common-mode gain is low.

In bipolar implementations, the output transistors must be large enough to handle the peak current without experiencing high-level injection. In the MOS counterpart, the transistors must be wide enough to (a) sustain a gate-source voltage commensurate with the

[9]The stage of Fig. 10.34 draws part of the current from V_{DD2}, creating some switching activity on V_{DD1}. This issue is resolved later.

supply voltage and (b) steer most of the tail current to one side with the limited voltage swing provided by the preceding stage.

Owing to large device dimensions, the output stage suffers from a substantial input capacitance. In bipolar technology, emitter followers often precede the stage to provide buffering. As shown in Fig. 10.35, the followers lower the output impedance of the

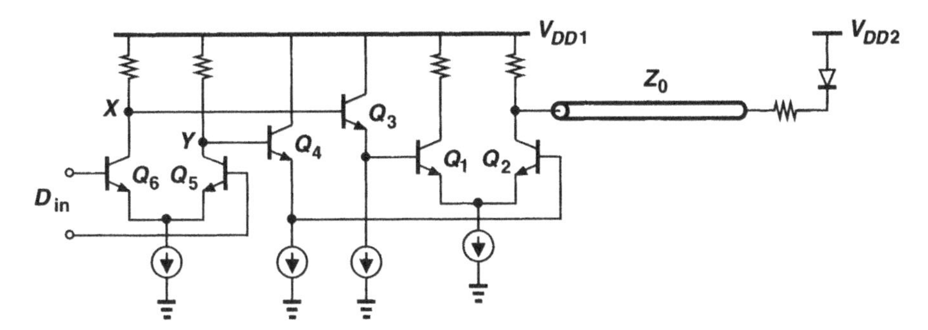

Figure 10.35 Example of a laser driver circuit.

predriver circuit, improving the speed at nodes X and Y. In addition to consuming voltage headroom, followers exhibit an inductive output impedance that, along with the input capacitance of Q_1 and Q_2, may yield considerable ringing in the step response. This behavior can be quantified with the aid of Fig. 10.36, where the base-collector capacitance and the

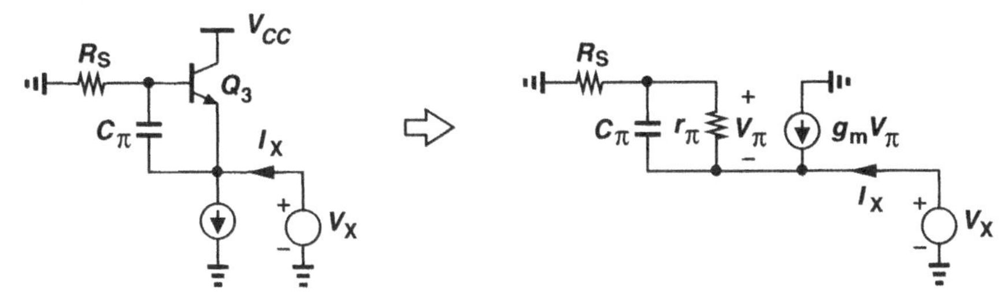

Figure 10.36 Output impedance of an emitter follower driven by a source resistance.

Early effect are neglected. We have

$$V_\pi(C_\pi s + \frac{1}{r_\pi}) + g_m V_\pi = -I_X, \tag{10.3}$$

and hence

$$V_\pi = \frac{-I_X}{C_\pi s + \frac{1}{r_\pi} + g_m}. \tag{10.4}$$

Also

$$-R_S(I_X + g_m V_\pi) + V_\pi = -V_X. \tag{10.5}$$

Substituting from (10.4) for V_π, we obtain

$$Z_{out} = \frac{r_\pi C_\pi R_S s + r_\pi + R_S}{r_\pi C_\pi s + \beta + 1}. \tag{10.6}$$

At $s = 0$, $V_X/I_X = (r_\pi + R_S)/(\beta + 1) \approx 1/g_m + R_S/(\beta + 1)$. On the other hand, at $s = \infty$, $V_X/I_X = R_S$. Thus, if $R_S > 1/g_m + R_S/(\beta + 1)$, the output impedance *increases* with frequency, displaying an inductive component. This condition usually holds because the follower is designed to lower the source impedance.

From Eq. (10.6), it is possible to construct an equivalent circuit representing the output impedance of emitter followers. Since Z_{out} contains a resistive component equal to $R_1 = (r_\pi + R_S)/(\beta + 1)$, we have

$$Z_{out} - \frac{r_\pi + R_S}{\beta + 1} = \frac{r_\pi C_\pi s(\beta R_S - r_\pi)}{(\beta + 1)(r_\pi C_\pi s + \beta + 1)}. \tag{10.7}$$

Inverting the result to obtain the admittance gives

$$\frac{1}{Z_{out} - R_1} = \frac{\beta + 1}{\beta R_S - r_\pi} + \frac{(\beta + 1)^2}{r_\pi C_\pi s(\beta R_S - r_\pi)}. \tag{10.8}$$

The first term is the inverse of a resistor $R_2 = (\beta R_S - r_\pi)/(\beta + 1)$. The second term yields an inductive value of

$$L_1 = \frac{r_\pi C_\pi(\beta R_S - r_\pi)}{(\beta + 1)^2}. \tag{10.9}$$

The overall output impedance therefore appears as shown in Fig. 10.37.

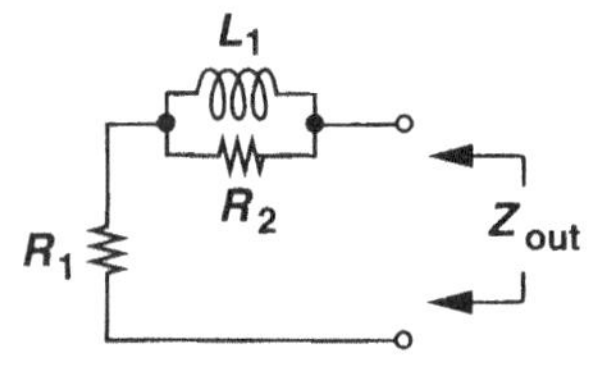

Figure 10.37 Equivalent output impedance of an emitter follower.

The value of L_1 in Eq. (10.9) can be rewritten as

$$L_1 \approx \frac{C_\pi \beta(R_S - 1/g_m)}{g_m(\beta + 1)} \tag{10.10}$$

$$\approx \frac{C_\pi}{g_m}(R_S - \frac{1}{g_m}), \tag{10.11}$$

which, for $R_S \gg 1/g_m$, reduces to

$$L_1 \approx \frac{C_\pi}{g_m} R_S. \tag{10.12}$$

At high bias currents, $C_\pi = C_{je} + C_b \approx C_b = g_m \tau_F$, and hence

$$L_1 \approx \tau_F R_S. \tag{10.13}$$

To reduce the ringing at the output of followers, two precautions can be taken. First, the bias current can be lowered, thereby increasing the value of R_1 in Fig. 10.37 and overdamping the step response. Second, the value of R_S can be reduced by preceding the circuit with another follower. Of course, the voltage headroom penalty may be prohibitive.

Another issue related to emitter followers arises if the interconnect line feeding the base exhibits substantial inductance. Illustrated in Fig. 10.38, such a topology is in fact used in

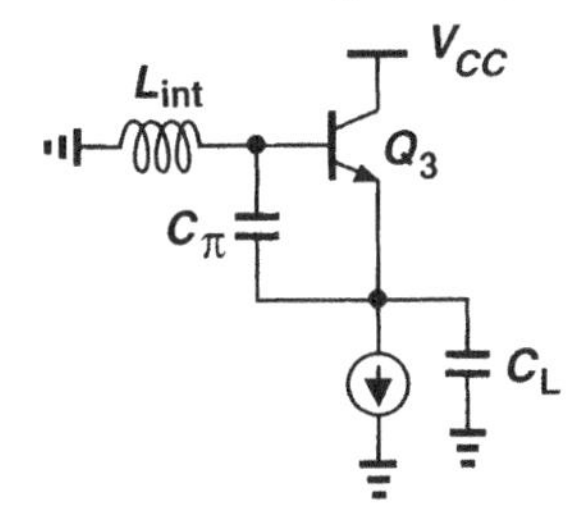

Figure 10.38 Effect of inductance in series with input of an emitter follower.

oscillator design (Chapter 6). With high-speed transistors, an interconnect having a length of a few hundreds of microns may introduce enough inductance to create heavy ringing at the output. Thus, accurate modeling of the interconnect lines is essential.

In CMOS technology, both the poor drive capability and the large voltage headroom consumption of source followers make them ill-suited to high-speed applications. As with output buffers (Chapter 5), inductive peaking and/or simply tapered differential pairs appear to be the only practical solutions here.

The large capacitance introduced by the transistors at the output nodes leads to a long time constant. This difficulty can be alleviated by inductive peaking at the output (Fig. 10.39) [4]. Here, inductors are placed in series with the on-chip termination resistors,

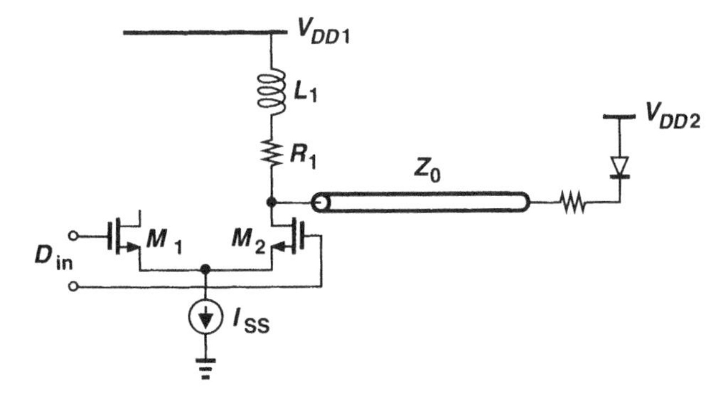

Figure 10.39 Inductive peaking at output.

sharpening the output pulses.

Even with differential drivers, the pulses delivered to the laser may suffer from significantly different rise and fall times. This is because the laser itself is single-ended and the turn-on and turn-off transients of each output transistor are asymmetric [4]. To understand this effect, consider the differential pair shown in Fig. 10.40 and recall that the large tail

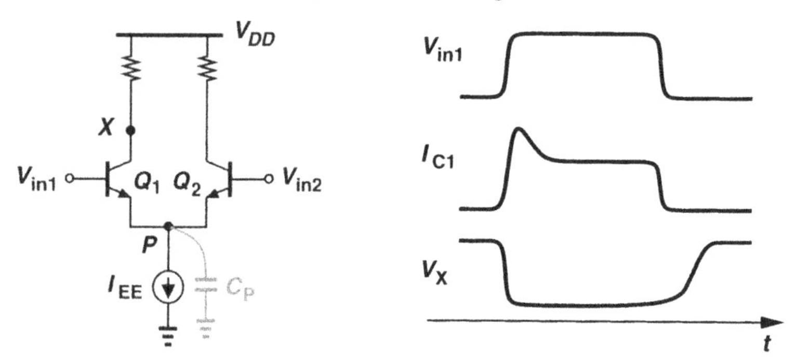

Figure 10.40 Effect of tail capacitance in a differential pair.

current source introduces substantial capacitance at node P. On the rising edge of V_{in1}, Q_1 draws a current equal to the sum of I_{EE} and the displacement current required to charge C_P. On the falling edge of V_{in1}, on the other hand, Q_1 simply turns off and the output node charges up with its natural time constant. As a result, the single-ended output voltage exhibits shorter fall time than rise time.

In order to suppress the output pulse asymmetry, the *input* rise time and fall time can be made asymmetric in the opposite direction. Illustrated in Fig. 10.41, the idea is to introduce a small difference between the output impedances of the followers, thereby slowing

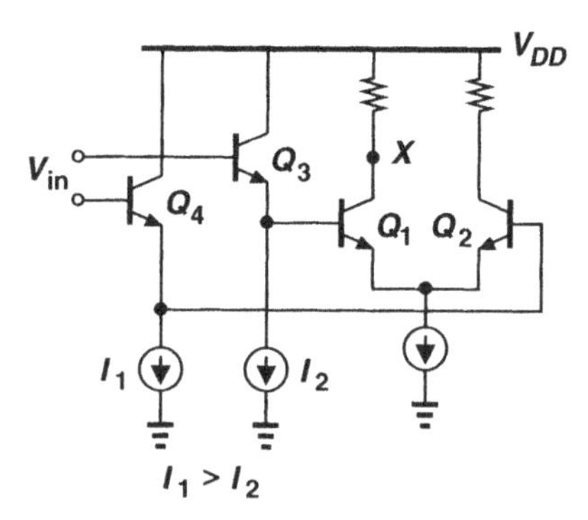

Figure 10.41 Asymmetric emitter followers to cancel pulsewidth distortion.

down the turn-on of Q_1. Alternatively, an offset voltage can be added at the input of the differential pair to adjust the switching speed of Q_1 and Q_2 [4].

While equalizing the single-ended output rise and fall times, both of the above techniques may yield significant duty cycle distortion. For this reason, another offset voltage must be introduced in the predriver circuit to intentionally distort the pulsewidth in the opposite direction [4].

Another important issue in the output stage relates to the effect of load asymmetries on the input pulse shape [Fig. 10.42(a)]. Due to the large gate-drain capacitance of M_1 and M_2, the single-ended signals at P and Q are affected differently. As a result, both pulsewidth distortion and pulse asymmetry may arise. For this reason, it is preferable to route both outputs through identical interconnects and terminate them into approximately equal impedance levels [Fig. 10.42(b)]. This also minimizes the transient currents drawn

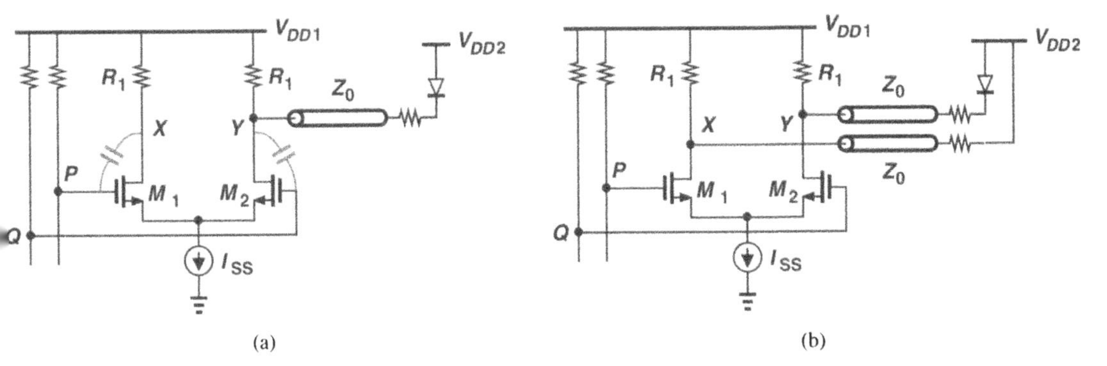

Figure 10.42 (a) Effect of coupling through gate-drain capacitance of output devices with asymmetric loads, (b) load symmetry to minimize the effect.

from both V_{DD1} and V_{DD2}.

Most lasers and modulators experience a voltage swing of more than 2 V when turning on and off. In low-voltage technologies, delivering such swings becomes difficult. This problem can be alleviated by adding a cascode transistor to the output stage, with the gate of the device connected to a supply voltage well above the device breakdown voltage (Fig. 10.43). As the output voltage rises, the cascode devices shield the input transistors.

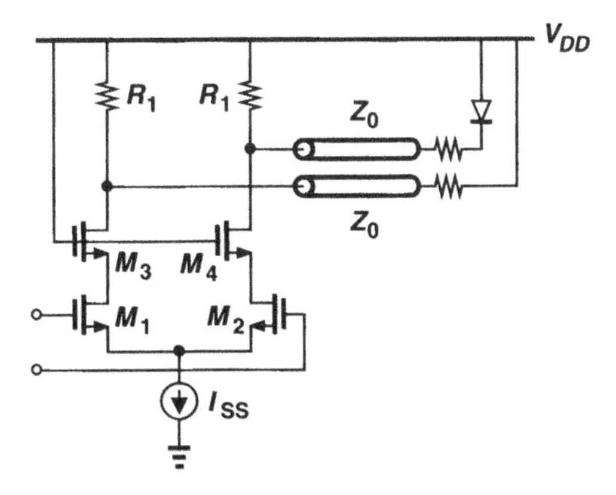

Figure 10.43 Laser driver employing cascode devices.

However, when the output falls, M_3 and M_4 enter the triode region, introducing a greater capacitance in the signal path. Note that this technique is less useful in bipolar realizations as it would not allow large swings *below* the base voltage of the cascode transistors.

10.4.1 Power Control

In order to maintain a constant optical power in the presence of temperature variations and aging, the laser driver must employ a means of controlling the bias current. Note that the monitor photodiode must exhibit stable characteristics with temperature and age, but its speed is not critical as it only measures the average optical power.

The power control circuit typically assumes the form shown in Fig. 10.44, where the

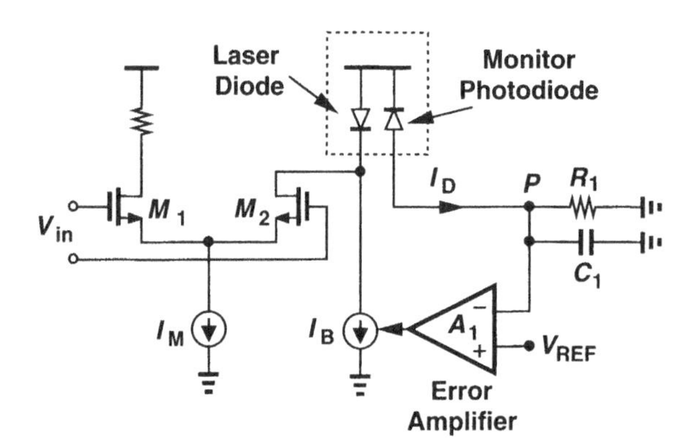

Figure 10.44 Power control in a laser driver.

current produced by the photodiode is converted to voltage, low-pass filtered, and compared with a reference. The resulting error then adjusts I_B such that V_P approaches V_{REF}. Current source I_M establishes the laser modulation current independently.

Interestingly, the power control loop performs high-pass filtering on the signal: the R_1C_1 network suppresses high-frequency components in the feedback loop, but, at sufficiently low frequencies, the negative feedback is strong enough to attenuate the output signal. Thus, the circuit displays a high-pass transfer function from V_{in} to P_{out}.

Let us examine the above effect in the time domain. If, as depicted in Fig. 10.45, the input contains random ONEs and ZEROs, then the average current produced by the photodiode is approximately equal to half of its peak value. The loop ensures that the product of this average current and R_1 remains close to V_{REF}. Now suppose a long run of ONEs is applied at the input, turning the laser nearly off. Then, the average photodiode current drops substantially, allowing V_P to fall toward zero. However, the loop attempts to maintain V_P in the vicinity of V_{REF} by raising I_B considerably. In fact, if the long run of ONEs continues indefinitely, the laser power eventually rises to half of the value produced for an optical ONE.

As with continuous-time offset cancellation in limiting amplifiers (Chapter 5), the high-pass filtering modulates the decision threshold. The value of R_1C_1 must therefore be sufficiently large to minimize the degradation. From calculations in Chapter 5, we write the corner frequency as

$$\omega_{-3dB} \approx \frac{\beta}{R_1C_1}, \tag{10.14}$$

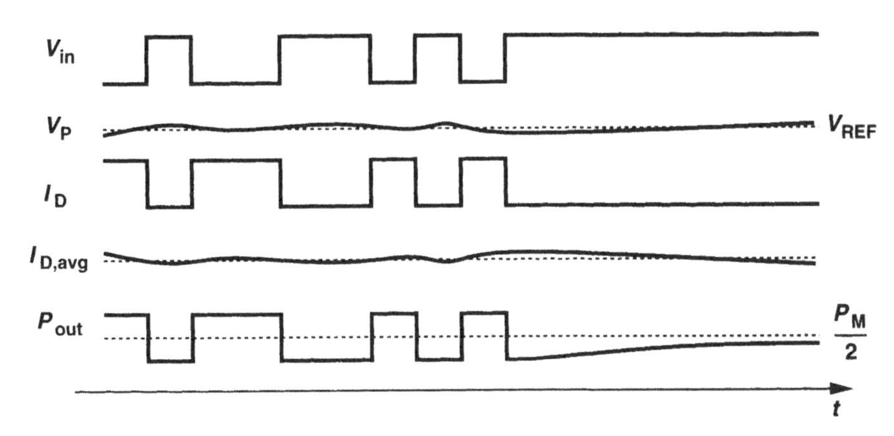

Figure 10.45 Response of power control circuit to a long run.

where β denotes the loop gain. For a laser efficiency of η, a photodiode responsivity of R_{PD}, the loop gain of the circuit in Fig. 10.44 is equal to

$$\beta = \eta \cdot R_{PD} \cdot R_1 \cdot A_R, \tag{10.15}$$

where A_R represents the transconductance gain from the input of the error amplifier to the output current of I_B. Thus,

$$\omega_{-3dB} \approx \frac{\eta R A_R}{C_1}, \tag{10.16}$$

a value independent of R_1 and suggesting that C_1 must be large enough to minimize degradation of the data. A corner frequency of a few tens of kilohertz may necessitate an off-chip capacitor.

10.5 New Developments in Laser Driver Design

A laser driver must deliver a high current with a large voltage swing and a broad bandwidth. These three requirements severely tighten the design space, making the driver the most challenging among high-speed circuits.

Figure 10.46 shows a generic laser driver architecture. The output stage must drive an off-chip load (e.g., 50 Ω) and also an on-chip "back-termination" resistor. As explained in Chapter 5, the back termination is necessary to absorb reflections due to slight mismatches in the package, the connectors, and the routing to the load. It is desirable to choose R_T equal to the characteristic impedance of the routing, e.g., 50 Ω, but this "wastes" half of the current swing produced by the differential pair. Thus, R_T is typically chosen in the range of 75 Ω to 100 Ω.

The design of the driver begins with the output stage and the required current swing [5]. Suppose we wish to deliver 40 mA to a 50-Ω load and hence another 27 mA to a 75-Ω

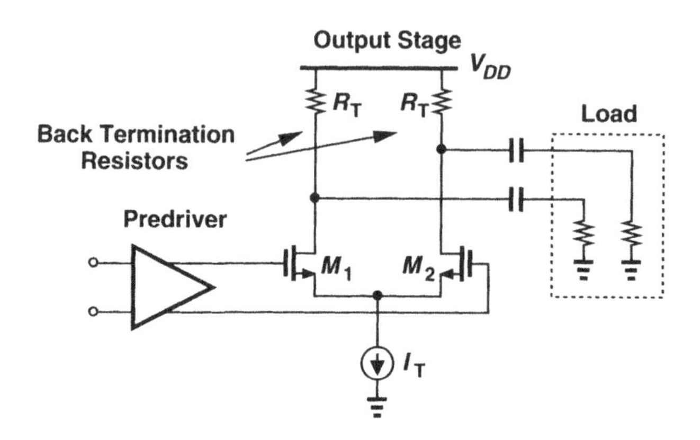

Figure 10.46 Generic laser diver cascade.

termination resistor. The tail current, I_T, of the output stage must therefore exceed 67 mA. However, the differential pair steers the tail current completely only if the transistors are very wide and/or the voltage swing provided by the predriver is large, both of which make the predriver design difficult. This means that in practice (a) a great deal of iteration is necessary between the predriver design and the output stage design, and (b) the tail current of the output stage is likely to be quite higher than 67 mA. For example, the design in [5] employs a W/L of 3300 for the output transistors along with $I_T = 90$ mA and a single-ended predriver swing of 600 mV_{pp}.

The input capacitance of the output stage makes the predriver design difficult, often more so than the output stage itself! For example, with the above choice of $W/L = 3300$, the design in [5] presents to the predriver a gate-source capacitance of 0.75 pF and a Mill-multiplied gate-drain capacitance of 0.88 pF (Fig. 10.47). For a bandwidth of 10 GHz at

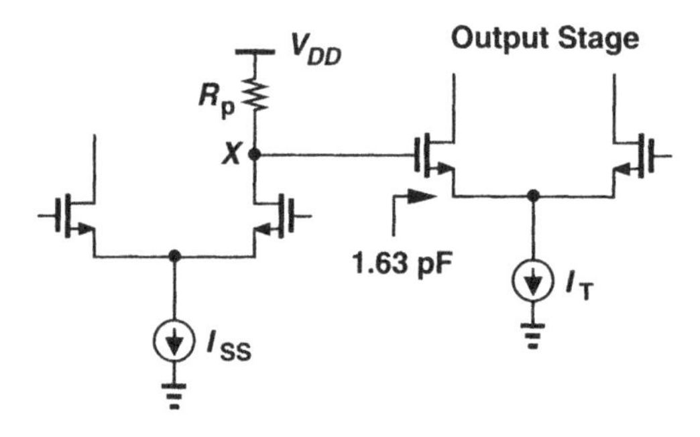

Figure 10.47 Example of laser driver circuit.

node X, $R_p \leq 10\ \Omega$, and for a voltage swing of 600 mV at this node, $I_{SS} \geq 62$ mA! We must therefore apply broadband techniques at X so as to allow a larger R_p and smaller I_{SS}.

We begin with adding negative capacitance at X [5]. Illustrated in Fig. 10.48(a), the cir-

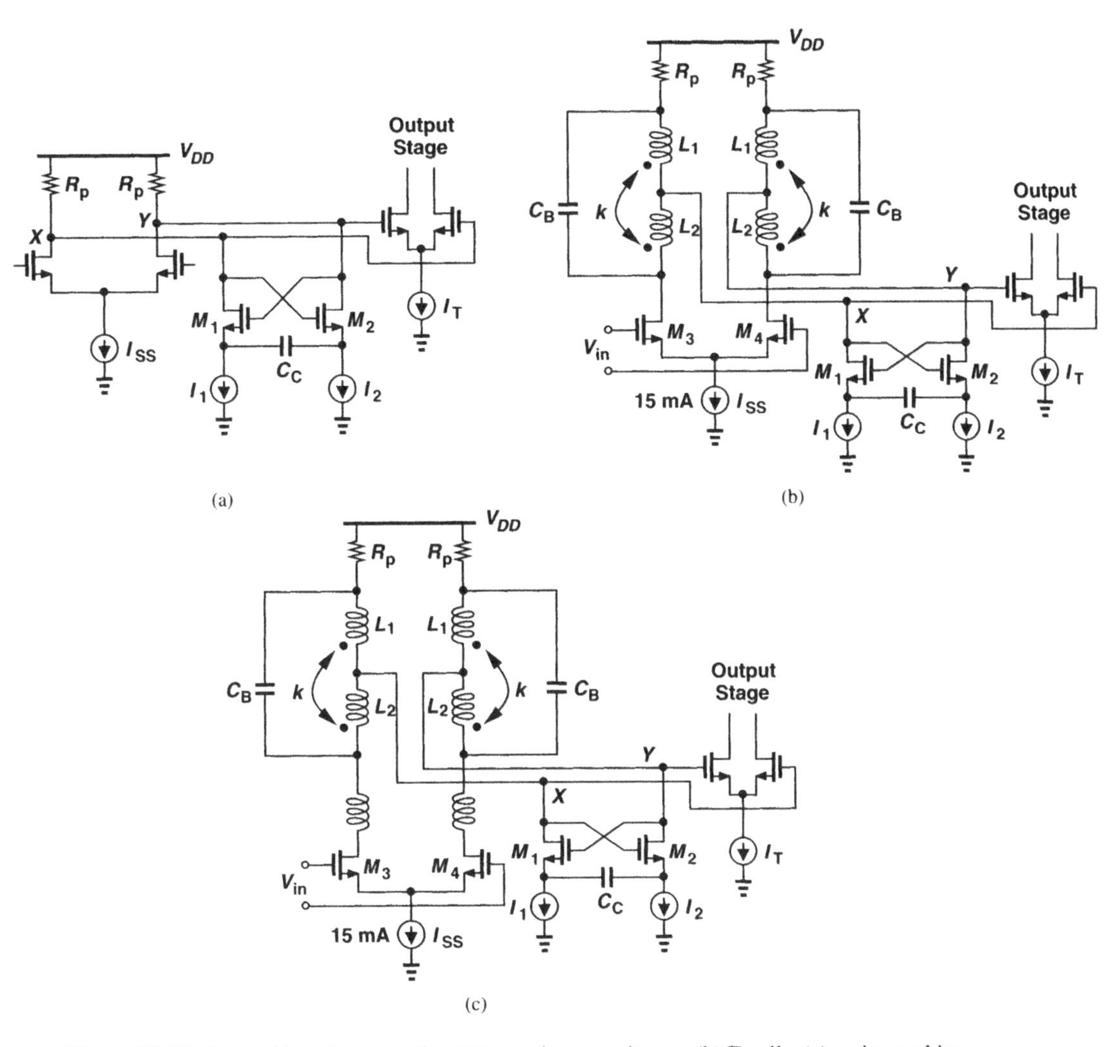

Figure 10.48 Laser driver incorporating (a) negative capacitance, (b) T-coils, (c) series peaking.

cuit incorporates the negative-impedance converter studied in Chapter 5. To avoid ringing in the step response, C_c is limited to 200 fF, presenting a single-ended negative capacitance of −400 fF at X (or Y). The remaining 1.23-pF capacitance must be tackled by other means. In this design, $I_1 = I_2 = 2.5$ mA.

For our next broadband technique, we can consider shunt peaking, T-coil peaking, etc. The low resistance values required at the interface between the predriver and the output stage make T-coil peaking attractive here. We thus arrive at the topology in Fig. 10.48(b). In this design, $R_p = 40\ \Omega$ and the total inductance of each T-coil is 3 nH [5].

While the negative capacitance and the T-coils in the above circuit deal with the capacitance at X and Y, the drain junction capacitance of M_3 and M_4 still limits the bandwidth. Chosen wide enough ($W/L = 444$) to steer a tail current of 15 mA with a single-ended input swing of 400 mV_{pp}, these transistors contribute significant capacitance. This issue is resolved by inserting two 1.75-nH inductors in series with the drains to create series peaking [Fig. 10.48(c)].

The design described above delivers a current of 100 mA at 10 Gb/s while consuming 675 mW [5]. The circuit is realized in 0.18-μm CMOS technology.

In order to deliver large voltage swings to lasers or optical modulators without stressing the output transistors, we can add cascode devices [6], as is often practiced in RF power amplifiers. Consider the cascode differential stage shown in Fig. 10.49(a), where V_{DD} is

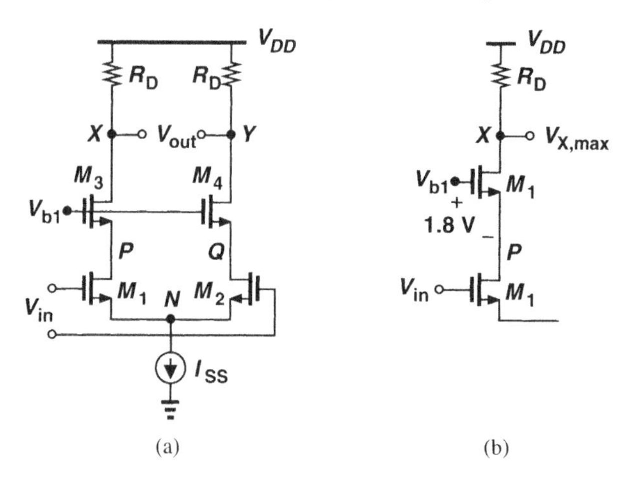

Figure 10.49 (a) Cascode laser driver, (b) detailed voltages.

chosen quite higher than the maximum allowable V_{DS} and V_{GS} of the transistors. We note that when M_1 turns off, $V_X = V_{DD}$ and $V_{DS3} = V_X - V_P \approx V_X - (V_{b1} - V_{TH}) = V_{DD} - (V_{b1} - V_{TH})$. Thus, to avoid stressing M_3, V_{b1} must be chosen *high* enough. Similarly, $V_{DS1} = V_P - V_N$ must also remain below the maximum allowable value. For example, in 0.18-μm technology, we have $V_{DS3}, V_{DS1} \leq 1.8$ V. Thus, allowing about 400 mV for the tail current source, we can choose V_{DD} equal to $2 \times 1.8\text{ V} + 0.4\text{ V} = 4$ V.

How much output voltage swing do we obtain in the above example? The voltage at node X can reach a maximum of V_{DD}. To determine the minimum value, we note that, as V_X falls and drives M_3 into the triode region, V_P also falls, increasing V_{GS3} [Fig. 10.49(b)]. To ensure M_3 is not stressed, we have $V_{GS3} = V_{b1} - V_P \leq 1.8$ V and hence $V_P \geq V_{b1} - 1.8$ V. Now, suppose V_{DS3} is small so that $V_X \approx V_P$. It follows that the maximum peak-to-peak swing at X is given by

$$V_{X,pp} \approx V_{DD} - (V_{b1} - 1.8\text{ V}). \tag{10.17}$$

Since $V_{DD} - (V_{b1} - V_{TH}) = 1.8$ V, we have

$$V_{X,pp} \approx 3.6\text{ V} - V_{TH}. \tag{10.18}$$

This, of course, is the upper bound because V_{DS3} may not be negligible when V_X reaches the lowest level. The reader is encouraged to compare this result with that of a simple differential pair.

At low supply voltages, it is desirable to eliminate the tail current source from the output stage, saving a headroom of a few hundred millivolts. However, it becomes difficult to define the bias current of this stage. To understand this point, consider the example illustrated in Fig. 10.50(a), where the output transistors sense an input common-mode level of

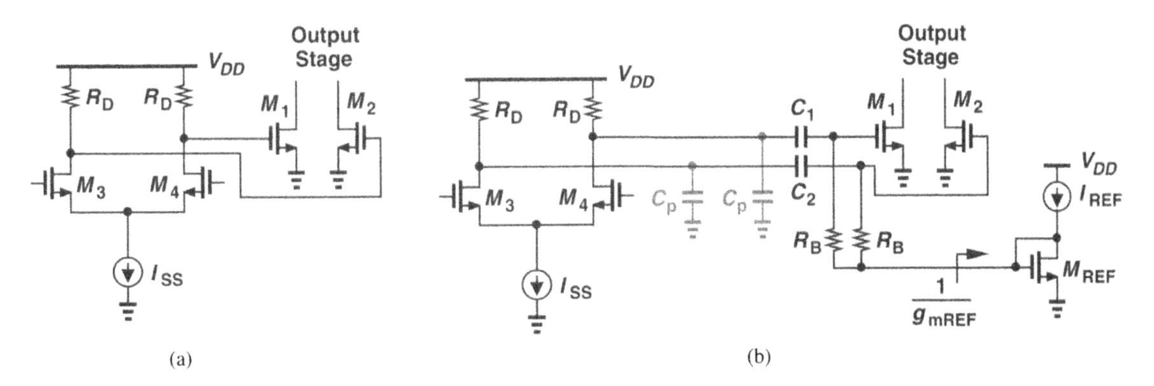

Figure 10.50 (a) Direct and (b) capacitive coupling between prediver and output stage .

$V_{DD} - (I_{SS}/2)R_D$. The bias current of M_1 and M_2 therefore depends on their threshold voltage and mobility and the supply voltage.

Can we precede the output stage with capacitive coupling to allow accurate definition of the bias current? Shown in Fig. 10.50(b), the idea is to form a current mirror for the output stage, i.e., copy I_{REF} onto M_1 and M_2. This approach must deal with two issues. First, the high-pass filter consisting of C_1 (or C_2) and R_B must exhibit a sufficiently long time constant so as to create negligible baseline wander (Chapter 2) in the presence of long runs. For example, for a maximum run length of 72 bits, R_BC_1 must reach several hundred bit periods. More importantly, some standards (e.g., SONET) specify a cut-off frequency as low as 40 kHz, demanding a time constant of about 4 μs and hence large resistor and capacitor values.

The second issue relates to the parasitics of the coupling capacitors in Fig. 10.50(b). To avoid significant signal attenuation, C_1 and C_2 must be about 5 to 10 times the input capacitance of the output stage, C_{in}. Since M_1 and M_2 are very large, so must be C_1 and C_2, thus introducing a considerable amount of bottom-plate parasitic capacitance to the substrate [C_p in Fig. 10.50(b)]. For example, if $C_1 = 5C_{in}$ and $C_p = 5\%C_1$, then $C_p = 0.25C_{in}$; i.e., capacitive coupling raises the load capacitance presented to the predriver by 25%.

In order to define the output stage bias without capacitive coupling, the predriver design can produce a CM level that, equivalently, biases the output transistors as in a current mirror [7]. To understand this technique, which has been realized in bipolar technology in [7], let us first consider the circuit shown in Fig. 10.51(a). Here, the supply voltage of the predriver, Q_3-Q_4,is set equal to $2V_{BE}$ rather than V_{CC}. Assuming for simplicity that

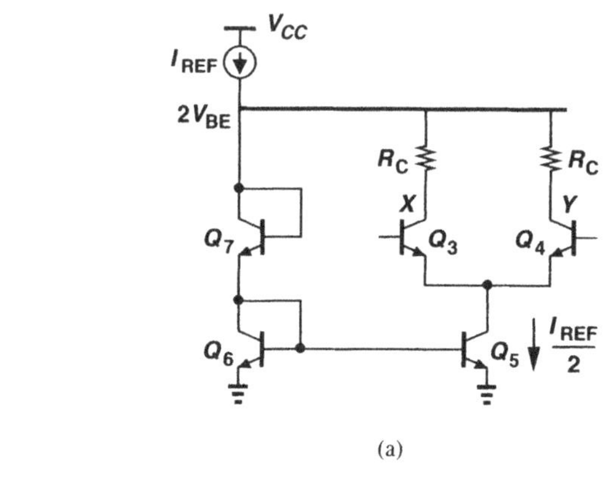

(a)

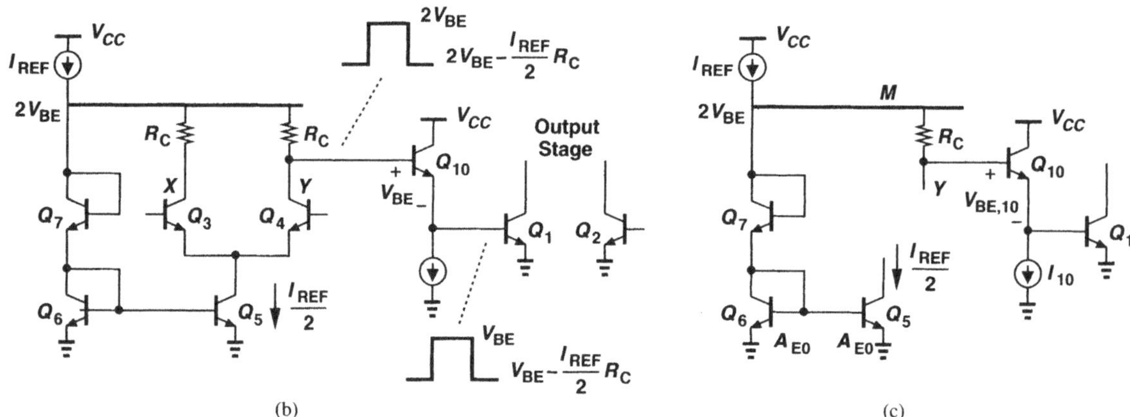

(b) (c)

Figure 10.51 (a) Biasing scheme to define voltage swings at X and Y in terms of V_{BE}, (b) laser driver cascade, (c) details of voltage swings in signal path.

Q_5-Q_7 are identical, we note that Q_5 draws a current of $I_{REF}/2$, creating high and low levels at X and Y given by $2V_{BE}$ and $2V_{BE} - (I_{REF}/2)R_C$, respectively. Of course, the input CM level of Q_3 and Q_4 must also be set properly to maintain them in the forward active region.

We now connect the above predriver through emitter followers to the output transistors [Fig. 10.51(b)]. Let us prove that the output current is a multiple of I_{REF}. Assuming a voltage gain of unity for the followers, we recognize that the base voltage of Q_1 and Q_2 is one V_{BE} lower than those at X and Y, respectively, and swings between V_{BE} and $V_{BE} - (I_{REF}/2)R_C$. Consequently, I_{C1} and I_{C2} swing between a well-defined multiple of I_{REF} and a small value if $(I_{REF}/2)R_C$ at least several hundred millivolts. To calculate the output current, we simplify the circuit as shown in Fig. 10.51(c), neglecting the base currents. We have

$$V_M = V_{BE7} + V_{BE6} \tag{10.19}$$

$$\approx V_{BE10} + V_{BE1}, \tag{10.20}$$

if Q_4 is off. These equations suggest that, to copy I_{REF} onto Q_1, we can choose $V_{BE10} = V_{BE7}$ and $V_{BE1} = V_{BE6}$. For example,, we assume Q_5 and Q_6 are identical and have an emitter area of A_{E0} and a saturation current of I_{S0}. That is, $V_{BE6} = V_T \ln[I_{REF}/(2I_{S0})]$ because Q_6 carries half of I_{REF}. Also, $V_{BE7} = V_T \ln[I_{REF}/(2I_{S7})]$ and $V_{BE10} = V_T \ln(I_{10}/I_{S10})$. It follows that

$$V_{BE1} = V_{BE7} + V_{BE6} - V_{BE10} \tag{10.21}$$

$$= V_T \ln \frac{I_{REF}}{2I_{S7}} + V_T \ln \frac{I_{REF}}{2I_{S0}} - V_T \ln \frac{I_{10}}{I_{S10}}. \tag{10.22}$$

Suppose $I_{10} = I_{REF}$ and $I_{S10} = 2I_{S7}$, then $V_{BE1} = V_T \ln[I_{REF}/(2I_{S0})] = V_T \ln(I_{C1}/I_{S1})$. We thus have

$$I_{C1} = \frac{I_{S1} I_{REF}}{2I_{S0}}, \tag{10.23}$$

revealing that the mirror factor from $I_{REF}/2$ to I_{C1} is equal to the ratio of the emitter areas of Q_1 and Q_6.

It is possible to improve the speed of the above circuit by replacing the bias current source of the emitter followers with diode-connected devices and also a cross-coupled pair [7]. Figure 10.52 depicts the final design. Note that, without the cross-coupled pair, the

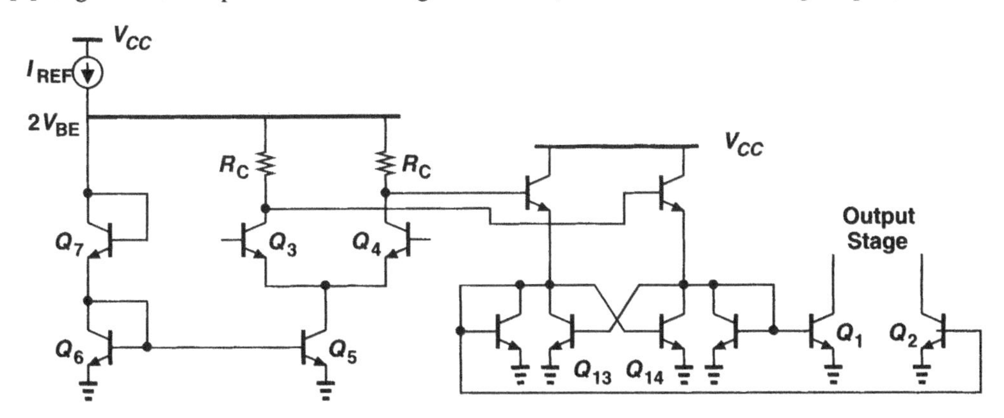

Figure 10.52 Complete laser driver.

diode-connected devices halve the followers' voltage gain. The negative resistance introduced by Q_{13} and Q_{14} can restore the gain. These four transistors also rapidly charge and discharge the very large capacitance seen at the bases of Q_1 and Q_2.

Realized in a 90-GHz BiCMOS technology, the above laser driver delivers a maximum modulation current of 85 mA with a maximum output voltage swing of 3.2 V. The circuit operates up to 4.25 Gb/s [7].

References

1. H.-I. Cong, et al., "A 10-Gb/s 16:1 Multiplexer and 10-GHz Clock Synthesizer in 0.25-μm SiGe BiCMOS," *IEEE J. Solid-State Circuits,* vol. 36, pp. 1946–1953, Dec. 2001.
2. C.-K. Yang, R. Farjad-Rad, and M. Horowitz, "A 0.5-μm CMOS 4-Gb/s Serial Link Transceiver," *IEEE J. Solid-State Circuits,* vol. 33, pp. 713–722, May 1998.
3. T. Otsuji, et al., "A Super-Dynamic Flip-Flop Circuit for Broadband Application up to 24 Gbit/s Utilizing Production-Level 0.2-μm GaAs MESFETs," *IEEE J. Solid-State Circuits,* vol. 32, pp. 1357–1362, Sept. 1997.
4. H.-M. Rein and M. Moller, "Design Considerations for Very High Speed Si Bipolar ICs Operating up to 50 Gb/s," *IEEE J. of Solid-State Circuits,* vol. 31, pp. 1076–1090, August 1996.
5. S. Galal and B. Razavi, "10-Gb/s limiting amplifier and laser/modulator driver in 0.18-um CMOS technology," *IEEE Journal of Solid-State Circuits,* vol. 38, pp. 2138-2146, Dec. 2003.
6. D.-U. Li and C.-M. Tsai, "10-13.6 Gbit/s 0.18 .m CMOS modulator drivers with 8 Vpp differential output swing," *Electronics Letters,* vol. 41, pp. 643-644, May 2005.
7. H. W. Fattaruso and B. Sheahan, "A 3-V 4.25-Gb/s laser driver with 0.4-V output voltage compliance," *IEEE Journal of Solid-State Circuits,* vol. 41, pp. 1930-1937, Aug. 2006.

Chapter 11

Burst-Mode Circuits

Optical communication has traditionally dealt with the transmission and reception of data over long distances (hundreds to thousands of miles) and, as such, has seen a small, niche market. In the past decade, however, a new type of optical communication has found widespread use in delivering data to homes and businesses. Employing "passive optical networks" (PONs), the new system poses interesting challenges in the design of transceiver circuits. This chapter deals with the "burst-mode" (BM) designs required in PONs. We begin with an overview of passive optical networks and their unique attributes. Next, we address the design of burst-mode transimpedance amplifiers and the problem of offset cancellation in BM limiting amplifiers. We then study BM CDR circuits and their jitter characteristics.

11.1 Passive Optical Networks

Suppose 20 homes in a neighborhood wish to find access to the Internet with data rates of several gigabits per second. (Today's DSL networks serving our homes support rates only up to about 1 or 2 megabits per second.) As shown in Fig. 11.1(a), the service provider can install 20 independent optical fibers (under the ground) along with 40 optical transceivers (20 at the provider end and one in each home). Such a network becomes extremely expensive if millions of homes must be served. It is also difficult to add one more home after the network is completed because a new fiber must be laid along the entire distance.

Can the homes share *one* fiber? Suppose a single fiber travels from the provider to the vicinity of the 20 homes [Fig. 11.1(b)] and is branched into 20 fibers for only a short distance. This topology requires only 21 optical transceivers and more easily lends itself to expansion. To reduce the installation and maintenance cost, a "passive" optical splitter (with no power supply) performs the branching function. This arrangement is called a passive optical network.

We must now discuss how exactly a PON delivers data. When the users transmit data (in the "upstream" direction), they time-share the main fiber. That is, user #1 transmits for T seconds and then turns off its laser; next, user #2 transmits for T seconds and then turns off

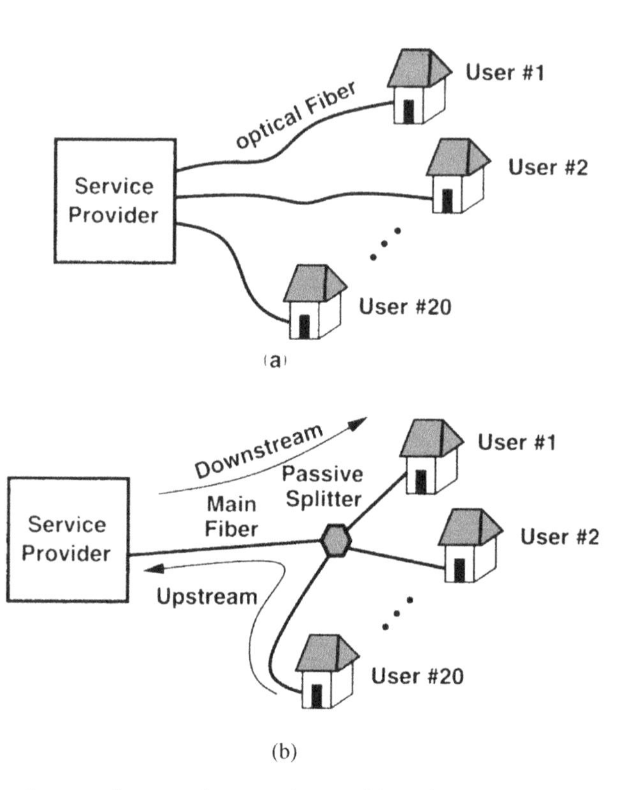

(b)

Figure 11.1 (a) Simple network connecting a service provider to homes, (b) use of a passive splitter to reduce the number of long fibers (PON).

its laser, etc. This time sharing is called "time-division multiple access" (TDMA). When the service provider transmits data to the users (in the "downstream" direction), each user is timed so as to receive only its own data and ignore others'.

The service provider station in Fig. 11.1(b) is called the "optical line unit" (OLT) and each user's device, an "optical network unit" (ONU). The OLT connects each PON to the "metro backbone," i.e., the main network in a city. The upstream and downstream data are modulated on different light wavelengths so that they can be distinguished.

While economically efficient for implementing the fiber-to-the-home (FTTH) technology, PONs present difficult circuit issues, especially at the OLT end. Since each user accesses the main fiber for a certain amount of time, the data received by the OLT is in the form of independent *bursts*. Moreover, due to aging of lasers and path loss differences, the data packets produced by different users may have considerably different *amplitudes*. Figure 11.2 illustrates the resulting data waveforms in the upstream path. The ratio of the highest and lowest amplitudes may reach 20 dB and is called the "soft/load" ratio.

The bursts of data arriving at the OLT exhibit different amplitudes and phases, impacting the design of the transimpedance amplifier, even the limiting amplifier, and the CDR circuit. "Burst-mode" optical circuits for rates as high as 10 Gb/s have thus been under extensive study.

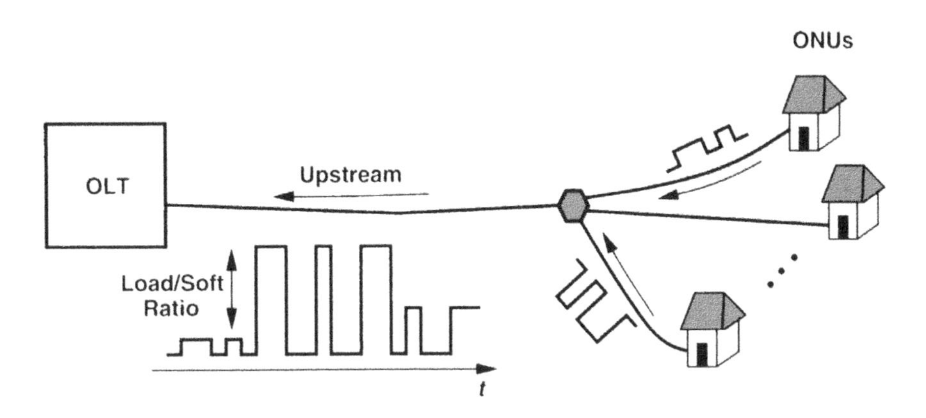

Figure 11.2 PON data waveforms.

We should mention that the ONUs in Fig. 11.2 also pose serious design problems. With a maximum of 32 ONUs defined by standards for one OLT, the passive splitting greatly reduces the light intensity received by each ONU, requiring a high-sensitivity TIA.

11.2 Burst-Mode TIAs

From our study of TIAs and limiting amplifiers (LAs) in previous chapters, we can construct the receiver front end shown in Fig. 11.3(a). Here, a TIA senses the input current, producing a single-ended output voltage. The TIA gain is set according to the input or output levels so as to avoid distortion of the binary data. In order to interface the TIA with the LA, a single-ended to differential (S/D) converter is necessary. Note that V_b must be chosen equal to the average value of the high and low voltage levels at the TIA output, V_H and V_L, respectively, and is typically derived from V_{TIA} by means of a low-pass filter [Fig. 11.3(b)]. The LPF time constant must be large enough to ensure V_b changes negligibly in the presence of a long sequence of ONEs or ZEROs.

We now address the issues facing the above front end in a burst-mode system. Depicted in Fig. 11.3(c), the data contains a "preamble" at the beginning of each burst to assist the receiver circuits. The preamble consists of a periodic sequence of ONEs and ZEROs (e.g., for a total of 10 bits), during which the receiver front end must adapt itself to the new signal amplitude (and the CDR must lock to the new signal phase).

How does the front end of Fig. 11.3(a) respond when a new burst arrives? First, the TIA may saturate because the AGC, which typically measures the average amplitude over a (large) number of bits, cannot adjust the gain immediately. As a result, the bits at the beginning of the burst may be distorted. and hence the speed, the data may experience heavy ISI. A BM TIA must therefore rapidly measure the data amplitude and command the AGC accordingly.

Second, the reference voltage, V_b, in Fig. 11.3(a) must also be adjusted. This is because the voltage level corresponding to ZEROs (i.e., zero light intensity) does not change (Fig.

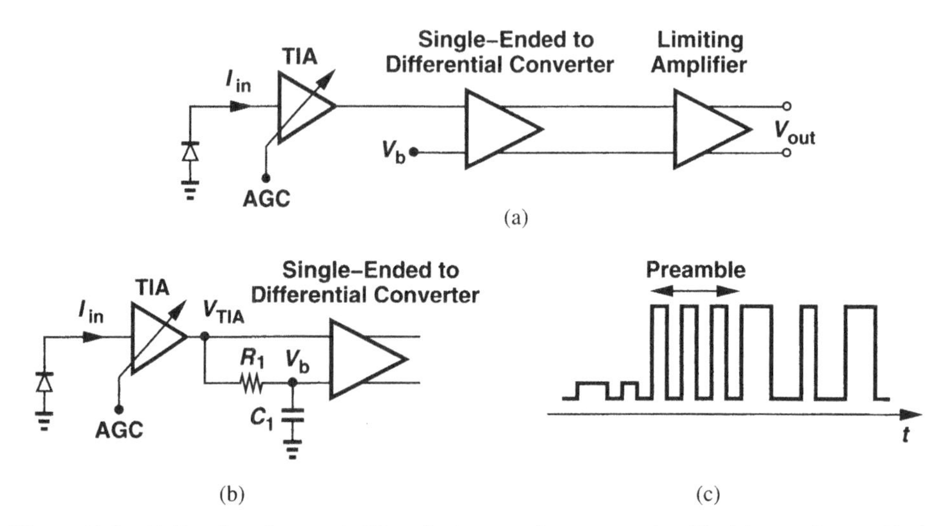

Figure 11.3 (a) Receiver front end, (b) typical approach to generating V_b, (c) use of preamble in data for dc level refresh.

11.4) but that corresponding to ONEs does. The difficulty here is that the circuit generat-

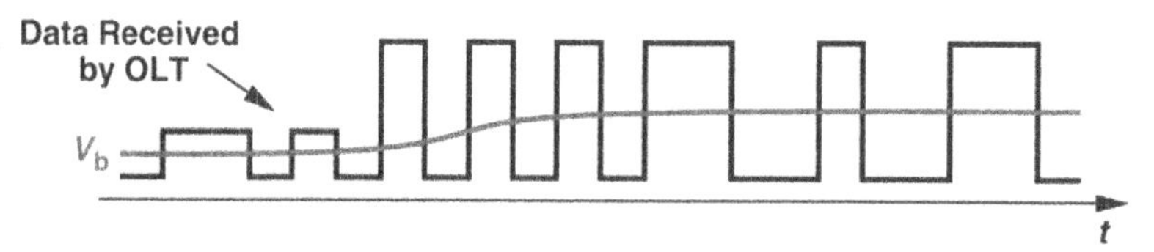

Figure 11.4 Data waveform received by OLT.

ing V_b, e.g., the low-pass filter in Fig. 11.3(b), must be fast enough to adjust V_b during the preamble, but slow enough not to change V_b significantly during long runs. These conflicting requirements cannot be met by a simple LPF.

Another attribute of PON systems exacerbates the problem of gain and reference adjustment. Some PON standards specify an extinction ratio as low as 6 dB for the ONU transmitter. For example, if the ONU laser produces 10 mW of light for each ONE, then the ZEROs are represented by 2.5 mW. Since the lasers in different ONUs may have different characteristics, the low light level received by the OLT may vary from one burst to the next (Fig. 11.5).

11.2.1 TIA with Top and Bottom Hold

In order to adjust both the gain and the reference, V_b, in the front end of Fig. 11.3(a), we must measure the high and low levels of the data waveform (V_H and V_L, respectively) at the TIA output. This can be accomplished through the use of peak detectors [1]. Called

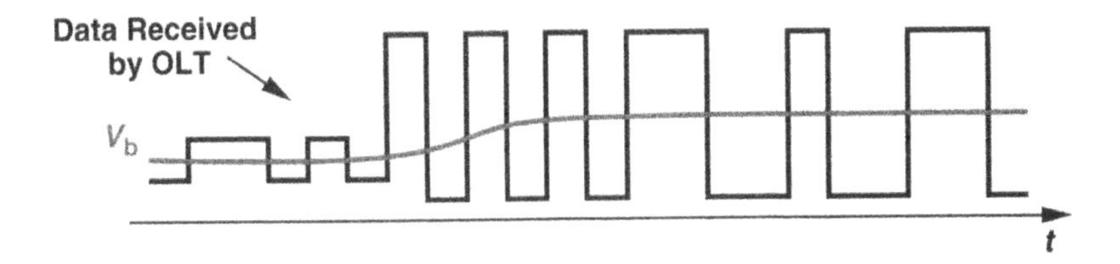

Figure 11.5 OLT data waveform when ONU lasers have different characteristics.

"top hold" and "bottom hold" in [1], the peak detectors rapidly measure V_H and V_L, respectively, providing information for the AGC and reference circuits.

Let us first construct a peak detector circuit. In CMOS technology, we employ a diode-connected MOSFET as a rectifier along with a capacitor [Fig. 11.6(a)]. Here, when V_{data}

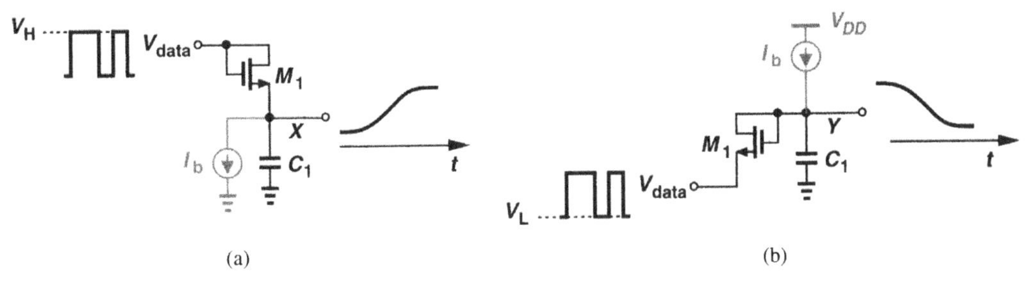

Figure 11.6 (a) Positive peak detector, (b) negative peak detector.

goes high, M_1 tuns on, charging C_1, and when V_{data} goes low, M_1 turns off. Thus, V_X carries information regarding the high data levels. Figure 11.6(b) depicts a similar arrangement for detecting the low levels.

The circuits of Fig. 11.6 entail several issues. First, they suffer from slow settling because the diode-connected MOSFET turns off gradually as V_X rises or V_Y falls. To alleviate this issue, a small bias current, I_b, can be applied to each transistor, but at the cost of corrupting V_X and V_Y during long runs. Second, V_X and V_Y contain a level shift approximately equal to the MOS threshold voltage, i.e., $V_X - V_Y$ is not equal to the difference between V_H and V_L. Third, if a new burst arrives with a smaller amplitude, then the peak detectors retain the previous levels because both MOSFETs are now off. To "refresh" the voltage across the capacitors, a large I_b can be added to each circuit so as to quickly lower V_X and raise V_Y, thus turning the diodes on. But such a current must be turned off automatically; hence the need for a reset command.

We now present a number of circuit techniques that deal with these issues [1]. Let us place the peak detector within a negative-feedback loop as shown in Fig. 11.7(a). Here, when M_1 is on, the loop resembles a two-stage op amp (comprising A_1 and A_2), acting as a unity-gain amplifier. Thus, $V_{out} \approx V_{in}$ even though M_1 introduces a level shift. When M_1 turns off, C_1 holds the voltage corresponding to the high data level, and $V_{out} \approx V_H$. But how do we ensure that the circuit responds properly to a new burst having a lower

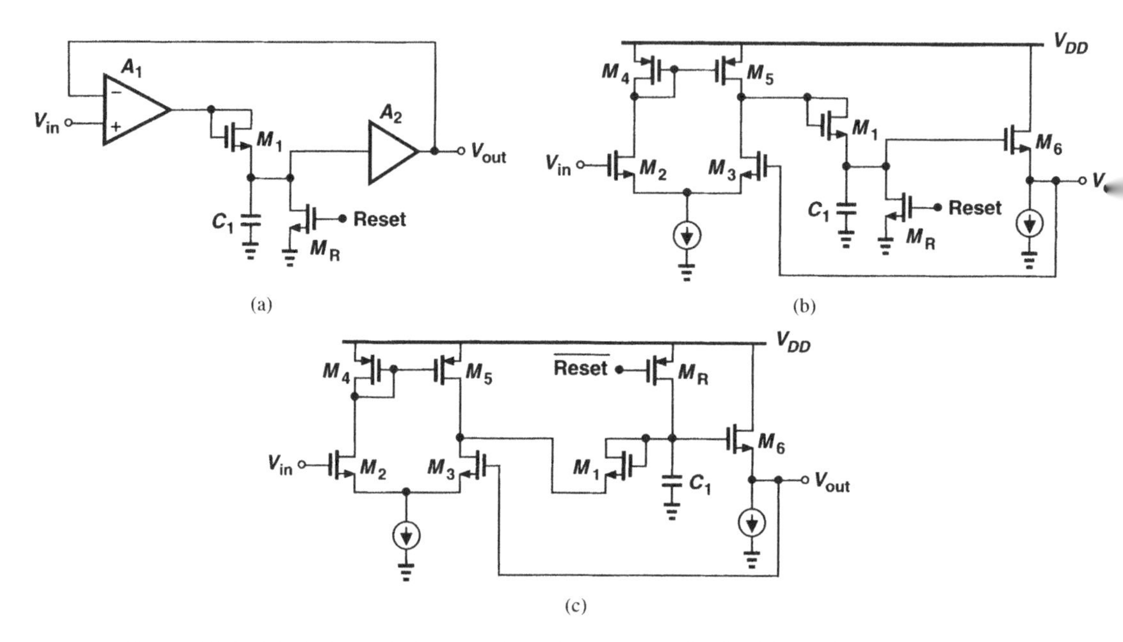

Figure 11.7 (a) Positive eak detector placed in a negative-feedback loop, (b) transistor implementation of (a), (c) implementation for negative peak detector.

amplitude? This is accomplished by generating a reset signal after the previous burst and thereby discharging C_1 [1] (as explained below).

The implementation of this "top hold" circuit is shown in Fig. 11.7(b) [1], where M_2-M_5 form A_1 and the source follower M_6 operates as A_2. In [1], the diode is in fact realized as a pn junction. While improving the settling speed, a forward-biased pn junction in CMOS technology may cause latchup, demanding a careful layout. Also, note that $V_{GS1} + V_{GS6}$ limits the voltage headroom, an issue greatly alleviated if M_6 is converted to a PMOS source follower. The "bottom hold" circuit is readily obtained as depicted in Fig. 11.7(c) [1]. With negligible device mismatches, the difference between these two circuits' outputs is approximately equal to $V_H - V_L$.

How do we exploit $V_H - V_L$ to rapidly adjust the TIA gain? The gain must be changed in proportion to $1/(V_H - V_L)$. Shown in Fig. 11.8 is an example of a gain adjustment circuit. A differential pair, M_1-M_2, converts $V_H - V_L$ to a differential current. Similarly, another pair, M_3-M_4, produces a differential current in proportion to a desired amplitude, V_1. The difference between these currents flows through an active current mirror, thereby generating a control voltage. As $V_H - V_L$ exceeds V_1, V_{cont} rises, reducing the on-resistance of M_F and hence the gain of the TIA. Resistor R_{F2} may be necessary to "soften" the TIA gain variation as a function of V_{cont}.

Two points merit attention in the above circuit. First, the input-referred offset voltage of the "differential difference" amplifier (M_1-M_6) must be much less than V_1. Second, since switching R_{F2} into the circuit may introduce peaking in the frequency response and

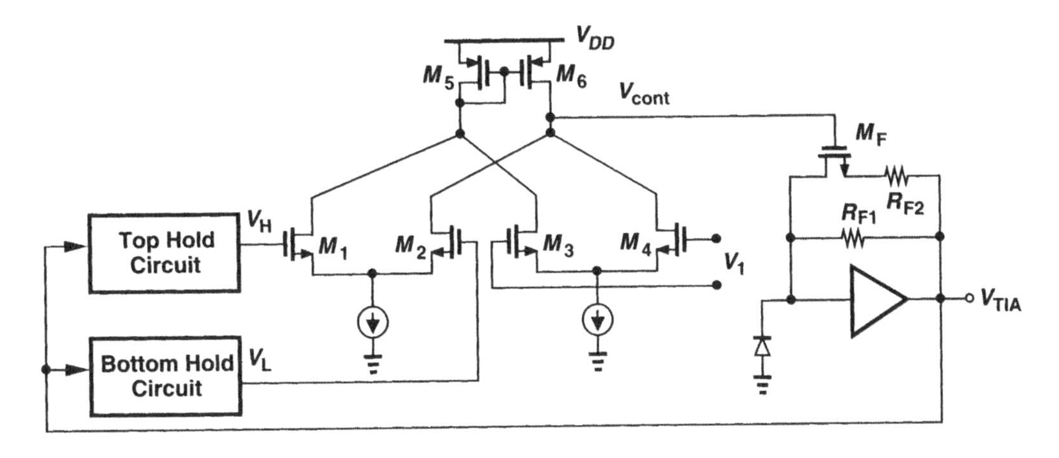

Figure 11.8 TIA gain adjustment using $V_H - V_L$.

hence substantial ISI, a small capacitor can be added in parallel with R_{F2} so as to avoid underdamped transient behavior.

The reset command in the circuits of Fig. 11.7 plays an essential role in peak detection. This reset can be afforded because PON standards in fact allow a "guard time" between consecutive bursts, during which the ONUs do not transmit. Figure 11.9 shows the details of the data waveform received by the OLT. During the guard time, the data swing collapses

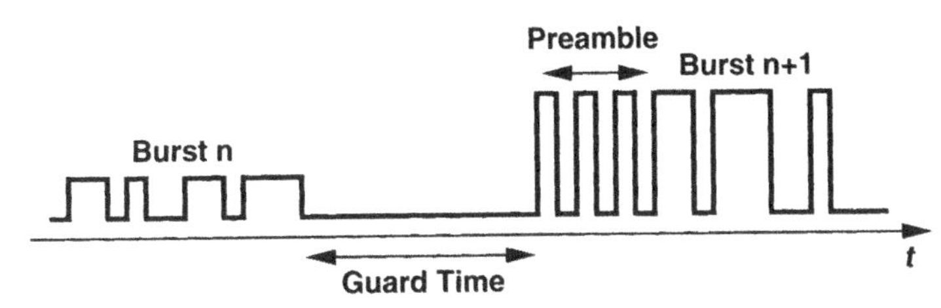

Figure 11.9 Use of guard time in data waveform to allow reset.

to zero, signaling the receiver to generate a reset command [1]. The reset remains active until the next burst arrives. To produce the reset signal, we can utilize the circuit of Fig. 11.8, choosing V_1 to be a small voltage, e.g., 10 mV. As shown in Fig. 11.10, if $V_H - V_L$ falls below this value, the differential difference amplifier creates a relatively low value at X and hence a high logical level at the output. If the gain of this amplifier is inadequate, another differential pair can precede the inverter. The design in [1] employs a different method of reset generation.

In addition to the TIA gain, we must also adjust the reference voltage of the single-ended to differential converter. We recognize that the average of V_H and V_L, $(V_H + V_L)/2$, can indeed serve this purpose [1]. As illustrated in Fig. 11.11, a resistive divider with $R_1 = R_2$ generates $V_b = (V_H + V_L)/2$. A small capacitor, C_b, reduces the ripple in the presence of long runs.

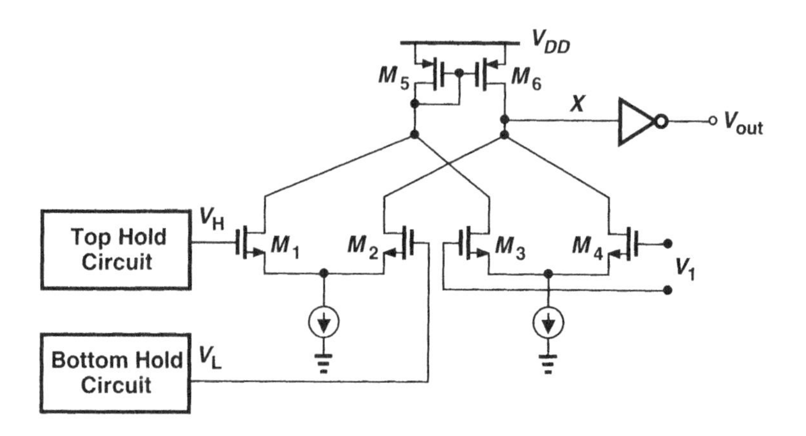

Figure 11.10 Reset generation circuit.

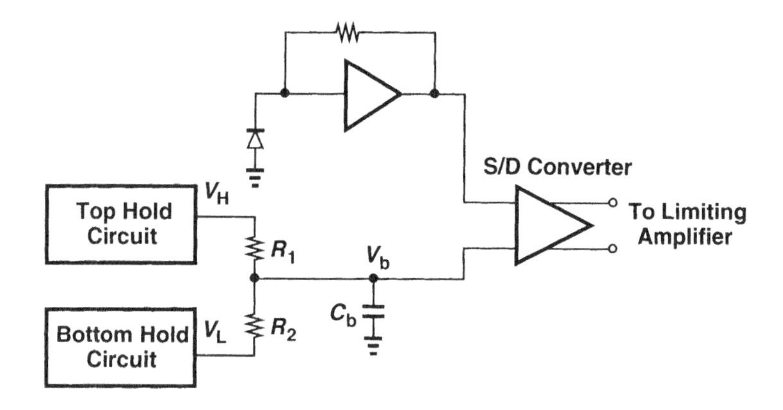

Figure 11.11 Generation of reference voltage, V_b, for S/D converter.

11.2.2 Burst-Mode TIA Variants

A number of other burst-mode TIAs have been developed that address the issues described in the previous section. We study some here.

It is possible to construct a bottom hold circuit similar to that of Fig. 11.7 but with no diode-connected device. Illustrated in Fig. 11.12(a) [2], the idea is to employ a common-source stage, M_1, to charge the capacitor. Here, a comparator compares the level stored on C_1 with the input. If $V_{in} < V_{out}$, the comparator generates a high logical level at X, turning M_1 on and drawing current from C_1. The output voltage thus falls until it crosses V_{in} and the comparator turns off M_1. This change is much faster than that afforded by a diode-connected device.

The circuit of Fig. 11.12(a) assumes a fast comparator. In reality, however, the com-

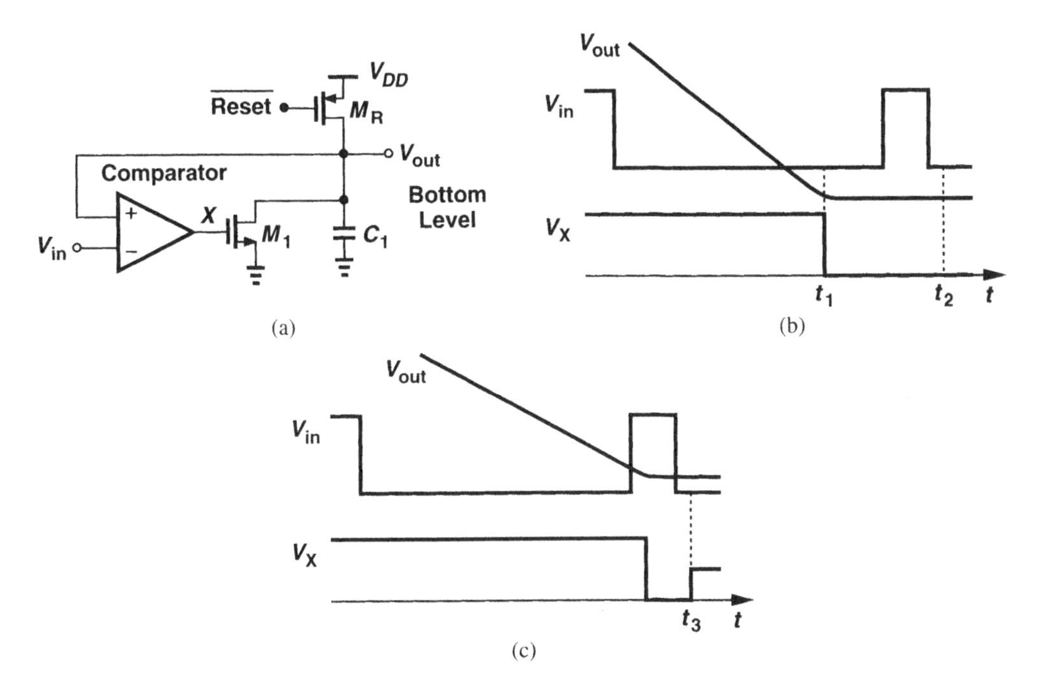

Figure 11.12 (a) BM TIA using a comparator, (b) effect of finite speed of comparator, (c) effect with slow discharge.

parator is realized as a high-gain amplifier, suffering from substantial delay.[1] Thus, as M_1 discharges C_1, the comparator does not turn M_1 off until V_{out} falls *well* below V_{in} [Fig. 11.12(b)]. Th error incurred by V_{out} in this case is *irreversible* because subsequent low input levels, e.g., that at t_2, are greater than V_{out} and hence unable to turn M_1 on. To avoid this undershoot, M_1 can be chosen a narrow MOSFET (or C_1 a large capacitor), thereby ensuring that V_{out} does not fall below V_{in} during one bit [Fig. 11.12(c)]. Unfortunately, if $V_{out} - V_{in}$ at t_3 is small (e.g., less than some tens of millivolts), so is the comparator output, failing to turn M_1 on. That is, V_{out} still incurs an error.

One may wonder whether the above error ultimately becomes a common-mode component if the *difference* between the bottom hold and top hold results is of interest. The reader is encouraged to investigate these issues for the top hold counterpart of this circuit and determine whether the error due to the comparator delay appears as a differential or common-mode component in $V_H - V_L$.

Illustrated in Fig. 11.13 is another TIA gain control technique [3]. In the simplified topology of Fig.11.13(a), a PMOS device compares V_X with V_{out}. For large input currents, the (inverting) TIA experiences a high V_X and a low V_{out}, turning on M_p. Consequently, M_p absorbs some of the input current and reduces the distortion. This clipping operation

[1] Unlike applications in which the comparator be can clocked to achieve a high speed, this circuit receives no clock and must resort to "asynchronous" comparison.

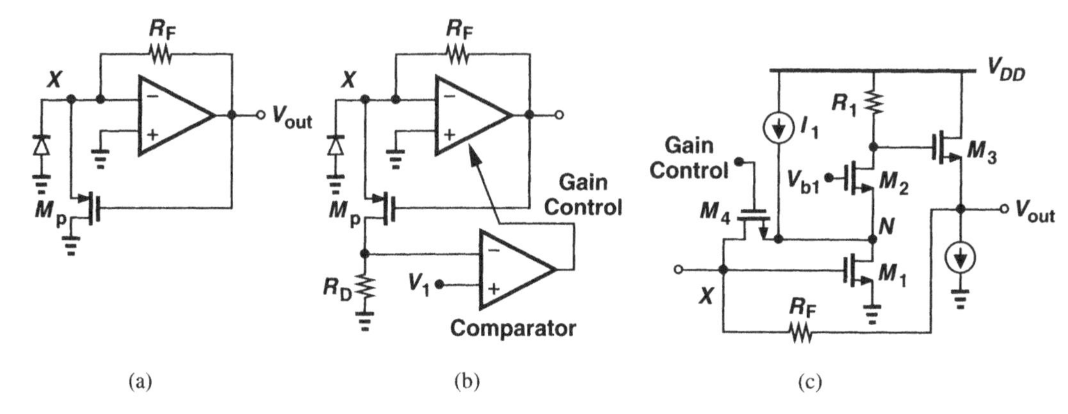

Figure 11.13 (a) TIA gain control by clipping, (b) use of current monitor to adjust core's gain, (c) core implementation.

proves effective if the delay from X to V_{out} is small enough to turn on M_p *before* the input stage of the TIA saturates. Of course, if V_{out} falls excessively, then the *output stage* saturates and the data is still distorted. In other words, these saturation effects must subside by the end of the preamble.

An interesting observation here is that transistor M_p in Fig. 11.13(a) can also monitor the input current, producing a proportional voltage [3]. As shown in Fig. 11.13(b), if the voltage across R_p exceeds a certain level, V_1, then the gain of the internal TIA circuit can be reduced. Figure 11.13(c) depicts the TIA circuit used in [3]. The amplifier consists of a cascode stage, M_1 and M_2, and a source follower, M_3. The gain is controlled by the voltage-dependent resistor M_4. This method of gain control affects the phase margin to a lesser extent than adjusting the feedback resistor, R_F [3]. This is because the *open-loop* gain of the amplifier is reduced is this case—but at the cost of poorly-defined closed-loop gain. Nonetheless, since M_4 turns on for only very high input currents (beyond what the PMOS clamp can absorb), the imprecise gain is not critical.

The circuit of Fig. 11.13(c) merits a few remarks. First, as M_4 turns on, creating a path between the gate of M_1 and the source of M_2, the dc level at the output is disturbed [3]. Specifically, since M_4 now injects the additional input current into node N, V_N tends to rise, and so does V_{out}. To resolve this issue, [3] adjusts the gate voltage of the cascode device, V_{b1}, by means of a replica circuit. Second, the current source I_1 is added to allow a high transconductance for M_1 without an excessive voltage drop across R_1. For example, if I_1 carries 75% of I_{D1}, then resistor R_1 can be quadrupled for a given voltage headroom, contributing less noise and raising the open-loop gain. However, the noise current of I_1 directly adds to that of M_1, requiring that I_1 be designed with a large overdrive voltage.

11.2.3 Offset Correction in Limiting Amplifiers

The foregoing studies reveal that the reference voltage necessary for single-ended to differential conversion incurs some error due to the top and bottom hold nonidealities. As a result, the S/D converter output exhibits a finite offset, which may saturate the limiting amplifier back-end stages (Fig. 11.14). Typical offset cancellation techniques used in

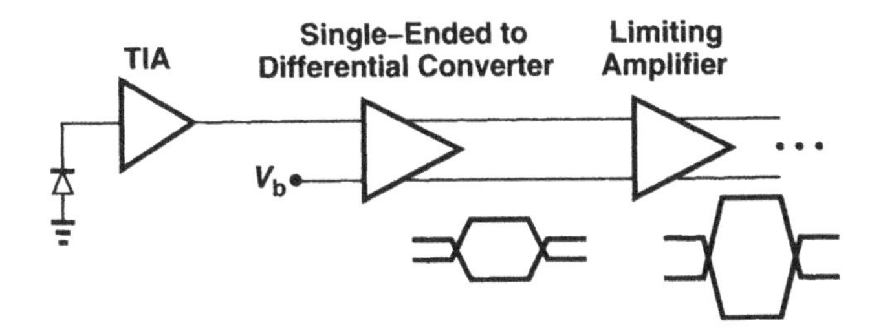

Figure 11.14 Saturation of a limiting amplifier due to dc offset.

continuous-mode circuits take a long time to settle and are thus ill-suited to a burst-mode environment.

It is possible to reduce the offset by applying feedforward or feedback to one or more stages in the limiting amplifier. The critical requirement here is fast settling of the offset. Consider the arrangement shown in Fig. 11.15(a) [4], where a feedforward path measures

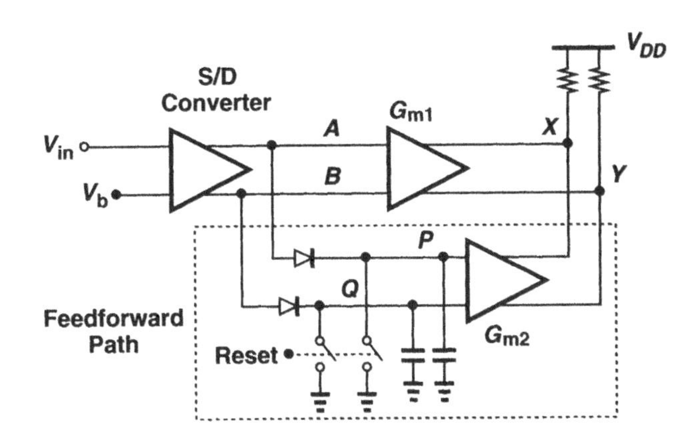

Figure 11.15 Offset cancellation by feedforward.

the peaks at A and B, amplifies the difference, and injects a differential current into nodes X and Y so as to remove the offset. Each of G_{m1} and G_{m2} stages can be implemented by a simple differential pair. If $V_P - V_Q$ is equal to the offset in $V_A - V_B$, then we choose $G_{m1} = G_{m2}$ for complete offset cancellation. Due to the peak detector nonidealities, the cancellation is not perfect, but the feedforward path can be added to the subsequent stages as well, progressively lowering the offset through the chain [4].

Let us extend the above concept to feedback configurations as well. Illustrated in Fig. 11.16, the idea is to measure the peaks at X and Y and adjust the offset between A and B. With sufficient speed in the peak detectors, the negative feedback loop quickly settles,

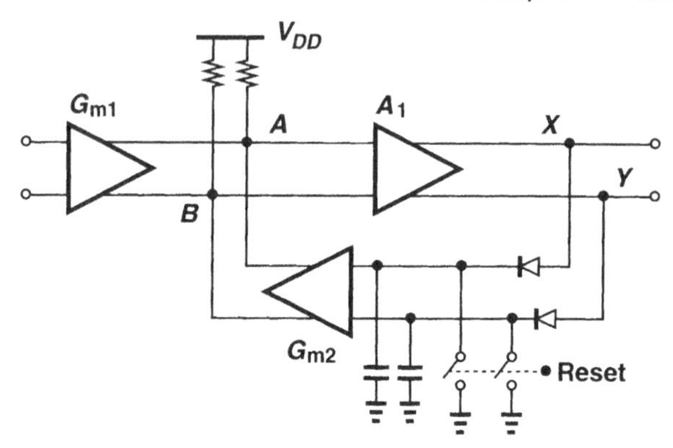

Figure 11.16 Offset cancellation by feedback.

yielding a small offset at the output. This technique can be applied to each stage of the limiting amplifier to ensure that the offset does not saturate the back-end stages.

11.3 Burst-Mode CDR Circuits

Our study of phase-locked CDR circuits reveals a lock time of at least several hundred input cycles. Burst-mode applications must therefore avoid transients in a PLL environment.

How can we recover the clock without a PLL? An interesting technique introduced by Banu and Dunlop [5] employs two "gated oscillators" that are directly activated and deactivated by the input data. Let us first see what we mean by a gated oscillator. Consider the ring oscillator shown in Fig. 11.17, where transistor M_1 operates as a switch [5]. Here,

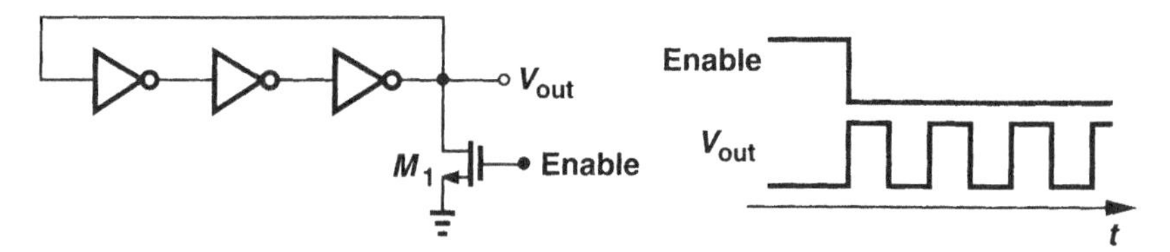

Figure 11.17 Gated ring oscillator.

if Enable is high and M_1 strong enough, then $V_{out} \approx 0$ and the circuit ceases to oscillate. When Enable falls to zero, V_{out} rises and oscillation begins. The key point in this gated oscillator is that, when activated, it produces a well-defined edge (a rising edge in this example) with minimal delay with respect to the enable command.

We now construct a BM CDR circuit using two gated oscillators. As illustrated in Fig. 11.18 [5], we enable Oscillator 1 by the low levels of the data and Oscillator 2 by the high levels. Thus, on the falling edge of D_{in}, V_{out1} goes up and continues as a periodic waveform. Similarly, on the rising edge of D_{in}, V_{out2} rises and toggles thereafter. If we "stitch" V_{out1} and V_{out2} using an OR gate, then we obtain a periodic clock at all times whose rising edges are aligned with the input data transitions. That is, V_{CK} can serve as

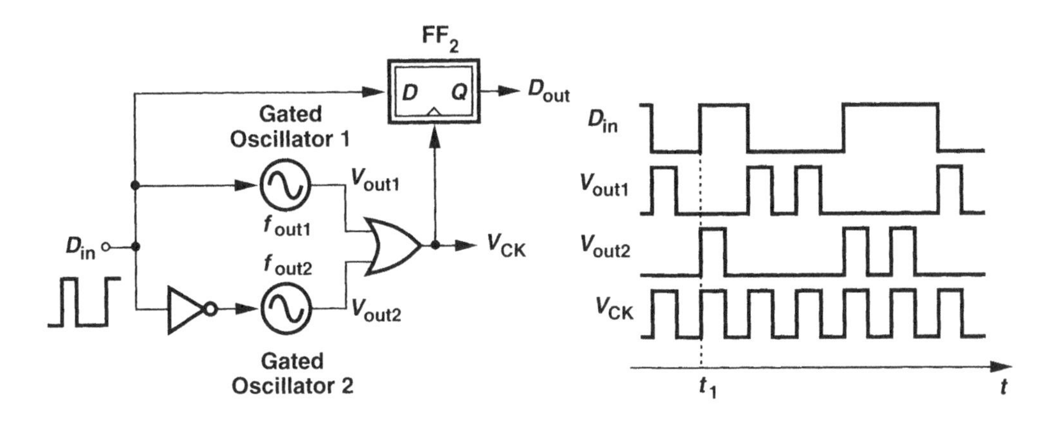

Figure 11.18 BM CDR circuit using two gated oscillators.

the recovered clock. As shown in Fig. 11.18, a flipflop samples D_{in} on the falling edges of this clock, producing the recovered data.

While simple and elegant, the above CDR circuit relies on several assumptions. First, the delay from D_{in} to V_{CK} (through each gated oscillator) is zero. Second, each oscillator period is exactly equal to the input bit period. Third, the two oscillators operate at exactly the same frequency, $f_{out1} = f_{out2}$. Fourth, each oscillator's jitter accumulation over long runs (e.g., 72 bits) is small enough. Since none of these assumptions may hold in practice, we must determine how various nonidealities impact the recovered clock and data.

11.3.1 Effect of Finite Delays

In the BM CDR architecture of Fig. 11.18, the finite delays produce an undesirable skew between the recovered clock, V_{CK}, and the input data. To investigate this effect, let us consider a more realistic implementation of the circuit. As shown in Fig. 11.19(a), we buffer each oscillator by an inverter and combine the two outputs using a NAND gate. Now, a data transition at the gate of M_1 experiences three gate delays: from D_{in} to X, from X to Y, and from Y to W. (We neglect the skew within the retiming flipflop.) This skew becomes a significant fraction of the input bit period at high speeds. For example, with a three-stage ring oscillator, the oscillation period is equal to six gate delays, and hence the skew (about three gate delays) reaches roughly 0.5 UI!

In order to remove the above skew, a path containing a similar delay can be inserted in series with the flipflop input. As shown in Fig. 11.19(b), we approximate the path from D_{in} to X by an inverter and follow it by another inverter to match Inv_1 in Fig. 11.19(a). A replica NAND gate completes the deskew path.

We must also consider the effect of the delay through the inverter at the enable input of Oscillator 2 in Fig. 11.18. The reader is encouraged to analyze this case, but since in practice D_{in} is available in differential (complementary) form, such an inverter is unnecessary.

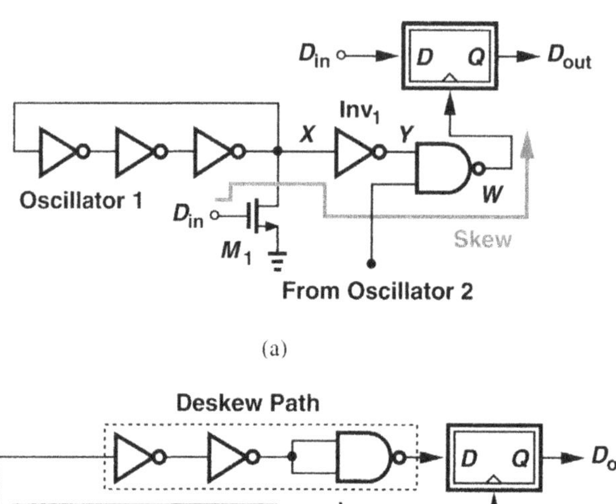

(a)

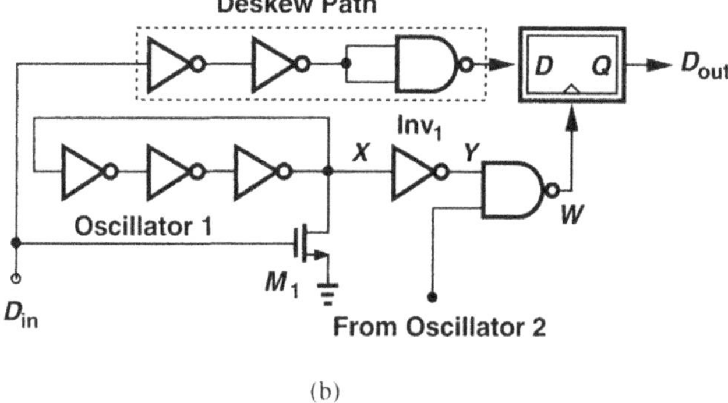

(b)

Figure 11.19 (a) Problem of skew in recovered clock path, (b) use of replica circuit to reduce skew.

11.3.2 Effect of Frequency Mismatch and Offset

Suppose the two oscillators in Fig. 11.18 exhibit a small frequency mismath of $\Delta f = f_{out1} - f_{out2} < 0$. For simplicity, we assume f_{out1} is equal to the input data rate, r_{in}. Consequently, when D_{in} is low, Oscillator 1 produces an integer number of cycles [Fig. 11.20(a)], but when D_{in} is high, Oscillator 2 runs slightly faster, creating a rising edge at t_1, just before D_{in} goes high. The corresponding clock pulse is therefore wider. That is, the recovered clock suffers from jitter. Interestingly, however, this jitter does not affect the recovered data because the *falling* edge of V_{CK} (at t_2) remains at its ideal point in time. This also holds if $\Delta f > 0$. Though benign in this case, excessive clock jitter due to frequency mismatch may still violate the optical standard's jitter requirements.

Let us now compute the maximum clock jitter as a result of frequency mismatch in Fig. 11.18. A maximum run length of N (e.g., 72 consecutive ONEs or ZEROs) corresponds to NT_{in} seconds, where $T_n = 1/r_{in}$ denotes the input bit period. With our assumption that f_{out1} is equal to r_{in}, we have $NT_{in} = N/f_{out1}$. Oscillator 2 generates N cycles in N/f_{out2} seconds [Fig. 11.20(b)]. Thus, the last rising edge of V_{out2} occurs $\Delta T = N/f_{out1} - N/f_{out2}$ seconds before D_{in} rises. This jitter can be written as

$$\Delta T = \frac{N}{f_{out1} - \Delta f} - \frac{N}{f_{out1}} \tag{11.1}$$

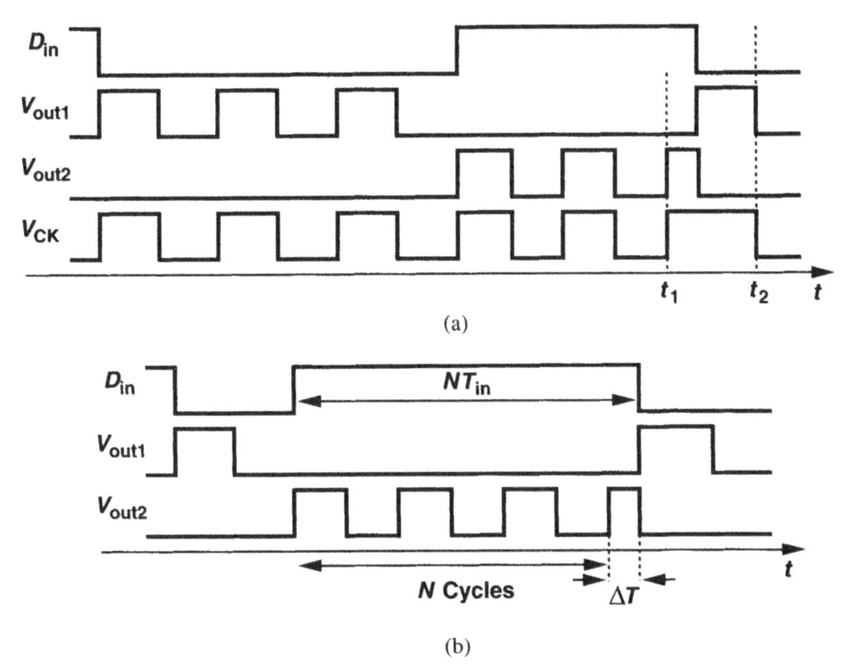

Figure 11.20 (a) Effect of frequency mismatch between two gated oscillators, (b) waveform examples to calculate resulting clock jitter.

$$\approx \frac{N}{f_{out1}} \cdot \frac{\Delta f}{f_{out1}}. \tag{11.2}$$

For example, if $\Delta f / f_{out1} = 1\%$ and $N = 72$, then $\Delta T = 0.72/f_{out1}$, i.e., the jitter is equal to 0.72 UI, an unacceptably large value. The two oscillators must therefore be designed and laid out for better matching.

In the above analysis, we have assumed abrupt transitions throughout the circuit. In reality, finite rise and fall times along with frequency offset may produce *glitches*. We study this phenomenon with the aid of the gated oscillators in Fig. 11.21, where the input data is assumed to be differential. Suppose V_{out2} goes high at $t = t_1$ due to frequency mismatch. Slightly later, D_{in} falls, forcing V_{out1} and V_{out2} to change in *opposite* directions and producing an indeterminate logical level at the inputs of the OR gate at $t = t_2$. The OR gate may therefore exihibit a glitch at its output.

How does the glitch affect the recovered data? In the CDR architecture of Fig. 11.18, the data is sampled on the falling edges of V_{CK}. As a result, in Fig. 11.21, D_{in} is still properly sampled at $t = t_3$. However, if the glitch is deep enough, the falling edge in V_{CK} just before $t = t_2$ attempts to sample D_{in} during its transition, possibly causing metastability.

Even though benign, the clock jitter may still not be acceptable to the standard. Moreover, since the stages following the CDR circuit must utilize this clock to ensure proper timing with respect to the recovered data, the effect of these glitches on their performance must be examined carefully.

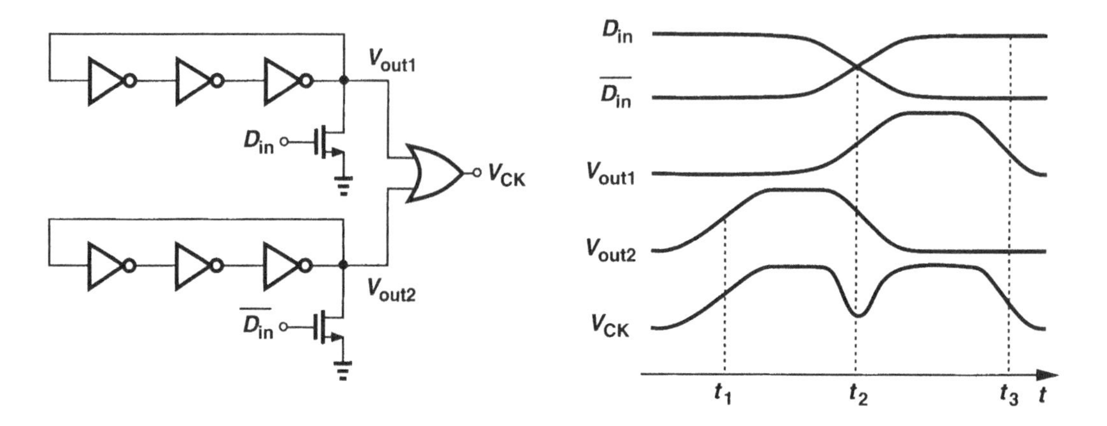

Figure 11.21 Glitch in recovered clock due to frequency mismatch.

Can we use gated LC oscillators to achieve a lower frequency mismatch? We must first decide how an LC oscillator can be gated so that, upon arrival of a data edge, it produces a predictable transition. Fig. 11.22(a) shows two gating methods that do *not* serve this purpose. This is because the nominal symmetry of the circuit implies that only mismatches

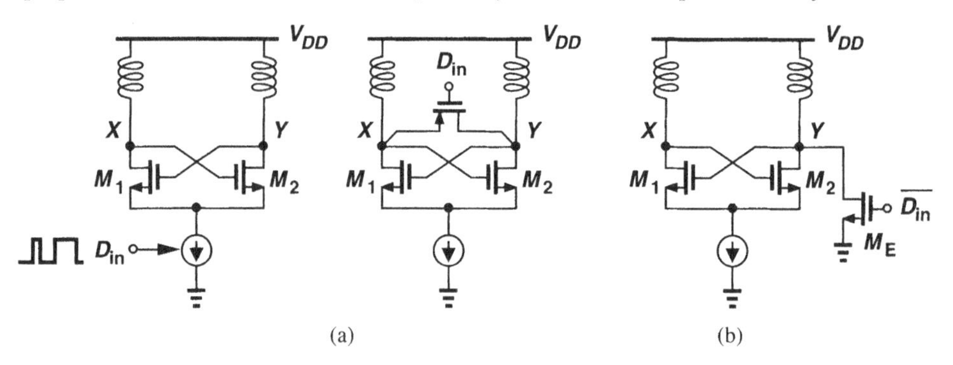

Figure 11.22 (a) Incorrect, and (b) correct gating of LC oscillators.

between the two sides determine whether X or Y goes high after D_{in} enables the oscillator. We therefore consider the arrangement of Fig. 11.22(b), where M_E pulls node Y low and disables the oscillator when $\overline{D_{in}}$ is high.[2] After $\overline{D_{in}}$ goes low, V_Y rises, V_X falls, and oscillation begins.

It is generally understood that an oscillator employing a resonator with a quality factor of Q takes roughly Q cycles to reach steady state. The reader may then wonder if the gating technique shown in Fig. 11.22(b) suffers from the same effect and hence fails to produce a proper clock edge immediately. However, simulations reveal that the large initial difference

[2]Here, M_E draws a large current through the inductor. We can tie the source of M_E to a moderate voltage, e.g., $\approx V_{DD}/2$, to reduce this current.

imposed between V_X and V_Y drives the oscillation toward steady state in less than one cycle.

If area, tuning range, unwanted coupling, or other issues prohibit the use of LC oscillators, we can incorporate ring oscillators and pay close attention to their layout so as to obtain good matching. Figure 11.23(a) shows an example of a highly-matched layout style.

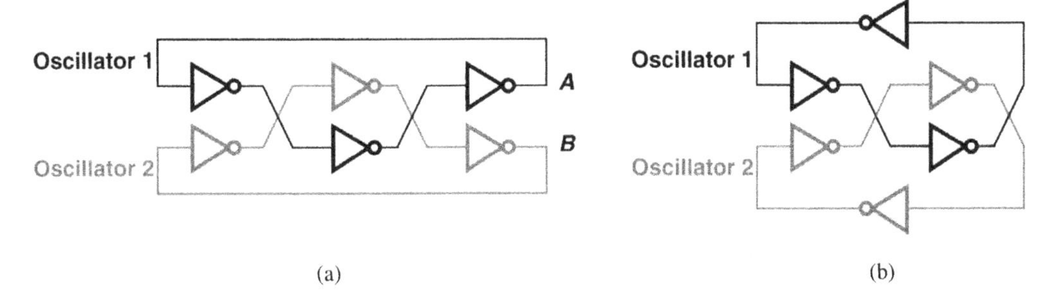

Figure 11.23 Interwoven ring oscillator layouts to improve matching.

Here, the two rings are "interwoven" to average out their mismatches. If the feedback lines introduce excessive capacitance at nodes A and B, the topology of Fig. 11.23(b) can be used but at the cost of less averaging between the two oscillators.[3]

We now study the effect of frequency *offset* in the architecture of Fig. 11.18. In this case, f_{out1} and f_{out2} are equal but unequal to the input data rate, r_{in}. Illustrated in Fig. 11.24 for $f_{out1} = f_{out2} = f_{out} > r_{in}$, the frequency offset displaces the rising clock edge

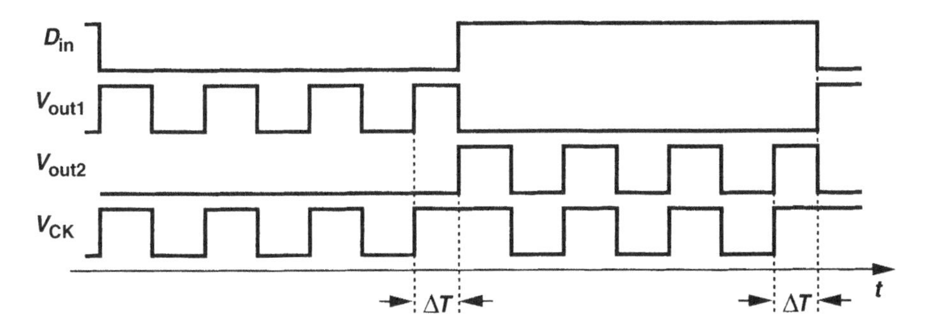

Figure 11.24 Effect of frequency offset in BM CDR circuit.

on both the falling and rising edges of the input data. The amount of jitter in the recovered clock can be written from Eq. (11.2) as

$$\Delta T = \frac{N}{f_{out}} \cdot \frac{\Delta f}{f_{out}}, \tag{11.3}$$

[3]This is because of the long distance between the top and bottom inverters.

where $\Delta f = f_{out} - r_{in}$. In this situation, too, the finite rise and fall times lead to glitches in the recovered clock.

The problem of frequency offset becomes particularly acute if the CDR circuit employs ring oscillators. This is because the center frequency of these oscillators may vary by almost a factor of 2 with process and temperature. Thus, some means of tuning is necessary to ensure that $f_{out1} \approx f_{out2} \approx r_{in}$. This can be accomplished through the use of a phase-locked loop. Depicted in Fig. 11.25, the idea is to compare f_{out}/M with an accurate

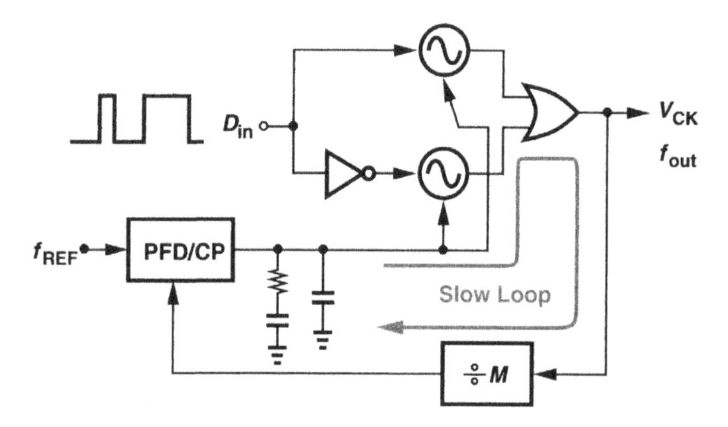

Figure 11.25 Use of PLL to reduce CDR frequency offset.

reference and tune the two oscillators so as to minimize the error. But what happens to the output *phase* here? We wish the gated oscillators to produce output edges aligned with D_{in} whereas the PLL attempts to align the output phase with the reference, f_{REF}. The phase adjustment by the PLL becomes noticeable in the absence of input data transitions because no input phase information is received. For this reason, the PLL time constant must be long enough to allow negligible shift in the output phase during long runs.

11.3.3 Jitter Characteristics

Th burst-mode CDR architecture of Fig. 11.18 exhibits jitter characteristics that are quite different from those of phase-locked circuits. We study jitter generation, transfer, and tolerance here.

Jitter Generation The jitter generated by the BM CDR circuit arises from the gated oscillators. From the time one oscillator is enabled until it is disabled, the recovered clock accumulates jitter [Fig. 11.26(a)]. For a run length of N bits, the clock jitter is given by

$$\Delta T_{CK} = \sqrt{N}\Delta T_{cc}, \tag{11.4}$$

where ΔT_{cc} denotes the cycle-to-cycle jitter. As explained in Chapter 9, ΔT_{cc} can be related to the free-running phase noise of the oscillator. Also, recall that the jitter in a phase-locked loop accumulates to a level given by $\Delta T_{cc}/\sqrt{2\pi f_u T_{in}}$, where f_u is the -3-

dB bandwidth. By analogy, we can say from Eq. (11.4) that the burst-mode circuit has an equivalent jitter accumulation bandwidth of $f_u = (2\pi N T_{in})^{-1}$.

How is the recovered data affected by the jitter generation? Interestingly, the oscillator jitter accumulation does not corrupt the data here! This is because each data transition starts one oscillator anew, and it is the very next clock edge that samples the data [e.g., at $t = t_1$ in Fig. 11.26(a)]. By contrast, phase-locked CDRs do not allow the input data transitions

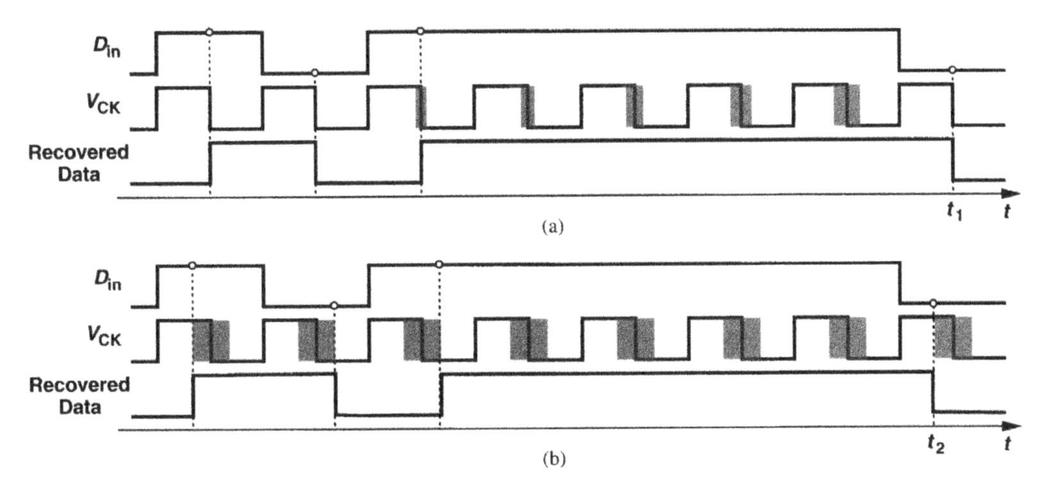

Figure 11.26 Jitter generation in (a) burst-mode, and (b) phase-locked CDR circuits.

to "jump start" the oscillator and hence impress the oscillator jitter upon the recovered data [e.g., at t_2 in Fig. 11.26(b)].

Jitter Transfer In the study of jitter generation, the reader may have noticed that the input data jitter travels to the BM CDR output *without attenuation*. Thus, if one input edge is suddenly displaced by ΔT, so is the corresponding output edge. This CDR architecture therefore negligibly filters the input data jitter, exhibiting a jitter transfer equal to unity up to high jitter frequencies. One may then wonder why we use this circuit at all! It appears that if the retiming flipflop in Fig. 11.18 is replaced with a wire, we still obtain the same data at the output. This is true, but the key role of the CDR circuit here is to generate a *clock* having a proper timing relationship with respect to the data (i.e., a clock edge in the middle of the data eye). This pair of recovered clock and data can be robustly used by subsequent stages, e.g., a demultiplexer.

Jitter Tolerance The lack of jitter filtration in the BM CDR architecture of Fig. 11.18 suggests a *high* tolerance of jitter. This is because a data edge displacement of ΔT (almost) immediately appears in the recovered clock, shifting the sampling point by the same amount (Fig. 11.27). This occurs regardless of the jitter frequency. In phase-locked CDR circuits, on the other hand, the loop time constant does not allow the recovered clock edges to closely track the input edges as the jitter frequency rises, eventually causing sampling errors.

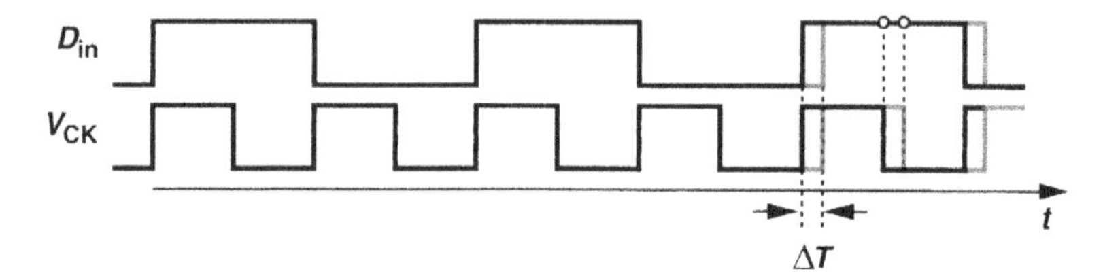

Figure 11.27 Jitter tolerance in a BM CDR circuit.

Let us study the jitter tolerance in the presence of long runs in the input data. Suppose the bit period is modulated by a sinusoid $J(t) = J_p \sin \omega_j t$, where J_p and ω_j denote the peak value and frequency of the jitter, respectively. We consider three cases. First, with no jitter, the falling edges of the recovered clock sample the input data at the proper point, producing D_{out} [Fig. 11.28(a)]. Second, with small, slow jitter, the input edge is displaced by a maximum amount equal to J_p and the clock may suffer from a glitch, but the sampling point remains in the middle of the data eye [Fig. 11.28(b)]. That is, the length of the run does not impact the sampling point.

In the third case, we assume a large, *fast* jitter [Fig. 11.28(c)]. Suppose the jitter changes so much that, starting from $t = t_1$, it reduces the bit period from T_{in} to $T_{in}/2$. Then, the falling edge of the recovered clock coincides with the next data transition (at $t = t_2$), causing an error. Such amount of jitter is the maximum theoretical jitter tolerance of the BM CDR circuit. This maximum can be calculated by setting $J(t)$ equal to $T_{in}/2$ at $t_2 = T_{in}/2$ in Fig. 11.28(c):

$$J_p \sin \omega_j \frac{T_{in}}{2} = \frac{T_{in}}{2}. \tag{11.5}$$

It follows that

$$J_p = \frac{T_{in}}{2 \sin \dfrac{\omega_j T_{in}}{2}}. \tag{11.6}$$

Since $T_{in} = 1$ UI,

$$J_p = \frac{1}{2 \sin \dfrac{\omega_j T_{in}}{2}} \text{ UI} \tag{11.7}$$

Note that the peak-to-peak input jitter is twice this value. If the jitter frequency is much lower than the data rate, i.e., if $\omega_j T_{in}/2 \ll 1$ rad, then

$$J_p \approx \frac{1}{\omega_j T_{in}} \text{ UI} \tag{11.8}$$

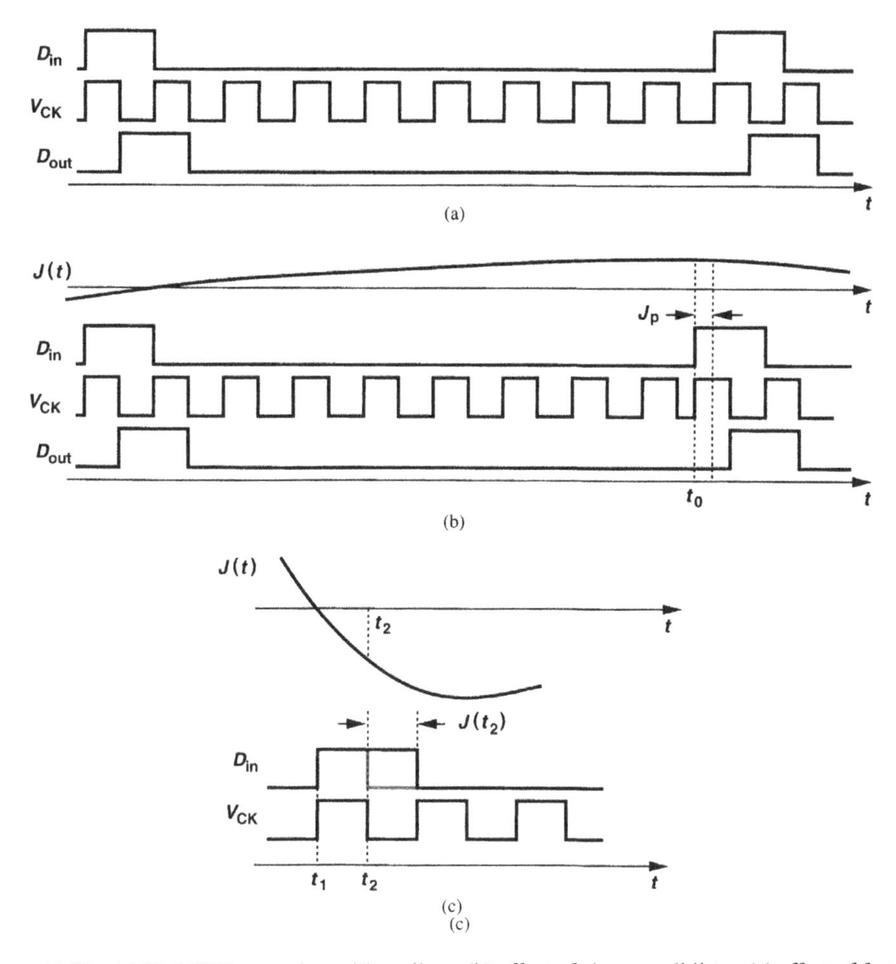

Figure 11.28 (a) BM CDR operation with no jitter, (b) effect of slow, small jitter, (c) effect of fast, large jitter.

11.4 Alternative BM CDR Architectures

In this section, we describe a number of BM CDR techniques that improve upon the basic architecture in Fig. 11.18.

Recall from Section 11.3.2 that frequency offset can lead to glitches in the clock. The PLL in Fig. 11.25 can correct the offset but it drives the phase of the clock toward that of the reference—unless the PLL bandwidth is small. An alternative approach employs only frequency comparison, thus allowing the clock phase to remain undisturbed during long runs. For example, as illustrated in Fig. 11.29, the recovered clock frequency can be compared to a reference, with the result driving a counter and a DAC [6]. The DAC adjusts the two oscillators' frequencies so as to minimize the error $|f_{CK}/M - f_{REF}|$. If

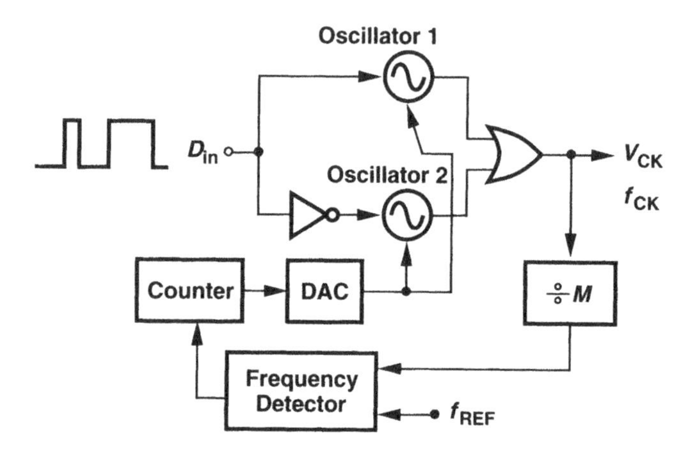

Figure 11.29 BM CDR circuit using a frequency-locked loop.

this adjustment occurs at the startup and the counter contents remain frozen afterwards, the phase of V_{CK} does not drift toward the reference.

In the above architecture, the DAC resolution is critical as it determines (a) the minimum frequency step and hence the residual frequency offset, and (b) the maximum frequency error that can be corrected. In order to estimate the necessary DAC resolution, let us assume that the oscillation frequency can vary with process and temperature from $(1-\alpha)f_{nom}$ to $(1+\alpha)f_{nom}$, where f_{nom} is the nominal value. The total change is therefore equal to $2\alpha f_{nom}$. We also note from Eq. (11.3) that the maximum frequency offset, Δf, for a jitter of ΔT is equal to $\Delta T f_{out}^2/N$. Thus, the DAC must provide a frequency step size of $\Delta T f_{out}^2/N$ across a total frequency range of $2\alpha f_{nom}$. In other words, the DAC resolution (dynamic range) is given by

$$\text{DAC Resolution} = \frac{2\alpha f_{nom}}{\Delta T f_{out}^2} N, \tag{11.9}$$

if a maximum jitter of ΔT and a maximum run length of N are required. In the locked condition, $f_{out} \approx f_{nom}$, and

$$\text{DAC Resolution} = \frac{2\alpha}{\Delta T f_{nom}} N. \tag{11.10}$$

As an example, suppose $\alpha = 30\%$, $N = 72$ bits, and $\Delta T = 0.1$ UI (i.e., $\Delta T f_{nom} = 0.1$). Then, the DAC resolution is 432, around 9 bits. The design in [6] incorporates a $\Sigma\Delta$ DAC to achieve a high resolution with low complexity. Of course, the DAC output must be heavily filtered so that its noise does not modulate the VCO significantly.

Our BM CDR studies have thus far focused on the architecture of Fig. 11.18, with two oscillators gated by the high and low levels of data. Alternatively, one can employ a single continuously-running oscillator that is controlled by data *edges* rather than data levels. As explained in Chapter 9, an XOR gate along with a delay stage can serve as an edge detector

[Fig. 11.30(a)]. The spectrum of such a full-wave-rectified output contains an impulse at

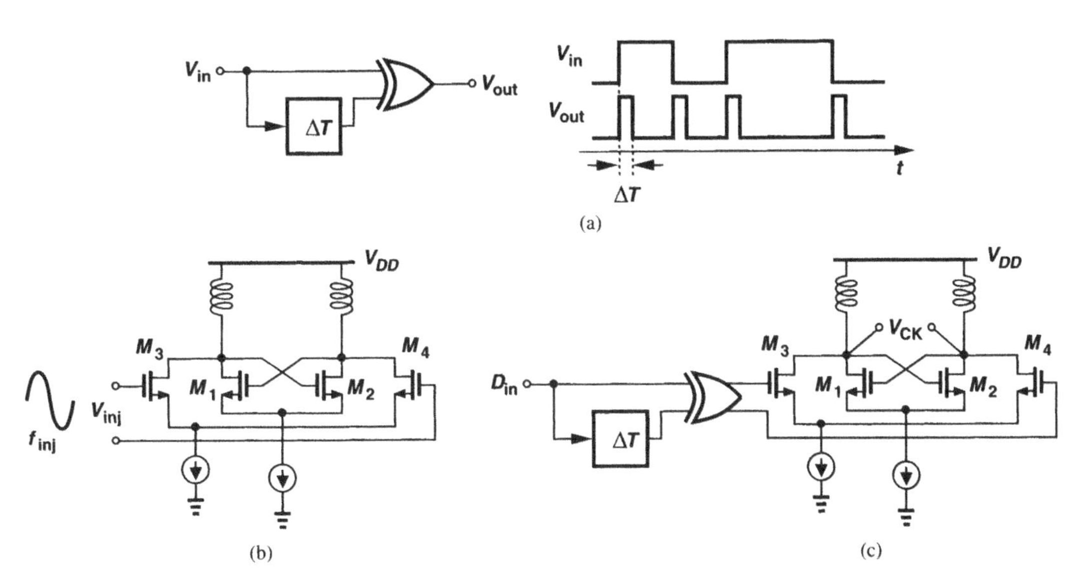

Figure 11.30 (a) Edge detector using an XOR gate, (b) injection into an oscillator, (c) injection of edge-detected data into an oscillator.

the data rate, $r_{in} = 1/T_{in}$.

Before utilizing an edge detector in a BM CDR environment, let us briefly study the notion of "injection locking." From the quadrature oscillator analysis in Chapter 7, we know that a signal can be unilaterally coupled (injected) into an oscillator as shown in Fig. 11.30(b). Suppose the oscillator operates at a frequency of f_0 before injection occurs. What happens if we now inject a sinusoid at a frequency $f_{inj} \neq f_0$? If the injection is strong enough and $|f_{inj} - f_0|$ small enough, the oscillation frequency shifts from f_0 to f_{inj} [7, 8]. We say the oscillator is "injection-locked" to the input. An interesting property of this arrangement is that the oscillator suppresses high-frequency jitter in the injected signal—in a manner similar to a PLL. But, unlike a PLL, an oscillator can injection-lock in a shorter time. The reader is encouraged to consult [7] and [8] to better understand these concepts.

Now, we inject the output of the edge detector into an oscillator [Fig. 11.30(c)]. The sinusoidal component at $f = r_{in}$ in edge-detected waveform can "pull" the oscillation frequency from f_0 and lock the oscillator to the input data. Thus, a *single* oscillator can provide the recovered clock with some filtering of the other components in the edge-detected waveform.

The use of injection locking in CDR circuits has been described by [9, 10, 11] and others. However, this approach entails a number of issues that manifest themselves in practice. First, it is unclear from Fig. 11.30(c) what the timing relationship is between the recovered clock, V_{CK}, and the input data, D_{in}. The following factors create this ambiguity: (1) the skew due to the XOR gate, (2) the phase of the sinusoidal component in the edge-detected waveform, and (3) the variation of the output phase of the oscillator as a function of $|f_0 - 1/T_{in}|$. As described in [7, 8], if f_0 happens to be equal to $1/T_{in}$, then the phase difference between the input and output of the injection-locked oscillator is zero. However, if $f_0 \neq 1/T_{in}$, after the oscillator locks to the input, a certain amount of phase difference arises. Thus, it is difficult to guarantee that the clock samples the middle of the input data eye under all process and temperature conditions.

Second, if the oscillator injection-locks quickly, it also loses lock quickly. That is, the oscillator may return to its free-running state during long runs, or at least experience substantial phase drift. That is, the injection locking approach may constrain the run length to be shorter than the preamble.

References

1. Q. Le et al, "A burst-mode receiver for 1.25-Gb/s ethernet PON with AGC and internally created reset signal," *IEEE Journal of Solid-State Circuits,* vol. 39, pp. 2379-2388, Dec. 2004.
2. H.-L. Chu and S.-I. Liu, "A 10Gb/s burst-mode transimpedance amplifier in 0.13-um CMOS," *Proc. ASSCC,* pp. 400-403, Nov. 2007.
3. K. Nishimura et al, "A 1.25-Gb/s CMOS burst-mode optical transceiver for ethernet PON system," *IEEE Journal of Solid-State Circuits,* vol. 40, pp. 1027-1034, April 2005.
4. M. Nakamura et al, "1.25-Gb/s burst-mode receiver ICs with quick response for PON systems," *IEEE Journal of Solid-State Circuits,* vol. 40, pp. 2680-2688, Dec. 2005.
5. M. Banu and A. Dunlop, "A 660Mb/s CMOS clock recovery circuit with instantaneous locking for NRZ data and burst-mode transmission," *ISSCC Dig. Tech. Papers,* pp 102-103, Feb. 1993.
6. J. Terada et al, "A 10.3 Gb/s burst-mode CDR using a $\Sigma\Delta$ DAC," *IEEE Journal of Solid-State Circuits,* vol. 43, pp. 2921-2928, Dec. 2008.
7. R. Adler, "A study of locking phenomena in oscillators," *Proc. of the IEEE*, vol. 61, No. 10, pp. 1380-1385, Oct. 1973.
8. B. Razavi, "A study of injection locking and pulling in oscillators," *IEEE J. of Solid-State Circuits,* vol. 39, pp. 1415-1424, Sep. 2004.
9. T. Gabara, "A 3.25 Gb/s injection locked CMOS clock recovery cell," *Proc. IEEE Custom Integrated Circuits Conf.*, pp. 521-524, Sep. 1999.
10. J. Lee and M. Liu, "A 20-Gb/s burst-mode clock and data recovery circuit using injection-locking technique," *IEEE J. of Solid-State Circuits,* vol. 43, pp. 619-630, March 2008.
11. J. Zhan et al., "A full-rate injection-locked 10.3 Gb/s clock and data recovery circuit in a 45 GHz," *Proc. IEEE Custom Integrated Circuits Conf.*, pp. 557-560, Sept. 2005.

Index

Printed and bound by CPI Group (UK) Ltd, Croydon, CR0 4YY

16/04/2025

14658426-0003